Marco Abate

Holomorphic Dynamics on Hyperbolic Riemann Surfaces

De Gruyter Studies in Mathematics

Volume 89

Marco Abate

Holomorphic Dynamics on Hyperbolic Riemann Surfaces

DE GRUYTER

Mathematics Subject Classification 2020
Primary: 37F99; Secondary: 30C80; 30F45; 30J99; 37F44

Author
Prof. Dr. Marco Abate
Università di Pisa
Dipartimento di Matematica
Largo Pontecorvo 5
56127 Pisa
Italy
marco.abate@unipi.it

ISBN 978-3-11-060105-3
e-ISBN (PDF) 978-3-11-060197-8
e-ISBN (EPUB) 978-3-11-059874-2
ISSN 0179-0986

Library of Congress Control Number: 2022944255

Bibliographic information published by the Deutsche Nationalbibliothek
The Deutsche Nationalbibliothek lists this publication in the Deutsche Nationalbibliografie; detailed bibliographic data are available on the Internet at http://dnb.dnb.de.

Typesetting: VTeX UAB, Lithuania
Printing and binding: CPI books GmbH, Leck

www.degruyter.com

Contents

Introduction

In the last 50 years, dynamical systems have become one of the main objects of study in mathematics, with many applications outside mathematics. It is a huge subject that can be considered from many points of view. There are discrete and continuous dynamical systems; there are local and global dynamical systems; there are one-dimensional dynamical systems and infinitely-dimensional dynamical systems; there are measurable dynamical systems, topological dynamical systems, smooth dynamical systems—and there are holomorphic dynamical systems.

This book is devoted to a (relatively small) portion of the (quite vast) area of holomorphic dynamical systems: one-dimensional dynamical systems on Riemann surfaces; more specifically, on hyperbolic Riemann surfaces. The investigation of one-dimensional holomorphic dynamical systems started in the second half of the nineteenth century, more or less in the same years when Poincaré began to understand the importance of dynamical systems and started to investigate them in earnest. About 150 years later, the field of one-dimensional holomorphic dynamical systems is still a very active area of research, both on nonhyperbolic Riemann surfaces (mainly the Riemann sphere $\widehat{\mathbb{C}}$ and the complex plane $\mathbb{C}$) and on hyperbolic Riemann surfaces (the unit disk $\mathbb{D}$ and all Riemann surfaces whose universal cover is the disk), with several new papers appearing every year. There are many books describing the basics of holomorphic dynamics on the Riemann sphere (see, e. g., [287]); on the other hand, the only book devoted to holomorphic dynamics on hyperbolic Riemann surfaces as far as I know is [3], that has been out of print since at least 20 years ago (more about this later). By the way, I do not know of any introductory book on the dynamics of holomorphic functions on the plane, a curious hole in the literature.

Let me now describe a bit more precisely what this book is about. A *discrete holomorphic dynamical system* is given by a holomorphic self-map f of a complex manifold M. (A *continuous holomorphic dynamical system* is instead usually given by a holomorphic vector field on a complex manifold; in this book, however, we shall take a slightly different point of view, as I shall explain below.) As often happens with dynamical systems, the object of study is classical, in this case holomorphic maps; it is the kind of questions that one asks on these objects that characterize the field. Namely, we associate to f the sequence $\{f^\nu\}$ of iterates of f, where f^ν is the composition of f with itself $\nu \in \mathbb{N}$ times; we also associate to each point $z \in X$ its *orbit* $\{f^\nu(z)\}$. In dynamical systems we are then interested in the *asymptotic behavior* of the sequence of iterates, that is, what happens as ν goes to infinity: is the sequence convergent? If it is not convergent, can we anyway describe the set of accumulation points? What about single orbits, do they all have the same behavior or different points that can behave differently? What about stability, that is, what happens if we perturb the starting point of the orbit or even the original functions? Do chaotic behaviors appear? And so on. This book shall try and give some answers for dynamical systems defined on hyperbolic Riemann surfaces. In this case, the Montel theorem prevents the appearance of chaos;

https://doi.org/10.1515/9783110601978-201

so, as we shall see, the flavor of the theory and the kind of results we shall obtain is quite different from the case of holomorphic dynamical systems on nonhyperbolic Riemann surfaces—even though recently it has been discovered that the theory developed for hyperbolic Riemann surfaces can be useful for understanding the behavior of dynamical systems in the complex plane (see, e. g., [38]).

As anticipated above, the investigation of this subject began with the works of Schröder [368, 369] in 1870 and Kœnigs [244] in 1883. They were mainly interested in the local situation for holomorphic functions of one variable. Let z_0 be a point of the complex plane $\mathbb{C}$ and f a holomorphic function defined in a neighborhood of z_0 such that $f(z_0) = z_0$. Then the behavior of the sequence of iterates of f near z_0 depends on the value of the derivative of f at z_0. More specifically, if $|f'(z_0)| < 1$ every point z sufficiently close to z_0 is attracted by z_0 (i. e., $f^\nu(z) \to z_0$ as $\nu \to +\infty$) while if $|f'(z_0)| > 1$ the points are repelled away from z_0—or, if you prefer, they are attracted by z_0 under the action of f^{-1}, which is defined in a neighborhood of z_0. Finally, if $|f'(z_0)| = 1$ (and there is a bounded neighbourhood of z_0 sent into itself by f; otherwise more complicated things can happen), the behavior of $\{f^\nu\}$ is cyclic, with a finite period if $f'(z_0)$ is a root of unity. As we shall see in Chapter 4, this local behavior has global repercussions; in a very precise sense, if $0 < |f'(z_0)| < 1$ then the linear map given by the multiplication by $f'(z_0)$ is a good model for the dynamics of f. In 1904, Böttcher [71] was able to give a model also when $f'(z_0) = 0$. Furthermore, for global maps in hyperbolic Riemann surfaces necessarily $|f'(z_0)| \leq 1$ and when $|f'(z_0)| = 1$ then f is an automorphism with simple dynamics; so in our context, the more interesting case is when f has no fixed points.

The first really deep work on global holomorphic dynamical systems has been done by Julia [215] in 1918. He investigated the dynamics of rational functions defined on the Riemann sphere $\widehat{\mathbb{C}}$ and discovered that the global behavior of the sequence of iterates is both complicated and fascinating. Near fixed points it is possible to adapt and clarify the local description, but new phenomena arise, linked for instance to the distribution of periodic points (i. e., fixed points of f^ν with $\nu > 1$). A main problem was the description of the *Julia set* of f, that is, of the set of points $z_0 \in \widehat{\mathbb{C}}$ such that the sequence of iterates $\{f^\nu\}$ is not equicontinuous in any neighborhood of z_0. The idea is that if $\{f^\nu\}$ is equicontinuous in a neighborhood of z_0 then there is a subsequence $\{f^{\nu_k}\}$ converging uniformly near z_0 and then the behavior of the sequence of iterates is somehow under control. In other words, the Julia set is in some sense the singular set for the asymptotic behavior of $\{f^\nu\}$; it is the set where chaotic behavior appears.

Slightly later, in a series of papers Fatou [146–148] extended and deepened Julia's work, also investigating holomorphic dynamical systems on the complex plane generated by transcendental entire functions [149]. Again, a main role is played by the Julia set, defined replacing the notion of equicontinuity by the notion of normality: the *Fatou set*, the complement of the Julia set, is the largest open subset where the sequence of iterates is normal in the sense of Montel.

After Fatou, the study of dynamical systems generated by rational and entire functions momentarily lost its impetus. Besides the works of Cremer [132, 133], Siegel [378], Töpfer [391], and Baker [29–32], mainly devoted to the study of periodic points, both locally and globally by using Nevanlinna's distribution value theory, and Brolin [86], devoted to a deep investigation of the iteration of polynomials of low degree, and a few others, nothing really new appeared.

The situation changed completely in the 1970s and 1980s when the work of Brjuno, Hermann, Sullivan, Douady, Hubbard, and many others shed a completely new light on the topic, showing its deep relationship with the theory of quasi-conformal mappings and opening the gates for a flood of exciting new and deep results that is still going on nowadays, thanks to so many mathematicians (including some Fields medalists) that it is impossible to list their names here.

However, this is not the subject of this book. As hinted above, a main source of complexity in the study of holomorphic dynamical systems on $\widehat{\mathbb{C}}$ and $\mathbb{C}$ is that the sequence of iterates is not normal everywhere, and thus chaos appears. On the other hand, the Montel theorem implies that on hyperbolic Riemann surfaces the whole sequence of iterates is normal everywhere. This completely changes the situation. In fact, by normality, the sequence of iterates is relatively compact in a suitable function space and the compactness has strong consequences on the dynamics of f. For instance, as mentioned before, if f is a holomorphic self-map of a hyperbolic Riemann surface with a fixed-point z_0, then $|f'(z_0)| \le 1$; moreover, $|f'(z_0)| = 1$ if and only if f is an automorphism and $f'(z_0) = 1$ if and only if f is the identity. This can be obtained by noticing that the sequence of iterates should have a converging subsequence and, therefore, the coefficients of the Taylor expansion of f^ν at z_0 cannot tend to infinity as $\nu \to +\infty$; since $(f^\nu)'(z_0) = f'(z_0)^\nu$, we get $|f'(z_0)| \le 1$ and from this it is not too difficult to prove the rest of the assertion (see Theorem 3.1.10). It should be remarked that the strength of this approach was completely understood only after its application (due to H. Cartan [104, 105] and to Carathéodory [98] in the 1930s) to the theory of holomorphic maps of several complex variables, probably because in one variable it was initially somehow concealed by the Schwarz–Pick lemma.

Thus we have the hope to be able to understand the holomorphic dynamics on hyperbolic Riemann surfaces by using the Montel theorem and the Schwarz–Pick lemma. As already remarked by Julia [215], if f is a holomorphic function of $\mathbb{D}$ into itself with a fixed-point $z_0 \in \mathbb{D}$, then the behavior of $\{f^\nu\}$ can be easily derived by the Schwarz–Pick lemma: if $|f'(z_0)| < 1$, then z_0 is globally attractive (and not just locally attractive as already proved by Kœnigs) and if $|f'(0)| = 1$ then f is a non-Euclidean rotation about z_0.

The new ideas needed to study what happens when f has no fixed points were provided by Wolff [414–416] and Denjoy [135] in 1926. Let $\tau \in \partial\mathbb{D}$; then as $z \in \mathbb{D}$ tends to τ, the Poincaré disks of center z and fixed Euclidean radius tend to a *horocycle* at τ, that is to an Euclidean disk internally tangent to $\partial\mathbb{D}$ at τ. Then Wolff proved a sort of Schwarz lemma for holomorphic functions without fixed points, using the horocycles:

if f sends $\mathbb{D}$ into itself without fixed points, then there exists a unique point $\tau \in \partial\mathbb{D}$ (called the *Wolff point* of f) such that f sends every horocycle at τ into itself. Knowing this, it is then not to difficult to prove, using the Montel theorem, that the sequence of iterates $\{f^\nu\}$ converges, uniformly on compact sets, to the constant map sending all $\mathbb{D}$ in τ; this is the *Wolff–Denjoy theorem*.

For a multiply connected domain $D \subset \mathbb{C}$ different from $\mathbb{C}^*$ and, more generally, for multiply connected hyperbolic Riemann surfaces the dynamics has been described by Heins [184, 191] first in 1941 and then with more details in 1988. If f has a fixed point, the local picture forces the global one, exactly as in $\mathbb{D}$. If instead f has no fixed points, then the sequence of iterates tends to the boundary. In particular, if the boundary of D is sufficiently regular, then either the sequence of iterates converges, uniformly on compact sets, to a constant map $\tau \in \overline{D}$ or f is an automorphism.

With a few partial exceptions, the study of continuous holomorphic dynamical systems on Riemann surfaces started much later. A sequence of iterates can be interpreted as a homomorphism from the semigroup $\mathbb{N}$ endowed with the sum to the semigroup of holomorphic self-maps of the Riemann surface endowed with the composition. From this point of view, a continuous holomorphic dynamical system is a *one-parameter semigroup*, that is a continuous homomorphism from the semigroup $\mathbb{R}^+$ endowed with the sum to the semigroup of holomorphic self-maps of the Riemann surface endowed with the composition. Again, we are interested in the asymptotic behavior. On the unit disk, Berkson and Porta [61] in 1978 showed that one-parameter semigroups can be recovered as the flow of a semicomplete holomorphic vector field, the *infinitesimal generator* of the semigroup. Shortly later, Heins [190] in 1981 has been able to classify one-parameter semigroups on all Riemann surfaces, showing that the only interesting ones are on $\mathbb{D}$; furthermore, using the results of Berkson and Porta, he was able to give neat geometric representations of one-parameter semigroups on $\mathbb{D}$.

This was more or less the state of the art in 1989 when [3] was published. The first part of that book was devoted to holomorphic dynamical systems in one complex variable; the second part of the book dealt with the theory of holomorphic dynamical systems in several complex variables, a subject that (with a few notable exceptions) was just starting to be developed at that time. Thirty years have passed; also considering that [3] went soon out of print, a few years ago I thought that it was time for a new updated edition. My expectation was that the second part of the book would have needed a thorough rewriting, because the landscape of the field in several variables has changed a lot in the intervening years; but I also thought that the updating of the first part should have been a much easier affair, because in one variable the theory seemed to be already more or less complete at the end of 1980s.

Well, I was wrong. The book you have in your hands is the updated version of only the first part of [3] and it is about three times longer than the original, going from about 100 pages to more than 300 pages. What happened is that in the last 30 years, even though the basic of the subject of course remained the same, many new exciting developments have appeared and many new applications have been discovered;

moreover, a new light has been shed on results that were already known in 1989 but that I then left out because it was not yet clear (at least to me) how important they were. A partial list of the new results included here is: the multi-point Schwarz–Pick lemma discovered by Beardon and Minda [48] in 2004; the Burns–Krantz theorem [91] published in 1994 on the boundary rigidity of holomorphic self-maps of $\mathbb{D}$, with the generalization given by Bracci, Kraus, and Roth [82] in 2020; the study of random iteration on hyperbolic Riemann surfaces, started essentially in the 1990s by Gill and others but whose main results were obtained by Beardon, Carne, Minda, and Ng [52] in 2004, by Keen and Lakic [224, 223, 225] in 2006 and by Short, Christodoulou and myself [116, 10] in 2021; the whole theory of models, starting from the fundamental work of Pommerenke [338], Baker–Pommerenke [34] and Cowen [126] at the beginning of the 1980s and then revised and completed by Arosio and Bracci [22] in 2016; the study of backward dynamics done by Bracci and Poggi-Corradini [76, 327] in 2003; and so on. Furthermore, the study of continuous holomorphic dynamical systems literally exploded, producing so many new results that to present even just the most important ones would need yet another book—that, luckily, has already appeared [80]. (And yes, in a few years you will also get the updated edition of the second part of [3]. I hope.)

Some comments and remarks about the structure of this work are in order. I have written this book keeping in mind two different goals (and audiences). First of all, this is intended as a reference book on holomorphic dynamical systems on hyperbolic Riemann surfaces and related topics. During my own investigations, I found many beautiful theorems never presented in book form; furthermore, the whole theory seemed to me requiring a comprehensive exposition collecting several results scattered around in the literature. This allowed a unified exposition of the main results and a clearer discussion of the threads connecting them.

So, the first audience of this book is mainly composed by researchers in holomorphic dynamical systems; they will find an up-to-date description of the field, open problems to solve and many references to several topics not discussed here. But, as already anticipated, I also had another goal in mind. This book would also like to be an introduction to this area for, say, first-year Ph. D. students (or for good master students, too), giving them both a sample of typical features and techniques, presented from scratch starting from the Schwarz lemma, as well as motivations provided by the historical development of the theory.

Also for this reason, I tried to keep prerequisites to a minimum. Besides a good knowledge of the basics of function theory of one complex variable, only a good topological background (up to covering spaces and the fundamental group) is needed. Sometimes we shall use notions or results from ordinary differential equations, differential geometry or measure theory, but whenever an external result is needed I have tried to always give a precise statement and a reference to a place where a proof can be found. Moreover, the Appendix contains statements and proofs of a few classical results not always covered by standard introductory courses in complex analysis.

Let us now briefly describe the actual content of this book; more details can be found in the introductions to each chapter.

Chapter 1 is a thorough introduction to geometric function theory on hyperbolic Riemann surfaces. We shall discuss the Schwarz–Pick lemma, including the multi-point version; the Poincaré metric and distance; the structure of the automorphism group of $\mathbb{D}$ and, more generally, of hyperbolic Riemann surfaces; the Montel, Vitali, and Picard theorems in full generality; the classification of Riemann surfaces; the boundary behavior of the universal covering map of multiply connected hyperbolic domains; the Ahlfors–Schwarz–Pick lemma; and much more.

In Chapter 2, we introduce the horocycles in $\mathbb{D}$ and their main properties, the Julia and Wolff lemmas that are boundary versions of the Schwarz lemma. We use them to study the angular derivative of holomorphic self-maps of $\mathbb{D}$ into itself, proving the Julia–Wolff–Carathéodory theorem, and then to investigate the structure of the automorphism group of a hyperbolic Riemann surface. We shall also prove the Lindelöf theorem on the existence of nontangential limits and the Burns–Krantz theorem on the boundary rigidity of holomorphic self-maps of $\mathbb{D}$.

Chapter 3 is devoted to discrete dynamics on hyperbolic Riemann surfaces. We start by describing the theory for holomorphic functions with a fixed point; we present two proofs of the Wolff–Denjoy theorem; we develop the iteration theory on hyperbolic Riemann surfaces and its version in finitely connected hyperbolic domains; we study the stability of the Wolff point; we introduce the notion of model for a holomorphic self-map, proving its existence and uniqueness; and we study random iteration on hyperbolic Riemann surfaces.

In Chapter 4, we concentrate our attention on the dynamics in the unit disk, where we can get deeper results. We study in detail how the orbits approach the Wolff point; we prove a complete classification of the possible models that can arise for holomorphic self-maps of $\mathbb{D}$, including ways to detect the model just looking at the map; we study the backward dynamics, understanding what happens to the orbits in the past and not just in the future; and we get a few results about the existence of common fixed points for commuting maps.

In Chapter 5, we investigate the one-parameter semigroups of holomorphic functions on a Riemann surface. In particular, we present the results of Berkson–Porta and Heins cited above about the existence and properties of the infinitesimal generator, the classification of one-parameter semigroups on Riemann surfaces other than $\mathbb{D}$ and the geometrical realization of one-parameter semigroups on $\mathbb{D}$.

Finally, as anticipated, to help the reader the Appendix contains the statement and proofs of a number of less standard results in real and complex analysis of one variable that we happen to use in the book.

Each section of each chapter ends with notes, containing history, comments, remarks, indications of related topics, and references to the bibliography. I tried to systematically trace who did what when and indeed the list of references includes more than 400 entries from 1826 to 2022. However, I am painfully aware that this list is not complete and I apologize in advance to anybody I forgot to mention or that did not receive a correct attribution.

Let us end this (long) introduction with the pleasant duty of acknowledgments. First and foremost, I would like to deeply thank my wife, Adele, and my sons, Leonardo, Jacopo, and Niccolò that supported me (in both the Italian meanings of the word: sustain and endure) during all these years, not complaining too much when their respectively husband and father disappeared in a mathematical hole with a faraway look clearly showing that he was not listening to the much more important things that they were saying to him. But when I emerge from the hole, I look at them and I am proud of who they are and have become. And I apologize, at least most of the time, and they forgive me, most of the times.

I thankfully and fondly remember the late Edoardo Vesentini, my Ph. D. advisor, that so many years ago trusted me when I told him that I wanted to study holomorphic dynamical systems (a field that at the time I called iteration theory). If I had a good start in my mathematical career, it is because of him.

The complete list of friends and colleagues that helped me and accompanied me in this 30-year long journey would occupy too much space to be printed here—and I consider myself a very lucky person for having such an extensive list. A special role in my mathematical and personal life has been played by Filippo Bracci, Chiara de Fabritiis, Graziano Gentili, Giorgio Patrizio, Jasmin Raissy, Tamara Servi, and Francesca Tovena but I extend a heartfelt thanks to all of you on the list. I could not have done it without you, really.

A special thanks goes to my editors at de Gruyter, in particular to Apostolos Damialis, Steve Elliot, and Nadja Schedensack that put up with all my delays for so many years. At last, this book is completed; let us see how many years I will need for the next one.

Last but not least, I would like to thank my mother, Silvana, 88 years old but still going strong and still trying to understand what I actually do for a living. Teaching, is clear. Administration, is understandable. Mathematical research... but she always trusted me no matter what. Thanks mom, this book is for you.

1 The Schwarz lemma and Riemann surfaces

A characteristic feature of the theory of holomorphic functions is a very strong relationship between analytical properties of functions and geometrical properties of domains. A striking example of this phenomenon is the path connecting the Schwarz lemma to the Montel theorem passing through the Poincaré distance on hyperbolic Riemann surfaces. One of the main goals of this chapter is to unwind this thread starting from the very beginning, both for its own interest and because it will provide us with a number of tools we shall need later on.

The main connection between analytical and geometrical aspects of the theory is the invariant version of the Schwarz lemma devised by Pick, stating that any holomorphic function of the unit disk $\mathbb{D}$ of $\mathbb{C}$ into itself is a weak contraction for the Poincaré distance. In other words, the geometry, i. e., the Poincaré distance imposes a strong functional constraint on the space of holomorphic self-maps of $\mathbb{D}$: equicontinuity. Using the universal covering map, we can then carry over the construction of the Poincaré distance to any hyperbolic Riemann surface and, by means of the Ascoli–Arzelà theorem, this will eventually lead to a geometrical, and slightly unusual, proof of the Montel theorem.

From a differential geometric point of view, the contraction properties of holomorphic maps are due to the fact that the Poincaré metric has (constant) negative curvature. This is expressed by the Ahlfors lemma, that in turn has important consequences on the analytical properties of holomorphic functions.

In this chapter, we shall then present the Schwarz–Pick and the Ahlfors lemmas and their interpretations in terms of the Poincaré distance and metric, and we shall use them to study in detail geometrical and analytical properties of hyperbolic Riemann surfaces, providing a firm foundation for the discussion of holomorphic dynamics on these surfaces.

More precisely, in Section 1.1 we shall introduce the Schwarz–Pick lemma while in Section 1.2 we shall describe the geometry of the Poincaré distance. In Section 1.3, we shall introduce the upper half-plane $\mathbb{H}^+$, a domain canonically biholomorphic to the unit disk where sometimes computations are easier. The classical Schwarz lemma is traditionally used to compute the automorphism group of the unit disk; Section 1.4 is devoted to a detailed study of these automorphisms and their fixed points. Section 1.5 contains more recent materials, including a multipoint version of the Schwarz–Pick lemma that we shall use to get some inequalities that shall be needed in Chapter 2.

Section 1.6 is an introduction to the basic theory of Riemann surfaces. Our starting point is the Riemann uniformization theorem; from there on, the exposition is almost self-contained, but relies heavily on the theory of covering spaces. The necessary background can be found in the first few sections of [154] and in [260]. In Section 1.7, which is the main core of this chapter, we shall prove the Montel theorem and deduce many important facts about hyperbolic Riemann surfaces, including the Picard theorems.

https://doi.org/10.1515/9783110601978-001

In Section 1.8, we shall study the boundary behavior of the universal covering map of hyperbolic domains, proving a number of results that we shall need in Chapter 3 to study dynamics in these domains. Section 1.9 is devoted to the Poincaré metric, the infinitesimal version of the Poincaré distance, and Section 1.10 to the Ahlfors lemma, a far-reaching differential geometric generalization of the Schwarz lemma. Warning: these two sections need some basic knowledge of differential geometry to be fully understood. Finally, the last section of this chapter is devoted to a characterization, expressed in terms of the Poincaré metric, of a class of domains, the Bloch domains, that we shall need in Chapter 3 to study random dynamics.

1.1 The Schwarz–Pick lemma

In this section, we shall prove the basic result underlying all the material covered in this book: the Schwarz–Pick lemma. As a first application, we shall use it to compute the automorphism group of the unit disk in the complex plane.

Definition 1.1.1. The (Euclidean) *open disk* $D(z_0, r)$ of *center* $z_0 \in \mathbb{C}$ and *radius* $r > 0$ is given by

$$D(z_0, r) = \{z \in \mathbb{C} \mid |z - z_0| < r\}.$$

If $z_0 = 0$, we shall sometimes write $\mathbb{D}_r$ instead of $D(0, r)$; moreover, $\mathbb{D}$ will denote the open *unit disk* of center 0 and radius 1 in the complex plane. The *unit circle* $\partial\mathbb{D}$ will sometimes be denoted by $\mathbb{S}^1$. Finally, if $A \subset \mathbb{C}$ and $z \in \mathbb{C}$ we shall denote by $d(z, A)$ the Euclidean distance of z from A, i. e.,

$$d(z, A) = \inf_{w \in A} |z - w|.$$

We can now state and prove the fundamental and classical *Schwarz lemma*.

Theorem 1.1.2 (Schwarz lemma, 1905). *Let $f: \mathbb{D} \to \mathbb{D}$ be a holomorphic function such that $f(0) = 0$. Then*

$$|f(z)| \leq |z| \tag{1.1}$$

for all $z \in \mathbb{D}$; moreover,

$$|f'(0)| \leq 1. \tag{1.2}$$

Furthermore, equality in (1.1) for some $z \neq 0$ or in (1.2) occurs if and only if f is a rotation, i. e., if and only if there is $\theta \in \mathbb{R}$ such that $f(z) = e^{i\theta} z$ for all $z \in \mathbb{D}$.

Proof. Since $f(0) = 0$, we can define a holomorphic function $g\colon \mathbb{D} \to \mathbb{C}$ by setting

$$g(z) = \begin{cases} \frac{f(z)}{z} & \text{if } z \neq 0, \\ f'(0) & \text{if } z = 0. \end{cases}$$

Given $z \in \mathbb{D}$, pick $|z| < r < 1$; then, by the maximum principle, we get

$$|g(z)| \leq \sup_{|w|=r} |g(w)| = \sup_{|w|=r} \frac{|f(w)|}{r} \leq \frac{1}{r}.$$

Letting $r \to 1$ we get $|g(z)| \leq 1$, i. e., (1.1) and (1.2).

If equality holds in (1.2), or in (1.1) for some nonzero z, then, again by the maximum principle, g must be a constant, necessarily of modulus one, and the last assertion follows. □

As a first example of the power of the Schwarz lemma, we use it to prove the classical *Liouville theorem.*

Theorem 1.1.3 (Liouville, 1844). *A bounded entire function is necessarily constant.*

Proof. Let $f\colon \mathbb{C} \to \mathbb{D}$ be a bounded holomorphic function. Composing with a translation, we can assume that $f(0) = 0$; multiplying by a suitable constant, we can assume that the image of f is contained in $\mathbb{D}$.

For any $R > 0$, let $f_R\colon \mathbb{D} \to \mathbb{D}$ be given by $f_R(\zeta) = f(R\zeta)$. Since $f_R(0) = 0$, the Schwarz lemma yields

$$|f_R(\zeta)| \leq |\zeta| \tag{1.3}$$

for all $\zeta \in \mathbb{D}$.

Choose now $w \in \mathbb{C}$. For R large enough, we have $|w| < R$; hence $w/R \in \mathbb{D}$ and (1.3) yields

$$|f(w)| = |f_R(w/R)| \leq \frac{|w|}{R}.$$

Letting $R \to +\infty$, we then get $f(w) = 0$. Since w was arbitrary, we conclude that $f \equiv 0$, and we are done. □

Historically, one of the first applications of the Schwarz lemma has been the computation of the automorphism group of $\mathbb{D}$.

Definition 1.1.4. Given two open sets $D_1, D_2 \subseteq \mathbb{C}$, we denote by $\mathrm{Hol}(D_1, D_2)$ the space of holomorphic functions from D_1 into D_2, endowed with the topology of uniform convergence on compact subsets (or equivalently, with the compact-open topology). A *biholomorphism* is an invertible holomorphic function; it is well known that the inverse is automatically holomorphic. If there exists a biholomorphism $f\colon D_1 \to D_2$ we shall say that D_1 and D_2 are *biholomorphic*. An *automorphism* of D is a biholomorphism of D with itself. The group of all automorphisms of D will be denoted by $\mathrm{Aut}(D) \subset \mathrm{Hol}(D, D)$.

Proposition 1.1.5. *Every automorphism $\gamma \colon \mathbb{D} \to \mathbb{D}$ of $\mathbb{D}$ is of the form*

$$\gamma(z) = e^{i\theta} \frac{z - a}{1 - \bar{a}z} \tag{1.4}$$

for some $\theta \in \mathbb{R}$ and $a = \gamma^{-1}(0) \in \mathbb{D}$. In particular, every $\gamma \in \mathrm{Aut}(\mathbb{D})$ extends continuously to a homeomorphism of $\overline{\mathbb{D}}$ onto itself.

Proof. First of all, every γ of the form (1.4) is an automorphism of $\mathbb{D}$. Indeed,

$$1 - \left| \frac{z - a}{1 - \bar{a}z} \right|^2 = \frac{(1 - |a|^2)(1 - |z|^2)}{|1 - \bar{a}z|^2} \tag{1.5}$$

for all $z \in \mathbb{D}$, and so $\gamma(\mathbb{D}) \subseteq \mathbb{D}$. Furthermore, the inverse of γ is

$$\gamma^{-1}(z) = e^{-i\theta} \frac{z + ae^{i\theta}}{1 + \bar{a}e^{-i\theta}z},$$

and thus it is again of the form (1.4). In particular, every automorphism of this form extends continuously to a homeomorphism of $\overline{\mathbb{D}}$ onto itself.

Let Γ be the set consisting of the automorphisms of type (1.4). It is easy to check that Γ is a group under composition and that for every $z_0 \in \mathbb{D}$ there is $\gamma \in \Gamma$ such that $\gamma(z_0) = 0$.

Let ϕ be an automorphism of $\mathbb{D}$; we have to prove that $\phi \in \Gamma$. Put $z_0 = \phi(0)$ and choose $\gamma_0 \in \Gamma$ such that $\gamma_0(z_0) = 0$. Then $\phi_0 = \gamma_0 \circ \phi$ is an automorphism of $\mathbb{D}$ such that $\phi_0(0) = 0$. If we prove that $\phi_0 \in \Gamma$ we are done, because then $\phi = \gamma_0^{-1} \circ \phi_0 \in \Gamma$, too.

Applying (1.2) to ϕ_0 and ϕ_0^{-1}, we see that $|\phi_0'(0)| = 1$. Therefore, the uniqueness assertion in the Schwarz lemma implies that ϕ_0 is a rotation; in particular, $\phi_0 \in \Gamma$, and the proof is completed. □

Definition 1.1.6. For each $a \in \mathbb{D}$, let $\gamma_a \in \mathrm{Aut}(\mathbb{D})$ be given by

$$\gamma_a(z) = \frac{z - a}{1 - \bar{a}z}.$$

In particular, every automorphism of $\mathbb{D}$ is given by some γ_a followed by a rotation.

Remark 1.1.7. For later use, we record here that $\gamma_a(a) = 0$ and that

$$\gamma_a'(z) = \frac{1 - |a|^2}{(1 - \bar{a}z)^2};$$

in particular,

$$\gamma_a'(0) = 1 - |a|^2 \quad \text{and} \quad \gamma_a'(a) = \frac{1}{1 - |a|^2}.$$

Thus every automorphism of $\mathbb{D}$ extends continuously to a neighborhood of $\overline{\mathbb{D}}$; in particular, every automorphism of $\mathbb{D}$ sends $\partial\mathbb{D}$ into itself. It turns out that the corresponding action of $\mathrm{Aut}(\mathbb{D})$ on $\partial\mathbb{D}$ is fairly good.

Definition 1.1.8. We say that a group Γ acting on a set X is *transitive* if for every $x_1, x_2 \in X$ there is $\gamma \in \Gamma$ such that $\gamma(x_1) = x_2$; *simply transitive* if the previous γ is unique; *doubly transitive* if for every $x_1, x_2, y_1, y_2 \in X$ with $x_1 \neq x_2$ and $y_1 \neq y_2$ there is $\gamma \in \Gamma$ such that $\gamma(x_1) = y_1$ and $\gamma(x_2) = y_2$.

Then we have the following result.

Corollary 1.1.9. $\mathrm{Aut}(\mathbb{D})$ *acts transitively on* $\mathbb{D}$ *and doubly transitively on* $\partial\mathbb{D}$. *Furthermore, for each pair* $(z_0, v_0) \in \mathbb{D} \times \mathbb{C}$ *with* $v_0 \neq 0$ *there is a unique* $\gamma \in \mathrm{Aut}(\mathbb{D})$ *such that* $\gamma(0) = z_0$ *and* $\gamma'(0)$ *is a positive multiple of* v_0.

Proof. Given $z_0, w_0 \in \mathbb{D}$, the automorphism $\gamma_{w_0}^{-1} \circ \gamma_{z_0}$ sends z_0 in w_0, and thus $\mathrm{Aut}(\mathbb{D})$ is transitive on $\mathbb{D}$.

Given $\sigma_0, \tau_0 \in \partial\mathbb{D}$, the rotation $\gamma(z) = \tau_0\overline{\sigma_0}z$ sends σ_0 in τ_0, and thus $\mathrm{Aut}(\mathbb{D})$ is transitive on $\partial\mathbb{D}$. To prove that it is doubly transitive, choose $\sigma_1, \sigma_2, \tau_1, \tau_2 \in \partial\mathbb{D}$ with $\sigma_1 \neq \sigma_2$ and $\tau_1 \neq \tau_2$; we seek $\gamma \in \mathrm{Aut}(\mathbb{D})$ such that $\gamma(\sigma_1) = \tau_1$ and $\gamma(\sigma_2) = \tau_2$. Obviously, it is enough to show that such a γ exists when $\sigma_1 = 1$ and $\sigma_2 = -1$. Moreover, we can also assume $\tau_1 = 1$, because $\mathrm{Aut}(\mathbb{D})$ contains the rotations. Summing up, given $\tau \in \partial\mathbb{D}$ we want $\gamma \in \mathrm{Aut}(\mathbb{D})$ such that $\gamma(1) = 1$ and $\gamma(-1) = \tau$. Let $\sigma \in \partial\mathbb{D}$ be the square root of $-\tau$ with positive real part; set $a = (\sigma - 1)/(\sigma + 1)$ and $\alpha = (\sigma + 1)/(\overline{\sigma} + 1) = (1 - \overline{a})/(1 - a)$. Then $|a| < 1 = |\alpha|$, and

$$\gamma(z) = \alpha \frac{z - a}{1 - \overline{a}z}$$

behaves as required.

Finally, choose a $\gamma_0 \in \mathrm{Aut}(\mathbb{D})$ such that $\gamma_0(0) = z_0$. All other automorphisms of $\mathbb{D}$ sending 0 to z_0 are of the form $\gamma(z) = \gamma_0(e^{i\theta}z)$ for some $\theta \in \mathbb{R}$. Since $\gamma'(0) = e^{i\theta}\gamma_0'(0) \neq 0$, it follows that there exists a unique $\gamma \in \mathrm{Aut}(\mathbb{D})$ such that $\gamma(0) = z_0$ and $\gamma'(0) = \lambda v_0$ with $\lambda > 0$. □

The transitivity of $\mathrm{Aut}(\mathbb{D})$ has the useful consequence that if we have to prove something (invariant under automorphisms) about one point $z_0 \in \mathbb{D}$ we can assume without loss of generality that $z_0 = 0$. Analogously, if we have to prove something invariant under automorphisms about two points $\sigma_1, \sigma_2 \in \partial\mathbb{D}$ we can assume without loss of generality that $\sigma_1 = 1$ and $\sigma_2 = -1$.

A first application of this principle allows us to prove that no holomorphic self-map of $\mathbb{D}$ different from the identity can have more than one fixed point.

Definition 1.1.10. Two self-maps $f: X \to X$ and $g: Y \to Y$ are *conjugated* if there exists a bijection $\gamma: Y \to X$ such that $g = \gamma^{-1} \circ f \circ \gamma$. If f and g are continuous (resp., holomorphic), we shall always assume that γ is a homeomorphism (resp., a biholomorphism). Notice that the only map conjugated to the identity is the identity itself.

Remark 1.1.11. In the sequel, the expression "up to conjugation" will mean that we are replacing a map f by a conjugated map $g = \gamma^{-1} \circ f \circ \gamma$.

Definition 1.1.12. A *fixed point* of a function $f: X \to X$ is a point $x_0 \in X$ such that $f(x_0) = x_0$. We shall denote by $\mathrm{Fix}(f)$ the set of fixed points of f.

Remark 1.1.13. If $g = \gamma^{-1} \circ f \circ \gamma$ is conjugated to f, then $\mathrm{Fix}(f) = \gamma(\mathrm{Fix}(g))$ because $\gamma^{-1} \circ f \circ \gamma(x) = x$ if and only if $f(\gamma(x)) = \gamma(x)$.

Corollary 1.1.14. *Let $f \in \mathrm{Hol}(\mathbb{D}, \mathbb{D})$ be with two distinct fixed points $z_0 \neq z_1 \in \mathbb{D}$. Then $f \equiv \mathrm{id}_{\mathbb{D}}$.*

Proof. Thanks to Corollary 1.1.9, up to a conjugation we can assume $z_0 = 0$. Then f satisfies $f(0) = 0$ and $f(z_1) = z_1$; from the equality assertion in Theorem 1.1.2, we get $f(z) = e^{i\theta} z$ for a suitable $\theta \in \mathbb{R}$. But $e^{i\theta} z_1 = z_1$ implies $e^{i\theta} = 1$, and so $f \equiv \mathrm{id}_{\mathbb{D}}$ as claimed. □

A sufficient condition for the existence of fixed points, that will later be considerably generalized and proved in a different way (see Proposition 3.3.4), is given by the *Ritt theorem*.

Corollary 1.1.15 (Ritt theorem, 1920). *Let $f: \mathbb{D} \to \mathbb{D}$ be holomorphic and such that $f(\mathbb{D})$ is relatively compact in $\mathbb{D}$. Then f has a unique fixed point in $\mathbb{D}$.*

Proof. Let $r < 1$ be such that $|f(z)| < r$ for all $z \in \mathbb{D}$. Then on $\partial D(0, r)$ we have

$$\left|(z - f(z)) - z\right| = \left|f(z)\right| < r = |z|.$$

Hence an application of the Rouché theorem (see, e. g., Theorem A.1.1) to $\mathrm{id}_{\mathbb{D}} - f$ and $\mathrm{id}_{\mathbb{D}}$ shows that the number of zeroes of $\mathrm{id}_{\mathbb{D}} - f$ (i. e., the number of fixed points of f) is equal to the number of zeroes of $\mathrm{id}_{\mathbb{D}}$ in $D(0, r)$, that is one. □

Since $\mathrm{Aut}(\mathbb{D})$ is transitive on $\mathbb{D}$, from the point of view of complex geometry all points of $\mathbb{D}$ must have the same properties. In particular, the origin is not a special point, and the special role played by it in the classical Schwarz lemma must be purely incidental. Indeed, using the automorphisms we can restate the Schwarz lemma in a more homogeneous form, our first version of the *Schwarz–Pick lemma*.

Corollary 1.1.16 (Schwarz–Pick lemma, 1915). *Let $f: \mathbb{D} \to \mathbb{D}$ be holomorphic. Then*

$$\left|\frac{f(z) - f(w)}{1 - \overline{f(w)}f(z)}\right| \leq \left|\frac{z - w}{1 - \overline{w}z}\right| \tag{1.6}$$

for all $z, w \in \mathbb{D}$; moreover,

$$\frac{|f'(z)|}{1-|f(z)|^2} \le \frac{1}{1-|z|^2} \tag{1.7}$$

for all $z \in \mathbb{D}$. Furthermore, equality in (1.6) for some $z \neq w \in \mathbb{D}$ or in (1.7) for some $z \in \mathbb{D}$ occurs if and only if $f \in \mathrm{Aut}(\mathbb{D})$, and then the equality holds for all $z, w \in \mathbb{D}$.

Proof. Fix $w \in \mathbb{D}$. Let $\gamma_1, \gamma_2 \in \mathrm{Aut}(\mathbb{D})$ be given by

$$\gamma_1(z) = \frac{z+w}{1+\overline{w}z} \quad \text{and} \quad \gamma_2(z) = \frac{z-f(w)}{1-\overline{f(w)}z},$$

and put $g = \gamma_2 \circ f \circ \gamma_1 \in \mathrm{Hol}(\mathbb{D},\mathbb{D})$. Then $g(0) = 0$ and the classical Schwarz lemma yields

$$\left|\frac{f(\gamma_1(\zeta)) - f(w)}{1-\overline{f(w)}f(\gamma_1(\zeta))}\right| \le |\zeta|$$

for all $\zeta \in \mathbb{D}$, and

$$\frac{1}{1-|f(w)|^2}|f'(w)|(1-|w|^2) \le 1,$$

where we used Remark 1.1.7. Taking $\zeta = \gamma_1^{-1}(z)$ we are done. □

Notes to Section 1.1

In 1869, Schwarz [371] proved the following result: let $f: \mathbb{D} \to \mathbb{D}$ be a holomorphic injective function continuous up to the boundary such that $f(0) = 0$ and $\rho_1 \le |f(\tau)| \le \rho_2$ for all $\tau \in \partial\mathbb{D}$; then $\rho_1|z| \le |f(z)| \le \rho_2|z|$ for all $z \in \overline{\mathbb{D}}$. A result closer to Theorem 1.1.2 has been proved by Poincaré [333] in 1884; he used it to compute $\mathrm{Aut}(\mathbb{D})$. Only in 1905, Carathéodory [92] recognized the real significance of the Schwarz lemma, giving it the statement and the proof (inspired by E. Schmidt) we know today, and christening it in [93].

The Schwarz lemma has been generalized in many ways. If $f \in \mathrm{Hol}(\mathbb{D},\mathbb{D})$ is not an automorphism, then it is a strict contraction with respect to the Poincaré distance and metric; it is natural to wonder whether it would be possible to strengthen the statement of the Schwarz lemma at least for some classes of functions, along the line of the original statement by Schwarz quoted above. For instance, this is possible assuming to know the image of a given point (see, e. g., [136, 290, 282, 48] and [139, Section 6.3]), or the value of the derivative at a given point (see, e. g., [355, 47, 283, 48]) or that the function is univalent (see, e. g., [20] and references therein).

Another line of research consists in proving Schwarz lemmas for higher order derivatives. The best estimate

$$\left(1-|z|\right)^{k-1}\left|\frac{f^{(k)}(z)}{k!}\right| \le \frac{1-|f(z)|^2}{1-|z|^2}$$

for all $z \in \mathbb{D}$ has been proved by Ruscheweyh [362] in 1985; see [19] for a proof with Hilbert space methods. See also [313, 274, 275, 57, 430, 134, 27] and references therein.

[90] and [62] contain versions of the Schwarz lemma expressed in terms of diameters, capacity, and area of the image of subdisks.

An interesting extension of the Schwarz lemma has been proved by Nehari [302] in 1947: take $f \in \mathrm{Hol}(\mathbb{D}, \mathbb{D})$ with a finite number of critical points in $\mathbb{D}$, and let $B \in \mathrm{Hol}(\mathbb{D}, \mathbb{D})$ be a Blaschke product (see Definition 1.5.5) having the same critical points as f with the same multiplicities (such a Blaschke product always exists and it is unique up to post-composition with an automorphism); then $|f^h(z)| \le |B^h(z)|$, where f^h (respectively, B^h) denotes the hyperbolic derivative of f (respectively, B) in the sense of Definition 1.5.1. Furthermore, equality at one point holds if and only if $f = \gamma \circ B$ with $\gamma \in \mathrm{Aut}(\mathbb{D})$, where at a critical point $z_0 \in \mathbb{D}$ the equality should be intended as $\lim_{z\to z_0} |f^h(z)|/|B^h(z)| = 1$. See also [251, 252, 356] for more details on the Nehari theorem.

More on the history of the Schwarz and Schwarz–Pick lemmas can be found in [359, 312, 68] and in the books [27, 141].

The Liouville theorem 1.1.3 has actually been first proved by Cauchy [107] in 1844; the proof we gave is taken from [100, § 69].

Corollary 1.1.15 is a particular case of a more general statement proved in [352]; see Proposition 3.3.4.

1.2 The Poincaré distance

The action of $\mathrm{Aut}(\mathbb{D})$ on $\mathbb{D}$ is transitive but it is *not* doubly transitive. Indeed, the Schwarz lemma implies that given $z_1, z_2 \in \mathbb{D}$ there exists $\gamma \in \mathrm{Aut}(\mathbb{D})$ such that $\gamma(0) = 0$ and $\gamma(z_1) = z_2$ if and only if $|z_1| = |z_2|$. This is the first clue to the existence of an underlying geometrical structure that must be preserved by $\mathrm{Aut}(\mathbb{D})$ and, therefore, should be strictly correlated to the holomorphic structure.

This geometrical structure is represented by the *Poincaré distance.*

Definition 1.2.1. The *Poincaré distance* $\omega\colon \mathbb{D} \times \mathbb{D} \to \mathbb{R}^+$ on the unit disk $\mathbb{D}$ is given by

$$\omega(z_1, z_2) = \frac{1}{2} \log \frac{1 + \left|\frac{z_2 - z_1}{1 - \overline{z_1} z_2}\right|}{1 - \left|\frac{z_2 - z_1}{1 - \overline{z_1} z_2}\right|}. \tag{1.8}$$

Remark 1.2.2. The function $t \mapsto \frac{1}{2} \log \frac{1+t}{1-t}$ is the inverse $\tanh^{-1}$ of the hyperbolic tangent function $\tanh t = \frac{e^t - e^{-t}}{e^t + e^{-t}}$. In particular, $\tanh^{-1}$ is a strictly increasing diffeomorphism between $[0, 1)$ and $\mathbb{R}^+ = [0, +\infty)$. We recall here the useful addition formula for the hyperbolic tangent:

$$\tanh(a + b) = \frac{\tanh a + \tanh b}{1 + (\tanh a)(\tanh b)}, \tag{1.9}$$

valid for all $a, b \in \mathbb{R}$.

By definition, for all $z \in \mathbb{D}$ we have

$$\omega(0, z) = \tanh^{-1}(|z|) = \frac{1}{2} \log \frac{1 + |z|}{1 - |z|}.$$

Moreover, if $\gamma \in \mathrm{Aut}(\mathbb{D})$ then Corollary 1.1.16 applied to γ and γ^{-1} yields

$$\omega(\gamma(z_1), \gamma(z_2)) = \omega(z_1, z_2)$$

for all z_1, $z_2 \in \mathbb{D}$; in other words, the Poincaré distance is invariant under the action of the automorphism group of $\mathbb{D}$.

Of course, we have to verify that (1.8) actually defines a distance. The symmetry is evident, as well as the fact that $\omega(z, w) = 0$ if and only if $z = w$. To prove the triangular inequality,

$$\omega(z_1, z_3) \leq \omega(z_1, z_2) + \omega(z_2, z_3),$$

using the transitivity of $\mathrm{Aut}(\mathbb{D})$ and the invariance of ω under the action of $\mathrm{Aut}(\mathbb{D})$, we can assume $z_1 = 0$, and thus applying tanh and recalling (1.9) we see that it suffices to prove that

$$|z_3| \leq \frac{|z_2| + \left|\frac{z_2 - z_3}{1 - \overline{z_3} z_2}\right|}{1 + |z_2|\left|\frac{z_2 - z_3}{1 - \overline{z_3} z_2}\right|}$$

for all $z_2, z_3 \in \mathbb{D}$. This inequality is equivalent to

$$\frac{|z_3| - |z_2|}{1 - |z_2||z_3|} \leq \left|\frac{z_2 - z_3}{1 - \overline{z_3} z_2}\right|,$$

which follows easily from $1 - |z_2||z_3| \leq |1 - \overline{z_3} z_2|$, and the triangular inequality is proved.

We may now rephrase Corollary 1.1.16, obtaining the (first part of the) invariant *Schwarz–Pick lemma.*

Theorem 1.2.3 (Schwarz–Pick lemma, 1915). *Let $f: \mathbb{D} \to \mathbb{D}$ be holomorphic. Then*

$$\omega(f(z), f(w)) \leq \omega(z, w) \tag{1.10}$$

for all z, $w \in \mathbb{D}$. Moreover, equality in (1.10) *for some $z \neq w \in \mathbb{D}$ occurs if and only if $f \in \mathrm{Aut}(\mathbb{D})$, and then the equality holds for all $z, w \in \mathbb{D}$.*

Proof. Since $\tanh^{-1}$ is strictly increasing, (1.10) is equivalent to (1.6), and the assertions follow. □

Remark 1.2.4. In Section 1.9, after recalling a few basic facts of differential geometry, we shall also give a geometrical interpretation to (1.7), expressed in terms of the Poincaré metric, the infinitesimal version of the Poincaré distance; see Theorem 1.9.7.

So, we have constructed a distance on $\mathbb{D}$ weakly contracted by holomorphic functions. This is exactly the geometrical structure whose existence we suggested before.

Corollary 1.2.5. *Let z_1, z_2, w_1, w_2 be four points of $\mathbb{D}$. Then there exists an automorphism γ of $\mathbb{D}$ such that $\gamma(z_1) = w_1$ and $\gamma(z_2) = w_2$ if and only if $\omega(z_1, z_2) = \omega(w_1, w_2)$.*

Proof. It is clear that, by Theorem 1.2.3, the existence of such an automorphism implies the equality of the Poincaré distances. Conversely, assume that $\omega(z_1, z_2) = \omega(w_1, w_2)$. Clearly, we can suppose $w_1 = 0$. Let $\gamma_1 \in \mathrm{Aut}(\mathbb{D})$ be such that $\gamma_1(z_1) = 0 = w_1$. Then

$$\omega(0, \gamma_1(z_2)) = \omega(z_1, z_2) = \omega(0, w_2),$$

i. e., $|\gamma_1(z_2)| = |w_2|$. Hence there exists a rotation γ_2 about the origin such that $\gamma_2(\gamma_1(z_2)) = w_2$, and $\gamma = \gamma_2 \circ \gamma_1$ is the automorphism we were seeking. □

Now we would like to describe the geometry of the Poincaré metric; in particular, we would like to know the shape of balls and of geodesics.

Definition 1.2.6. Given $z \in \mathbb{D}$ and $R > 0$, the *Poincaré ball* $B_{\mathbb{D}}(z, R)$ of *center* z and *radius* R is

$$B_{\mathbb{D}}(z_0, R) = \{z \in \mathbb{D} \mid \omega(z, z_0) < R\}.$$

A *geodesic* for ω is a continuous curve $\sigma\colon \mathbb{R} \to \mathbb{D}$ such that $\omega(\sigma(s), \sigma(t)) = |s - t|$ for all $s, t \in \mathbb{R}$; a *geodesic ray* is the restriction to $\mathbb{R}^+$ of a geodesic.

Proposition 1.2.7. (i) *The Poincaré ball $B_{\mathbb{D}}(z_0, R)$ of center $z_0 \in \mathbb{D}$ and radius $R > 0$ is an open Euclidean disk of center*

$$\zeta_0 = \frac{1 - (\tanh R)^2}{1 - (\tanh R)^2 |z_0|^2} z_0,$$

and radius

$$\rho = \frac{(\tanh R)(1 - |z_0|^2)}{1 - (\tanh R)^2 |z_0|^2}. \tag{1.11}$$

(ii) *If $z \in \partial B_{\mathbb{D}}(z_0, R)$, then*

$$\frac{||z_0| - \tanh R|}{1 - (\tanh R)|z_0|} \le |z| \le \frac{|z_0| + \tanh R}{1 + (\tanh R)|z_0|}.$$

(iii) *The Poincaré distance is a complete distance inducing on $\mathbb{D}$ the Euclidean topology.*

(iv) *The geodesics for the Poincaré metric are the (suitably parametrized) Euclidean diameters of $\mathbb{D}$ and the intersections with $\mathbb{D}$ of Euclidean circles orthogonal to $\partial\mathbb{D}$. In particular, geodesics are smooth curves and any two distinct points of $\mathbb{D}$ are connected by a unique geodesic.*

Proof. (i) Clearly,

$$B_{\mathbb{D}}(z_0, R) = \left\{w \in \mathbb{D} \;\middle|\; \left|\frac{w - z_0}{1 - \overline{z_0} w}\right| < \tanh R\right\}.$$

Let us put, for simplicity, $r = \tanh R \in (0,1)$. Then $w \in B_{\mathbb{D}}(z_0, R)$ if and only if $|w - z_0|^2 < r^2|1 - \overline{z_0}w|^2$ if and only if

$$|w|^2 - 2\frac{1-r^2}{1-r^2|z_0|^2}\operatorname{Re}(\overline{z_0}w) < \frac{r^2 - |z_0|^2}{1-r^2|z_0|^2}.$$

On the other hand,

$$|w-\zeta_0|^2 = |w|^2 - 2\frac{1-r^2}{1-r^2|z_0|^2}\operatorname{Re}(\overline{z_0}w) + \frac{(1-r^2)^2}{(1-r^2|z_0|^2)^2}|z_0|^2;$$

hence w belongs to the Euclidean disk of center ζ_0 and radius ρ if and only if

$$\begin{aligned}|w|^2 - 2\frac{1-r^2}{1-r^2|z_0|^2}\operatorname{Re}(\overline{z_0}w) &< \frac{r^2(1-|z_0|^2)^2}{(1-r^2|z_0|^2)^2} - \frac{(1-r^2)^2}{(1-r^2|z_0|^2)^2}|z_0|^2\\ &= \frac{r^2-|z_0|^2}{1-r^2|z_0|^2},\end{aligned}$$

and we are done.

(ii) We clearly have $||\zeta_0| - \rho| \le |z| \le |\zeta_0| + \rho$. The assertion then follows from the equalities

$$|\zeta_0| - \rho = \frac{1-r^2}{1-r^2|z_0|^2}|z_0| - \frac{r(1-|z_0|^2)}{1-r^2|z_0|^2} = \frac{|z_0|-r}{1-r|z_0|}$$

and

$$|\zeta_0| + \rho = \frac{1-r^2}{1-r^2|z_0|^2}|z_0| + \frac{r(1-|z_0|^2)}{1-r^2|z_0|^2} = \frac{|z_0|+r}{1+r|z_0|}.$$

(iii) The fact that ω induces the Euclidean topology follows immediately from (i). This also shows that the closure of a Poincaré ball is always compact in $\mathbb{D}$, and thus ω is a complete distance.

(iv) Given $\tau \in \partial\mathbb{D}$ define $\sigma_\tau\colon \mathbb{R} \to \mathbb{D}$ by

$$\sigma_\tau(t) = \tanh(t)\tau. \tag{1.12}$$

Then (1.9) yields

$$\omega(\sigma_\tau(s), \sigma_\tau(t)) = \omega\left(0, \left|\frac{\tanh(t) + \tanh(-s)}{1 + \tanh(t)\tanh(-s)}\right|\right) = \omega(0, \tanh|t-s|) = |t-s|,$$

and thus σ_τ is a geodesic with $\sigma_\tau(0) = 0$.

Conversely, assume that $\sigma\colon \mathbb{R} \to \mathbb{D}$ is a geodesic with $\sigma(0) = 0$. In particular, for every $t \in \mathbb{R}$ we have $\omega(0, \sigma(t)) = |t|$; hence $|\sigma(t)| = |\tanh(t)|$ for all $t \in \mathbb{R}$. Thus we can write $\sigma(t) = \tau(t)\tanh(t)$ for a suitable continuous curve $\tau\colon \mathbb{R} \to \partial\mathbb{D}$; we claim that τ is constant.

Take t_0, $t \in \mathbb{R}^*$ and put $\tau_0(t) = \overline{\tau(t_0)}\tau(t)$. Since σ is a geodesic, we have $\omega(\sigma(t), \sigma(t_0)) = |t - t_0|$, and hence

$$\left|\frac{\tau_0(t)\tanh(t) - \tanh(t_0)}{1 - \tau_0(t)\tanh(t)\tanh(t_0)}\right| = \left|\frac{\sigma(t) - \sigma(t_0)}{1 - \overline{\sigma(t_0)}\sigma(t)}\right| = \tanh(|t - t_0|)$$
$$= \left|\frac{\tanh(t) - \tanh(t_0)}{1 - \tanh(t)\tanh(t_0)}\right|,$$

again by (1.9). Taking the square of this equality, subtracting 1 and recalling (1.5) we get

$$\left|1 - \tau_0(t)\tanh(t)\tanh(t_0)\right|^2 = \left|1 - \tanh(t)\tanh(t_0)\right|^2.$$

Developing the modulus squared, we see that this can happen if and only if $\operatorname{Re}\tau_0(t) \equiv 1$. But this means that $\tau_0 \equiv 1$, and thus τ is constant, as claimed.

Summing up, we have proved that the geodesics issuing from the origin are all and only of the form $\sigma_\tau(t) = \tanh(t)\tau$ with $\tau \in \partial\mathbb{D}$; in particular, they are all smooth, and each $z \in \mathbb{D} \setminus \{0\}$ is connected to 0 by a unique geodesic. Since automorphisms are isometries for the Poincaré distance, and the automorphism group is transitive on $\mathbb{D}$, it follows that the geodesics are all and only of the form

$$\sigma(t) = e^{i\theta}\gamma_a(\tanh t) = e^{i\theta}\frac{\tanh t - a}{1 - \overline{a}\tanh t}$$

for suitable $a \in \mathbb{D}$ and $\theta \in \mathbb{R}$. In particular, they are all smooth, and two distinct points are connected by a unique geodesic.

We are left to prove that when $a \neq 0$ the support of σ is the intersection with $\mathbb{D}$ of an Euclidean circle orthogonal to ∂D. The support of σ is obtained by rotating the set $\gamma_a(C)$, where $C = (-1, 1)$. A point w belongs to $\gamma_a(C)$ if and only if $\gamma_a^{-1}(w) \in C$, i. e., if and only if

$$0 = \operatorname{Im}\frac{w + a}{1 + \overline{a}w} = \frac{1}{|1 + \overline{a}w|^2}\{(1 + |w|^2)\operatorname{Im} a - \operatorname{Im}[(1 - a^2)\overline{w}]\}.$$

If $\operatorname{Im} a = 0$, this happens if and only if $\operatorname{Im} w = 0$, and so in this case $\gamma_a(C) = C$. If $\operatorname{Im} a \neq 0$, then $w \in \mathbb{D}$ belongs to $\gamma_a(C)$ if and only if

$$|w|^2 - 2\operatorname{Re}\left[i\frac{a^2 - 1}{2\operatorname{Im} a}\overline{w}\right] + 1 = 0,$$

i. e., if and only if

$$|w - b|^2 = |b|^2 - 1, \tag{1.13}$$

where $b = i\frac{a^2-1}{2\operatorname{Im} a}$. Since $|b| > 1$, (1.13) is the equation of an Euclidean circle orthogonal to $\partial\mathbb{D}$, and we are done. □

Figure 1.1 shows four Poincaré balls with their Euclidean and Poincaré centers, and Figure 1.2 shows four Poincaré geodesics issuing from the same point.

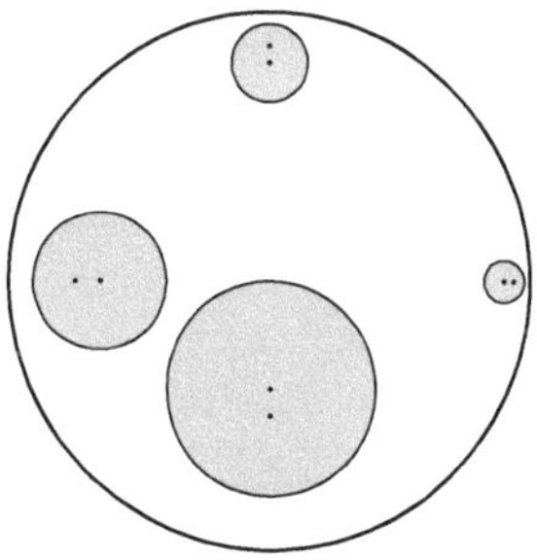

Figure 1.1: Poincaré balls with (Poincaré) radius 1/2. The Poincaré centers are the dots closer to $\partial\mathbb{D}$.

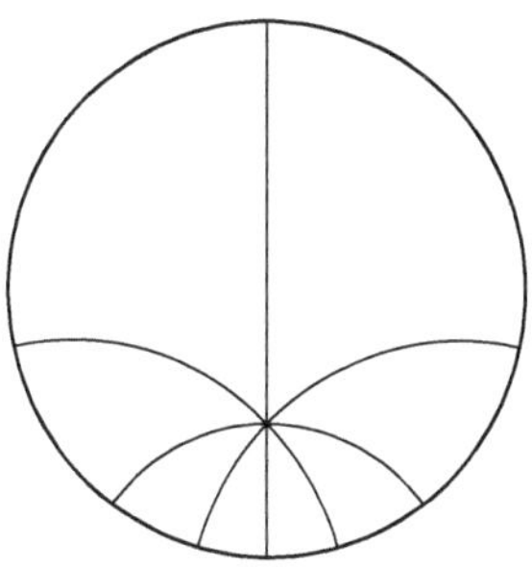

Figure 1.2: Geodesics for the Poincaré metric.

The Euclidean center ζ_0 of a Poincaré ball belongs to the radius passing through the Poincaré center z_0 but it is closer to the origin, and we have the neat relation

$$z_0 - \zeta_0 = \rho(\tanh R)z_0,$$

where ρ is the Euclidean radius of the ball $B_{\mathbb{D}}(z_0, R)$. Furthermore, if we let z_0 go to the boundary of $\mathbb{D}$ keeping the Poincaré radius R constant then the Euclidean radius ρ of $B_{\mathbb{D}}(z_0, R)$ goes to 0. In the next chapter, we shall discuss what happens to $B_{\mathbb{D}}(z_0, R)$ when we send z_0 to the boundary keeping the Euclidean radius constant.

Notes to Section 1.2

The Poincaré distance has been first studied by Beltrami [55, 56] in 1868 and then by Poincaré starting from 1882 in his work on Fuchsian groups [331, 332].

It is possible to give an axiomatic characterization of the Poincaré distance: it is the unique $\mathrm{Aut}(\mathbb{D})$-invariant function $d\colon \mathbb{D} \times \mathbb{D} \to \mathbb{R}$ such that $d(0,s) = d(0,t) + d(t,s)$ for all $0 \le t \le s < 1$ and such that

$$\lim_{t\to 0^+} \frac{d(0,t)}{t} = 1.$$

For a proof see, e. g., [210, Proposition 1.1.23].

One can also introduce the *Möbius distance* $\mu\colon \mathbb{D} \times \mathbb{D} \to \mathbb{R}^+$ by

$$\mu(z,w) = \left|\frac{w-z}{1-\bar{z}w}\right|.$$

It is easy to check that μ is a distance on $\mathbb{D}$; moreover, by Corollary 1.1.16, it is weakly contracted by holomorphic self-maps of D. However, it is not complete; furthermore, it cannot have geodesics because it satisfies a strict triangular inequality $\mu(z_1,z_3) < \mu(z_1,z_2) + \mu(z_2,z_3)$ for all triples of distinct points in $\mathbb{D}$. See, e. g., [210, Section 1.1] for this and more on the geometry of the Möbius distance.

The fully invariant version Theorem 1.2.3 expressed using the Poincaré metric has been suggested by Pick [322, 323] in 1915, though Carathéodory [93] in 1912 already knew Corollary 1.1.16.

It is possible to prove that the group of all isometries for the Poincaré distance consists of all holomorphic and antiholomorphic automorphisms of $\mathbb{D}$; see, e. g., [210, Proposition 1.1.20].

Related to Corollaries 1.1.9 and 1.2.5 there is the following theorem due to Pick [323, 324] and Nevanlinna [303, 304, 306]: given $n \ge 1$ distinct points $z_1,\dots,z_n \in \mathbb{D}$ and n more points, $w_1,\dots,w_n \in \mathbb{D}$, there exists a $f \in \mathrm{Hol}(\mathbb{D},\mathbb{D})$ such that $f(z_j) = w_j$ for all $j = 1,\dots,n$ if and only if the Hermitian matrix

$$Q_n = \left(\frac{1-w_h\overline{w_k}}{1-z_h\overline{z_k}}\right)_{h,k=1,\dots,n}$$

is positive semidefinite. Furthermore, f can be a Blaschke product, and it is unique if and only if $\det Q = 0$. See [279] for an elementary proof. This result can be considered as a generalization of the Schwarz–Pick lemma because when $n = 2$ the condition $\det Q_2 \ge 0$ is equivalent to (1.6).

More on the geometry of the Poincaré metric and distance can be found, e. g., in [65] and in [210, Chapter 1].

1.3 The upper half-plane

In this section we introduce a very useful domain biholomorphic to the unit disk. As we shall see, working in this other domain will sometimes be easier than working in the unit disk and the results obtained there will be easily transferred to the unit disk via a canonical biholomorphism.

Definition 1.3.1. The *upper half-plane* $\mathbb{H}^+$ is the domain in $\mathbb{C}$ given by

$$\mathbb{H}^+ = \{z \in \mathbb{C} \mid \operatorname{Im} z > 0\}.$$

Remark 1.3.2. The closure of $\mathbb{H}^+$ in $\mathbb{C}$ clearly is $\mathbb{H}^+ \cup \mathbb{R}$. However, for reasons that will become clearer in the sequel, it is natural to consider $\mathbb{H}^+$ as embedded in the

Riemann sphere $\widehat{\mathbb{C}} = \mathbb{C} \cup \{\infty\}$ (see Example 1.6.3). Thus for us the closure $\overline{\mathbb{H}^+}$ of $\mathbb{H}^+$ will be $\overline{\mathbb{H}^+} = \mathbb{H}^+ \cup \mathbb{R} \cup \{\infty\}$ and the boundary of $\mathbb{H}^+$ will be $\partial\mathbb{H}^+ = \mathbb{R} \cup \{\infty\}$.

Definition 1.3.3. The *Cayley transform* is the function $\Psi\colon \mathbb{D} \to \mathbb{H}^+$ given by

$$\Psi(z) = i\frac{1+z}{1-z}. \tag{1.14}$$

Since

$$\operatorname{Im}\left(i\frac{1+z}{1-z}\right) = \frac{1-|z|^2}{|1-z|^2},$$

it is easily verified that Ψ is a biholomorphism between $\mathbb{D}$ and $\mathbb{H}^+$, with inverse

$$\Psi^{-1}(w) = \frac{w-i}{w+i}.$$

Notice that

$$1 - \left|\Psi^{-1}(w)\right|^2 = \frac{4\operatorname{Im} w}{|w+i|^2}$$

and that

$$\Psi'(z) = \frac{2i}{(1-z)^2}, \qquad (\Psi^{-1})'(w) = \frac{2i}{(w+i)^2}$$

for all $z \in \mathbb{D}$ and $w \in \mathbb{H}^+$.

Remark 1.3.4. Using the convention $1/0 = \infty$ (see also Section 1.6) then Ψ extends to a homeomorphism of $\overline{\mathbb{D}}$ with $\overline{\mathbb{H}^+}$, sending 1 in ∞, 0 in i and -1 in 0.

Sometimes we shall need a homeomorphism of $\overline{\mathbb{D}}$ with $\overline{\mathbb{H}^+}$ sending a given point $\tau \in \partial\mathbb{D}$ to ∞.

Definition 1.3.5. Given $\tau \in \partial\mathbb{D}$, the *generalized Cayley transform* $\Psi_\tau\colon \mathbb{D} \to \mathbb{H}^+$ is given by

$$\Psi_\tau(z) = \Psi(\overline{\tau}z) = i\frac{\tau+z}{\tau-z}. \tag{1.15}$$

It is a biholomorphism extending continuously to the closures and so that $\Psi_\tau(\tau) = \infty$. Its inverse is given by

$$\Psi_\tau^{-1}(w) = \tau\frac{w-i}{w+i}.$$

Using the Cayley transform, we can transfer the Poincaré distance from $\mathbb{D}$ to $\mathbb{H}^+$. Easy computations show that

$$\omega(\Psi^{-1}(w_1), \Psi^{-1}(w_2)) = \frac{1}{2}\log\frac{1 + \left|\frac{w_2 - w_1}{w_2 - \overline{w_1}}\right|}{1 - \left|\frac{w_2 - w_1}{w_2 - \overline{w_1}}\right|}$$

for every $w_1, w_2 \in \mathbb{H}^+$.

Definition 1.3.6. The *Poincaré distance* $\omega_{\mathbb{H}^+}$ on $\mathbb{H}^+$ is given by

$$\omega_{\mathbb{H}^+}(w_1, w_2) = \tanh^{-1}\left|\frac{w_2 - w_1}{w_2 - \overline{w_1}}\right| = \frac{1}{2}\log\frac{1 + \left|\frac{w_2 - w_1}{w_2 - \overline{w_1}}\right|}{1 - \left|\frac{w_2 - w_1}{w_2 - \overline{w_1}}\right|}. \tag{1.16}$$

In particular, the Cayley transform Ψ is an isometry from the Poincaré distance of $\mathbb{D}$ to the Poincaré distance of $\mathbb{H}^+$.

Using the Cayley transform, we get a Schwarz–Pick lemma for holomorphic self-maps of the upper half-plane.

Proposition 1.3.7. *Let $F: \mathbb{H}^+ \to \mathbb{H}^+$ be holomorphic. Then F weakly contracts the Poincaré distance, i. e.,*

$$\omega_{\mathbb{H}^+}(F(w_1), F(w_2)) \leq \omega_{\mathbb{H}^+}(w_1, w_2)$$

or, equivalently,

$$\left|\frac{F(w_2) - F(w_1)}{F(w_2) - \overline{F(w_1)}}\right| \leq \left|\frac{w_2 - w_1}{w_2 - \overline{w_1}}\right| \tag{1.17}$$

for all w_1, $w_2 \in \mathbb{H}^+$. Equality in (1.17) *for some $w_1 \neq w_2 \in \mathbb{H}^+$ occurs if and only if $F \in \mathrm{Aut}(\mathbb{H}^+)$, and then it holds for all w_1, $w_2 \in \mathbb{H}^+$. Furthermore,*

$$\frac{|F'(w)|}{\operatorname{Im} F(w)} \leq \frac{1}{\operatorname{Im} w}$$

for all $w \in \mathbb{H}^+$, with equality in one point (and hence everywhere) if and only if $F \in \mathrm{Aut}(\mathbb{H}^+)$.

Proof. It suffices to apply Corollary 1.1.16 or Theorem 1.2.3 to $\Psi^{-1} \circ F \circ \Psi$. □

We can also compute $\mathrm{Aut}(\mathbb{H}^+)$.

Proposition 1.3.8. *Every automorphism $\gamma: \mathbb{H}^+ \to \mathbb{H}^+$ of $\mathbb{H}^+$ is of the form*

$$\gamma(w) = \frac{aw + b}{cw + d}, \tag{1.18}$$

for some $a, b, c, d \in \mathbb{R}$ such that $ad - bc = 1$. In particular, every $\gamma \in \mathrm{Aut}(\mathbb{H}^+)$ extends to a homeomorphism of $\overline{\mathbb{H}^+}$ onto itself. Furthermore, $\mathrm{Aut}(\mathbb{H}^+)$ is isomorphic to $\mathbf{PGL}(2, \mathbb{R}) = \mathbf{SL}(2, \mathbb{R})/\{\pm I_2\}$.

Proof. The function γ is an automorphism of $\mathbb{H}^+$ if and only if $\Psi^{-1} \circ \gamma \circ \Psi$ is an automorphism of $\mathbb{D}$. Plugging the Cayley transform into (1.4) we find exactly (1.18), with $a, b, c, d \in \mathbb{R}$ satisfying $D = ad - bc > 0$. But now if we divide a, b, c, and d by $\sqrt{D}$ we can express γ in the form (1.18) with coefficients satisfying $ad - bc = 1$. Notice that if $c \neq 0$ then $\gamma(\infty) = a/c$ and $\gamma^{-1}(\infty) = -d/c$; if $c = 0$, then $\gamma(\infty) = \gamma^{-1}(\infty) = \infty$.

As a consequence of (1.18), the map $\mu \colon \mathbf{SL}(2, \mathbb{R}) \to \mathrm{Aut}(\mathbb{H}^+)$ given by

$$\mu \begin{pmatrix} a & b \\ c & d \end{pmatrix} (w) = \frac{aw + b}{cw + d}$$

is surjective, and it is easily seen to be a group homomorphism. Its kernel is equal to $\{\pm I_2\}$, and the assertion follows. □

The upper half-plane model is sometimes useful to understand questions regarding the behavior of geometrical objects at a point of the boundary. For instance, the Cayley transform sends the geodesics ending at $1 \in \partial\mathbb{D}$ into the vertical lines in $\mathbb{H}^+$, quite a simpler object. On the other hand, the study of objects linked to internal points may be formally easier in $\mathbb{D}$ than in $\mathbb{H}^+$. For instance, the subgroup of automorphisms of $\mathbb{H}^+$ fixing $i = \Psi(0)$ is composed by the automorphisms having the somewhat cumbersome form

$$\gamma(w) = \frac{w \cos\theta - \sin\theta}{w \sin\theta + \cos\theta}.$$

For these reasons, from now on in the proofs we shall often move back and forth from $\mathbb{D}$ to $\mathbb{H}^+$, according to the current situation.

Remark 1.3.9. The automorphisms of both $\mathbb{D}$ and $\mathbb{H}^+$ are *fractional linear maps*, i. e., maps of the form

$$f(z) = \frac{az + b}{cz + d}$$

where $a, b, c, d \in \mathbb{C}$ are such that $ad - bc \neq 0$. The Cayley transform too is a fractional linear map. This is not a coincidence, because we shall see (Proposition 1.6.19) that the fractional linear maps are exactly the automorphisms of the Riemann sphere.

By the way, it is easy to check that the image of an Euclidean circle or line through a fractional linear maps is an Euclidean line or circle (taking into account the fact that a circle might be sent in a line or a line in a circle).

We end this section with a couple of formulas that will be useful later on.

Proposition 1.3.10. (i) *For every $z_1, z_2 \in \mathbb{D}$ we have*

$$\begin{cases} \sinh \omega(z_1, z_2) = \dfrac{|z_2 - z_1|}{\sqrt{(1 - |z_1|^2)(1 - |z_2|^2)}}, \\ \cosh \omega(z_1, z_2) = \dfrac{|1 - \overline{z_1} z_2|}{\sqrt{(1 - |z_1|^2)(1 - |z_2|^2)}}. \end{cases} \tag{1.19}$$

(ii) *For every $w_1, w_2 \in \mathbb{H}^+$ we have*

$$\begin{cases} \sinh \omega_{\mathbb{H}^+}(w_1, w_2) = \dfrac{|w_2 - w_1|}{2\sqrt{\operatorname{Im} w_1 \operatorname{Im} w_2}}, \\ \cosh(2\omega_{\mathbb{H}^+}(w_1, w_2)) = 1 + \dfrac{|w_2 - w_1|^2}{2 \operatorname{Im} w_1 \operatorname{Im} w_2}. \end{cases} \tag{1.20}$$

Proof. (i) Formulas (1.19) follow using the fact that

$$\tanh \omega(z_1, z_2) = \left| \frac{z_2 - z_1}{1 - \overline{z_1} z_2} \right|$$

(see Remark 1.2.2), equation (1.5) and the easily verified formulas

$$(\sinh t)^2 = \frac{(\tanh t)^2}{1 - (\tanh t)^2} \quad \text{and} \quad (\cosh t)^2 = \frac{1}{1 - (\tanh t)^2},$$

valid for all $t \in \mathbb{R}$.

(ii) This time we use the formulas

$$\tanh \omega_{\mathbb{H}^+}(w_1, w_2) = \left| \frac{w_2 - w_1}{w_2 - \overline{w_1}} \right|, \quad 1 - \left| \frac{w_2 - w_1}{w_2 - \overline{w_1}} \right|^2 = \frac{4 \operatorname{Im} w_1 \operatorname{Im} w_2}{|w_2 - \overline{w_1}|^2}$$

and

$$(\sinh t)^2 = \frac{(\tanh t)^2}{1 - (\tanh t)^2}, \quad \cosh(2t) = 1 + \frac{(\sinh t)^2}{2}.$$ □

Notes to Section 1.3

The Cayley transform has been introduced by Cayley [108] in 1846 as a map sending skew-symmetric matrices into special orthogonal matrices: if A is a skew-symmetric $n \times n$ matrix with real coefficients then $\Psi(A) = (I - A)^{-1}(I + A)$ is an orthogonal matrix with determinant 1.

The relevance of the upper half-plane as model for the hyperbolic geometry has been first recognised by Beltrami [56] in 1868 and then has been thoroughly exploited by Poincaré starting from 1882 (see [331, 332]).

1.4 Fixed points of automorphisms

In this section we collect several facts about automorphisms of $\mathbb{D}$ and $\mathbb{H}^+$, mainly regarding their fixed points.

We recall (Proposition 1.1.5) that an automorphism γ of $\mathbb{D}$ extends continuously to $\overline{\mathbb{D}}$, and the extension (still denoted by γ) sends $\overline{\mathbb{D}}$ into itself; in particular, it makes sense to look for fixed points in $\overline{\mathbb{D}}$.

Proposition 1.4.1. *Let $\gamma \in \mathrm{Aut}(\mathbb{D})$, $\gamma \neq \mathrm{id}_{\mathbb{D}}$. Then either*
(i) *γ has a unique fixed point in $\mathbb{D}$, or*
(ii) *γ has a unique fixed point in $\partial\mathbb{D}$, or*
(iii) *γ has two distinct fixed points in $\partial\mathbb{D}$.*

Proof. Write

$$\gamma(z) = e^{i\theta}\frac{z - z_0}{1 - \overline{z_0}z},$$

for some $\theta \in \mathbb{R}$ and $z_0 \in \mathbb{D}$. The equation satisfied by the fixed points of γ is

$$\overline{z_0}z^2 + (e^{i\theta} - 1)z - e^{i\theta}z_0 = 0. \tag{1.21}$$

If $z_0 = 0$, then $\gamma(z) = e^{i\theta}z$ and we are in the first case. If $z_0 \neq 0$, then (1.21) has (counting multiplicities) exactly two roots, $z_1, z_2 \in \mathbb{C}$, with possibly $z_1 = z_2$. Moreover,

$$|z_1||z_2| = |z_0/\overline{z_0}| = 1.$$

Therefore, either just one of them is in $\mathbb{D}$—and we are again in case (i)—or both are in $\partial\mathbb{D}$, and we are either in case (ii) or (iii). □

Definition 1.4.2. An automorphism of $\mathbb{D}$ different from the identity is called *elliptic* if it has a (unique) fixed point in $\mathbb{D}$, *parabolic* if it has a unique fixed point on $\partial\mathbb{D}$ and *hyperbolic* if it has two distinct fixed points on $\partial\mathbb{D}$. The *axis* of a hyperbolic automorphism γ is the unique geodesic of $\mathbb{D}$ connecting the two fixed points of γ.

Analogously, by Proposition 1.3.8 an automorphism of $\mathbb{H}^+$ extends continuously to $\overline{\mathbb{H}^+}$ and sends $\overline{\mathbb{H}^+}$ into itself. Therefore, the same definitions make sense for automorphisms of $\mathbb{H}^+$ (recalling to include the point at infinity). In this case, we may tell elliptic, parabolic and hyperbolic automorphisms apart directly from their representation as elements of $\mathbf{PGL}(2, \mathbb{R})$.

Definition 1.4.3. The *trace* of $\gamma \in \mathrm{Aut}(\mathbb{H}^+)$ is given by $\operatorname{tr}\gamma = |\operatorname{tr} A_\gamma|$, where $A_\gamma \in \mathbf{SL}(2, \mathbb{R})$ is any representative of γ (see Proposition 1.3.8) and $\operatorname{tr} A$ is the usual trace of the matrix A. It is easy to check that $\operatorname{tr}\gamma$ does not depend on the chosen representative and that it is invariant under conjugation.

Then we have the following result.

Proposition 1.4.4. *Let $\gamma \in \operatorname{Aut}(\mathbb{H}^+)$, $\gamma \neq \mathrm{id}_{\mathbb{H}^+}$. Then γ is elliptic (resp., parabolic, hyperbolic) if and only if its trace is less than 2 (resp., equal to 2, greater than 2).*

Proof. Let γ be represented by (1.18), so that $\operatorname{tr}\gamma = |a + d|$. The fixed-points equation for γ is

$$cw^2 + (d - a)w - b = 0.$$

If $c \neq 0$, then γ is elliptic (resp., parabolic, hyperbolic) if and only if this equation has two distinct complex roots (resp., one double real root, two distinct real roots), i. e., if and only if $D = (d-a)^2 + 4bc < 0$ (resp., $D = 0$, $D > 0$). Using the constraint $ad - bc = 1$, we easily compute

$$D = (a + d)^2 - 4,$$

and the assertion follows in this case.

If $c = 0$, then $d = a^{-1}$ and γ has a fixed point at ∞; in particular, γ cannot be elliptic. Hence it is hyperbolic if and only if it has a fixed point different from ∞, i. e., if and only if $d \neq a$, which is equivalent to $|a + d| > 2$, and we are done. □

Remark 1.4.5. To better understand the different kinds of automorphisms, we use Corollary 1.1.9. If $\gamma \in \operatorname{Aut}(\mathbb{H}^+)$ is hyperbolic, up to conjugation we can assume that the fixed points of γ are 0 and ∞. Hence $\gamma(w) = aw$ for some $a > 0$, $a \neq 1$, and its trace is $a + a^{-1} > 2$. In particular, the set of hyperbolic automorphisms fixing two given points of $\partial\mathbb{H}^+$ (or $\partial\mathbb{D}$) is a group isomorphic to $(\mathbb{R}^+_*, \cdot)$, where $\mathbb{R}^+_* = (0, +\infty)$.

Analogously, if γ is parabolic we can assume that its fixed point is ∞; thus $\gamma(w) = w + b$ for some $b \in \mathbb{R}^*$. In particular, the set of parabolic automorphisms fixing a given point of $\partial\mathbb{H}^+$ (or $\partial\mathbb{D}$) is a group isomorphic to $(\mathbb{R}, +)$.

Finally, if γ is elliptic we can assume its fixed point is i, and write

$$\gamma(w) = \frac{w\cos\theta - \sin\theta}{w\sin\theta + \cos\theta} \tag{1.22}$$

for some $\theta \in \mathbb{R}$; the trace of γ is then $2|\cos\theta|$. In particular, the set of elliptic automorphisms fixing a given point of $\mathbb{H}^+$ (or $\mathbb{D}$) is a group isomorphic to $(\mathbb{S}^1, \cdot)$. This becomes even clearer considering the automorphisms of $\mathbb{D}$ fixing the origin, because they are all of the form $\gamma(z) = e^{i\theta}z$ with $e^{i\theta} \in \mathbb{S}^1$.

Remark 1.4.6. In $\mathbb{D}$, an elliptic automorphism γ fixing 0 is just a rotation about the origin. For this reason, elliptic automorphisms are sometimes called *non-Euclidean rotations*.

Remark 1.4.7. For later use, we remark that the parabolic automorphisms of $\mathbb{D}$ with fixed point $\tau \in \partial\mathbb{D}$ are of the form

$$\gamma(z) = \sigma_0 \frac{z - a}{1 - \bar{a}z}, \tag{1.23}$$

where

$$\sigma_0 = \frac{2 - ib}{2 + ib} \quad \text{and} \quad a = \frac{ib}{ib - 2}\tau$$

for some $b \in \mathbb{R}$. This follows because every such γ must be of the form

$$\gamma(z) = \tau\Psi^{-1}(\Psi(\bar{\tau}z) + b),$$

where Ψ is the Cayley transform.

Not surprisingly, the elliptic automorphisms of trace 0 have peculiar properties.

Proposition 1.4.8. *Let $\gamma \in \mathrm{Aut}(\mathbb{H}^+)$. Then γ has trace zero if and only if $\gamma^2 = \mathrm{id}_{\mathbb{H}^+}$.*

Proof. Write γ as in (1.18). Then a computation yields

$$\gamma^2(w) = \frac{(a + d)(aw + b) - w}{(a + d)(cw + d) - 1},$$

where we have used $ad - bc = 1$, and the assertion follows. □

We shall see in Section 4.6 that the behavior of holomorphic self-maps of the unit disk will be somewhat modeled on the behavior of the automorphisms. For this reason, we now prove a couple of results that will turn out to be baby versions of several theorems we shall see in this book. We start with a computation (that can be compared with Corollaries 2.3.16 and 2.5.5).

Lemma 1.4.9. *Let $\gamma \in \mathrm{Aut}(\mathbb{D})$ be a hyperbolic automorphism with fixed points $\sigma_1, \sigma_2 \in \partial\mathbb{D}$. Then $\gamma'(\sigma_1)$ and $\gamma'(\sigma_2)$ are positive real numbers different from 1 with $\gamma'(\sigma_1)\gamma'(\sigma_2) = 1$. If instead $\gamma \in \mathrm{Aut}(\mathbb{D})$ is a parabolic automorphism with fixed point $\tau \in \partial\mathbb{D}$ then $\gamma'(\tau) = 1$.*

Proof. First of all, if $\sigma \in \partial\mathbb{D}$ is a fixed point of γ then for any $\gamma_1 \in \mathrm{Aut}(\mathbb{D})$ the point $\tilde{\sigma} = \gamma_1(\sigma)$ is a fixed point of $\tilde{\gamma} = \gamma_1 \circ \gamma \circ \gamma_1^{-1}$ so that $\tilde{\gamma}'(\tilde{\sigma}) = \gamma'(\sigma)$. Therefore, by Corollary 1.1.9, in the hyperbolic case we can assume $\sigma_1 = 1$ and $\sigma_2 = -1$. Then Remark 1.4.5 gives

$$\gamma(z) = \Psi^{-1}(a\Psi(z)) = \frac{a - 1 + (a + 1)z}{a + 1 + (a - 1)z}$$

for some $a \in \mathbb{R}^+ \setminus \{0, 1\}$. An easy computation yields $\gamma'(1) = 1/a$ and $\gamma'(-1) = a$ as required.

Finally, if γ is parabolic we can assume $\tau = 1$ and that γ is given by (1.23). Then

$$\gamma'(1) = \sigma_0 \frac{1 - |a|^2}{(1 - \overline{a})^2} = \frac{2 - ib}{2 + ib} \frac{4}{4 + b^2} \frac{(2 + ib)^2}{4} = 1,$$

as claimed. □

As a consequence, we have our very first dynamical result (that should be compared with Theorem 3.2.1).

Proposition 1.4.10. *Given $\gamma \in \mathrm{Aut}(\mathbb{D})$ put $\gamma^0 = \mathrm{id}_{\mathbb{D}}$ and $\gamma^k = \gamma \circ \gamma^{k-1}$ for $k \in \mathbb{N}^*$. Then:*
(i) *if γ is hyperbolic or parabolic then $\gamma^k \to \tau$ uniformly on compact subsets, where $\tau \in \partial\mathbb{D}$ is a fixed point of γ; in particular, if γ is hyperbolic then τ is the only fixed point of γ such that $|\gamma'(\tau)| < 1$;*
(ii) *if $\gamma \neq \mathrm{id}_{\mathbb{D}}$ is elliptic then the sequence $\{\gamma^k(z)\}$ converges if and only if z is the fixed point of γ.*

Proof. If γ is hyperbolic, we have seen that without loss of generality we can transfer everything to $\mathbb{H}^+$ and assume that $\gamma(w) = aw$ with $a \in \mathbb{R}^+ \setminus \{1\}$. Then clearly $\gamma^k(w) = a^k w$; hence $\{\gamma^k\}$ converges to either to ∞ or to 0 according to whether $a > 1$ or $0 < a < 1$. We are then done because 0 and ∞ are the fixed points of γ and $a = \gamma'(0)$ (the derivative at the other fixed point is $1/a$, by Lemma 1.4.9).

If γ is parabolic, again transferring everything to $\mathbb{H}^+$, we can assume $\gamma(w) = w + b$ for some $b \in \mathbb{R}^*$. Then $\gamma^k(w) = w + kb$; so $\gamma^k \to \infty$ and we are done in this case too.

Finally, if γ is elliptic without loss of generality we can assume that the fixed point of γ is the origin. Thus $\gamma(z) = e^{2\pi i\theta} z$ for some $\theta \in \mathbb{R} \setminus \mathbb{Z}$, so that $\gamma^k(z) = e^{2\pi ik\theta} z$. Clearly, $\gamma^k(0) = 0$ for all $k \in \mathbb{N}$, and hence $\{\gamma^k(0)\}$ trivially converges. Conversely, assume, by contradiction, that $\gamma^k(z) \to z_0 \in \mathbb{C}$ for some $z \neq 0$. Since $|\gamma^k(z)| = |z|$, it follows that $|z_0| = |z| < 1$, i. e., $z_0 \in \mathbb{D}$. Then we can compute $\gamma(z_0)$; however,

$$\gamma(z_0) = \gamma\Big(\lim_{k \to +\infty} \gamma^k(z)\Big) = \lim_{k \to +\infty} \gamma^{k+1}(z) = z_0,$$

i. e., z_0 is a fixed point of γ. But the only fixed point of γ is the origin, and from $|z| = |z_0|$ we infer $z = 0$, contradiction. □

As we shall often see in this book, there are relations between commuting functions and fixed points.

Definition 1.4.11. We shall say that two self-maps $f, g: X \to X$ of a set X *commutes* (or that they are *commuting functions*) if $f \circ g = g \circ f$.

A first instance of the relationships mentioned above is the following (compare with Theorem 3.1.20 and Section 4.10).

Proposition 1.4.12. *Let $\gamma_1, \gamma_2 \in \mathrm{Aut}(\mathbb{D})$, both different from the identity. Then $\gamma_1 \circ \gamma_2 = \gamma_2 \circ \gamma_1$ if and only if γ_1 and γ_2 have the same fixed points in $\overline{\mathbb{D}}$.*

Proof. Transfer everything on $\mathbb{H}^+$. If γ_1 is parabolic, without loss of generality we can assume $\gamma_1(w) = w + \beta$, with $\beta \in \mathbb{R}^*$. Write γ_2 as in (1.18); then $\gamma_1 \circ \gamma_2 = \gamma_2 \circ \gamma_1$ yields

$$\begin{pmatrix} a + \beta c & b + \beta d \\ c & d \end{pmatrix} = \pm \begin{pmatrix} a & \beta a + b \\ c & \beta c + d \end{pmatrix}.$$

Hence $c = 0$, $a = d$ and $ad = 1$, showing that $\gamma_2(w) = w + b$, and thus γ_2 has the same fixed points as γ_1. Conversely, if $\mathrm{Fix}(\gamma_2) = \mathrm{Fix}(\gamma_1) = \{\infty\}$ then $\gamma_2(w) = w + b$ for a suitable $b \in \mathbb{R}^*$ and so γ_1 and γ_2 commute.

If γ_1 is hyperbolic, without loss of generality we can assume $\gamma_1(w) = \lambda w$ for some $\lambda > 0$, $\lambda \neq 1$. Now $\gamma_1 \circ \gamma_2 = \gamma_2 \circ \gamma_1$ yields

$$\begin{pmatrix} \lambda a & \lambda b \\ c & d \end{pmatrix} = \pm \begin{pmatrix} \lambda a & b \\ \lambda c & d \end{pmatrix}.$$

Hence $b = c = 0$ and $ad = 1$, showing that $\gamma_2(w) = a^2 w$, and thus in this case, too, γ_2 has the same fixed points as γ_1. Conversely, if $\mathrm{Fix}(\gamma_2) = \mathrm{Fix}(\gamma_1) = \{0, \infty\}$ then $\gamma_2(w) = aw$ for a suitable $a \in \mathbb{R}^+ \setminus \{0, 1\}$ and so γ_1 and γ_2 commute.

Finally, if γ_1 is elliptic we can come back to $\mathbb{D}$ and assume, without loss of generality, that $\mathrm{Fix}(\gamma_1) = \{0\}$, so that $\gamma_1(z) = e^{i\phi} z$ for a suitable $\phi \in \mathbb{R} \setminus (2\pi\mathbb{Z})$. Write γ_2 as in (1.4); then $\gamma_1 \circ \gamma_2 = \gamma_2 \circ \gamma_1$ yields

$$e^{i(\phi+\theta)} \frac{z - a}{1 - \bar{a} z} = e^{i\theta} \frac{e^{i\phi} z - a}{1 - \bar{a} e^{i\phi} z}.$$

Putting $z = a$, we obtain $(e^{i\phi} - 1)a = 0$, and hence $a = 0$, because $e^{i\phi} \neq 1$. Therefore, $\gamma_2(z) = e^{i\theta} z$, i. e., it has the same fixed points as γ_1. Conversely, if $\mathrm{Fix}(\gamma_2) = \mathrm{Fix}(\gamma_1) = \{0\}$ then $\gamma_2(z) = e^{i\theta} z$ for a suitable $\theta \in \mathbb{R}$ and so γ_1 and γ_2 commute. □

In this book, we shall encounter several results of this kind: commuting holomorphic maps must have common fixed points (under suitable hypotheses, of course). For the moment, we shall content ourselves with two corollaries about subgroups of $\mathrm{Aut}(\mathbb{D})$ that we shall need in our study of Riemann surfaces. Since we have seen that $\mathrm{Aut}(\mathbb{D})$ contains subgroups isomorphic to $(\mathbb{R}, +)$, $(\mathbb{R}^+_*, \cdot)$ and $(\mathbb{S}^1, \cdot)$ we start with a description of the subgroups of these groups.

Proposition 1.4.13. (i) *The subgroups of $(\mathbb{R}, +)$ either are of the form $\mathbb{Z}x_0$ for some $x_0 \geq 0$, and hence discrete, closed, and cyclic, or they are dense in $\mathbb{R}$.*

(ii) *The subgroups of $(\mathbb{R}^+_*, \cdot)$ either are discrete, closed, and cyclic or they are dense in $\mathbb{R}^+_*$.*

(iii) *The subgroups of $(\mathbb{S}^1, \cdot)$ either are finite cyclic, generated by a root of unity, or they are dense. In particular, if $\theta \notin \mathbb{Q}$ then the sequence $\{e^{2\pi k i\theta}\}_{k\in\mathbb{N}}$ accumulates 1.*

(iv) *The proper closed subgroups of* $(\mathbb{C}, +)$ *are of the form* $\mathbb{Z}x_0$ *or* $\mathbb{R}x_0$ *for some* $x_0 \in \mathbb{C}^*$ *or of the form* $\mathbb{Z}x_1 \oplus \mathbb{Z}x_2$ *or* $\mathbb{R}x_1 \oplus \mathbb{Z}x_2$ *where* $x_1, x_2 \in \mathbb{C}$ *are* $\mathbb{R}$*-linearly independent.*

Proof. (i) Let Γ be a subgroup of $(\mathbb{R}, +)$. If $\Gamma = \{0\}$, we are done; otherwise $\Gamma \cap \mathbb{R}^+ \neq \emptyset$. Put $x_0 = \inf\{x \in \Gamma \mid x > 0\}$; we claim that if $x_0 > 0$ then $\Gamma = \mathbb{Z}x_0$, whereas if $x_0 = 0$ then Γ is dense in $\mathbb{R}$.

Assume $x_0 > 0$, and suppose for the moment that Γ is closed, so that $x_0 \in \Gamma$. Take $x \in \Gamma$, and choose $k \in \mathbb{Z}$ such that $kx_0 \leq x < (k+1)x_0$. Then $x - kx_0 \in \Gamma$ and satisfies $0 \leq x - kx_0 < x_0$; the minimality of x_0 forces $x - kx_0 = 0$, i. e., $x = kx_0 \in \mathbb{Z}x_0$, as desired. If Γ were not closed, the same argument applied to the closure $\overline{\Gamma}$ of Γ would give $\overline{\Gamma} = \mathbb{Z}x_0$, and thus $\Gamma = \overline{\Gamma}$.

Assume instead $x_0 = 0$; this means that Γ contains arbitrarily small positive elements. Fix $\varepsilon > 0$, and choose $x_\varepsilon \in \Gamma$ such that $0 < x_\varepsilon < \varepsilon$. Given $x \in \mathbb{R}$, there exists $k \in \mathbb{Z}$ such that $kx_\varepsilon \leq x < (k+1)x_\varepsilon$; in particular, $0 \leq x - kx_\varepsilon < x_\varepsilon < \varepsilon$. Since $kx_\varepsilon \in \Gamma$ and ε is arbitrary, this exactly means that Γ is dense in $\mathbb{R}$.

(ii) The function $\exp\colon \mathbb{R} \to \mathbb{R}^+_*$ is an isomorphism of topological groups, and hence the assertion follows from (i).

(iii) If $\Gamma = \{1\}$, we are done. Otherwise, put $\theta_0 = \inf\{\theta \in [0,1) \mid e^{2\pi i\theta} \in \Gamma\}$; we claim that if $\theta_0 > 0$, then $\theta_0 \in \mathbb{Q}$ and Γ is finite cyclic generated by $e^{2\pi i\theta_0}$, whereas if $\theta_0 = 0$ then Γ is dense in $\mathbb{S}^1$.

Assume $\theta_0 > 0$ and, as in (i), momentarily suppose that Γ is closed, so that $e^{2\pi i\theta_0} \in \Gamma$. Take $e^{2\pi i\theta} \in \Gamma$; then necessarily $\theta \geq \theta_0$ and we can find $k \in \mathbb{N}$ such that $k\theta_0 \leq \theta < (k+1)\theta_0$. Arguing as in (i) we then get $e^{2\pi i\theta} = e^{2\pi ik\theta_0}$, and thus Γ is cyclic generated by $e^{2\pi i\theta_0}$. In particular, there must exists $k_0 \in \mathbb{N}$ such that $e^{2\pi ik_0\theta_0}$ is the inverse of $e^{2\pi i\theta_0}$; but this means $(k_0+1)\theta_0 \in \mathbb{Z}$, and thus $\theta_0 \in \mathbb{Q}$. In particular, $e^{2\pi i\theta_0}$ is a root of unity and Γ is finite cyclic.

If $\theta_0 > 0$ and Γ is not necessarily closed, the same argument applied to $\overline{\Gamma}$ shows that $\overline{\Gamma}$ is finite, and so it should coincide with Γ.

Assume now that $\theta_0 = 0$. Given $0 < \theta \leq 1$ and $\varepsilon > 0$, as in (i) we can find $0 < \theta_\varepsilon < \min\{\theta, \varepsilon\}$ and $k \in \mathbb{N}$ such that $e^{2\pi i\theta_\varepsilon} \in \Gamma$ and $k\theta_\varepsilon \leq \theta < (k+1)\theta_\varepsilon$; therefore, being ε arbitrary we can find elements of Γ arbitrarily close to $e^{2\pi i\theta}$, and this means that Γ is dense in $\mathbb{S}^1$.

Finally, take $\theta \in [0,1) \setminus \mathbb{Q}$, and consider $\Gamma = \{e^{2\pi ik\theta}\}_{k\in\mathbb{Z}}$. Clearly, Γ is a subgroup of $\mathbb{S}^1$; moreover, it is infinite, because $e^{2\pi ik_1\theta} = e^{2\pi ik_2\theta}$ for two $k_1, k_2 \in \mathbb{Z}$ if and only if $(k_1 - k_2)\theta \in \mathbb{Z}$, and this can happen if and only if $k_1 = k_2$ because $\theta \notin \mathbb{Q}$. It follows that Γ is dense in $\mathbb{S}^1$; to conclude it suffices to show that 1 can be accumulated by elements of the form $e^{2\pi ik\theta}$ with $k \in \mathbb{N}$. But indeed, being Γ dense, given $\varepsilon > 0$ there exist $k_1, k_2 \in \mathbb{Z}$ such that $|k_1 - k_2\theta| < \varepsilon$; since $|k_1 - k_2\theta| = |-k_1 - (-k_2)\theta|$ we can actually assume that $k_2 \geq 0$, and we are done.

(iv) Let Γ be a proper closed subgroup of $(\mathbb{C}, +)$. If Γ is contained in a line $\mathbb{R}x_0$, the assertion follows from (i).

Assume first that Γ is not contained in any line, but contains a line $\Gamma_1 = \mathbb{R}x_1$. Then Γ/Γ_1 is a subgroup of $\mathbb{C}/\Gamma_1$, which is isomorphic to $\mathbb{R}$; hence, by (i), either Γ/Γ_1 is isomorphic to $\mathbb{R}$—and then $\Gamma = \mathbb{C}$, against the assumption that Γ were a proper subgroup of $\mathbb{C}$—or Γ/Γ_1 is discrete and generated by $x_2 + \Gamma_1$, which means that $\Gamma = \mathbb{R}x_1 \oplus \mathbb{Z}x_2$.

So assume that Γ does not contain any line $\mathbb{R}x$; we claim that Γ is discrete. If not, there is a sequence $\{x_\nu\} \subset \Gamma$ converging to some $x_0 \in \mathbb{C}$. Since Γ is closed, $x_0 \in \Gamma$; up to replace x_ν by $x_\nu - x_0$ we can assume $x_0 = 0$. Without loss of generality, we can assume that $x_\nu \neq 0$ for all $\nu \in \mathbb{N}$ and that $x_\nu/|x_\nu| \to \tau \in \mathbb{S}^1$. If we prove that the line $\mathbb{R}\tau$ is contained in Γ, we have the contradiction we were looking for. Given $t \neq 0$ and $\nu \in \mathbb{N}$ choose $m_\nu \in \mathbb{Z}$ so that $|m_\nu - \frac{t}{|x_\nu|}| < 1$. Then

$$|m_\nu x_\nu - t\tau| \leq \left|\left(m_\nu - \frac{t}{|x_\nu|}\right)x_\nu\right| + \left|\frac{tx_\nu}{|x_\nu|} - t\tau\right| < |x_\nu| + |t|\left|\frac{x_\nu}{|x_\nu|} - \tau\right|.$$

Since $x_\nu \to 0$ and $\frac{x_\nu}{|x_\nu|} \to \tau$, this shows that $t\tau \in \overline{\Gamma} = \Gamma$, and hence $\mathbb{R}\tau \subseteq \Gamma$, contradiction.

Thus we can assume Γ discrete. Set $\delta_1 = \inf\{|x| \mid x \in \Gamma \setminus \{0\}\} > 0$. Since Γ is discrete and closed, there exists $x_1 \in \Gamma$ such that $|x_1| = \delta_1$. If $\Gamma \subseteq \mathbb{R}x_1$, then $\Gamma = \mathbb{Z}x_1$ by (i); so assume that Γ is not contained in $\mathbb{R}x_1$. Since $\Gamma \setminus \mathbb{R}x_1$ is still discrete and closed, we can find $x_2 \in \Gamma \setminus \mathbb{R}x_1$ such that $|x_2| = \delta_2$, where

$$\delta_2 = \inf\{|x| \mid x \in \Gamma \setminus \mathbb{R}x_1\} > 0;$$

we claim that $\Gamma = \mathbb{Z}x_1 \oplus \mathbb{Z}x_2$. Indeed, take $x \in \Gamma$. Since x_1 and x_2 are $\mathbb{R}$-linearly independent there exist $\lambda_1, \lambda_2 \in \mathbb{R}$ such that $x = \lambda_1 x_1 + \lambda_2 x_2$; up to replace x_2 by $-x_2$ without loss of generality we can assume $\lambda_1\lambda_2 \geq 0$. Summing elements of $\mathbb{Z}x_1 \oplus \mathbb{Z}x_2$, we can reduce to the case $\lambda_1, \lambda_2 \in [-1/2, 1/2)$; it suffices to prove that then $x = 0$. Since x_1 and x_2 are $\mathbb{R}$-linearly independent, we have $\operatorname{Re}(x_1\overline{x_2}) < |x_1||x_2| \leq \delta_2^2$; therefore,

$$|x|^2 = \lambda_1^2|x_1|^2 + \lambda_2^2|x_2|^2 + 2\lambda_1\lambda_2 \operatorname{Re}(x_1\overline{x_2}) < (\lambda_1 + \lambda_2)^2\delta_2^2 < \delta_2^2.$$

The minimality of $|x_2|$ thus yields $\lambda_2 = 0$ but then the minimality of $|x_1|$ yields also $\lambda_1 = 0$, as claimed, and we are done. □

Corollary 1.4.14. *Let Γ be a nontrivial discrete Abelian subgroup of* $\operatorname{Aut}(\mathbb{D})$. *Then Γ is cyclic. In particular, if Γ does not contain elliptic elements then Γ is isomorphic to $\mathbb{Z}$.*

Proof. By Proposition 1.4.12, all elements of Γ (except the identity) have the same fixed points. Hence Γ is (topologically) isomorphic to a discrete subgroup of $(\mathbb{R}^+_*, \cdot)$, $(\mathbb{R}, +)$ or $(\mathbb{S}^1, \cdot)$, according to whether Γ consists of hyperbolic, parabolic, or elliptic automorphisms. Hence it is cyclic, and infinite cyclic in the hyperbolic and parabolic cases, by Proposition 1.4.13. □

Corollary 1.4.15. *Let Γ be a non-Abelian subgroup of* $\mathrm{Aut}(\mathbb{D})$ *without elliptic elements. Then Γ contains a hyperbolic automorphism.*

Proof. Transfer everything on $\mathbb{H}^+$, as usual. If all elements of Γ are hyperbolic we are clearly done. Assume then that there is $\gamma \in \Gamma$ parabolic; up to conjugation, we can suppose $\gamma(w) = w + \beta$ for some $\beta \in \mathbb{R}^*$. Since Γ is non-Abelian, by Proposition 1.4.12 it contains some other element γ_1 with different fixed points. If γ_1 is hyperbolic, we are done. If γ_1 is parabolic, write it as in (1.18); clearly, c must be nonzero. Then $\gamma^k \circ \gamma_1$ has trace $|a + d + k\beta c|$ and so it is hyperbolic for k sufficiently large by Proposition 1.4.4. □

We end this section with two technical results we shall need later on.

Lemma 1.4.16. (i) *Given two distinct points $z_0, z_1 \in \mathbb{D}$ let ℓ be the unique geodesic passing through z_0 and z_1. Then there is a unique hyperbolic automorphism $\gamma \in$* $\mathrm{Aut}(\mathbb{D})$ *with axis ℓ and such that $\gamma(z_0) = z_1$.*

(ii) *Let $\gamma \in$* $\mathrm{Aut}(\mathbb{D})$ *be a hyperbolic automorphism with axis ℓ. Then*

$$\omega(c,\gamma(c)) \le \omega(z,\gamma(z)) \le e^{2\omega(z,c)}\omega(c,\gamma(c)) \tag{1.24}$$

for all $c \in \ell$ and $z \in \mathbb{D}$. Furthermore, $\omega(c,\gamma(c))$ does not depend on $c \in \ell$.

(iii) *Let $\ell\colon \mathbb{R} \to \mathbb{D}$ be a geodesic. Then there exists a unique hyperbolic automorphism $\gamma \in$* $\mathrm{Aut}(\mathbb{D})$ *such that $\gamma(\ell(t)) = \ell(t+1)$ for all $t \in \mathbb{R}$.*

Proof. (i) Since the conjugate of a hyperbolic automorphism is still a hyperbolic automorphism, up to a conjugation we can assume $z_0 = 0$ and $z_1 \in \mathbb{R} \cap \mathbb{D}$; in particular, ℓ is the diameter connecting -1 and 1.

Using the Cayley transform, we can transfer everything to $\mathbb{H}^+$; so ℓ is the positive imaginary axis, $z_0 = i$ and $z_1 = it$ for some $t > 0$. Then the unique hyperbolic automorphism γ of $\mathbb{H}^+$ with axis ℓ and sending z_0 into z_1 is $\gamma(w) = tw$.

(ii) Using the Cayley transform, we can replace $\mathbb{D}$ by $\mathbb{H}^+$. Up to a conjugation, we can assume that $\gamma(w) = \lambda w$ and, up to replacing γ by γ^{-1}, we can assume $\lambda > 1$. In particular, the axis ℓ of γ is the positive imaginary axis.

Take $c = it \in \ell$ and $w = |w|e^{i\theta} \in \mathbb{H}^+$, with $t > 0$ and $\theta \in (0,\pi)$. Then

$$\tanh \omega_{\mathbb{H}^+}(c,\gamma(c)) = \left|\frac{\lambda - 1}{\lambda + 1}\right| \le \left|\frac{\lambda - 1}{\lambda - e^{-2i\theta}}\right| = \left|\frac{\lambda w - w}{\lambda w - \overline{w}}\right| = \tanh \omega_{\mathbb{H}^+}(w,\gamma(w))$$

because $|\lambda + 1| \ge |\lambda - e^{-2i\theta}|$ for all $\theta \in \mathbb{R}$ and $\lambda > 1$. So, the left inequality in (1.24) is proved; moreover, we have

$$\omega_{\mathbb{H}^+}(c,\gamma(c)) = \frac{1}{2}|\log \lambda|$$

for all $c \in \ell$.

Now, since the function $t \mapsto \frac{\sinh t}{t}$ is strictly increasing for $t > 0$, from the left inequality we get

$$\omega_{\mathbb{H}^+}(w, \gamma(w)) \le \frac{\sinh \omega_{\mathbb{H}^+}(w, \gamma(w))}{\sinh \omega_{\mathbb{H}^+}(c, \gamma(c))} \omega_{\mathbb{H}^+}(c, \gamma(c)) = \frac{|w|}{\operatorname{Im} w} \omega_{\mathbb{H}^+}(c, \gamma(c)),$$

where the last equality follows from the first formula in (1.20). The second formula in (1.20) yields

$$\frac{|w|}{\operatorname{Im} w} = \cosh(2\omega_{\mathbb{H}^+}(w, i|w|)),$$

and the right inequality in (1.24) follows because $\cosh t \le e^t$ for all $t \ge 0$ and because $\omega_{\mathbb{H}^+}(w, i|w|) \le \omega_{\mathbb{H}^+}(w, c)$ for all $c \in \ell$, as it is easily verified.

(iii) We can again transfer everything to $\mathbb{H}^+$; up to a conjugation, recalling (1.12) we can assume that $\ell(t) = e^{2t}i$. Then $\gamma(w) = e^2 w$ is a hyperbolic automorphism of $\mathbb{H}^+$ with axis ℓ satisfying $\gamma(\ell(t)) = \ell(t+1)$ for all $t \in \mathbb{R}$, and it is unique by part (i). □

Lemma 1.4.17. (i) *Let $\gamma \in \operatorname{Aut}(\mathbb{D})$ be a hyperbolic automorphism with fixed points $\sigma, \tau \in \partial\mathbb{D}$. Then γ sends any circular arc or diameter $C \subset \overline{\mathbb{D}}$ connecting σ with τ into itself.*

(ii) *Let $\gamma \in \operatorname{Aut}(\mathbb{D})$ be a parabolic automorphism with fixed point $\tau \in \partial\mathbb{D}$. Then γ sends any circle $C \subset \overline{\mathbb{D}}$ passing through τ into itself.*

Proof. We already noticed in Remark 1.3.9 that automorphisms of $\mathbb{D}$ send circular arcs or diameters into circular arcs or diameters.

(i) By Corollary 1.1.9, without loss of generality we can assume $\sigma = 1$ and $\tau = -1$. Transfer everything on $\mathbb{H}^+$, via the Cayley transform Ψ. It is easy to check that the image through Ψ of a circular arc (or diameter) connecting -1 with 1 is a half-line issuing from 0; since there is $\lambda > 0$ such that $\Psi \circ \gamma \circ \Psi^{-1}(w) = \lambda w$ for all $w \in \mathbb{H}^+$, the assertion follows.

(ii) Without loss of generality, we can assume $\tau = 1$. Transfer everything on $\mathbb{H}^+$, via the Cayley transform Ψ. It is easy to check that the image through Ψ of a circumference passing through 1 is a horizontal line. Since there is $b \in \mathbb{R}$ such that $\Psi \circ \gamma \circ \Psi^{-1}(w) = w + b$ for all $w \in \mathbb{H}^+$, the assertion follows. □

Notes to Section 1.4

Most of the content of this section, as well as the approach we shall use in our study of Riemann surfaces, is essentially taken from the theory of Fuchsian and Kleinian groups, as developed by Poincaré [331–333]. A modern introduction to this beautiful theory is [249].

Concerning Proposition 1.4.13, $(\mathbb{C}, +)$ has many curious not closed subgroups. For instance, there are connected dense subgroups of $\mathbb{C}$ whose complement is still dense [213, 276]; on the other hand, it is known [423, 173] that the only path-connected subgroups of $\mathbb{C}$ are of $\{0\}$, $\mathbb{R}x_0$, and $\mathbb{C}$.

The left inequality in Lemma 1.4.16 is classical; the right inequality comes from [115].

1.5 Multipoint Schwarz–Pick lemmas

In this section, we shall prove a few generalizations of the Schwarz–Pick lemma, that we shall need in the next chapter, involving more than two points. We start introducing two useful notions.

Definition 1.5.1. Let $f \in \mathrm{Hol}(\mathbb{D},\mathbb{D})$. The *hyperbolic derivative* of f is the function $f^h\colon \mathbb{D} \to \mathbb{C}$ given by

$$f^h(z) = f'(z)\frac{1-|z|^2}{1-|f(z)|^2}.$$

The *hyperbolic difference quotient* of f is the function $f^*\colon \mathbb{D}\times\mathbb{D} \to \mathbb{C}$ given by

$$f^*(z,w) = \begin{cases} \dfrac{f(z)-f(w)}{1-\overline{f(w)}f(z)} \Big/ \dfrac{z-w}{1-\overline{w}z} & \text{if } z \neq w;\\ f^h(z) & \text{if } z = w.\end{cases}$$

The hyperbolic difference quotient is clearly continuous on $\mathbb{D}\times\mathbb{D}$.

The hyperbolic difference quotient is built so as to encode the information given by the Schwarz–Pick lemma.

Lemma 1.5.2. *Let $f \in \mathrm{Hol}(\mathbb{D},\mathbb{D})$. Then:*
(i) $|f^*(z,w)| \leq 1$ *for all* $z, w \in \mathbb{D}$;
(ii) *for every* $w_0 \in \mathbb{D}$ *the function* $z \mapsto f^*(z,w_0)$ *is holomorphic;*
(iii) *there exist* $z_0, w_0 \in \mathbb{D}$ *such that* $|f^*(z_0,w_0)| = 1$ *if and only if* $f \in \mathrm{Aut}(\mathbb{D})$ *if and only if* $|f^*| \equiv 1$.

Proof. (i) and (iii) follow from Corollary 1.1.16. Furthermore, $f^*(\cdot,w_0)$ is clearly holomorphic on $\mathbb{D}\setminus\{w_0\}$, and thus (ii) follows from (i) and the Riemann extension theorem. □

Remark 1.5.3. If $\gamma \in \mathrm{Aut}(\mathbb{D})$ is given by

$$\gamma(z) = e^{i\theta}\frac{z-a}{1-\overline{a}z}$$

then it is easy to check that for all z, $w \in \mathbb{D}$ we have

$$\gamma^*(z,w) = e^{i\theta}\frac{1-\overline{a}w}{1-a\overline{w}}.$$

The hyperbolic difference quotient and the hyperbolic derivative satisfy a natural chain rule.

Lemma 1.5.4. *Let $f, g \in \mathrm{Hol}(\mathbb{D}, \mathbb{D})$. Then*

$$(f \circ g)^h(z) = f^h(g(z))g^h(z)$$

for all $z \in \mathbb{D}$ and

$$(f \circ g)^*(z, w) = f^*(g(z), g(w))g^*(z, w)$$

for all $z, w \in \mathbb{D}$. In particular if $\gamma_1, \gamma_2 \in \mathrm{Aut}(\mathbb{D})$ then

$$|(\gamma_1 \circ f \circ \gamma_2)^*(z, w)| = |f^*(\gamma_2(z), \gamma_2(w))|$$

for all $z, w \in \mathbb{D}$.

Proof. The first two formulas follow immediately from

$$(f \circ g)'(z)\frac{1 - |z|^2}{1 - |(f \circ g)(z)|^2} = f'(g(z))g'(z)\frac{1 - |z|^2}{1 - |f(g(z))|^2}\frac{1 - |g(z)|^2}{1 - |g(z)|^2}$$

and from

$$\begin{aligned}&\frac{(f \circ g)(z) - (f \circ g)(w)}{1 - \overline{(f \circ g)(w)}(f \circ g)(z)} \Big/ \frac{z - w}{1 - \overline{w}z}\\ &\quad = \left[\frac{f(g(z)) - f(g(w))}{1 - \overline{f(g(w))}f(g(z))} \Big/ \frac{g(z) - g(w)}{1 - \overline{g(w)}g(z)}\right]\left[\frac{g(z) - g(w)}{1 - \overline{g(w)}g(z)} \Big/ \frac{z - w}{1 - \overline{w}z}\right].\end{aligned}$$

The last formula is a consequence of the previous two and of Lemma 1.5.2(iii). □

We have seen that the equality in the usual Schwarz–Pick lemma is realized by the automorphisms of $\mathbb{D}$. The next definition introduces the class of functions realizing the equality in the multipoint Schwarz–Pick lemmas we are going to discuss.

Definition 1.5.5. A (finite) *Blaschke product* is a $B \in \mathrm{Hol}(\mathbb{D}, \mathbb{D})$ continuous up to the boundary of $\mathbb{D}$ with $B(\partial\mathbb{D}) \subseteq \partial\mathbb{D}$. Since a Blaschke product B cannot vanish in a neighborhood of $\partial\mathbb{D}$, it must have a finite number $d \geq 0$ of zeroes in $\mathbb{D}$, counted with respect to their multiplicity. This number d is the *degree* of B.

Lemma 1.5.6. *A function $B \in \mathrm{Hol}(\mathbb{D}, \mathbb{D})$ is a Blaschke product of degree $d \geq 0$ if and only if there are $\theta \in \mathbb{R}$ and $a_1, \dots, a_d \in \mathbb{D}$ such that*

$$B(z) = e^{i\theta} \prod_{j=1}^{d} \frac{z - a_j}{1 - \overline{a_j}z}.$$

Furthermore, if $\gamma \in \mathrm{Aut}(\mathbb{D})$ then $B \circ \gamma$ and $\gamma \circ B$ are still Blaschke products of the same degree d.

Proof. If B has degree 0, then the maximum principle applied to $1/B$ implies that $|B| \equiv 1$, and hence $B \equiv e^{i\theta}$ for a suitable $\theta \in \mathbb{R}$.

Assume $d \geq 1$, and let $a_1, \dots, a_d \in \mathbb{D}$ be the zeroes of B, listed accordingly to their multiplicities. Put

$$B_0(z) = \prod_{j=1}^{d} \frac{z - a_j}{1 - \overline{a_j} z}.$$

Then B/B_0 is holomorphic without zeroes in $\mathbb{D}$ and $|B/B_0| = |B_0/B| \equiv 1$ on $\partial\mathbb{D}$. By the maximum principle, we get $|B/B_0|, |B_0/B| \leq 1$, and thus B/B_0 is a constant of modulus 1, as required.

If $\gamma \in \mathrm{Aut}(\mathbb{D})$, then $B \circ \gamma$ clearly is a Blaschke product of the same degree. On the other hand, also $\gamma \circ B$ is still a Blaschke product; we should check that it has the same degree as B. If $a = \gamma^{-1}(0)$, then $(\gamma \circ B)(z) = 0$ if and only if $B(z) = a$ if and only if

$$e^{i\theta} \prod_{j=1}^{d} (z - a_j) = a \prod_{j=1}^{d} (1 - \overline{a_j} z).$$

This is a polynomial equation of degree exactly d; moreover, all roots are in $\mathbb{D}$, because the maximum principle applied to $z \mapsto 1/B(\frac{1}{z})$ shows that $B(\mathbb{C} \setminus \overline{\mathbb{D}}) \subseteq (\mathbb{C} \cup \{\infty\}) \setminus \overline{\mathbb{D}}$. Thus $\gamma \circ B$ has exactly d zeroes in $\mathbb{D}$, counted with multiplicities, and we are done. □

We can now use the hyperbolic difference quotient to prove a *three-point Schwarz–Pick lemma*.

Proposition 1.5.7 (Beardon-Minda, 2004). *Let $f \in \mathrm{Hol}(\mathbb{D}, \mathbb{D}) \setminus \mathrm{Aut}(\mathbb{D})$. Then*

$$\omega(f^*(z, v), f^*(w, v)) \leq \omega(z, w) \tag{1.25}$$

for all $z, v, w \in \mathbb{D}$. Furthermore, equality holds for some $z_0 \neq w_0$ and v_0 if and only if it holds everywhere if and only if f is a Blaschke product of degree 2.

Proof. Given $v \in \mathbb{D}$, Lemma 1.5.2 implies that $f^*(\cdot, v) \in \mathrm{Hol}(\mathbb{D}, \mathbb{D})$, and thus (1.25) follows from the usual Schwarz–Pick lemma.

By the same argument, if we have equality in (1.25) for some $z_0 \neq w_0$ and v_0 it follows that $f^*(\cdot, v_0) \in \mathrm{Aut}(\mathbb{D})$. Now, we clearly have

$$\gamma_{v_0}(z) f^*(z, v_0) = \gamma_{f(v_0)}(f(z)). \tag{1.26}$$

If $f^*(\cdot, v_0) \in \mathrm{Aut}(\mathbb{D})$, then $B = \gamma_{v_0} f^*(\cdot, v_0)$ is a Blaschke product of degree 2; then Lemma 1.5.6 implies that $f = \gamma_{f(v_0)}^{-1} \circ B$ is a Blaschke product of degree 2, as claimed.

Conversely, if f is a Blaschke product of degree 2, then for any $v \in \mathbb{D}$ we have that $\gamma_{f(v)} \circ f$ is a Blaschke product of degree 2 vanishing at v. Therefore, γ_v divides $\gamma_{f(v)} \circ f$; so (1.26) implies that $f^*(\cdot, v)$ is a Blaschke product of degree 1, i. e., an automorphism of $\mathbb{D}$, and then in (1.25) the equality holds everywhere. □

Remark 1.5.8. Let B be a Blaschke product of degree 2. Since B is onto but not an automorphism, it must have a critical point $c \in \mathbb{D}$. Then $\gamma_{B(c)} \circ B \circ \gamma_{-c}$ is a Blaschke product of degree 2 having 0 as fixed critical point. Therefore, $\gamma_{B(c)} \circ B \circ \gamma_{-c}(z) = e^{i\theta}z^2$ for a suitable $\theta \in \mathbb{R}$, and thus

$$B(z) = \gamma_{-B(c)}\big(e^{i\theta}\gamma_c(z)^2\big).$$

Now, if $B_0(z) = e^{i\theta}z^2$ then it is easy to check that

$$B_0^*(z,w) = e^{i\theta}\frac{z+w}{1+\overline{w}z} = e^{i\theta}\gamma_{-w}(z).$$

Using Lemma 1.5.4 and Remark 1.5.3, we then obtain

$$B^*(z,w) = \frac{1+\overline{B(c)}e^{i\theta}\gamma_c(w)^2}{1+B(c)e^{-i\theta}\overline{\gamma_c(w)}^2}e^{i\theta}\gamma_{-\gamma_c(w)}\big(\gamma_c(z)\big)\frac{1-\overline{c}w}{1-c\overline{w}}. \tag{1.27}$$

For later use, given $c \in \mathbb{D}$ we define $R_c \in \mathrm{Aut}(\mathbb{D})$ by $R_c(z) = \gamma_{-c}(-\gamma_c(z))$. In other words, R_c is a non-Euclidean rotation of angle π about the point c.

We can now prove a sort of *four-point Schwarz–Pick lemma.*

Corollary 1.5.9. *Let $f \in \mathrm{Hol}(\mathbb{D},\mathbb{D}) \setminus \mathrm{Aut}(\mathbb{D})$. Then*

$$\omega(0,f^*(z,v)) \le \omega(0,f^*(u,w)) + \omega(u,v) + \omega(w,z) \tag{1.28}$$

for all $z,u,v,w \in \mathbb{D}$, with equality for some $(z,u) \neq (w,v)$ if and only if f is a Blaschke product of degree 2 and $R_c(v)$, $R_c(u)$, w and z lie in this order on a geodesic, where c is the critical point of f.

Proof. Take any $z,w,v \in \mathbb{D}$. Proposition 1.5.7 yields

$$\begin{aligned}\omega(0,f^*(z,v)) &\le \omega(0,f^*(w,v)) + \omega(f^*(w,v),f^*(z,v))\\ &\le \omega(0,f^*(w,v)) + \omega(w,z).\end{aligned} \tag{1.29}$$

Now $|f^*(w,v)| = |f^*(v,w)|$; therefore, arguing as above we get

$$\begin{aligned}\omega(0,f^*(w,v)) = \omega(0,f^*(v,w)) &\le \omega(0,f^*(u,w)) + \omega(f^*(u,w),f^*(v,w))\\ &\le \omega(0,f^*(u,w)) + \omega(u,v),\end{aligned} \tag{1.30}$$

and hence

$$\omega(0,f^*(z,v)) \le \omega(0,f^*(u,w)) + \omega(u,v) + \omega(w,z)$$

for all $z,u,v,w \in \mathbb{D}$, as claimed.

If we have equality for some $(z,u) \neq (w,v)$, then we should necessarily have $\omega(f^*(w,v),f^*(z,v)) = \omega(w,z)$ and $\omega(f^*(u,w),f^*(v,w)) = \omega(u,v)$, and thus we get that f is a Blaschke product of degree 2, by Proposition 1.5.7. Furthermore, we must have

$$\omega(0,f^*(z,v)) = \omega(0,f^*(w,v)) + \omega(f^*(w,v),f^*(z,v)),$$

that can happen only if 0, $f^*(w,v)$ and $f^*(z,v)$ lie in this order on a geodesic; since the only geodesics issuing from 0 are rays, recalling (1.27) this happens if and only if 0, $\gamma_{-\gamma_c(v)}(\gamma_c(w))$ and $\gamma_{-\gamma_c(v)}(\gamma_c(z))$ are on the same geodesic. Applying $\gamma_{\gamma_c(v)}$, we see that this happens if and only if $-\gamma_c(v)$, $\gamma_c(w)$ and $\gamma_c(z)$ lie on the same geodesic in this order; applying γ_{-c} we finally get that $R_c(v)$, w and z lie in this order on a geodesic.

Analogously, we must have

$$\omega(0,f^*(v,w)) = \omega(0,f^*(u,w)) + \omega(f^*(u,w),f^*(v,w));$$

arguing as before we see that 0, $\gamma_{-\gamma_c(w)}(\gamma_c(u))$ and $\gamma_{-\gamma_c(w)}(\gamma_c(v))$ are on the same geodesic in this order. Applying first $\gamma_{\gamma_c(w)}$, then multiplying by -1 (which reverses the order on a geodesic) and lastly applying γ_{-c} we see that this is equivalent to having $R_c(v)$, $R_c(u)$ and w in this order on the same geodesic. Since there is a unique geodesic passing through two distinct points, it follows that $R_c(v)$, $R_c(u)$, w and z lie in this order on the same geodesic, as claimed.

Conversely, if f is a Blaschke product of degree 2 with critical point $c \in \mathbb{D}$ and $u,v,z,w \in \mathbb{D}$ are such that $R_c(v)$, $R_c(u)$, w, and z lie in this order on a geodesic then the previous arguments show that we have equality in all steps of (1.29) and of (1.30) and thus in (1.28), and we are done. □

Corollary 1.5.10. *Let $f \in \mathrm{Hol}(\mathbb{D},\mathbb{D}) \setminus \mathrm{Aut}(\mathbb{D})$. Then*

$$\omega(|f^h(z)|,|f^h(w)|) \leq 2\omega(z,w) \tag{1.31}$$

for all $z,w \in \mathbb{D}$. Furthermore, equality holds for some $z \neq w$ if and only if f is a Blaschke product of degree 2 and z and w lie on a geodesic issuing from the critical point of f.

Proof. Take $z,w \in \mathbb{D}$; without loss of generality, we can assume $|f^h(z)| \geq |f^h(w)|$. Then

$$\omega(|f^h(z)|,|f^h(w)|) = \omega(0,|f^h(z)|) - \omega(0,|f^h(w)|) = \omega(0,f^h(z)) - \omega(0,f^h(w)).$$

Now (1.28) applied with $u = w$ and $v = z$ yields

$$\omega(0,f^h(z)) \leq \omega(0,f^h(w)) + 2\omega(z,w),$$

and (1.31) follows. By Corollary 1.5.9, the equality holds if and only if f is a Blaschke product of degree 2 and $R_c(z)$, $R_c(w)$, w and z lie in this order on the same geodesic, where c is the critical point of f. But this can happen if and only if w and z lie on a geodesic issuing from c, and we are done. □

Take now $f \in \mathrm{Hol}(\mathbb{D},\mathbb{D}) \setminus \mathrm{Aut}(\mathbb{D})$. Corollary 1.5.9 with $u = v = w$ becomes

$$\omega(0, f^*(z,w)) \le \omega(0, f^h(w)) + \omega(z,w) \tag{1.32}$$

with equality for some $z \neq w$ if and only if f is a Blaschke product of degree 2 and w and z lie in this order on a geodesic ray emanating from the critical point c of f. As a consequence, we get the following sharpening of the Schwarz lemma.

Corollary 1.5.11. *Let $f \in \mathrm{Hol}(\mathbb{D},\mathbb{D}) \setminus \mathrm{Aut}(\mathbb{D})$. Then*

$$|f^*(z,w)| \le \frac{\left|\frac{z-w}{1-\overline{w}z}\right| + |f^h(w)|}{1 + |f^h(w)|\left|\frac{z-w}{1-\overline{w}z}\right|} \tag{1.33}$$

for all $z, w \in \mathbb{D}$, with equality for some $z \neq w$ if and only if f is a Blaschke product of degree 2 and w and z lie in this order on a geodesic ray emanating from the critical point c of f.

Moreover, if $f(0) = 0$ we get

$$|f(z)| \le |z| \frac{|z| + |f'(0)|}{1 + |f'(0)||z|}, \tag{1.34}$$

with equality for some $z \neq 0$ if and only if f is a Blaschke product of degree 2 and $z = -tc$ for some $t > 0$, where c is the critical point of f.

Proof. Applying tanh to both sides of (1.32) and recalling the addition formula (1.9), we get (1.33) as well as the assertion about the equality. The second part of the statement follows taking $w = 0$ and assuming $f(0) = 0$. □

We end this section with another multipoint estimate that we shall need in Section 3.7.

Theorem 1.5.12 (Christodoulou-Short, 2019). *Given $a, b, z \in \mathbb{D}$ with $a \neq b$, let*

$$K_{a,b}(z) = \frac{1}{2\sinh \omega(a,b)} \exp\big[2(\omega(z,a) + \omega(a,b) + \omega(b,z))\big].$$

Then we have

$$\omega(f(z), \gamma(z)) \le K_{a,b}(z)\big[\omega(f(a), \gamma(a)) + \omega(f(b), \gamma(b))\big] \tag{1.35}$$

for all $f \in \mathrm{Hol}(\mathbb{D},\mathbb{D})$ and $\gamma \in \mathrm{Aut}(\mathbb{D})$.

Proof. Up to replacing f by $\gamma^{-1} \circ f$, we can assume that $\gamma = \mathrm{id}_{\mathbb{D}}$.

Let us first assume that $f(b) = b$. If $f = \mathrm{id}_{\mathbb{D}}$, there is nothing to prove; so without loss of generality, we can assume that $f(a) \neq a$ (see Corollary 1.1.14). Furthermore, it suffices to prove the assertion for $b = 0$. Indeed, putting $\tilde{f} = \gamma_b \circ f \circ \gamma_b^{-1}$ we have $\tilde{f}(0) = 0$; setting $\tilde{z} = \gamma_b(z)$ and $\tilde{a} = \gamma_b(a)$ we have $\omega(\tilde{f}(\tilde{z}), \tilde{z}) = \omega(f(z), z)$, $\omega(\tilde{f}(\tilde{a}), \tilde{a}) = \omega(f(a), a)$

and $K_{\tilde{a},0}(\tilde{z}) = K_{a,b}(z)$, and thus the assertion for f and b follows from the assertion from $\tilde{f}$ and 0.

Furthermore, it suffices to prove the assertion for $f \notin \mathrm{Aut}(\mathbb{D})$. Indeed, if $f \in \mathrm{Aut}(\mathbb{D})$ we can write (1.35) for $f_r = rf$ and then take the limit as $r \to 1^-$; since $K_{a,0}(z)$ does not depend on f, in this way we get (1.35) for automorphisms knowing it for nonsurjective maps.

Since $K_{a,0}(z) \geq 1$ always, (1.35) clearly holds if $\omega(f(z),z) \leq \omega(f(a),a)$; so assume that $\omega(f(z),z) > \omega(f(a),a)$. Since $t \mapsto \frac{\sinh t}{t}$ is strictly increasing when $t > 0$, we then have

$$\frac{\omega(f(z),z)}{\omega(f(a),a)} < \frac{\sinh \omega(f(z),z)}{\sinh \omega(f(a),a)};$$

so it suffices to prove that the latter ratio is bounded by $K_{a,0}(z)$.

We clearly can assume $z \neq 0$. To simplify the notation, from now on we put $s(z,w) = \sinh \omega(z,w)$ and $c(z,w) = \cosh \omega(z,w)$. Furthermore, set $g(z) = f^*(z,0)$; since f is not an automorphism we know that g is a holomorphic self-map of $\mathbb{D}$. The triangular inequality $|1 - g(z)| \leq |1 - g(a)| + |g(a) - g(z)|$ yields

$$\frac{|z - f(z)|}{|z|} \leq \frac{|a - f(a)|}{|a|} + |g(a) - g(z)|. \tag{1.36}$$

Recalling (1.19) we can rewrite (1.36) in the following form:

$$\frac{s(z,f(z))}{s(z,0)c(f(z),0)} \leq \frac{s(a,f(a))}{s(a,0)c(f(a),0)} + \frac{s(g(a),g(z))}{c(g(a),0)c(g(z),0)}.$$

Since $c(f(z),0) \leq c(z,0)$, we obtain

$$\frac{s(z,f(z))}{s(a,f(a))} \leq \frac{s(z,0)c(z,0)}{s(a,0)}\left(\frac{1}{c(f(a),0)} + \frac{s(g(a),g(z))s(a,0)}{c(g(a),0)c(g(z),0)s(a,f(a))}\right).$$

Now, $s(z,0)c(z,0) = \frac{1}{2}\sinh(2\omega(0,z)) \leq \frac{1}{4}e^{2\omega(0,z)}$; therefore, we have

$$\frac{s(z,f(z))}{s(a,f(a))} \leq \frac{e^{2\omega(0,z)}}{4s(a,0)}\left(\frac{1}{c(f(a),0)} + \frac{s(g(a),g(z))s(a,0)}{c(g(a),0)c(g(z),0)s(a,f(a))}\right);$$

we must now estimate the term in the large brackets.

Since $\omega(g(z),0) \geq |\omega(g(a),0) - \omega(g(a),g(z))|$ and cosh is an even function, recalling the formula for the hyperbolic cosine of a sum we get

$$c(g(z),0) \geq c(g(a),0)c(g(a),g(z)) - s(g(a),0)s(g(a),g(z)).$$

Multiplying both sides by $c(g(a),0)$ and recalling again (1.19), we obtain

$$c(g(a),0)c(g(z),0) \geq \frac{c(g(a),g(z)) - |g(a)|s(g(a),g(z))}{1 - |g(a)|^2}$$

$$\geq \frac{c(g(a),g(z)) - |g(a)|s(g(a),g(z))}{1+|g(a)|}\frac{|a|}{|a-f(a)|}$$
$$= \frac{c(g(a),g(z)) - |g(a)|s(g(a),g(z))}{1+|g(a)|}\frac{s(a,0)c(f(a),0)}{s(a,f(a))}.$$

Therefore,

$$\begin{aligned}
&\frac{1}{c(f(a),0)} + \frac{s(g(a),g(z))s(a,0)}{c(g(a),0)c(g(z),0)s(a,f(a))}\\
&\leq \frac{1}{c(f(a),0)}\left[1+\frac{(1+|g(a)|)s(g(a),g(z))}{c(g(a),g(z)) - |g(a)|s(g(a),g(z))}\right]\\
&= \frac{1}{c(f(a),0)}\left[\frac{c(g(a),g(z))+s(g(a),g(z))}{c(g(a),g(z)) - |g(a)|s(g(a),g(z))}\right]\\
&\leq \frac{c(g(a),g(z))+s(g(a),g(z))}{c(g(a),g(z)) - s(g(a),g(z))} = e^{2\omega(g(a),g(z))}\\
&\leq e^{2\omega(a,z)}.
\end{aligned}$$

Putting all together, we finally get

$$\begin{aligned}
\omega(f(z),z) &\leq \frac{1}{4\sinh\omega(a,0)}\exp[2(\omega(0,z)+\omega(a,z))]\omega(f(a),a)\\
&\leq K_{a,0}(z)\omega(f(a),a),
\end{aligned} \tag{1.37}$$

and we have proved (1.35) when $f(b) = b$.

Let us now assume that $f(b) \neq b$; then there is a unique geodesic ℓ passing through b and $f(b)$. By Lemma 1.4.16(i), there is a unique hyperbolic automorphism γ of $\mathbb{D}$ with axis ℓ and sending $f(b)$ in b. Then $\tilde{f} = \gamma\circ f$ fixes b, and the first inequality in (1.37) yields

$$\omega(\tilde{f}(z),z) \leq M_{a,b}(z)\omega(\tilde{f}(a),a),$$

where $M_{a,b}(z) = \frac{1}{4\sinh\omega(a,b)}\exp[2(\omega(b,z)+\omega(a,z))]$. Now

$$\begin{aligned}
\omega(f(z),z) &\leq \omega(f(z),\tilde{f}(z)) + \omega(\tilde{f}(z),z)\\
&\leq \omega(f(z),\tilde{f}(z)) + M_{a,b}(z)\omega(\tilde{f}(a),a)\\
&\leq \omega(f(z),\tilde{f}(z)) + M_{a,b}(z)\omega(\tilde{f}(a),f(a)) + M_{a,b}(z)\omega(f(a),a).
\end{aligned}$$

Since $f(b)$ belongs by construction to ℓ, we can apply Lemma 1.4.16(ii) twice with $c = f(b)$ obtaining

$$\begin{aligned}
\omega(f(z),z) &\leq [e^{2\omega(f(z),f(b))} + M_{a,b}(z)e^{2\omega(f(a),f(b))}]\omega(f(b),b) + M_{a,b}(z)\omega(f(a),a)\\
&\leq [e^{2\omega(z,b)} + M_{a,b}(z)e^{2\omega(a,b)}](\omega(f(a),a)+\omega(f(b),b))\\
&\leq \frac{e^{2(\omega(z,b)+\omega(a,b)+\omega(a,z))}}{2\sinh\omega(a,b)}(\omega(f(a),a)+\omega(f(b),b)),
\end{aligned}$$

where the last inequality follows from the fact that $e^{2t} \geq 4\sinh t$ for all $t \geq 0$, and we are done. □

To give a first example of the usefulness of this estimate, we prove the following result.

Corollary 1.5.13. *Let* $\{f_\nu\} \subset \mathrm{Hol}(\mathbb{D}, \mathbb{D})$ *be a sequence of holomorphic self-maps of* $\mathbb{D}$. *Assume that there exist* $z_0, z_1 \in \mathbb{D}$ *with* $z_0 \neq z_1$ *such that* $f_\nu(z_j) \to w_j \in \mathbb{D}$ *as* $\nu \to +\infty$ *for* $j = 0, 1$, *with* $\omega(w_0, w_1) = \omega(z_0, z_1)$. *Then* $\{f_\nu\}$ *converges, uniformly on compact subsets, to the unique* $\gamma \in \mathrm{Aut}(\mathbb{D})$ *such that* $\gamma(z_j) = w_j$ *for* $j = 1, 2$.

Proof. Corollary 1.2.5 implies that there exists a (unique, by Corollary 1.1.14) automorphism $\gamma \in \mathrm{Aut}(\mathbb{D})$ such that $\gamma(z_j) = w_j$ for $j = 0, 1$. Then we can apply Theorem 1.5.12 with $a = z_0$ and $b = z_1$, and it follows that if $f_\nu(z_j) \to w_j = \gamma(z_j)$ for $j = 0, 1$ then $f_\nu \to \gamma$ uniformly on compact subsets, as claimed. □

Notes to Section 1.5

This section is mostly taken from the elegant paper [48] where Beardon and Minda, in 2004, found a unified way to obtain and generalize many estimates scattered in the literature. In particular, results due to Dieudonné [136], Rogosinski [355], Goluzin [168, 169], Beardon and Carne [47], Yamashita [425, 426], Beardon [43], Osserman [313], Kaptanoğlu [221], and Mercer [282–284] can be recovered as special cases of the material described in this section (in particular, we shall prove Goluzin's estimate in Lemma 2.7.1). See also [49] for other applications, and [39, 114] for further generalizations.

For instance, an immediate consequence of Proposition 1.5.7 applied with $v = z$ and $w = 0$ is the classical *Dieudonné lemma* [136]: if $f \in \mathrm{Hol}(\mathbb{D}, \mathbb{D})$ is such that $f(0) = 0$ then

$$\left| f'(z) - \frac{f(z)}{z} \right| \leq \frac{|z|^2 - |f(z)|^2}{|z|(1 - |z|^2)}$$

for all $z \in \mathbb{D} \setminus \{0\}$. See also [114] for further generalizations of Dieudonné lemma and [51] for what happens in domains different from $\mathbb{D}$.

Corollary 1.5.11 is in [282, 313], but it was possibly known before.

Theorem 1.5.12 and Corollary 1.5.13 come from [115], where it is also possible to find the following statement: for all $z_0, z \in \mathbb{D}^*$ we have $\omega_{\mathbb{D}^*}(f(z), z) \leq L_{z_0}(z)^3 \omega_{\mathbb{D}^*}(f(z_0), z_0)$ for all $f \in \mathrm{Hol}(\mathbb{D}^*, \mathbb{D}^*)$ with $f(0) = 0$, where $L_{z_0}(z) = \frac{-8}{|z_0| \log |z_0|} \exp \omega_{\mathbb{D}^*}(z_0, z)$.

1.6 Riemann surfaces

In this section, we collect a few fundamental facts concerning Riemann surfaces, up to the classification of Riemann surfaces with Abelian fundamental group. As mentioned in the Introduction to this chapter, our main tool will be the theory of covering spaces.

Definition 1.6.1. A *Riemann surface* is a one-dimensional connected complex manifold. More precisely, giving a Riemann surface amounts to specifying a connected

Hausdorff topological space X with a countable basis and a family $\mathcal{A} = \{(U_\alpha, \varphi_\alpha)\}$ satisfying the following properties:
(i) $\{U_\alpha\}$ is an open cover of X;
(ii) each *chart* $\varphi_\alpha: U_\alpha \to \mathbb{C}$ is a homeomorphism between U_α and an open subset of $\mathbb{C}$;
(iii) each *change of coordinates* $\varphi_\beta \circ \varphi_\alpha^{-1}$ is holomorphic where defined.

The family $\mathcal{A}$ is an *atlas*; two atlases on the same topological space define the same Riemann surface if and only if their union is still an atlas.

See, e. g., [154] for more information on Riemann surfaces.

Definition 1.6.2. A continuous map $f: X \to Y$ between two Riemann surfaces is *holomorphic* if for any atlas $\{(U_\alpha, \varphi_\alpha)\}$ of X and any atlas $\{(V_\beta, \psi_\beta)\}$ of Y the function $\psi_\beta \circ f \circ \varphi_\alpha^{-1}$ is holomorphic where defined as a function of one complex variable. We shall denote by $\mathrm{Hol}(X, Y) \subset C^0(X, Y)$ the space of holomorphic functions from X to Y, endowed with the compact-open topology (equivalently, with the topology of uniform convergence on compact subsets; see, e. g., [227, p. 230]). A *biholomorphism* is an invertible holomorphic function; it is easy to check that the inverse is automatically holomorphic. If there exists a biholomorphism $f: X \to Y$ we shall say that the Riemann surfaces X and Y are *biholomorphic*. An *automorphism* of X is a biholomorphism of X with itself; we shall denote by $\mathrm{Aut}(X)$ the group of automorphisms of X. Finally, the *isotropy group* $\mathrm{Aut}_{z_0}(X)$ of $z_0 \in X$ is the subgroup of automorphisms of X fixing z_0.

Example 1.6.3. The *Riemann sphere* $\widehat{\mathbb{C}} = \mathbb{C} \cup \{\infty\}$ is a prototypical example of Riemann surface. As a topological space, it is homeomorphic to the unit sphere $\mathbb{S}^2$ in $\mathbb{R}^3$. More precisely, it is endowed with the standard topology on $\mathbb{C}$, and it has as fundamental system of neighborhoods of ∞ the complements of compact sets in $\mathbb{C}$; in particular, a sequence $\{z_\nu\} \subset \mathbb{C}$ converges to ∞ in $\widehat{\mathbb{C}}$ if and only if $|z_\nu| \to +\infty$. It becomes a Riemann surface when considered with the atlas $\{(U_0, \varphi_0), (U_1, \varphi_1)\}$, where $U_0 = \mathbb{C}$, $\varphi_0 = \mathrm{id}_{\mathbb{C}}$, $U_1 = \mathbb{C}^* \cup \{\infty\}$, and $\varphi_1: U_1 \to \mathbb{C}$ is given by

$$\varphi_1(w) = \begin{cases} \frac{1}{w} & \text{if } w \neq \infty, \\ 0 & \text{if } w = \infty. \end{cases}$$

The Riemann sphere is also sometimes called *extended complex plane* or *projective complex line*, and in the latter case it is often denoted by $\mathbb{P}^1(\mathbb{C})$.

Remark 1.6.4. It is well known that $\mathrm{Hol}(\widehat{\mathbb{C}}, \widehat{\mathbb{C}})$ coincides with the set of rational maps; see, e. g., [294, pp. 30, 41].

The main fact we shall use without proof is the fundamental *Riemann uniformization theorem.*

Theorem 1.6.5 (Riemann uniformization theorem, 1907). *Every simply connected Riemann surface is biholomorphic either to the Riemann sphere $\widehat{\mathbb{C}}$, or to the complex plane $\mathbb{C}$, or to the unit disk $\mathbb{D}$.*

A proof can be found, e. g., in [154] or in [143].

The importance of this theorem lies in the fact that, as we shall see momentarily, every Riemann surface is the quotient of a simply connected Riemann surface by a group of automorphisms; therefore, the study of Riemann surfaces is reduced to the investigation of particular subgroups of $\mathrm{Aut}(\widehat{\mathbb{C}})$, $\mathrm{Aut}(\mathbb{C})$, and $\mathrm{Aut}(\mathbb{D})$. We shall consistently use this approach throughout this section, reverting as needed to a more geometric point of view in subsequent sections.

Our first aim then is to describe any Riemann surface as quotient of a simply connected Riemann surface. We begin with the following basic result.

Lemma 1.6.6. *Let X be a Riemann surface. Then there exists a holomorphic covering map $\pi: \widetilde{X} \to X$ such that $\widetilde{X}$ is a simply connected Riemann surface. Moreover, if $\hat{\pi}: \widehat{X} \to X$ is another holomorphic covering map with $\widehat{X}$ simply connected then there exists a biholomorphism $\Phi: \widehat{X} \to \widetilde{X}$ such that $\hat{\pi} = \pi \circ \Phi$.*

Proof. Let $\pi: \widetilde{X} \to X$ be the topological universal covering map of X, and let $\{(U_\alpha, \varphi_\alpha)\}$ be an atlas for X; without loss of generality, we can assume that each U_α is connected and well covered by π, i. e., each $\pi^{-1}(U_\alpha)$ is the disjoint union of open subsets $\widetilde{U}_{\alpha\beta}$ of $\widetilde{X}$ such that $\pi|_{\widetilde{U}_{\alpha\beta}}$ is a homeomorphism between $\widetilde{U}_{\alpha\beta}$ and U_α for each α and β. Then it is easy to see that setting $\tilde{\varphi}_{\alpha\beta} = \varphi_\alpha \circ \pi|_{\widetilde{U}_{\alpha\beta}}$ we get an atlas $\tilde{\mathcal{A}} = \{(\widetilde{U}_{\alpha\beta}, \tilde{\varphi}_{\alpha\beta})\}$ on $\widetilde{X}$ such that π is holomorphic.

Assume now that $\hat{\pi}: \widehat{X} \to X$ is another holomorphic covering map with $\widehat{X}$ simply connected. It is well known (see, e. g., [260, Proposition 12.6]) that there exists a homeomorphism $\Phi: \widehat{X} \to \widetilde{X}$ such that $\hat{\pi} = \pi \circ \Phi$; it suffices to prove that Φ is holomorphic.

Fix $\hat{z}_0 \in \widehat{X}$, and choose a chart $(U_\alpha, \varphi_\alpha)$ centered in $\hat{\pi}(z_0)$ with $U_\alpha \subset X$ connected and well covered by both $\hat{\pi}$ and π. Denote by $\widehat{U}_0 \subset \widehat{X}$ the connected component of $\hat{\pi}^{-1}(U_\alpha)$ containing $\hat{z}_0$, and by $\widetilde{U}_0 \subset \widetilde{X}$ the connected component of $\pi^{-1}(U_\alpha)$ containing $\tilde{z}_0 = \Phi(\hat{z}_0)$. Since Φ is a homeomorphism satisfying $\hat{\pi} = \pi \circ \Phi$ we immediately get $\Phi(\widehat{U}_0) = \widetilde{U}_0$, and thus $\Phi|_{\widehat{U}_0} = (\pi|_{\widetilde{U}_0})^{-1} \circ \hat{\pi}|_{\widehat{U}_0}$ is holomorphic. Since $\hat{z}_0$ was arbitrary, it follows that Φ is holomorphic everywhere, and we are done. □

Definition 1.6.7. Let X be a Riemann surface. We shall denote by $\pi_X: \widetilde{X} \to X$ the unique (up to biholomorphisms) holomorphic universal covering map whose existence is proved in Lemma 1.6.6; the simply connected Riemann surface $\widetilde{X}$ will be called the *universal covering surface* of X. Moreover, X is said *elliptic* (resp., *parabolic*, *hyperbolic*) if its universal covering surface is $\widehat{\mathbb{C}}$ (resp., $\mathbb{C}$, $\mathbb{D}$).

Definition 1.6.8. Let $\pi: \widetilde{X} \to X$ be a holomorphic (not necessarily universal) covering map of a Riemann surface X. An *automorphism of the covering* (or *deck transformation*) is an automorphism γ of $\widetilde{X}$ such that $\pi \circ \gamma = \pi$. If $\pi_X : \widetilde{X} \to X$ is the universal covering map we shall denote by Γ_X the group of automorphisms of π_X.

The automorphisms of a covering clearly form a group. Standard covering spaces theory (see, e. g., [260, Chapter 11] and [154, Section 1.5]) gives the following result.

Proposition 1.6.9. *Let $\pi_X : \widetilde{X} \to X$ be the universal covering map of a Riemann surface X. Then:*
(i) *the group Γ_X of automorphisms of π_X is isomorphic to the fundamental group $\pi_1(X)$ of X;*
(ii) *Γ_X is simply transitive on the fibers of the covering, i. e., for any point $z \in X$ and any pair of points $w_1, w_2 \in \pi_X^{-1}(z)$ there exists a unique $\gamma \in \Gamma_X$ such that $\gamma(w_1) = w_2$.*

Remark 1.6.10. For later use, we recall here the definition of the isomorphism μ between $\pi_1(X)$ and Γ_X. Choose base points $q \in X$ and $\tilde{q} \in \widetilde{X}$ such that $\pi_X(\tilde{q}) = q$. Let $[\sigma]$ be an element of $\pi_1(X, q)$, and $\sigma : [0, 1] \to X$ a loop representing $[\sigma]$ with $\sigma(0) = q$. Let $\tilde{\sigma} : [0, 1] \to \widetilde{X}$ be the unique curve such that $\pi_X \circ \tilde{\sigma} = \sigma$ and $\tilde{\sigma}(0) = \tilde{q}$. Since $\tilde{\sigma}(1)$ belongs to $\pi_X^{-1}(q)$ there exists a unique $\gamma \in \Gamma_X$ such that $\gamma(\tilde{q}) = \tilde{\sigma}(1)$; we put $\mu([\sigma]) = \gamma$. It is clear that μ is well-defined, and it is not difficult to check that it is an isomorphism (see, e. g., [154, p. 34] or [260, p. 250]).

It turns out that the automorphism group of a universal covering map cannot be any subgroup of the automorphism group of the covering surface, but it must satisfy two important properties.

Definition 1.6.11. Let X be a topological space and Γ a group of homeomorphisms of X. We say that Γ acts *freely* on X if no element of Γ other than the identity has a fixed point; that Γ is *properly discontinuous* at a point $z \in X$ if there exists a neighborhood U of z such that the set $\{\gamma \in \Gamma \mid \gamma(U) \cap U \neq \emptyset\}$ is finite; and that Γ is *properly discontinuous* tout-court if it is so at every point.

Then we have the following.

Proposition 1.6.12. *Let $\widetilde{X}$ be a simply connected Riemann surface and Γ a subgroup of* $\mathrm{Aut}(\widetilde{X})$. *Then:*
(i) *If Γ is properly discontinuous and acts freely on $\widetilde{X}$ then the orbit space $\widetilde{X}/\Gamma$ has a natural structure of Riemann surface such that the canonical projection $\tilde{\pi} : \widetilde{X} \to \widetilde{X}/\Gamma$ is its holomorphic universal covering map.*
(ii) *If Γ_X is the automorphism group of the universal covering map $\pi_X : \widetilde{X} \to X$ of a Riemann surface X, then Γ_X is a closed subgroup of* $\mathrm{Aut}(\widetilde{X})$, *properly discontinuous and acting freely on $\widetilde{X}$. Moreover, X is canonically biholomorphic to $\widetilde{X}/\Gamma_X$.*

Proof. (i) Given $z_0 \in \widetilde{X}$, since Γ is properly discontinuous there is a neighborhood $\widetilde{U}$ of z_0 such that $\{\gamma \in \Gamma \mid \gamma(\widetilde{U}) \cap \widetilde{U} \neq \emptyset\}$ is finite; we can also assume that $\widetilde{U}$ is the domain of a chart $\tilde{\varphi}$. Furthermore, since Γ acts freely, we have $\gamma(z_0) \neq z_0$ for all $\gamma \neq \mathrm{id}_{\widetilde{X}}$;

therefore, up to shrinking $\widetilde{U}$, we can assume that $\{\gamma \in \Gamma \mid \gamma(\widetilde{U}) \cap \widetilde{U} \neq \emptyset\} = \{\mathrm{id}_{\widetilde{X}}\}$. This implies that $\gamma_1(\widetilde{U}) \cap \gamma_2(\widetilde{U}) = \emptyset$ as soon as $\gamma_1 \neq \gamma_2$. In particular, $\tilde{\pi}$ restricted to $\gamma(\widetilde{U})$ is a homeomorphism between $\gamma(\widetilde{U})$ and its image $U \subset \widetilde{X}/\Gamma$ for all $\gamma \in \Gamma$. Since

$$\tilde{\pi}^{-1}(\pi(\widetilde{U})) = \bigcup_{\gamma \in \Gamma} \gamma(\widetilde{U}),$$

it follows that U is well covered by $\tilde{\pi}$, and we can define a chart φ on U by setting $\varphi = \tilde{\varphi} \circ (\tilde{\pi}|_{\widetilde{U}})^{-1}$. It is then easy to check that in this way we have endowed $\widetilde{X}/\Gamma$ with a natural structure of Riemann surface such that the canonical projection $\tilde{\pi}$ is a holomorphic covering map.

(ii) If $\{\gamma_\nu\} \subset \Gamma_X$ is a sequence converging to some $\gamma \in \mathrm{Aut}(\widetilde{X})$, then from $\pi_X \circ \gamma_\nu = \pi_X$ for all ν we infer $\pi_X \circ \gamma = \pi_X$, i. e., $\gamma \in \Gamma_X$, and thus Γ_X is closed in $\mathrm{Aut}(\widetilde{X})$.

Fix $z_0 \in \widetilde{X}$ and a well covered connected neighborhood U of $\pi(z_0)$. Let $\widetilde{U}$ be the connected component of $\pi^{-1}(U)$ containing z_0; then $\pi_X|_{\widetilde{U}}$ is a biholomorphism between $\widetilde{U}$ and U. Take $\gamma \in \Gamma_X$ such that $\gamma(\widetilde{U}) \cap \widetilde{U} \neq \emptyset$. If $z \in \gamma(\widetilde{U}) \cap \widetilde{U}$, choose $w \in \widetilde{U}$ so that $\gamma(w) = z$. From $\pi_X \circ \gamma = \pi_X$, we get $\pi_X(z) = \pi_X(w)$; since $\pi_X|_{\widetilde{U}}$ is injective we get $w = z = \gamma(w)$. So γ is the identity on an open subset, and hence everywhere. In other words, $\{\gamma \in \Gamma_X \mid \gamma(\widetilde{U}) \cap \widetilde{U} \neq \emptyset\} = \{\mathrm{id}_{\widetilde{X}}\}$. Since z_0 is an arbitrary point of $\widetilde{X}$, this means exactly that Γ_X is properly discontinuous and acts freely on $\widetilde{X}$.

Finally, Proposition 1.6.9(ii) implies that the Γ_X-orbit $\{\gamma(\tilde{z}) \mid \gamma \in \Gamma_X\}$ of a point $\tilde{z} \in \widetilde{X}$ coincides with the fiber $\pi_X^{-1}(\pi_X(\tilde{z}))$ of $\pi_X(\tilde{z})$; hence there is a canonical bijection χ between X and $\widetilde{X}/\Gamma_X$ associating to each $z \in X$ its fiber $\pi_X^{-1}(z)$. In particular, the canonical projection $\tilde{\pi}: \widetilde{X} \to \widetilde{X}/\Gamma_X$ is given by $\tilde{\pi} = \chi \circ \pi_X$. Then from the definition of the Riemann surface structure on $\widetilde{X}/\Gamma_X$ given in (i) it follows immediately that χ is holomorphic—and hence a biholomorphism. □

A Riemann surface X is thus given by its universal covering surface $\widetilde{X}$ and by a properly discontinuous subgroup Γ of $\mathrm{Aut}(\widetilde{X})$ acting freely on $\widetilde{X}$ and isomorphic to $\pi_1(X)$. As a consequence, in principle analytic and geometric properties of X should follow from analytic and geometric properties of $\widetilde{X}$ and algebraic properties of Γ. One of the main tools for this reduction is given by the lifting of maps, a typical construction in covering spaces that we now recall to fix the terminology.

Definition 1.6.13. Let X and Y be two Riemann surfaces, and $\pi_X: \widetilde{X} \to X$ and $\pi_Y: \widetilde{Y} \to Y$ holomorphic covering maps. A *lifting* of a map $f \in \mathrm{Hol}(X, Y)$ is a holomorphic map $\tilde{f} \in \mathrm{Hol}(\widetilde{X}, \widetilde{Y})$ such that $f \circ \pi_X = \pi_Y \circ \tilde{f}$.

It turns out that when π_X and π_Y are universal covering maps then liftings always exist and are uniquely determined by the value at one point.

Proposition 1.6.14. *Let X and Y be two Riemann surfaces, $\pi_X: \widetilde{X} \to X$ and $\pi_Y: \widetilde{Y} \to Y$ their universal covering maps, and let Γ_X and Γ_Y be the corresponding automorphism groups. Take $f \in \mathrm{Hol}(X, Y)$. Then:*

(i) *for every* $\tilde{z}_0 \in \widetilde{X}$ *and* $\tilde{w}_0 \in \widetilde{Y}$ *such that* $\pi_Y(\tilde{w}_0) = f(\pi_X(\tilde{z}_0))$ *there exists a unique lifting* $\tilde{f} \in \mathrm{Hol}(\widetilde{X}, \widetilde{Y})$ *such that* $\tilde{f}(\tilde{z}_0) = \tilde{w}_0$*;*
(ii) *if* $\tilde{f} \in \mathrm{Hol}(\widetilde{X}, \widetilde{Y})$ *is a lifting of* f *then all other liftings are of the form* $\gamma \circ \tilde{f}$ *with* $\gamma \in \Gamma_Y$*;*
(iii) *if* $\tilde{f} \in \mathrm{Hol}(\widetilde{X}, \widetilde{Y})$ *is a lifting of* f *then*

$$\forall \gamma \in \Gamma_X \ \exists \gamma_1 \in \Gamma_Y : \quad \tilde{f} \circ \gamma = \gamma_1 \circ \tilde{f}; \tag{1.38}$$

(iv) f *is a biholomorphism if and only if any lifting* $\tilde{f} \in \mathrm{Hol}(\widetilde{X}, \widetilde{Y})$ *is a biholomorphism such that* $\tilde{f}^{-1}$ *satisfies* (1.38)*;*
(v) *a map* $\tilde{f} \in \mathrm{Hol}(\widetilde{X}, \widetilde{Y})$ *is a lifting of a holomorphic map from* X *to* Y *if and only if* (1.38) *is satisfied.*

Proof. These statements for continuous maps are well known (see, e. g., [260, Chapters 11, 12]). To get the statements for holomorphic maps, it suffices to remark that a lifting $\tilde{f}$ of a continuous map f is holomorphic if and only if f is holomorphic, because the universal covering maps are holomorphic and locally invertible. □

Another kind of lifting problem is the following: given $f \in \mathrm{Hol}(X, Y)$ find $\hat{f} \in \mathrm{Hol}(X, \widetilde{Y})$ such that $f = \pi_Y \circ \hat{f}$. This time the problem is not always solvable; $\hat{f}$ exists if and only if $f_*(\pi_1(X))$ is trivial, where $f_*: \pi_1(X) \to \pi_1(Y)$ is the homomorphism induced by f on the fundamental groups (see, e. g., [260, Theorem 11.15]). The relations among f, $\tilde{f}$, and $\hat{f}$ are summarized by the following commutative diagram:

$$\begin{array}{ccc} \widetilde{X} & \xrightarrow{\tilde{f}} & \widetilde{Y} \\ {\scriptstyle \pi_X}\downarrow & \overset{\hat{f}}{\nearrow} & \downarrow{\scriptstyle \pi_Y} \\ X & \xrightarrow{f} & Y \end{array} .$$

Another way of looking at this question makes use of the notion of automorphic function.

Definition 1.6.15. Let X, Y be Riemann surfaces, and Γ a subgroup of $\mathrm{Aut}(X)$. A function $g \in \mathrm{Hol}(X, Y)$ is *automorphic* under Γ if $g \circ \gamma = g$ for all $\gamma \in \Gamma$.

Remark 1.6.16. Let $\pi_X: \widetilde{X} \to X$ be the universal covering map of a Riemann surface X, with automorphism group Γ_X. If Y is another Riemann surface, a function $g \in \mathrm{Hol}(\widetilde{X}, Y)$ is automorphic under Γ_X if and only if there is a function $g_o \in \mathrm{Hol}(X, Y)$ such that $g = g_o \circ \pi_X$. This time the explicative commutative diagram is

$$\begin{array}{ccc} \widetilde{X} & \xrightarrow{\tilde{g}_o} & \widetilde{Y} \\ {\scriptstyle \pi_X}\downarrow & \overset{g}{\searrow} & \downarrow{\scriptstyle \pi_Y} \\ \widetilde{X}/\Gamma_X = X & \xrightarrow{g_o} & Y \end{array} .$$

Coming back to our $f \in \mathrm{Hol}(X,Y)$, we see that there exists $\hat{f} \in \mathrm{Hol}(X,\widetilde{Y})$ such that $f = \pi_Y \circ \hat{f}$ if and only if any lifting (and hence all liftings) $\tilde{f} \in \mathrm{Hol}(\widetilde{X},\widetilde{Y})$ is automorphic under Γ_X. For future reference, we summarize the situation in the following statement.

Proposition 1.6.17. *Let X and Y be two Riemann surfaces, $\pi_X: \widetilde{X} \to X$ and $\pi_Y: \widetilde{Y} \to Y$ their universal covering maps, and $\Gamma_X \subset \mathrm{Aut}(\widetilde{X})$ and $\Gamma_Y \subset \mathrm{Aut}(\widetilde{Y})$ the corresponding automorphism groups. Choose a function $f \in \mathrm{Hol}(X,Y)$. Then the following statements are equivalent:*
(i) *$f_*(\pi_1(X))$ is trivial;*
(ii) *there exists $\hat{f} \in \mathrm{Hol}(X,\widetilde{Y})$ such that $f = \pi_Y \circ \hat{f}$;*
(iii) *a lifting (and hence any lifting) of f is automorphic under Γ_X.*

Proof. Again, the statement is well known for continuous functions (see, e. g., [260, Chapters 11, 12]); we get the statement for holomorphic functions by remarking (as in the proof of Proposition 1.6.14) that $\hat{f}$ is holomorphic if and only if f is holomorphic. □

Clearly, two different subgroups of $\mathrm{Aut}(\widetilde{X})$ can give rise to the same Riemann surface. As a first application of the theory of liftings, we show when this happens.

Proposition 1.6.18. *Let $\pi_1: \widetilde{X}_1 \to X_1$ and $\pi_2: \widetilde{X}_2 \to X_2$ be the universal covering maps of two Riemann surfaces, and denote by Γ_j the automorphism group of π_j for $j = 1, 2$. Then X_1 and X_2 are biholomorphic if and only if there is a biholomorphism $\Phi: \widetilde{X}_1 \to \widetilde{X}_2$ such that $\Gamma_2 = \Phi\Gamma_1\Phi^{-1}$. In particular, if $\widetilde{X}$ is a simply connected Riemann surface and Γ_1, Γ_2 are properly discontinuous subgroups of $\mathrm{Aut}(\tilde{X})$ acting freely then $\widetilde{X}/\Gamma_1$ is biholomorphic to $\widetilde{X}/\Gamma_2$ if and only if Γ_1 is conjugated to Γ_2 in $\mathrm{Aut}(\widetilde{X})$.*

Proof. Let $\varphi: X_1 \to X_2$ be a biholomorphism. By Proposition 1.6.14(iv), we can lift φ to a biholomorphism $\Phi: \widetilde{X}_1 \to \widetilde{X}_2$ such that $\pi_2 \circ \Phi = \varphi \circ \pi_1$. Then, for every $\gamma_1 \in \Gamma_1$ we have

$$\pi_2 \circ \Phi \circ \gamma_1 = \varphi \circ \pi_1 \circ \gamma_1 = \varphi \circ \pi_1 = \pi_2 \circ \Phi,$$

i. e., $\Phi \circ \gamma_1 \circ \Phi^{-1} \in \Gamma_2$, and thus $\Phi\Gamma_1\Phi^{-1} \subseteq \Gamma_2$. Reversing the argument, we get $\Phi^{-1}\Gamma_2\Phi \subseteq \Gamma_1$, and one implication follows.

Conversely, assume that $\Gamma_2 = \Phi\Gamma_1\Phi^{-1}$ for a biholomorphism $\Phi: \widetilde{X}_1 \to \widetilde{X}_2$. Then Proposition 1.6.14(v) and (iv) imply that Φ is the lifting of a biholomorphism $\varphi: X_1 \to X_2$, and we are done. □

Of course, now we would like to know which subgroups of $\mathrm{Aut}(\widehat{\mathbb{C}})$, $\mathrm{Aut}(\mathbb{C})$, and $\mathrm{Aut}(\mathbb{D})$ are properly discontinuous and act freely. The first step is describing the automorphism group of the three simply connected Riemann surfaces. We already know $\mathrm{Aut}(\mathbb{D})$; we next compute the other two groups.

Proposition 1.6.19. (i) *Every automorphism γ of $\widehat{\mathbb{C}}$ is of the form*

$$\gamma(z) = \frac{az + b}{cz + d}, \tag{1.39}$$

where $a, b, c, d \subset \mathbb{C}$ are such that $ad - bc = 1$. The representation is unique up to sign, and $\mathrm{Aut}(\widehat{\mathbb{C}})$ is isomorphic to $\mathbf{PGL}(2, \mathbb{C}) = \mathbf{SL}(2, \mathbb{C})/\{\pm I_2\}$.

(ii) *Every automorphism γ of $\mathbb{C}$ is of the form*

$$\gamma(z) = az + b, \tag{1.40}$$

for some $a, b \in \mathbb{C}$ with $a \neq 0$.

Proof. (i) Let denote by Γ the set of maps of the form (1.39). It is easy to check that Γ is a subgroup of $\mathrm{Aut}(\widehat{\mathbb{C}})$ and that, moreover, Γ is doubly transitive on $\widehat{\mathbb{C}}$. Take $\gamma \in \mathrm{Aut}(\widehat{\mathbb{C}})$. Up to post-composing with an element of Γ, without loss of generality we can assume that $\gamma(0) = 0$ and $\gamma(\infty) = \infty$. We claim that then $h(z) = \gamma(z)/z$ is a bounded holomorphic function from $\mathbb{C}^*$ into itself. It suffices to show that h is bounded in a neighborhood of ∞ and in a neighborhood of 0. Let $g(\zeta) = 1/\gamma(1/\zeta)$. Since γ is an automorphism fixing ∞ we have $g(0) = 0$ and $g'(0) \in \mathbb{C}^*$; in particular, $\zeta/g(\zeta)$ tends to the finite nonzero value $1/g'(0)$ when $\zeta \to 0$. Setting $z = 1/\zeta$, we have $h(z) = \zeta/g(\zeta)$, and thus $h(z) \to 1/g'(0)$ as $z \to \infty$; in particular, h is bounded near ∞. Analogously, $h(z) \to \gamma'(0) \neq 0$ as $z \to 0$; so h is also bounded in a neighborhood of 0, and the claim is proved.

By the Riemann removable singularity theorem, it follows that h extends to a bounded entire holomorphic function; by the Liouville theorem, $h \equiv a$ for a suitable constant $a \in \mathbb{C}^*$. But then $\gamma(z) = az$ belongs to Γ, and thus we have proved that $\mathrm{Aut}(\widehat{\mathbb{C}}) = \Gamma$. The last statement follows as in the proof of Proposition 1.3.8.

(ii) Since γ is a homeomorphism of $\mathbb{C}$ with itself, it is easy to see that $|\gamma(z)| \to \infty$ as $z \to \infty$. It follows that the singularity of $1/\gamma(1/\zeta)$ at $\zeta = 0$ is removable, and thus γ extends to an automorphism of $\widehat{\mathbb{C}}$ fixing ∞. But the maps of the form (1.39) fixing ∞ are all of the form (1.40), and we are done. □

We can now show that $\mathrm{Aut}(\mathbb{C})$ is doubly transitive and that $\mathrm{Aut}(\widehat{\mathbb{C}})$ is even triply transitive.

Corollary 1.6.20. (i) *For any pair of triples $\{z_0, z_1, z_2\}$, $\{w_0, w_1, w_2\}$ of distinct points in $\widehat{\mathbb{C}}$, there exists a unique $\gamma \in \mathrm{Aut}(\widehat{\mathbb{C}})$ such that $\gamma(z_j) = w_j$ for $j = 0, 1, 2$.*

(ii) *For any pair of couples $\{z_0, z_1\}$, $\{w_0, w_1\}$ of distinct points in $\mathbb{C}$, there exists a unique $\gamma \in \mathrm{Aut}(\mathbb{C})$ such that $\gamma(z_j) = w_j$ for $j = 0, 1$.*

Proof. (i) It suffices to prove the statement for $z_0 = 0$, $z_1 = 1$, and $z_2 = \infty$. Assume first $\{w_0, w_1, w_2\} \subset \mathbb{C}$. We need to find $a, b, c, d \in \mathbb{C}$ such that

$$\frac{b}{d} = w_0, \quad \frac{a + b}{c + d} = w_1, \quad \frac{a}{c} = w_2, \quad ad - bc = 1.$$

A quick computation shows that this system has only the following solutions:

$$d = \pm\left[\frac{w_2 - w_1}{(w_2 - w_0)(w_1 - w_0)}\right]^{1/2}, \quad c = \frac{w_1 - w_0}{w_2 - w_1}d, \quad b = w_0 d, \quad a = w_2\frac{w_1 - w_0}{w_2 - w_1}d.$$

Since the two solutions differ only by a sign they determine a unique element of $\mathrm{Aut}(\widehat{\mathbb{C}})$, as claimed.

If instead $w_2 = \infty$, a similar argument yields

$$d = \pm\left[\frac{1}{w_1 - w_0}\right]^{1/2}, \quad c = 0, \quad b = w_0 d, \quad a = (w_1 - w_0)d,$$

and again we get a unique element of $\mathrm{Aut}(\widehat{\mathbb{C}})$, as claimed. In an analogous way, we prove the assertion when $w_0 = \infty$ or $w_1 = \infty$.

(ii) It suffices to prove the statement for $z_0 = 0$ and $z_1 = 1$, and the unique solution is $\gamma(z) = (w_1 - w_0)z + w_0$. □

We can also determine the number of fixed points of automorphisms of $\mathbb{C}$ and $\widehat{\mathbb{C}}$, and which automorphisms commute with each other.

Corollary 1.6.21. (i) *Every nontrivial automorphism of $\widehat{\mathbb{C}}$ has exactly one fixed point or two distinct fixed points.*

(ii) *If $\gamma \in \mathrm{Aut}(\widehat{\mathbb{C}})$ has exactly one fixed point, then there exist $\gamma_0 \in \mathrm{Aut}(\widehat{\mathbb{C}})$ and $\beta \in \mathbb{C}^*$ such that*

$$\gamma_0^{-1} \circ \gamma \circ \gamma_0(z) = z + \beta; \tag{1.41}$$

if $\gamma \in \mathrm{Aut}(\widehat{\mathbb{C}})$ has exactly two distinct fixed points, then there exist $\gamma_0 \in \mathrm{Aut}(\widehat{\mathbb{C}})$ and $\alpha \in \mathbb{C} \setminus \{0, 1\}$ such that

$$\gamma_0^{-1} \circ \gamma \circ \gamma_0(z) = \alpha z. \tag{1.42}$$

(iii) *An automorphism $\eta \in \mathrm{Aut}(\widehat{\mathbb{C}})$ commutes with an automorphism of the form (1.41) if and only if there is $\lambda \in \mathbb{C}$ such that*

$$\gamma_0 \circ \eta \circ \gamma_0(z) = z + \lambda;$$

an automorphism $\eta \in \mathrm{Aut}(\widehat{\mathbb{C}})$ commutes with an automorphism of the form (1.42) with $\alpha \neq -1$ if and only if there is $\mu \in \mathbb{C}^*$ such that

$$\gamma_0^{-1} \circ \eta \circ \gamma_0(z) = \mu z;$$

if instead $\alpha = -1$ then we may also have

$$\gamma_0^{-1} \circ \eta \circ \gamma_0(z) = \mu z^{-1}.$$

(iv) *Every nontrivial automorphism of* $\mathbb{C}$ *has none or exactly one fixed point.*

(v) *If* $\gamma \in \mathrm{Aut}(\mathbb{C})$ *has no fixed points, then there is* $\beta \in \mathbb{C}^*$ *such that*

$$\gamma(z) = z + \beta; \tag{1.43}$$

if $\gamma \in \mathrm{Aut}(\mathbb{C})$ *has exactly one fixed point, then there are* $\alpha \in \mathbb{C} \setminus \{0, 1\}$ *and* $\beta \in \mathbb{C}$ *such that*

$$\gamma(z) = \alpha z + \beta. \tag{1.44}$$

(vi) *An automorphism* $\eta \in \mathrm{Aut}(\mathbb{C})$ *commutes with an automorphism of the form* (1.43) *if and only if there is* $b \in \mathbb{C}$ *such that*

$$\eta(z) = z + b;$$

an automorphism $\eta \in \mathrm{Aut}(\mathbb{C})$ *commutes with an automorphism of the form* (1.44) *if and only if there is* $a \in \mathbb{C}^*$ *such that*

$$\eta(z) = az + \frac{a-1}{\alpha-1}\beta.$$

Proof. (i) A point $z \in \mathbb{C}$ is a fixed point of a $\gamma \in \mathrm{Aut}(\widehat{\mathbb{C}})$ of the form (1.39) if and only if it is a root of the equation

$$cz^2 + (d-a)z - b = 0.$$

If $c \neq 0$, this equation has exactly two solutions, counted with multiplicities; since $\gamma(\infty) = a/c \neq \infty$ when $c \neq 0$ we have the assertion in this case. If $c = 0$, then this equation has at most one solution in $\mathbb{C}$; but since $\gamma(\infty) = \infty$ when $c = 0$ the assertion follows in this case, too.

(ii) Assume that γ has a unique fixed point; up to conjugation, we can assume that it is ∞. Then the analysis done in (i) shows that we must have $c = 0$ and $d = a$, and the assertion follows.

Assume instead that γ has two distinct fixed points; up to conjugation we can assume that $\mathrm{Fix}(\gamma) = \{0, \infty\}$. The analysis done in (i) then implies $b = c = 0$, and we are done.

(iii) In the first case, up to conjugating with γ_0 we can assume that $\gamma(z) = z + \beta$ with $\beta \in \mathbb{C}^*$. If we write η in the form (1.39), we see that η commutes with γ if and only if

$$\frac{a(z+\beta)+b}{c(z+\beta)+d} = \frac{az+b}{cz+d} + \beta.$$

Taking $z = \infty$, we immediately see that $c = 0$; then, taking $z = 0$, we get $a = d$. Therefore, $\eta(z) = z + b/d$, as claimed.

In the second case, up to conjugating with γ_0 we can assume that $\gamma(z) = \alpha z$ with $\alpha \in \mathbb{C} \setminus \{0, 1\}$. If we write η in the form (1.39), we see that η commutes with γ if and only if

$$\frac{a\alpha z + b}{c\alpha z + d} = \alpha \frac{az + b}{cz + d}.$$

Taking $z = \infty$, we immediately see that $c = 0$ or $a = 0$; taking $z = 0$, we see that $b = 0$ or $d = 0$. If $c = 0$, then $d \neq 0$; thus $b = 0$ and $\eta(z) = a^2 z$ as required. If $a = 0$, then $b \neq 0$; thus $d = 0$ and $\eta(z) = -b^2 z^{-1}$. But to satisfy $\eta(\alpha z) = \alpha\eta(z)$ in this case, we must also have $\alpha^2 = 1$; hence $\alpha = -1$ and we are done.

(iv) A point $z_0 \in \mathbb{C}$ is a fixed point of $\gamma(z) = az + b$ if and only if it is a root of the equation $(a - 1)z + b = 0$. Therefore, if $a \neq 1$ we have exactly one fixed point, and if $a = 1$ there are no fixed points.

(v) It follows immediately from the analysis made in (iv).

(vi) An automorphism $\eta(z) = az + b$ commutes with $\gamma(z) = z + \beta$ if and only if $a(z + \beta) + b = az + b + \beta$, i. e., if and only if $a = 1$.

An automorphism $\eta(z) = az + b$ commutes with $\gamma(z) = \alpha z + \beta$ with $\alpha \neq 0, 1$ if and only if $a\beta + b = \alpha b + \beta$, i. e., $b = \frac{a-1}{\alpha-1}\beta$, and we are done. □

The next lemma gives a necessary condition for a subgroup of $\mathrm{Aut}(\widetilde{X})$ to be properly discontinuous.

Lemma 1.6.22. *Let Γ be a closed group of automorphisms of a Riemann surface X, properly discontinuous at some point of X. Then Γ is discrete.*

Proof. Assume Γ is not discrete. Then there is an infinite sequence of distinct elements $\gamma_\nu \in \Gamma$ converging to an element $\gamma \in \Gamma$. Therefore, $\gamma^{-1} \circ \gamma_\nu \to \mathrm{id}_X$, and Γ cannot be properly discontinuous at any point of X. □

In the next section, we shall see (Proposition 1.7.22) that if X is a hyperbolic Riemann surface then every discrete subgroup of $\mathrm{Aut}(X)$ is everywhere properly discontinuous. For the moment, Lemma 1.6.22 is enough to describe the properly discontinuous subgroups of $\mathrm{Aut}(\widehat{\mathbb{C}})$ and $\mathrm{Aut}(\mathbb{C})$ acting freely on $\widehat{\mathbb{C}}$, respectively, $\mathbb{C}$; as a consequence, we get the list of all elliptic and parabolic Riemann surfaces.

Corollary 1.6.23. (i) *No nontrivial subgroup of* $\mathrm{Aut}(\widehat{\mathbb{C}})$ *acts freely on* $\widehat{\mathbb{C}}$. *In particular, the unique elliptic Riemann surface is* $\widehat{\mathbb{C}}$.

(ii) *The closed properly discontinuous subgroups of* $\mathrm{Aut}(\mathbb{C})$ *acting freely on* $\mathbb{C}$ *are, up to conjugation,* $\{\mathrm{id}_{\mathbb{C}}\}$, $\{\gamma(z) = z + n \mid n \in \mathbb{Z}\}$ *and*

$$\Gamma_\tau = \{\gamma(z) = z + m + n\tau \mid m, n \in \mathbb{Z}\}, \tag{1.45}$$

where $\tau \in \mathbb{H}^+$. *In particular, the parabolic Riemann surfaces are* $\mathbb{C}$, $\mathbb{C}^*$, *and the quotients* $\mathbb{C}/\Gamma_\tau$.

Proof. (i) Every element of $\mathrm{Aut}(\widehat{\mathbb{C}})$ has at least one fixed point in $\widehat{\mathbb{C}}$, by Corollary 1.6.21(i), and thus no subgroup of $\mathrm{Aut}(\widehat{\mathbb{C}})$ can act freely on $\widehat{\mathbb{C}}$.

(ii) By Corollary 1.6.21(v), the only elements of $\mathrm{Aut}(\mathbb{C})$ without fixed points are the translations $\gamma(z) = z + \beta$, which forms a subgroup isomorphic to $(\mathbb{C}, +)$. Then by Lemma 1.6.22, we need to consider only the discrete subgroups of $\mathbb{C}$, that are $\{0\}$, $\mathbb{Z}x_0$, and $\mathbb{Z}x_1 \oplus \mathbb{Z}x_2$, where $x_0 \in \mathbb{C}^*$ and $x_1, x_2 \in \mathbb{C}$ are $\mathbb{R}$-linearly independent (Proposition 1.4.13). They act properly discontinuously on $\mathbb{C}$, and the assertion follows in the first and third cases, because up to conjugation in $\mathrm{Aut}(\mathbb{C})$ we can assume $x_1 = 1$ and $x_2 \in \mathbb{H}^+$.

In the second case, up to a conjugation in $\mathrm{Aut}(\mathbb{C})$ we can assume $x_0 = 1$. Let $\pi_{\mathbb{C}^*}: \mathbb{C} \to \mathbb{C}^*$ be given by

$$\pi_{\mathbb{C}^*}(z) = \exp(2\pi i z). \tag{1.46}$$

It is easy to check that $\pi_{\mathbb{C}^*}$ is a covering map having $\mathbb{Z}$ as an automorphism group, and hence $\mathbb{C}/\mathbb{Z}$ is biholomorphic to $\mathbb{C}^*$. □

Definition 1.6.24. A *complex torus* is a Riemann surface of the form $\mathbb{C}/\Gamma_\tau$ where $\tau \in \mathbb{H}^+$ and $\Gamma_\tau \subset \mathrm{Aut}(\mathbb{C})$ is given by (1.45). Topologically, a complex torus is homeomorphic to $\mathbb{S}^1 \times \mathbb{S}^1$.

Remark 1.6.25. To each $\tau \in \mathbb{H}^+$, we can associate both the group $\Gamma_\tau \subset \mathrm{Aut}(\mathbb{C})$ given by (1.45) and the subgroup $\Lambda_\tau = \mathbb{Z} \oplus \tau\mathbb{Z} \subset \mathbb{C}$. It is then clear that X_τ can be obtained also as the quotient $\mathbb{C}/\Lambda_\tau$ of topological Abelian groups, and hence it has a natural structure of topological Abelian group. Denoting by $\pi_\tau: \mathbb{C} \to X_\tau = \mathbb{C}/\Gamma_\tau$ the natural projection, the operation $+$ on X_τ is defined by $\pi_\tau(z) + \pi_\tau(w) = \pi_\tau(z + w)$ for all $z, w \in \mathbb{C}$.

It turns out that the holomorphic self-maps of a torus are relatively few.

Proposition 1.6.26. *Let $X_\tau = \mathbb{C}/\Gamma_\tau$ be a complex torus, with $\tau \in \mathbb{H}^+$, and set $\Lambda_\tau = \mathbb{Z} \oplus \mathbb{Z}\tau$. Then a map $\tilde{f} \in \mathrm{Hol}(\mathbb{C}, \mathbb{C})$ is the lift of a map $f \in \mathrm{Hol}(X_\tau, X_\tau)$ if and only if there exist $c \in \mathbb{C}$ and $\lambda_0 \in \Lambda_\tau$ with $\lambda_0\tau \in \Lambda_\tau$ such that $\tilde{f}(z) = \lambda_0 z + c$. Furthermore, if this happens then $|\lambda_0|^2 \in \mathbb{N}$; moreover, if $\lambda_0 = m_0 + n_0\tau \notin \mathbb{Z}$ then τ is a solution of an equation of the form $n_0\tau^2 = m_1 + n_1\tau$ for suitable $m_1, n_1 \in \mathbb{Z}$ such that $n_1^2 + 4n_0m_1 < 0$.*

Proof. Denote by $\pi_\tau: \mathbb{C} \to X_\tau$ the universal covering map, and for $\lambda \in \mathbb{C}$ put $\gamma_\lambda(z) = z + \lambda$; then the automorphism group Γ_τ of π_τ is composed by the maps γ_λ with $\lambda \in \Lambda_\tau$. Clearly, Γ_τ is generated by γ_1 and γ_τ.

Assume that $\tilde{f}(z) = \lambda_0 z + c$ with $\lambda_0, \lambda_0\tau \in \Lambda_\tau$. Then it is easy to check that $\tilde{f} \circ \gamma_1 = \gamma_{\lambda_0} \circ \tilde{f}$ and $\tilde{f} \circ \gamma_\tau = \gamma_{\lambda_0\tau} \circ \tilde{f}$, and thus $\tilde{f}$ is the lifting of a $f \in \mathrm{Hol}(X_\tau, X_\tau)$ by Proposition 1.6.14(v).

Conversely, take $f \in \mathrm{Hol}(X_\tau, X_\tau)$, and let $\tilde{f}: \mathbb{C} \to \mathbb{C}$ be a lifting of f, so that $\pi_\tau \circ \tilde{f} = f \circ \pi_\tau$. By Proposition 1.6.14(iii), there exists $\lambda_0, \lambda_1 \in \Lambda_\tau$ such that $\tilde{f}(z + 1) = \tilde{f}(z) + \lambda_0$ and $\tilde{f}(z + \tau) = \tilde{f}(z) + \lambda_1$ for all $z \in \mathbb{C}$.

Set $\tilde{g}(z) = \tilde{f}(z) - \lambda_0 z$; we have $\tilde{g}(z+1) = \tilde{g}(z)$ and $\tilde{g}(z+\tau) = \tilde{g}(z) + \eta$, where $\eta = \lambda_1 - \lambda_0\tau \in \mathbb{C}$. If $\eta = 0$, then $\tilde{g}$ induces a holomorphic function $g\colon X_\tau \to \mathbb{C}$; since X_τ is compact, the maximum modulus principle implies that g is constant. Therefore, $\tilde{g}$ must be constant, too; hence there exists $c \in \mathbb{C}$ such that $\tilde{f}(z) = \lambda_0 z + c$, as claimed.

If $\eta \neq 0$, then $\tilde{g}$ induces a holomorphic function from X_τ into the Riemann surface $\mathbb{C}/\Gamma$, where Γ is the subgroup of $\mathrm{Aut}(\mathbb{C})$ generated by $\gamma(z) = z+\eta$. But the previous corollary says that $\mathbb{C}/\Gamma$ is biholomorphic to $\mathbb{C}^*$; therefore, we can argue as in the case $\eta = 0$ and we get again $\tilde{f}(z) = \lambda_0 z + c$ for a suitable $c \in \mathbb{C}$. In both cases, we have $\lambda_0\tau = \tilde{f}(\tau) - c = \tilde{f}(0) - c + \lambda_1 = \lambda_1$, and thus $\lambda_0\tau \in \Lambda_\tau$.

Now write $\lambda_0 = m_0 + n_0\tau$ with $m_0, n_0 \in \mathbb{Z}$. If $n_0 = 0$, then clearly $|\lambda_0|^2 \in \mathbb{N}$. If instead $n_0 \neq 0$, then $\lambda_0\tau$ belongs to Λ_τ if and only if there are $m_1, n_1 \in \mathbb{Z}$ such that $n_0\tau^2 = m_1 + n_1\tau$. Since $\operatorname{Im}\tau > 0$, we must have $n_1^2 + 4n_0m_1 < 0$ and

$$\tau = \frac{n_1 + i\sqrt{-n_1^2 - 4n_0m_1}}{2n_0}.$$

In particular, $\operatorname{Re}\tau = \frac{n_1}{2n_0}$ and

$$|\tau|^2 = \frac{n_1^2 + (-n_1^2 - 4n_0m_1)}{4n_0^2} = -\frac{m_1}{n_0}.$$

Therefore,

$$|\lambda_0|^2 = m_0^2 + 2m_0n_0\operatorname{Re}\tau + n_0^2|\tau|^2 = m_0^2 + m_0n_1 - m_1n_0 \in \mathbb{N}, \tag{1.47}$$

and we are done. □

Later on we shall see (Corollary 1.7.23) that the properly discontinuous subgroups of $\mathrm{Aut}(\mathbb{D})$ acting freely on $\mathbb{D}$ are exactly the discrete subgroups without elliptic elements.

A scrupulous reader may now ask for examples of hyperbolic Riemann surfaces. A first list of examples is provided by topology. Riemann surfaces with a non-Abelian fundamental group must be hyperbolic, for the nonhyperbolic Riemann surfaces always have an Abelian fundamental group. Thus all compact Riemann surfaces of genus at least 2 are hyperbolic; furthermore, it is clear that the noncompact Riemann surfaces not biholomorphic to a plane domain must be hyperbolic.

How about plane domains? First of all, again by topological considerations, a domain $D \subset \widehat{\mathbb{C}}$ with a fundamental group nontrivial and not biholomorphic to $\mathbb{Z}$ must be hyperbolic; for instance, $\widehat{\mathbb{C}}$ minus three points is hyperbolic. Furthermore, every bounded domain must be hyperbolic by the Liouville theorem.

At this point, the above mentioned scrupulous reader may begin suspecting that most of the plane domains are hyperbolic. In fact, almost all of them are so; this is a corollary of the following observation showing how nonhyperbolic and hyperbolic Riemann surfaces live in completely separated realms.

Proposition 1.6.27. *Every holomorphic function $f: X \to Y$ from an elliptic or parabolic Riemann surface X into a hyperbolic Riemann surface Y is constant.*

Proof. Let $\tilde{f}: \widetilde{X} \to \mathbb{D}$, where $\widetilde{X} = \widehat{\mathbb{C}}$ or $\mathbb{C}$, be a lifting of f to the universal covering. By the Liouville theorem, $\tilde{f}$ is constant, and thus f itself is constant. □

By the way, if we take $X = \mathbb{C}$ in this proposition we recover the *little Picard theorem.*

Corollary 1.6.28 (Little Picard theorem, 1879). *An entire function missing two values is constant.*

Proof. We have noticed that $\mathbb{C}$ minus two points (i. e., $\widehat{\mathbb{C}}$ minus three points) is hyperbolic; therefore, any entire holomorphic function with values in $\mathbb{C}$ minus two points is constant by Proposition 1.6.27. □

Another immediate consequence is the following.

Corollary 1.6.29. *Every domain $D \subset \widehat{\mathbb{C}}$ such that $\widehat{\mathbb{C}} \setminus D$ contains at least three points is a hyperbolic Riemann surface.*

Proof. In fact, by definition there exists a holomorphic immersion of D into $\widehat{\mathbb{C}}$ minus three points; therefore, D must be hyperbolic by Proposition 1.6.27. □

This result suggests the following.

Definition 1.6.30. A plane domain $D \subset \widehat{\mathbb{C}}$ such that $\widehat{\mathbb{C}} \setminus D$ contains at least three points will be called *hyperbolic.* More generally, a *hyperbolic domain* of a compact Riemann surface $\widehat{X}$ is a noncompact domain $D \subset \widehat{X}$ which is hyperbolic as a Riemann surface. For the sake of brevity, we shall often say "a hyperbolic domain $D \subset \widehat{X}$" instead of "a hyperbolic domain D in a compact Riemann surface $\widehat{X}$."

Another corollary is the commonest form of the *Riemann mapping theorem.*

Corollary 1.6.31. *Every simply connected domain $D \subset \mathbb{C}$ different from $\mathbb{C}$ is biholomorphic to $\mathbb{D}$.*

Proof. Indeed, if $D \neq \mathbb{C}$ then $\widehat{\mathbb{C}} \setminus D$ must contain at least three points and the assertion follows from Corollary 1.6.29. □

By the way, later on we shall need an important extension of the Riemann mapping theorem, the *Osgood–Taylor–Carathéodory theorem.*

Theorem 1.6.32 (Osgood–Taylor–Carathéodory, 1913). *Let $D \subset \mathbb{C}$ be a simply connected bounded domain. Then a biholomorphism $f: \mathbb{D} \to D$ extends continuously to a homeomorphism between $\overline{\mathbb{D}}$ and $\overline{D}$ if and only if ∂D is a Jordan curve.*

A proof can be found, e. g., in [88, Theorem 9.14] or in [125, Theorem 14.5.6].

Now it has become clear that most of Riemann surfaces are hyperbolic, and starting from the next section most of our theorems will be about hyperbolic Riemann surfaces. On the other hand, elliptic and parabolic Riemann surfaces share a particular feature: they all have Abelian fundamental groups. This suggests to look for hyperbolic Riemann surfaces with Abelian fundamental group; it turns out that they are very few.

Theorem 1.6.33. *Let X be a Riemann surface with Abelian fundamental group. Then either:*

(i) *$\pi_1(X)$ is trivial, and X is either $\widehat{\mathbb{C}}$, $\mathbb{C}$ or $\mathbb{D}$, or*

(ii) *$\pi_1(X) \cong \mathbb{Z}$, and X is either $\mathbb{C}^*$, $\mathbb{D}^* = \mathbb{D} \setminus \{0\}$ or an annulus*

$$A(r,1) = \{z \in \mathbb{C} \mid r < |z| < 1\}$$

for some $0 < r < 1$, or

(iii) *$\pi_1(X) \cong \mathbb{Z} \oplus \mathbb{Z}$ and X is a complex torus.*

Proof. By Corollary 1.6.23, we have to consider only hyperbolic Riemann surfaces. By Proposition 1.6.9 and Lemma 1.6.22, the automorphism group Γ of the universal covering map $\pi\colon \mathbb{D} \to X$ is an Abelian discrete subgroup of $\mathrm{Aut}(\mathbb{D})$ without elliptic elements (because it acts freely on $\mathbb{D}$); therefore, by Corollary 1.4.14, Γ is isomorphic to $\mathbb{Z}$ and it is generated by either a parabolic or a hyperbolic element γ.

Transfer everything to $\mathbb{H}^+$, as usual. If γ is parabolic, without loss of generality we can assume $\gamma(w) = w + 1$. Let $\pi_{\mathbb{D}^*}\colon \mathbb{H}^+ \to \mathbb{D}^*$ be given by

$$\pi_{\mathbb{D}^*}(w) = \exp(2\pi i w). \tag{1.48}$$

It is easy to check that $\pi_{\mathbb{D}^*}$ is a covering map having Γ as automorphism group, and hence $\mathbb{H}^+/\Gamma$ is biholomorphic to $\mathbb{D}^*$.

If γ is hyperbolic, without any loss of generality we can assume $\gamma(w) = aw$ for some $a > 1$. Set $0 < r = \exp(-2\pi^2/\log a) < 1$ and let $\pi_{A(r,1)}\colon \mathbb{H}^+ \to A(r,1)$ be given by

$$\pi_{A(r,1)}(w) = \exp\left(2\pi i \frac{\log w}{\log a}\right), \tag{1.49}$$

where $\log w$ is the principal branch of the logarithm in $\mathbb{H}^+$. Then it is easy to see that $\pi_{A(r,1)}$ is a covering map having Γ as automorphism group, and hence $\mathbb{H}^+/\Gamma$ is biholomorphic to $A(r,1)$. □

Remark 1.6.34. Note that if $a_1 \neq a_2$, with both a_1 and a_2 greater than 1, then the group generated by $\gamma_1(w) = a_1 w$ is not conjugated in $\mathrm{Aut}(\mathbb{H}^+)$ to the group generated by $\gamma_2(w) = a_2 w$. As a consequence, if $0 < r_1 \neq r_2 < 1$ then $A(r_1,1)$ and $A(r_2,1)$ are not biholomorphic (by Proposition 1.6.18).

Definition 1.6.35. A Riemann surface X, which is not simply connected is called *multiply connected*; if $\pi_1(X) \cong \mathbb{Z}$, it is called *doubly connected.*

Then we have the following.

Corollary 1.6.36. *Every doubly connected Riemann surface X is biholomorphic either to $\mathbb{C}^*$, to $\mathbb{D}^*$ or to an annulus $A(r,1)$ for some $0 < r < 1$.*

We can also use the automorphism group of the universal covering map to describe the automorphism group of a Riemann surface.

Proposition 1.6.37. *Let $\pi: \widetilde{X} \to X$ be the universal covering map of a Riemann surface X, let Γ be the automorphism group of π, and denote by $N(\Gamma)$ the normalizer of Γ in $\mathrm{Aut}(\widetilde{X})$. Then $\mathrm{Aut}(X) \cong N(\Gamma)/\Gamma$.*

Proof. Arguing exactly as in Proposition 1.6.18, we see that an automorphism of X gives rise to an automorphism of $\widetilde{X}$ normalizing Γ, and *vice versa.* □

As an application, we conclude this section by computing the automorphism group of the Riemann surfaces listed in Theorem 1.6.33.

Proposition 1.6.38. (i) *Every $\gamma \in \mathrm{Aut}(\mathbb{C}^*)$ is of the form $\gamma(z) = \lambda z^{\pm 1}$, for some $\lambda \in \mathbb{C}^*$. In particular, the connected component at the identity of $\mathrm{Aut}(\mathbb{C}^*)$ is isomorphic to $\mathbb{C}^*$.*

(ii) *Let $\tau \in \mathbb{H}^+$, and let X_τ be the torus $\mathbb{C}/\Gamma_\tau$. Then:*
 (a) *$\mathrm{Aut}(X_\tau)$ has a normal subgroup G isomorphic to X_τ and transitive on X_τ;*
 (b) *if $\tau = \frac{p+i}{q}$ with $p \in \mathbb{Z}$ and $q \in \mathbb{N}^*$ such that q divides $p^2 + 1$ then $\mathrm{Aut}(X_\tau)/G$ is isomorphic to $\mathbb{Z}_4$, and all elements of $\mathrm{Aut}(X_\tau) \setminus G$ are periodic of period 2 or 4;*
 (c) *if $\tau = \frac{2p+1+i\sqrt{3}}{2q}$ with $p \in \mathbb{Z}$ and $q \in \mathbb{N}^*$ such that q divides p^2+p+1 then $\mathrm{Aut}(X_\tau)/G$ is isomorphic to $\mathbb{Z}_6$, and all elements of $\mathrm{Aut}(X_\tau) \setminus G$ are periodic of period 2, 3, or 6;*
 (d) *in all other cases, $\mathrm{Aut}(X_\tau)/G$ is isomorphic to $\mathbb{Z}_2$ and all elements of $\mathrm{Aut}(X_\tau) \setminus G$ are periodic of period 2.*

 In particular, the connected component at the identity of $\mathrm{Aut}(X_\tau)$ is always isomorphic to X_τ. Furthermore, a function $f \in \mathrm{Hol}(X_\tau, X_\tau)$ belongs to $\mathrm{Aut}(X_\tau)$ if and only if it has a lifting $\tilde{f} \in \mathrm{Hol}(\mathbb{C}, \mathbb{C})$ of the form $\tilde{f}(z) = \lambda z + c$ with $c \in \mathbb{C}$ and $|\lambda| = 1$ so that $\lambda, \lambda\tau \in \mathbb{Z} \oplus \tau\mathbb{Z}$.

(iii) *Every $\gamma \in \mathrm{Aut}(\mathbb{D}^*)$ is of the form $\gamma(z) = e^{i\theta} z$ for some $\theta \in \mathbb{R}$.*

(iv) *Every $\gamma \in \mathrm{Aut}(A(r,1))$ is either of the form $\gamma(z) = e^{i\theta} z$ or of the form $\gamma(z) = e^{i\theta} r z^{-1}$ for some $\theta \in \mathbb{R}$.*

Proof. (i) The automorphism group Γ of the universal covering map (1.46) of $\mathbb{C}^*$ is generated by $\gamma_0(z) = z+1$. Then the normalizer of Γ in $\mathrm{Aut}(\mathbb{C})$ is composed by the functions $\gamma(z) = \pm(z+b)$ with $b \in \mathbb{C}$. Using the universal covering map to read the automorphisms in $\mathbb{C}^*$, the assertion follows.

(ii) Let $\pi_\tau\colon \mathbb{C} \to X_\tau$ be the universal covering map of X_τ, and let Γ_τ be the automorphism group of π_τ; as usual, Γ_τ is composed by the maps of the form $\gamma_\lambda(z) = z + \lambda$ with $\lambda \in \Lambda_\tau = \mathbb{Z} \oplus \mathbb{Z}\tau$.

In Remark 1.6.25, we noticed that X_τ has a natural structure of topological group. In particular, the translations with respect to the operation in X_τ form a group G of automorphisms of X_τ transitive on X_τ; clearly, G is isomorphic to X_τ. Recalling Proposition 1.6.14 we see that the liftings of these automorphisms are all the maps of the form γ_b with $b \in \mathbb{C}$; moreover, γ_{b_1} and γ_{b_2} are liftings of the same automorphism if and only if $b_1 - b_2 \in \Lambda_\tau$.

Let $f \in \mathrm{Aut}(X_\tau)$. By Proposition 1.6.26, we know that a lifting of f is of the form $\gamma_{c_0}^{\lambda_0}(z) = \lambda_0 z + c_0$, where $\lambda_0 \in \Lambda_\tau$ is such that $\lambda_0\tau \in \Lambda_\tau$ and $|\lambda_0|^2 \in \mathbb{N}$. If $\gamma_{c_1}^{\lambda_1}(z) = \lambda_1 z + c_1$ is a lifting of f^{-1}, then $\gamma_{c_0}^{\lambda_0} \circ \gamma_{c_1}^{\lambda_1}$ must be a lifting of the identity of X_τ. Since the liftings of the identity are exactly the γ_λ with $\lambda \in \Lambda_\tau$, we must have $\lambda_0\lambda_1 = 1$; from $|\lambda_0|^2, |\lambda_1|^2 \in \mathbb{N}$, we infer $|\lambda_0| = |\lambda_1| = 1$. In particular, $\lambda_1 = \overline{\lambda_0}$ and $\overline{\lambda_0}, \overline{\lambda_0}\tau \in \Lambda_\tau$. Furthermore, $(\gamma_{c_0}^{\lambda_0})^{-1} = \gamma_{-\overline{\lambda_0}c_0}^{\overline{\lambda_0}}$; then

$$(\gamma_{c_0}^{\lambda_0})^{-1} \circ \gamma_b \circ \gamma_{c_0}^{\lambda_0} = \gamma_{\overline{\lambda_0}b}$$

for all $b \in \mathbb{C}$, and thus G is normal in $\mathrm{Aut}(X_\tau)$, as claimed.

Our next aim is to determine $\mathrm{Aut}(X_\tau)/G$. Notice that $\gamma_c^\lambda = \gamma_0^\lambda \circ \gamma_c$; therefore, it suffices to find all γ_0^λ that are a lifting of an automorphism of X_τ to determine both $\mathrm{Aut}(X_\tau)/G$ and $\mathrm{Aut}(X_\tau) \setminus G$. For simplicity, we denote by $\tilde{\Gamma}$ the group of maps γ_0^λ, which are liftings of automorphisms of X_τ; what we have seen implies that $\mathrm{Aut}(X_\tau)/G$ is isomorphic to $\tilde{\Gamma}$.

We know that $\gamma_0^\lambda \in \tilde{\Gamma}$ if and only if $|\lambda| = 1$ and $\lambda, \lambda\tau, \overline{\lambda}, \overline{\lambda}\tau \in \Lambda_\tau$. Take $\lambda \in \Lambda_\tau$ such that $\lambda\tau \in \Lambda_\tau$ and $|\lambda| = 1$; we shall show that this is enough to have $\gamma_0^\lambda \in \tilde{\Gamma}$, thus proving the last assertion in the statement. Write $\lambda = m_0 + n_0\tau$; there are a few cases to consider.

Assume first $n_0 = 0$. Then $m_0 = \pm 1$, and clearly $\gamma_0^{\pm 1} \in \tilde{\Gamma}$. Notice that $(\gamma_c^{-1})^2 = \mathrm{id}_{\mathbb{C}}$; therefore, all automorphisms in the lateral G-class determined by γ_0^{-1} are periodic of period 2.

Now assume $n_0 \neq 0$. By Proposition 1.6.26 and (1.47), we know that there exist $m_1, n_1 \in \mathbb{Z}$ such that $n_0\tau^2 = m_1 + n_1\tau$; moreover, $n_1^2 + 4n_0m_1 < 0$ and $1 = |\lambda_0|^2 = m_0^2 + m_0n_1 - m_1n_0$. In particular, $n_0m_1 = m_0^2 + m_0n_1 - 1$ and

$$0 > n_1^2 + 4n_0m_1 = n_1^2 + 4m_0^2 + 4m_0n_1 - 4 = (n_1 + 2m_0)^2 - 4.$$

It follows that $n_1 + 2m_0 = 0$ or $n_1 + 2m_0 = \pm 1$.

If $n_1 + 2m_0 = 0$, we get $n_1 = -2m_0$ and $n_1^2 + 4n_0m_1 = -4$; therefore,

$$\tau = \frac{n_1 \pm \sqrt{n_1^2 + 4n_0m_1}}{2n_0} = \frac{-m_0 \pm i}{n_0},$$

where the sign of the square root is chosen so that the imaginary part of τ is positive. Since in this case $n_0 m_1 = -m_0^2 - 1$, we have that n_0 must divide $m_0^2 + 1$, and thus we are in case (b) with $q = |n_0|$ and $p = -\operatorname{sgn}(n_0)m_0$. Furthermore, $\lambda = m_0 + n_0\tau = \pm i$; in particular $\bar{\lambda} = -\lambda$ and $\bar{\lambda}\tau = -\lambda\tau$ automatically belong to Λ_τ. So $\tilde{\Gamma}$ is generated by γ_0^i, and hence it is isomorphic to $\mathbb{Z}_4$. Furthermore, $(\gamma_c^{\pm i})^2 = \gamma_{\pm ic+c}^{-1}$; so all automorphisms in the lateral G-classes determined by $\gamma_0^{\pm i}$ are periodic of period 4.

Assume $n_1 + 2m_0 = \pm 1$. We get $n_1 = \pm 1 - 2m_0$ and $n_1^2 + 4n_0 m_1 = -3$; therefore,

$$\tau = \frac{n_1 \pm \sqrt{n_1^2 + 4n_0 m_1}}{2n_0} = \frac{\pm 1 - 2m_0 \pm i\sqrt{3}}{2n_0},$$

where again the sign of the square root is chosen so that the imaginary part of τ is positive. Since in this case $n_0 m_1 = \pm m_0 - m_0^2 - 1$, we have that n_0 must divide $m_0^2 \mp m_0 + 1$, and thus we are in case (c) with $q = |n_0|$ and $p = -\operatorname{sgn}(n_0)m_0$ if $\pm\operatorname{sgn}(n_0) = 1$ or $p = -\operatorname{sgn}(n_0)m_0 - 1$ otherwise. Furthermore,

$$\lambda = m_0 + n_0\tau = \frac{\pm 1 \pm i\sqrt{3}}{2},$$

and all four choices of sign are allowable; in particular $\bar{\lambda}$ and $\bar{\lambda}\tau$ automatically belong to Λ_τ. Notice that

$$\frac{1 + i\sqrt{3}}{2} = e^{i\pi/3}, \quad \frac{-1 + i\sqrt{3}}{2} = e^{2i\pi/3}, \quad -1 = e^{3\pi i/3},$$
$$\frac{-1 - i\sqrt{3}}{2} = e^{4\pi i/3}, \quad \frac{1 - i\sqrt{3}}{2} = e^{5i\pi/3};$$

therefore, $\tilde{\Gamma}$ is generated by $\gamma_0^{e^{i\pi/3}}$, and hence it is isomorphic to $\mathbb{Z}_6$. Furthermore, $(\gamma_c^{e^{\pi i/3}})^3 = \gamma_{c'}^{-1}$, with $c' = (1 + i\sqrt{3})c$, and $(\gamma_c^{e^{2\pi i/3}})^3 = \mathrm{id}_{\mathbb{C}}$; so all automorphisms in the lateral G-classes determined by $\gamma_0^{e^{k\pi i/3}}$ are periodic of period 2, 3, or 6, and we are done.

(iii) The automorphism group Γ of the universal covering map (1.48) of $\mathbb{D}^*$ is generated by $\gamma_0(w) = w + 1$. Then the normalizer of Γ in $\operatorname{Aut}(\mathbb{H}^+)$ is given by $N(\Gamma) = \{\gamma(w) = w + b \mid b \in \mathbb{R}\}$. Using the universal covering map (1.48) as in (i) the assertion follows. Alternatively, notice that, by the Riemann extension theorem, every automorphism of $\mathbb{D}^*$ is the restriction of an automorphism of $\mathbb{D}$ leaving the origin fixed, and apply the Schwarz lemma.

(iv) The automorphism group Γ of the universal covering map (1.49) of $A(r, 1)$ is generated by $\gamma_0(w) = a_0 w$, where

$$a_0 = \exp(-2\pi^2/\log r).$$

Then $N(\Gamma)$ is generated by the functions $\gamma(w) = aw$ for $a > 0$ and by $\gamma(w) = -1/w$. Using the universal covering map (1.49), the assertion again follows. □

Notes to Section 1.6

The Uniformization Theorem 1.6.5 was first stated by Riemann [349] in 1851, but his proof had some gaps. The complete proof evolved in more than half a century, together with the modern concepts of Riemann surface and of n-dimensional manifold, through the works of many people. The most important ones are Poincaré [332] (who introduced new powerful methods and gave a partial proof of the existence of the universal covering map), Osgood [310] (who proved the theorem for plane domains), Hilbert [197] (who rigorously proved the Dirichlet principle, a main tool), again Poincaré [335], and Kœbe [240] (who in 1907 proved the complete statement), and Weyl [413] (who put the result in today's perspective). A direct proof of the Riemann mapping theorem Corollary 1.6.31 can be found, e. g., in [360].

The little Picard theorem was first proved by Picard [319] in 1879.

The Osgood–Taylor–Carathéodory Theorem 1.6.32 was conjectured by Osgood in 1901, and proved almost simultaneously in 1913 by Osgood and Taylor [311] and by Carathéodory [96]. Carathéodory himself, in his papers [94, 95], with his theory of prime ends made definitive investigations about the boundary behavior of the universal covering map of arbitrary simply connected domains; in particular, he proved that a biholomorphism $\varphi\colon \mathbb{D} \to D \subset \widehat{\mathbb{C}}$ extends continuously to the boundary if and only if ∂D is locally connected. A modern account can be found in [337] or in [125].

A natural question is when a biholomorphism $f\colon \mathbb{D} \to D$ has a C^k invertible extension to $\partial\mathbb{D}$ for some $k \in \mathbb{N}^* \cup \{\infty, \omega\}$. A necessary condition is that ∂D should be a simple closed C^k curve. That this condition is also sufficient when $k = \omega$ is an easy consequence of the Schwarz reflection principle. The fact that it is also sufficient if $k = \infty$ is much less trivial but still true; see, e. g., [54]. Surprisingly, this condition is not sufficient if $k < \infty$. An example for $k = 1$, due to Webster (see [412] and [54, p. 39]), is given by $f(z) = (z-1)/\log(z-1)$ and $D = f(\mathbb{D})$. Indeed, f is a biholomorphism between $\mathbb{D}$ and D such that ∂D is a closed simple curve of class C^1 but f does not admit a C^1 extension to $\partial\mathbb{D}$. Conditions for the C^k extension were given by Painlevé [315, 316], Kellogg [228], Seidel [372] and Warschawski [408–410]; see [411] for a simpler exposition of the main results. See also [169, Chapter X.1].

We remarked (Remark 1.6.34) that two annuli $A(r_1, 1)$ and $A(r_2, 1)$ are biholomorphic if and only if $r_1 = r_2$. On the other hand, two complex tori $X_{\tau_1} = \mathbb{C}/\Gamma_{\tau_1}$ and $X_{\tau_2} = \mathbb{C}/\Gamma_{\tau_2}$ can be biholomorphic even when $\tau_1 \neq \tau_2$. Setting $\Lambda_\tau = \mathbb{Z} \oplus \tau\mathbb{Z}$, using Proposition 1.6.18 it is easy to see that X_{τ_1} is biholomorphic to X_{τ_2} if and only if there is $\alpha \in \mathbb{C}^*$ such that $\alpha\Lambda_{\tau_1} = \Lambda_{\tau_2}$. Assume now, as we may, that $\tau_1, \tau_2 \in \mathbb{H}^+$. Then using the fact that 1 and τ_j are $\mathbb{Z}$-linearly independent for $j = 1, 2$, it is not difficult to check that there exists $\alpha \in \mathbb{C}^*$ such that $\alpha\Lambda_{\tau_1} = \Lambda_{\tau_2}$ if and only if there is $\gamma \in \mathrm{SL}(2, \mathbb{Z})$ such that $\gamma(\tau_1) = \tau_2$. Therefore, the set of complex tori modulo biholomorphisms can be identified with $\mathcal{X} = \mathbb{H}^+/\mathrm{SL}(2, \mathbb{Z})$, the *moduli space* of complex tori. It is then possible to prove (see, e. g., [373, p. 78]) that $\mathcal{X}$ is homeomorphic to the quotient of the set $D = \{w \in \mathbb{H}^+ \mid |w| \geq 1, |\operatorname{Re}(w)| \leq 1/2\}$ with respect to the equivalence relation identifying w with $w' = w \mp 1$ if $\operatorname{Re}(w) = \pm 1/2$ and w with $w' = -1/w$ if $|w| = 1$.

The complex tori listed in Proposition 1.6.38(ii)(b) and (c) are said to admit a *complex multiplication*.

1.7 Hyperbolic Riemann surfaces and the Montel theorem

In this section, we shall use the geometrical structure induced by the Poincaré distance on any hyperbolic Riemann surface to derive a direct proof of the Montel theorem. The significance of this approach is twofold. On one side, it is a beautiful example of the correlation between geometrical and functional aspects of the theory of holomorphic functions. On the other side, the constructions involved here will be fundamental for the rest of the book.

Our first aim is to transfer the Poincaré distance from $\mathbb{D}$ to any hyperbolic Riemann surface.

Definition 1.7.1. Let X be a hyperbolic Riemann surface and denote by $\pi_X\colon \mathbb{D} \to X$ its universal covering map. Then the *Poincaré distance* $\omega_X\colon X \times X \to \mathbb{R}^+$ on X is given for all $z, w \in X$ by

$$\omega_X(z,w) = \inf\{\omega(\tilde{z},\tilde{w}) \mid \tilde{z} \in \pi_X^{-1}(z),\ \tilde{w} \in \pi_X^{-1}(w)\}. \tag{1.50}$$

We shall denote by $B_X(z,r)$ the *Poincaré ball* of *center* $z \in X$ and *radius* $r > 0$ for ω_X. A *geodesic* for ω_X is a continuous curve $\sigma\colon I \to X$, where $I \subseteq \mathbb{R}$ is a (possibly infinite) interval, such that for each $t_0 \in I$ there exists $\delta > 0$ such that $\omega_X(\sigma(s),\sigma(t)) = |s-t|$ for all $s, t \in (t_0 - \delta, t_0 + \delta)$.

Remark 1.7.2. When $X = \mathbb{D}$, this definition of geodesic coincides with the one given in Definition 1.2.6. Indeed, a geodesic in the latter sense clearly is a geodesic in the former sense. Conversely, let $\sigma\colon I \to \mathbb{D}$ be a geodesic in the sense of Definition 1.7.1; without loss of generality, we can assume that $I = (-\ell, \ell)$ with $\ell \in \mathbb{R}^+ \cup \{\infty\}$. The argument used to prove Proposition 1.2.7(iv) shows that there is a $\delta > 0$ such that $\sigma|_{(-\delta,\delta)}$ is the restriction of a classical geodesic (that is, of a geodesic according to Definition 1.2.6) in $\mathbb{D}$. Let δ_0 be the supremum of the $\delta > 0$ such that $\sigma|_{(-\delta,\delta)}$ is the restriction of a classical geodesic. If $\delta_0 = \ell$, we are done. If, by contradiction, $\delta_0 < \ell$ we can apply the same argument to show that $\sigma|_{(\delta_0-\varepsilon,\delta_0+\varepsilon)}$ is the restriction of a geodesic for some $0 < \varepsilon < \ell - \delta_0$; by the uniqueness of the geodesic connecting two points, it should be the restriction of the same geodesic already giving $\sigma|_{(-\delta_0,\delta_0)}$. Repeating the same argument about $-\delta_0$, we see that σ is the restriction of a classical geodesic on an interval strictly larger than $(-\delta_0, \delta_0)$, against the maximality of δ_0.

We called ω_X a distance, but this requires a proof, that will be provided by the next proposition together with a few geometrical facts about ω_X.

Proposition 1.7.3. *Let $\pi_X\colon \mathbb{D} \to X$ be the universal covering map of a hyperbolic Riemann surface X. Then:*

(i) *for every $z, w \in X$, we have*

$$\omega_X(z,w) = \inf\{\omega(\tilde{z},\tilde{w}) \mid \tilde{w} \in \pi_X^{-1}(w)\}, \tag{1.51}$$

where $\tilde{z}$ is any point in the fiber $\pi_X^{-1}(z)$.

(ii) *$\pi_X(B_{\mathbb{D}}(\tilde{z},r)) = B_X(\pi_X(\tilde{z}),r)$ for all $\tilde{z} \in \mathbb{D}$ and $r > 0$;*

(iii) *ω_X is a complete distance on X inducing the standard topology; moreover, the closed balls $\overline{B_X(z,r)}$ are compact and connected;*

(iv) *given $z_1, z_2 \in X$, for each $\tilde{z}_1 \in \pi_X^{-1}(z_1)$ we can find $\tilde{z}_2 \in \pi_X^{-1}(z_2)$ such that $\omega_X(z_1,z_2) = \omega(\tilde{z}_1,\tilde{z}_2)$;*

(v) *π_X is a local isometry, i. e., each $\tilde{z}_0 \in \mathbb{D}$ has a neighborhood $\widetilde{U} \subset \mathbb{D}$ such that $\omega_X(\pi_X(\tilde{z}_1),\pi_X(\tilde{z}_2)) = \omega(\tilde{z}_1,\tilde{z}_2)$ for all $\tilde{z}_1,\tilde{z}_2 \in \widetilde{U}$;*

(vi) *a curve $\sigma\colon I \to X$ is a geodesic for ω_X if and only if a lifting $\tilde{\sigma}\colon I \to \mathbb{D}$ is a geodesic for ω. In particular, every geodesic is of class C^∞ and two points in X can be connected by a geodesic.*

Proof. (i) Indeed, (1.51) follows from the facts that the automorphism group of the universal covering is transitive on the fibers (Proposition 1.6.9) and that automorphisms of $\mathbb{D}$ are isometries for the Poincaré distance (Theorem 1.2.3).

(ii) Put $z = \pi_X(\tilde{z})$. If $w \in B_X(z,r)$, then by (1.51) there exists $\tilde{w} \in \pi_X^{-1}(w)$ such that $\omega(\tilde{z},\tilde{w}) < r$; therefore, $\tilde{w} \in B_{\mathbb{D}}(\tilde{z},r)$ and $w = \pi_X(\tilde{w}) \in \pi_X(B_{\mathbb{D}}(\tilde{z},r))$, i. e., $B_X(\pi_X(\tilde{z}),r) \subseteq \pi_X(B_{\mathbb{D}}(\tilde{z},r))$. Conversely, if $\tilde{w} \in B_{\mathbb{D}}(\tilde{z},r)$ then by definition $\omega_X(\pi_X(\tilde{z}),\pi_X(\tilde{w})) < r$, and so $\pi_X(B_{\mathbb{D}}(\tilde{z},r)) \subseteq B_X(\pi_X(\tilde{z}),r)$.

(iii) Using (1.50) and (1.51), it is a routine matter to prove that ω_X is a distance. Furthermore, notice that (i) and (ii) imply that

$$\pi_X^{-1}(B_X(z,r)) = \bigcup_{\tilde{z}\in\pi_X^{-1}(z)} B_{\mathbb{D}}(\tilde{z},r).$$

From this and (ii), it immediately follows that π_X is continuous and open from $\mathbb{D}$ to (X,ω_X), and hence ω_X induces the standard topology.

To prove that ω_X is complete, it suffices to prove that the closed balls $\overline{B_X(z,r)}$ are compact. Choose $\tilde{z} \in \mathbb{D}$ such that $\pi_X(\tilde{z}) = z$ and fix $\varepsilon > 0$. By (ii), $\overline{B_X(z,r)}$ is contained in $\pi_X(\overline{B_{\mathbb{D}}(\tilde{z},r+\varepsilon)})$. Since $\overline{B_{\mathbb{D}}(\tilde{z},r+\varepsilon)}$ is compact and $\overline{B_X(z,r)}$ is closed, the assertion follows. Finally, $\overline{B_X(z,r)}$ is connected because $B_X(z,r)$ is connected being the image via π_X of a connected set.

(iv) Put $M = \omega_X(z_1,z_2)$. Since $\pi_X^{-1}(z_2)$ is discrete and $\overline{B} = \overline{B_{\mathbb{D}}(\tilde{z}_1,M+1)}$ is compact, the intersection $\pi_X^{-1}(z_2) \cap \overline{B}$ is finite; then (1.51) implies that there exists $\tilde{z}_2 \in \pi_X^{-1}(z_2)$ such that $\omega_X(z_1,z_2) = \omega(\tilde{z}_1,\tilde{z}_2)$.

(v) Given $\tilde{z}_0 \in \mathbb{D}$, we can find $r > 0$ such that π_X is injective on $B_{\mathbb{D}}(\tilde{z}_0,3r)$. In particular, for each $\tilde{z}_1,\tilde{z}_2 \in B_{\mathbb{D}}(\tilde{z}_0,r)$ we have $\pi_X^{-1}(z_j) \cap B_D(\tilde{z}_0,3r) = \{\tilde{z}_j\}$ for $j = 1,2$, where $z_j = \pi_X(\tilde{z}_j)$. Therefore, if $\tilde{z} \in \pi_X^{-1}(z_1) \setminus \{\tilde{z}_1\}$ we have

$$\omega(\tilde{z},\tilde{z}_2) \ge \omega(\tilde{z},\tilde{z}_0) - \omega(\tilde{z}_0,\tilde{z}_2) > 3r - r = 2r > \omega(\tilde{z}_1,\tilde{z}_2).$$

By (ii), it follows that $\omega_X(z_1,z_2) = \omega(\tilde{z}_1,\tilde{z}_2)$, and so π_X is an isometry when restricted to $B_{\mathbb{D}}(\tilde{z}_0,r)$.

(vi) It follows immediately from the fact that π_X is a local isometry (see also Remark 1.7.2), and by the fact that two points in $\mathbb{D}$ can be connected by a geodesic, which is always of class C^∞ (Proposition 1.2.7). □

Corollary 1.7.4. *Let X be a Riemann surface. Then there exists an increasing sequence $\{K_\nu\}$ of compact connected subsets of X such that K_ν is contained in the interior part of $K_{\nu+1}$ for all $\nu \ge 1$ and*

$$\bigcup_{\nu\ge1} K_\nu = X.$$

Proof. If X is compact, it suffices to take $K_\nu = X$ for all $\nu \geq 1$. If $X = \mathbb{C}$, then the closed disks $K_\nu = \overline{D(0,\nu)}$ do the job. If $X = \mathbb{C}^*$, it suffices to take $K_\nu = \{z \in \mathbb{C}^* \mid 1/\nu \leq |z| \leq \nu\}$ for all $\nu \geq 1$.

Finally, if X is hyperbolic it suffices to take $K_\nu = \overline{B_X(z_0,\nu)}$, for any $z_0 \in X$ fixed; the K_ν are compact and connected by Proposition 1.7.3(iii). □

Example 1.7.5. Let us compute the Poincaré distance of the pointed unit disk $\mathbb{D}^*$. Since the Cayley transform is an isometry for the Poincaré distances, we can use the universal covering map $\pi_{\mathbb{D}^*}\colon \mathbb{H}^+ \to \mathbb{D}^*$ given by (1.48) instead of the equivalent covering map defined on $\mathbb{D}$.

Take $z_0 = r_0 e^{2\pi i\theta_0}$, $z_1 = r_1 e^{2\pi i\theta_1} \in \mathbb{D}^*$, with $0 < r_0, r_1 < 1$ and $\theta_0, \theta_1 \in [0,1)$; without loss of generality, we can assume $\theta_1 \geq \theta_0$. Since $z_0 = \pi_{\mathbb{D}^*}(\theta_0 + \frac{i}{2\pi}\log\frac{1}{r_0})$ and $\pi_{\mathbb{D}^*}^{-1}(z_1) = \{\theta_1 + m + \frac{i}{2\pi}\log\frac{1}{r_1} \mid m \in \mathbb{Z}\}$, thanks to (1.51) we have

$$\omega_{\mathbb{D}^*}(z_0,z_1) = \inf\left\{\omega_{\mathbb{H}^+}\left(\theta_0 + \frac{i}{2\pi}\log\frac{1}{r_0}, \theta_1 + m + \frac{i}{2\pi}\log\frac{1}{r_1}\right) \,\middle|\, m \in \mathbb{Z}\right\}.$$

Recalling (1.16) we see that we must find the infimum with respect to $m \in \mathbb{Z}$ of the quantity

$$f(m) = \left|\frac{\theta_1 + m + \frac{i}{2\pi}\log\frac{1}{r_1} - \theta_0 - \frac{i}{2\pi}\log\frac{1}{r_0}}{\theta_1 + m + \frac{i}{2\pi}\log\frac{1}{r_1} - \theta_0 + \frac{i}{2\pi}\log\frac{1}{r_0}}\right|.$$

Taking the square, we obtain

$$|f(m)|^2 = \frac{(\theta_1 - \theta_0 + m)^2 + \frac{1}{4\pi^2}|\log\frac{r_0}{r_1}|^2}{(\theta_1 - \theta_0 + m)^2 + \frac{1}{4\pi^2}|\log\frac{1}{r_0 r_1}|^2}.$$

The even function $x \mapsto \frac{x^2+a^2}{x^2+b^2}$ with $b^2 > a^2$ has a minimum at $x = 0$ and it is strictly increasing for $x > 0$. Hence the minimum of f is reached when $|\theta_1 - \theta_0 + m|$ is minimum. Since $0 \leq \theta_1 - \theta_0 < 1$, the latter minimum is achieved in $m = 0$ if $0 \leq \theta_1 - \theta_0 \leq 1/2$, and in $m = -1$ if $1/2 \leq \theta_1 - \theta_0 < 1$. Summing up, we have obtained the following formula:

$$\omega_{\mathbb{D}^*}(z_0,z_1) = \begin{cases} \frac{1}{2}\log\dfrac{1+\sqrt{\dfrac{(\theta_1-\theta_0)^2+\frac{1}{4\pi^2}|\log\frac{r_0}{r_1}|^2}{(\theta_1-\theta_0)^2+\frac{1}{4\pi^2}|\log\frac{1}{r_0r_1}|^2}}}{1-\sqrt{\dfrac{(\theta_1-\theta_0)^2+\frac{1}{4\pi^2}|\log\frac{r_0}{r_1}|^2}{(\theta_1-\theta_0)^2+\frac{1}{4\pi^2}|\log\frac{1}{r_0r_1}|^2}}} & \text{if } 0 \leq \theta_1 - \theta_0 \leq 1/2, \\[2ex] \frac{1}{2}\log\dfrac{1+\sqrt{\dfrac{(\theta_1-\theta_0-1)^2+\frac{1}{4\pi^2}|\log\frac{r_0}{r_1}|^2}{(\theta_1-\theta_0-1)^2+\frac{1}{4\pi^2}|\log\frac{1}{r_0r_1}|^2}}}{1-\sqrt{\dfrac{(\theta_1-\theta_0-1)^2+\frac{1}{4\pi^2}|\log\frac{r_0}{r_1}|^2}{(\theta_1-\theta_0-1)^2+\frac{1}{4\pi^2}|\log\frac{1}{r_0r_1}|^2}}} & \text{if } 1/2 \leq \theta_1 - \theta_0 < 1. \end{cases}$$

In particular, if $\theta_0 = \theta_1 = \theta$ and $r_0 \le r_1$ we get

$$\omega_{\mathbb{D}^*}(r_0 e^{2\pi i\theta}, r_1 e^{2\pi i\theta}) = \frac{1}{2}\log\frac{\log(1/r_0)}{\log(1/r_1)}.$$

If instead we take $r_0 = r_1 = r$ and $\theta_0 \le \theta_1$, we get

$$\omega_{\mathbb{D}^*}(re^{2\pi i\theta_0}, re^{2\pi i\theta_1}) = \begin{cases} \frac{1}{2}\log\dfrac{1+\sqrt{\frac{(\theta_1-\theta_0)^2}{(\theta_1-\theta_0)^2+\frac{1}{\pi^2}(\log r)^2}}}{1-\sqrt{\frac{(\theta_1-\theta_0)^2}{(\theta_1-\theta_0)^2+\frac{1}{\pi^2}(\log r)^2}}} & \text{if } 0 \le \theta_1 - \theta_0 \le 1/2, \\ \frac{1}{2}\log\dfrac{1+\sqrt{\frac{(\theta_1-\theta_0-1)^2}{(\theta_1-\theta_0-1)^2+\frac{1}{\pi^2}(\log r)^2}}}{1-\sqrt{\frac{(\theta_1-\theta_0-1)^2}{(\theta_1-\theta_0-1)^2+\frac{1}{\pi^2}(\log r)^2}}} & \text{if } 1/2 \le \theta_1 - \theta_0 < 1, \end{cases}$$

and so

$$\max_{\theta_1,\theta_2\in\mathbb{R}} \omega_{\mathbb{D}^*}(re^{2\pi i\theta_0}, re^{2\pi i\theta_1}) = \frac{1}{2}\log\frac{\sqrt{1+4(\log r)^2/\pi^2}+1}{\sqrt{1+4(\log r)^2/\pi^2}-1}. \tag{1.52}$$

The main property of the Poincarè distance on $\mathbb{D}$ was the Schwarz–Pick lemma. Accordingly, the main property of the Poincaré distance on an arbitrary Riemann surface is the following general *Schwarz–Pick lemma.*

Theorem 1.7.6. *Let X and Y be two hyperbolic Riemann surfaces, and $f\colon X \to Y$ a holomorphic function. Then*

$$\omega_Y(f(z_1), f(z_2)) \le \omega_X(z_1, z_2)$$

for all $z_1, z_2 \in X$. Furthermore, equality at some $z_1 \ne z_2$ implies that f is a covering map. Conversely, if f is a covering map then for every $w_1, w_2 \in Y$ and $z_1 \in X$ with $f(z_1) = w_1$ we can find $z_2 \in X$ so that $f(z_2) = w_2$ and $\omega_Y(w_1, w_2) = \omega_X(z_1, z_2)$.

Proof. Lift f to a holomorphic function $\tilde{f}\colon \mathbb{D} \to \mathbb{D}$ such that $\pi_Y \circ \tilde{f} = f \circ \pi_X$. Now take $z_1, z_2 \in X$. By Proposition 1.7.3(iv), we can find $\tilde{z}_1, \tilde{z}_2 \in \mathbb{D}$ so that $\pi_X(\tilde{z}_j) = z_j$ for $j = 1, 2$ and $\omega(\tilde{z}_1, \tilde{z}_2) = \omega_X(z_1, z_2)$. Since $\pi_Y(\tilde{f}(\tilde{z}_j)) = f(z_j)$ for $j = 1, 2$, the Schwarz–Pick lemma and (1.50) yield

$$\omega_Y(f(z_1), f(z_2)) \le \omega(\tilde{f}(\tilde{z}_1), \tilde{f}(\tilde{z}_2)) \le \omega(\tilde{z}_1, \tilde{z}_2) = \omega_X(z_1, z_2),$$

as claimed.

Now assume we have equality for some $z_1 \ne z_2$. Fix $\tilde{z}_1 \in \pi_X^{-1}(z_1)$; Proposition 1.7.3(iv) implies that there exists $\tilde{z}_2 \in \pi_X^{-1}(z_2)$ such that $\omega_X(z_1, z_2) = \omega(\tilde{z}_1, \tilde{z}_2)$. Therefore, we have

$$\begin{aligned}\omega(\tilde{z}_1, \tilde{z}_2) &= \omega_X(z_1, z_2) = \omega_Y(f(z_1), f(z_2)) \\ &= \omega_Y(\pi_Y(\tilde{f}(\tilde{z}_1)), \pi_Y(\tilde{f}(z_2))) \le \omega(\tilde{f}(\tilde{z}_1), \tilde{f}(\tilde{z}_2)) \le \omega(\tilde{z}_1, \tilde{z}_2).\end{aligned}$$

Therefore, $\omega(\tilde{f}(\tilde{z}_1),\tilde{f}(\tilde{z}_2)) = \omega(\tilde{z}_1,\tilde{z}_2)$; hence $\tilde{f} \in \mathrm{Aut}(\mathbb{D})$ by the Schwarz–Pick lemma and from $\pi_Y \circ \tilde{f} = f \circ \pi_X$ we deduce that f is a covering map, because $\pi_Y \circ \tilde{f}$ and π_X are covering maps.

Conversely, assume that f is a covering map. Then $\tilde{f}$ is a covering map (see, e. g., [260, Lemma 12.1]), and hence an automorphism of $\mathbb{D}$. Take $w_1, w_2 \in Y$ and $z_1 \in X$ such that $f(z_1) = w_1$. Choose $\tilde{z}_1 \in \mathbb{D}$ such that $\pi_X(\tilde{z}_1) = z_1$ and let $\tilde{w}_1 = \tilde{f}(\tilde{z}_1)$; in particular, we have

$$\pi_Y(\tilde{w}_1) = (\pi_Y \circ \tilde{f})(\tilde{z}_1) = (f \circ \pi_X)(\tilde{z}_1) = f(z_1) = w_1.$$

By Proposition 1.7.3(iv), there is $\tilde{w}_2 \in \pi_Y^{-1}(w_2)$ so that $\omega_Y(w_1,w_2) = \omega(\tilde{w}_1,\tilde{w}_2)$. Put $\tilde{z}_2 = \tilde{f}^{-1}(\tilde{w}_2)$ and $z_2 = \pi_X(\tilde{z}_2)$. By construction, we have $f(z_j) = w_j$ for $j = 1,2$, and $\omega(\tilde{z}_1,\tilde{z}_2) = \omega(\tilde{w}_1,\tilde{w}_2)$. Therefore, we get

$$\omega_Y(w_1,w_2) = \omega(\tilde{w}_1,\tilde{w}_2) = \omega(\tilde{z}_1,\tilde{z}_2) \geq \omega_X(z_1,z_2) \geq \omega_Y(w_1,w_2),$$

and hence $\omega_X(z_1,z_2) = \omega_Y(w_1,w_2)$ as desired. □

Remark 1.7.7. It should be noticed that if we set $\omega_X \equiv 0$ for a nonhyperbolic Riemann surface X, then (recalling Proposition 1.6.27) Theorem 1.7.6 holds for all holomorphic functions between any pair of Riemann surfaces.

An important consequence of Theorem 1.7.6 is that the family of holomorphic functions between two hyperbolic Riemann surfaces is equicontinuous. In particular, we have the following.

Corollary 1.7.8. *Let X and Y be hyperbolic Riemann surfaces. Then the topology of pointwise convergence on* $\mathrm{Hol}(X,Y)$ *coincides with the compact-open topology.*

Proof. Since $\mathrm{Hol}(X,Y)$ is equicontinuous, we can quote [227, p. 232]. □

From now on, then, when X and Y are hyperbolic we shall indifferently consider $\mathrm{Hol}(X,Y)$ endowed with the topology of pointwise convergence or with the compact-open topology. Notice that with the compact-open topology the space $\mathrm{Hol}(X,Y)$ is metrizable; see, e. g., [281, p. 68] or [17, p. 220].

We now describe the concepts involved in the Montel theorem.

Definition 1.7.9. Let X_1 and X_2 be two topological spaces. A sequence of continuous maps $\{f_\nu\} \subset C^0(X_1,X_2)$ is *compactly divergent* if for every pair of compact sets $K_1 \subset X_1$ and $K_2 \subset X_2$ there is $\nu_0 \in \mathbb{N}$ such that $f_\nu(K_1) \cap K_2 = \emptyset$ for all $\nu \geq \nu_0$. We shall sometimes say that $\{f_\nu\}$ *diverges to infinity* uniformly on compact sets.

If X_1 and X_2 are Riemann surfaces, a family $\mathcal{F} \subseteq \mathrm{Hol}(X_1,X_2)$ of holomorphic maps is called *normal* if every sequence in $\mathcal{F}$ admits a subsequence which is either convergent to a map in $\mathrm{Hol}(X_1,X_2)$ or compactly divergent.

Example 1.7.10. For instance, $\{z^k\}_{k\in\mathbb{N}}$ is not a normal family in $\mathrm{Hol}(\mathbb{C},\mathbb{C})$, whereas it is a normal family in $\mathrm{Hol}(\mathbb{D},\mathbb{D})$. Indeed,

$$\lim_{k\to+\infty}|z^k| = \begin{cases} 0 & \text{if } |z|<1,\\ 1 & \text{if } |z|=1,\\ +\infty & \text{if } |z|>1,\end{cases}$$

and so no subsequence of $\{z^k\}$ can be convergent or compactly divergent on $\mathbb{C}$, whereas the whole sequence $\{z^k\}$ converges to 0 in $\mathbb{D}$.

Normality is sort of a compactness condition; for instance, if Y is a compact Riemann surface, then a family $\mathcal{F}\subseteq\mathrm{Hol}(X,Y)$ is normal if and only if it is relatively compact, because the compactness of Y excludes the existence of compactly divergent sequences. So normality is naturally linked to the *Ascoli–Arzelá theorem*, that we are now going to state.

Theorem 1.7.11 (Ascoli–Arzelà, 1895). *Let X be a locally compact metric space and Y a metric space. Then a family $\mathcal{F}\subseteq C^0(X,Y)$ is relatively compact (with respect to the compact-open topology) if and only if:*
(i) *$\mathcal{F}(x)=\{f(x)\mid f\in\mathcal{F}\}$ is relatively compact in Y for every $x\in X$, and*
(ii) *$\mathcal{F}$ is equicontinuous.*

For a proof see, e. g., [227, p. 233].
Finally, we can state and prove the *Montel theorem.*

Theorem 1.7.12 (Montel, 1912). *Let X and Y be hyperbolic Riemann surfaces. Then* $\mathrm{Hol}(X,Y)$ *is a normal family.*

Proof. Let $\{f_\nu\}$ be a sequence of holomorphic functions from X to Y; we shall prove that if $\{f_\nu\}$ is not compactly divergent then it admits a convergent subsequence.

Assume $\{f_\nu\}$ is not compactly divergent; then there are compact sets $K_1\subset X$ and $K_2\subset Y$ such that $f_\nu(K_1)\cap K_2\neq\emptyset$ for infinitely many ν. Up to a subsequence, we can assume $f_\nu(K_1)\cap K_2\neq\emptyset$ for all ν.

We claim that $\{f_\nu(z_0)\}$ is relatively compact in Y for any $z_0\in X$. In fact, fix $z_0\in X$ and $w_0\in K_2$; furthermore, let d_1 denote the ω_X-diameter of K_1 and d_2 the ω_Y-diameter of K_2. Choose $z_\nu\in K_1$ such that $f_\nu(z_\nu)\in K_2$. Then Theorem 1.7.6 yields

$$\begin{aligned}\omega_Y(f_\nu(z_0),w_0) &\le \omega_Y(f_\nu(z_0),f_\nu(z_\nu))+\omega_Y(f_\nu(z_\nu),w_0)\\ &\le \omega_X(z_0,z_\nu)+\omega_Y(f_\nu(z_\nu),w_0)\\ &\le \omega_X(z_0,z)+\omega_X(z,z_\nu)+\omega_Y(f_\nu(z_\nu),w_0),\end{aligned}$$

where z is any point of K_1. Hence for every $\nu\in\mathbb{N}$ we have

$$\omega_Y(f_\nu(z_0),w_0)\le\min\{\omega_X(z_0,z)\mid z\in K_1\}+d_1+d_2.$$

Therefore, $\{f_\nu(z_0)\}$ is ω_Y-bounded, and thus relatively compact in Y, by Proposition 1.7.3(iii), and the claim is proved.

To complete the proof, it suffices now to invoke the Ascoli–Arzelà theorem: its hypotheses are fulfilled (again thanks to Theorem 1.7.6), and so $\{f_\nu\}$ is relatively compact. In particular, we can extract a converging subsequence and we are done. □

Corollary 1.7.13. *Let Y be a compact hyperbolic Riemann surface and X any Riemann surface. Then every sequence in* $\mathrm{Hol}(X, Y)$ *admits a converging subsequence.*

Proof. If X is not hyperbolic, then $\mathrm{Hol}(X, Y)$ contains only constant functions (Proposition 1.6.27) and the assertion follows because Y is compact. If X is hyperbolic, the assertion follows from Theorem 1.7.12 because the compactness of Y prevents the existence of compactly divergent subsequences. □

Actually, Theorem 1.7.12 characterizes hyperbolic Riemann surfaces.

Proposition 1.7.14. *Let X be a Riemann surface. Then X is hyperbolic if and only if* $\mathrm{Hol}(X, X)$ *is normal.*

Proof. By Theorem 1.7.12, it suffices to show that if X is not hyperbolic then $\mathrm{Hol}(X, X)$ is not normal.

Since $\mathrm{Hol}(\widehat{\mathbb{C}}, \widehat{\mathbb{C}})$, $\mathrm{Hol}(\mathbb{C}, \mathbb{C})$ and $\mathrm{Hol}(\mathbb{C}^*, \mathbb{C}^*)$ contain the sequence $\{z^k\}$, it is clear that neither of them is normal (see Example 1.7.10).

Let now consider a complex torus $X_\tau = \mathbb{C}/\Lambda_\tau$, where $\Lambda_\tau = \mathbb{Z} \oplus \tau\mathbb{Z}$ and $\tau \in \mathbb{H}^+$. Since X_τ is compact, to prove that $\mathrm{Hol}(X_\tau, X_\tau)$ is not normal it suffices to find a sequence of holomorphic self-maps of X_τ without converging subsequences.

Given $p \in \mathbb{N}^*$, let $\tilde{\mu}_p\colon \mathbb{C} \to \mathbb{C}$ be the linear map given by $\tilde{\mu}_p(z) = pz$. Since $\tilde{\mu}_p(\Lambda_\tau) \subseteq \Lambda_\tau$, every $\tilde{\mu}_p$ induces a holomorphic function $\mu_p \in \mathrm{Hol}(X_\tau, X_\tau)$. Clearly, $\pi_\tau(0) \in X_\tau$ is a fixed point of each μ_p, where π_τ denotes the canonical projection from $\mathbb{C}$ to X_τ. At a fixed point, we can compute the derivative; we have $\mu_p'(\pi_\tau(0)) = p$. So $\lim_{p\to+\infty} \mu_p'(\pi_\tau(0)) = +\infty$; if there would exist a subsequence $\{\mu_{p_k}\}$ of $\{\mu_p\}$ converging to some $g \in \mathrm{Hol}(X_\tau, X_\tau)$ we would have $g(\pi_\tau(0)) = \pi_\tau(0)$ and $\mu_{p_k}'(\pi_\tau(0)) \to g'(\pi_\tau(0)) \in \mathbb{C}$, contradiction. Therefore, $\{\mu_p\}$ has no converging subsequences and we are done. □

Another useful consequence of the Montel theorem is the following.

Proposition 1.7.15. *Let $D \subset \widehat{X}$ be a hyperbolic domain, and Y another Riemann surface. Then* $\mathrm{Hol}(Y, D)$ *is relatively compact in* $\mathrm{Hol}(Y, \widehat{X})$.

Proof. If Y is not hyperbolic, then $\mathrm{Hol}(Y, D)$ consists only of constant maps (Proposition 1.6.27) and the assertion follows from the compactness of $\widehat{X}$.

Assume then Y hyperbolic, and let $D_1 \subseteq \widehat{X}$ be a hyperbolic domain containing D and such that $\widehat{X} \setminus D_1$ is finite (consisting of three points if $\widehat{X}$ is the Riemann sphere, consisting of one point if $\widehat{X}$ is a torus, empty if $\widehat{X}$ is hyperbolic); it suffices to show that $\mathrm{Hol}(Y, D_1)$ is relatively compact in $\mathrm{Hol}(Y, \widehat{X})$.

Let $\{f_k\}$ be a sequence in $\mathrm{Hol}(Y, D_1)$; we need to find a subsequence converging in $\mathrm{Hol}(Y, \widehat{X})$. If $\{f_k\}$ admits a subsequence converging in $\mathrm{Hol}(Y, D_1)$, we are done; in particular, if $D_1 = \widehat{X}$ this always happens (Corollary 1.7.13) and the proof is completed.

By the Montel Theorem 1.7.12, we are left with the case $\{f_k\}$ compactly divergent in $\mathrm{Hol}(Y, D_1)$. Write $\widehat{X} \setminus D_1 = \{x_1, \ldots, x_p\}$, and fix a compact connected subset K_0 of Y. We claim that there exist $x_{j_0} \in \widehat{X} \setminus D_1$, a subsequence $\{f_{k_\nu}\}$ of $\{f_k\}$ and a connected neighborhood U_{j_0} of x_{j_0} in $\widehat{X}$ such that for every compact connected subset K of Y containing K_0 and every neighborhood U of x_{j_0} contained in U_{j_0} we have $f_{k_\nu}(K) \subset U$ for all ν large enough. Since (Corollary 1.7.4) Y is the increasing union of compact connected subsets, that without loss of generality we can assume to contain K_0, this will imply that $f_{k_\nu} \to x_{j_0}$ in $\mathrm{Hol}(Y, \widehat{X})$, and the assertion will follow.

For $j = 1, \ldots, p$, choose disjoint connected neighborhoods U_j of x_j in $\widehat{X}$. Since $\{f_k\}$ is compactly divergent, we have $f_k(K_0) \subset \bigcup_j U_j$ for k large enough. But K_0 is connected; hence there is a subsequence $\{f_{k_\nu}\}$ and an index j_0 such that $f_{k_\nu}(K_0) \subset U_{j_0}$ for all $\nu \in \mathbb{N}$.

Now let K be any compact connected subset of Y containing K_0, and $U \subset U_{j_0}$ any neighborhood of x_{j_0} in $\widehat{X}$. Since $\{f_{k_\nu}\}$ is compactly divergent, we have

$$f_{k_\nu}(K) \subset U \cup \bigcup_{j \neq j_0} U_j$$

for ν large enough. But $K \supseteq K_0$ is connected, and $f_{k_\nu}(K) \cap U_{j_0} \neq \emptyset$ for all $\nu \in \mathbb{N}$; therefore, $f_{k_\nu}(K) \subset U$ for all ν sufficiently large, and we are done. □

So, we have completely traced the way from the seemingly innocuous Schwarz lemma to the all powerful Montel theorem. The novice reader will soon learn to appreciate the strength of Montel theorem (the experienced reader appreciates it already, we hope). A first sample is given by the following consequence on the topology of $\mathrm{Hol}(X, Y)$.

Corollary 1.7.16. *Let X and Y be two hyperbolic Riemann surfaces. Then* $\mathrm{Hol}(X, Y)$ *is locally compact.*

Proof. Take $f \in \mathrm{Hol}(X, Y)$, fix $z_0 \in X$, and let $U \subset Y$ be a relatively compact open neighborhood of $f(z_0)$; it suffices to show that the neighborhood

$$W(z_0, U) = \{g \in \mathrm{Hol}(X, Y) \mid g(z_0) \in U\}$$

of f is relatively compact in $\mathrm{Hol}(X, Y)$. But indeed no sequence in $W(z_0, U)$ can be compactly divergent, and the assertion follows from Theorem 1.7.12. □

The first really important consequence is that the convergence of a sequence of functions in the usual compact-open topology is assured by very weak hypotheses. In the proof, we shall use an elementary topological lemma that we recall here for the sake of completeness.

Definition 1.7.17. Let X be a Hausdorff topological space. A *limit point* (or *accumulation point*) of a sequence $\{x_\nu\} \subset X$ is a limit in X of a subsequence of $\{x_\nu\}$.

Lemma 1.7.18. *Let X be a Hausdorff and first countable topological space and let $\{x_\nu\} \subset X$ be a sequence whose closure is compact in X. Assume that $\{x_\nu\}$ has a unique limit point $\overline{x} \in X$. Then the whole sequence $\{x_\nu\}$ converges to $\overline{x}$.*

Proof. Assume, by contradiction, that $\{x_\nu\}$ does not converge to $\overline{x}$. Then there exists an open neighborhood U of $\overline{x}$ and a subsequence $\{x_{\nu_k}\}$ such that $x_{\nu_k} \notin U$ for all $k \in \mathbb{N}$. Since the closure of $\{x_\nu\}$ is compact, we can extract from $\{x_{\nu_k}\}$ a subsequence converging to a point $y \in X$. But clearly $y \notin U$, and hence $y \neq \overline{x}$, against the hypothesis that $\{x_\nu\}$ has only one limit point. □

With this we can prove the *Vitali theorem*.

Theorem 1.7.19 (Vitali, 1903). *Let X and Y be hyperbolic Riemann surfaces, and let $\{f_\nu\}$ be a sequence of functions in* $\mathrm{Hol}(X, Y)$. *Assume there is a set $A \subset X$ with at least one accumulation point such that $\{f_\nu(z)\}$ converges in Y for every $z \in A$. Then $\{f_\nu\}$ converges uniformly on compact subsets of X to a function $f \in$* $\mathrm{Hol}(X, Y)$.

Proof. Clearly, the sequence $\{f_\nu\}$ cannot contain compactly divergent subsequences; hence, by Theorem 1.7.12 and Lemma 1.7.18, it suffices to show that it has only one limit point in $\mathrm{Hol}(X, Y)$. Let $f, g \in \mathrm{Hol}(X, Y)$ be two limit points of $\{f_\nu\}$. Since $\{f_\nu(z)\}$ converges for every $z \in A$, it follows that $f \equiv g$ on A, and hence everywhere by the identity principle. □

We can also apply the Montel theorem to the investigation of the topological structure of the automorphism group of a hyperbolic Riemann surface.

Proposition 1.7.20. (i) *Let $\{f_\nu\} \subset$* $\mathrm{Aut}(X)$ *be a sequence of automorphisms of a hyperbolic Riemann surface X converging to a holomorphic function $f \in$* $\mathrm{Hol}(X, X)$. *Then $f \in$* $\mathrm{Aut}(X)$. *In particular,* $\mathrm{Aut}(X)$ *is closed in* $\mathrm{Hol}(X, X)$.
(ii) *Let $\{f_\nu\}$ be a sequence of automorphisms of a hyperbolic domain $D \subseteq \widehat{X}$ converging to a holomorphic function $f: D \to \widehat{X}$. Then either f is a constant belonging to ∂D or $f \in$* $\mathrm{Aut}(D)$.

Proof. To reduce part (ii) to part (i), it suffices to remark that if $f(D) \cap \partial D \neq \emptyset$, then f is a constant belonging to ∂D thanks to the Hurwitz theorem (see Corollary A.1.5). Therefore, we can directly assume $f \in \mathrm{Hol}(D, D)$ and prove only part (i).

Set $g_\nu = f_\nu^{-1}$; by Theorem 1.7.12, up to a subsequence we can assume that either $g_\nu \to g \in \mathrm{Hol}(X,X)$ or $\{g_\nu\}$ is compactly divergent. Since, for any $z \in X$, we have $f_\nu(z) \to f(z) \in X$ and $g_\nu(f_\nu(z)) = z$, the sequence $\{g_\nu\}$ cannot be compactly divergent. Then

$$g(f(z)) = \lim_{\nu\to\infty} g_\nu(f_\nu(z)) = z$$

for every $z \in X$ and, since $g(X) \subseteq X$,

$$f(g(z)) = \lim_{\nu\to\infty} f_\nu(g_\nu(z)) = z$$

for all $z \in X$; therefore, $g = f^{-1}$ and $f \in \mathrm{Aut}(X)$. □

Corollary 1.7.21. *Let X be a hyperbolic Riemann surface. Then $\mathrm{Aut}(X)$ is closed in $\mathrm{Hol}(X, X)$, and hence it is locally compact. Furthermore, the isotropy group $\mathrm{Aut}_{z_0}(X)$ is compact for all $z_0 \in X$.*

Proof. The closure is Proposition 1.7.20; the local compactness is Corollary 1.7.16; the compactness of the isotropy group follows remarking that no sequence of elements of $\mathrm{Aut}_{z_0}(X)$ can be compactly divergent and recalling that $\mathrm{Aut}_{z_0}(X)$ is closed in $\mathrm{Aut}(X)$. □

Now we can keep an old promise.

Proposition 1.7.22. *Let X be a hyperbolic Riemann surface. Then every discrete subgroup Γ of $\mathrm{Aut}(X)$ is everywhere properly discontinuous.*

Proof. If Γ is not properly discontinuous at some point $z_0 \in X$, then there exists an infinite sequence of distinct elements $\gamma_\nu \in \Gamma$ and a sequence $\{z_\nu\} \subset X$ converging to z_0 such that $\gamma_\nu(z_\nu) \to z_0$. Up to a subsequence, we can assume that γ_ν tends to a function $\gamma: X \to X$, for $\{\gamma_\nu\}$ cannot have compactly divergent subsequences. By Proposition 1.7.20, $\gamma \in \mathrm{Aut}(X)$; hence $\gamma \in \Gamma$, for Γ is closed (see Proposition A.5.1), and this is impossible because Γ is discrete. □

Corollary 1.7.23. *The closed properly discontinuous subgroups of $\mathrm{Aut}(\mathbb{D})$ acting freely on $\mathbb{D}$ are the discrete subgroups without elliptic elements.*

Proof. It follows from Lemma 1.6.22 and Proposition 1.7.22. □

We shall now describe another application of the Schwarz–Pick lemma for hyperbolic Riemann surfaces Theorem 1.7.6, leading to a proof of the big Picard theorem. We start with the following.

Lemma 1.7.24. *Let X be a hyperbolic Riemann surface and let $f \in \mathrm{Hol}(\mathbb{D}^*, X)$ be such that there is a sequence $\{z_\nu\} \subset \mathbb{D}^*$ converging to 0 so that $f(z_\nu) \to w_0 \in X$ as $\nu \to +\infty$. Then $f_*(\pi_1(\mathbb{D}^*))$ is trivial.*

Proof. Set $r_\nu = |z_\nu|$ and let $\sigma_\nu: [0,1] \to \mathbb{D}^*$ be the curve $\sigma_\nu(t) = r_\nu e^{2\pi i t}$. It suffices to show that the curve $f \circ \sigma_\nu$ is homotopic to a point in X for ν large enough.

Choose a contractible neighborhood U of w_0 and let $r_0 > 0$ be such that the ω_X-ball of radius $2r_0$ and center w_0 is contained in U. Since $f(z_\nu) \to w_0$, we have $\omega_X(f(z_\nu), w_0) < r_0$ eventually. On the other hand, since $z_\nu \to 0$, using (1.52) we see that the $\omega_{\mathbb{D}^*}$-diameter of the image of σ_ν is less than r_0 for ν large enough. But this implies that the ω_X-diameter of the image of $f \circ \sigma_\nu$ is eventually less than r_0; thus the image of $f \circ \sigma_\nu$ is contained in U for ν large enough and we are done. □

The astonishing fact is that the *big Picard theorem* now follows.

Theorem 1.7.25. *Let $D \subseteq \widehat{X}$ be a hyperbolic domain contained in a compact Riemann surface $\widehat{X}$. Then every $f \in \mathrm{Hol}(\mathbb{D}^*, D)$ extends holomorphically to a function $\tilde{f} \in \mathrm{Hol}(\mathbb{D}, \widehat{X})$.*

Proof. Clearly, we can assume $\widehat{X} \setminus D$ is a finite set (containing at least three points if $\widehat{X} = \widehat{\mathbb{C}}$, at least one point if $\widehat{X}$ is a torus, even empty if $\widehat{X}$ itself is hyperbolic). If f is constant, the conclusion is trivially true; so let assume that f is not constant.

Suppose first there is a sequence $\{z_\nu\} \subset \mathbb{D}^*$ converging to 0 such that $f(z_\nu) \to w_0 \in D$. By Lemma 1.7.24, $f_*(\pi_1(\mathbb{D}^*))$ is trivial; hence, by Proposition 1.6.17, f lifts to a function $\hat{f}: \mathbb{D}^* \to \mathbb{D}$ such that $f = \pi_D \circ \hat{f}$, where $\pi_D: \mathbb{D} \to D$ is the universal covering map of D. Then the Riemann removable singularity theorem implies that $\hat{f}$ extends holomorphically to a function $f_1 \in \mathrm{Hol}(\mathbb{D}, \mathbb{C})$. Clearly, $f_1(\mathbb{D}) \subseteq \overline{\mathbb{D}}$. If $f_1(0) \in \partial\mathbb{D}$, then by the maximum principle f_1, and hence $\hat{f}$, and hence f, would be constant, impossible. So, $f_1 \in \mathrm{Hol}(\mathbb{D}, \mathbb{D})$, and clearly $\tilde{f} = \pi_D \circ f_1 \in \mathrm{Hol}(\mathbb{D}, D)$ is the required extension of f.

So, assume now that for no sequence $\{z_\nu\} \subset \mathbb{D}^*$ converging to 0 the sequence $\{f(z_\nu)\}$ converges to a point of D. This means that for every neighborhood U of $\widehat{X} \setminus D$ in $\widehat{X}$ there is a small disk V about 0 such that $f(V \setminus \{0\}) \subset U$; indeed, if this were not the case we could construct a sequence $\{z_\nu\} \subset \mathbb{D}^*$ converging to 0 such that $\{f(z_\nu)\} \subset D \setminus U$, which is compact, and thus we could find a subsequence converging to a point of D.

Now take U biholomorphic to a (not necessarily connected) bounded open set in $\mathbb{C}$ (it is possible because $\widehat{X} \setminus X$ is finite). Then another application of the Riemann removable singularity theorem shows that f extends holomorphically across 0 to a function $\tilde{f} \in \mathrm{Hol}(\mathbb{D}, \widehat{X})$, and we are done. □

Corollary 1.7.26 (Big Picard theorem, 1879). *Let $f \in \mathrm{Hol}(\mathbb{D}^*, \mathbb{C})$ be a holomorphic function with an essential singularity at the origin. Then $\mathbb{C} \setminus f(\mathbb{D}^*)$ contains at most one point.*

Proof. Assume, by contradiction, that $\mathbb{C} \setminus f(\mathbb{D}^*)$ contains at least two distinct points, z_0 and z_1. Then f belongs to $\mathrm{Hol}(\mathbb{D}^*, D)$, where $D = \widehat{\mathbb{C}} \setminus \{z_0, z_1, \infty\} \subset \widehat{\mathbb{C}}$ is a hyperbolic domain. But then Theorem 1.7.25 implies that f extends to a holomorphic map from $\mathbb{D}$ into $\widehat{\mathbb{C}}$, which is impossible because 0 is an essential singularity. □

We end this section with a result about the boundary behavior of the Poincaré distance essentially saying that the Poincaré distance between two sequences converging to different boundary points of a hyperbolic domain must diverge.

Proposition 1.7.27. *Let $D \subset \widehat{X}$ be a hyperbolic domain. Let $\{z_\nu\} \subset D$ be a sequence converging to $\tau_0 \in \partial D$. Let $\{w_\nu\} \subset D$ be another sequence. If there is $M > 0$ so that $\omega_D(z_\nu, w_\nu) < M$ for all $\nu \in \mathbb{N}$, then $w_\nu \to \tau_0$ as $\nu \to +\infty$.*

Proof. Put $B_\nu = B_D(z_\nu, M)$; it suffices to show that the d-diameter δ_ν of B_ν goes to 0 as $\nu \to +\infty$, where d is any fixed distance function on $\widehat{X}$ inducing the given topology.

Let us denote by $\pi_\nu\colon \mathbb{D} \to D$ an universal covering map such that $\pi_\nu(0) = z_\nu$, and set $f_\nu = \pi_\nu|_{B_{\mathbb{D}}(0,M)}$. Proposition 1.7.3(ii) implies that $B_\nu = \pi_\nu(B_{\mathbb{D}}(0, M))$. Fix a base point $w_0 \in D$, and set $K_n = \overline{B_D(w_0, n)}$, with $n \in \mathbb{N}^*$. Proposition 1.7.3(iii) says that each K_n is compact; moreover, clearly $D = \bigcup_n K_n$.

The same proposition also implies that $\omega_D(w_0, z_\nu) \to +\infty$ as $\nu \to +\infty$; as a consequence, for every fixed n we have $B_\nu \cap K_n = \emptyset$ as soon as ν is large enough.

Since $f_\nu(0) \to \tau_0 \in \widehat{X} \setminus K_n$, no subsequence of $\{f_\nu\}$ can be compactly divergent in $\widehat{X} \setminus K_n$; thus, being $\widehat{X} \setminus K_n$ clearly hyperbolic, by Theorem 1.7.12 we know that $\{f_\nu\}$ is relatively compact in $\mathrm{Hol}(B_{\mathbb{D}}(0, M), \widehat{X} \setminus K_n)$.

Let $\{f_{\nu_k}\}$ be a subsequence converging to $g \in \mathrm{Hol}(B_{\mathbb{D}}(0, M), \widehat{X} \setminus K_n)$. By the Hurwitz theorem (see Corollary A.1.5) if g is not constant then its image is contained in D, and thus it must intersect K_{n_0} for some n_0 large enough. But since B_ν is eventually disjoint from any K_n, the image of g must be contained in $\widehat{X} \setminus K_n$ for all $n \in \mathbb{N}$, contradiction. So g must be constant, i. e., $g \equiv \tau_0$.

Since this happens for any converging subsequence of $\{f_\nu\}$, by Lemma 1.7.18 the whole sequence $\{f_\nu\}$ converges to τ_0. Notice that, up to arguing with $M + 1$ instead of M, we can assume that $f_\nu \to g$ uniformly on $B_{\mathbb{D}}(0, M)$.

As a consequence, the d-diameter of the image of f_ν must converge to 0; but the image of f_ν is B_ν, and the proof is complete. □

Notes to Section 1.7

The first version of the Ascoli–Arzelà Theorem 1.7.11 is due to Ascoli [25] and Arzelà [24]; their results have later been generalized by Fréchet [157] in 1906.

The Montel Theorem 1.7.12 was first proved by Montel in 1912 [297], at least for hyperbolic domains in the complex plane. Montel's original proof used the modular function, i. e., the universal covering map of $\mathbb{C} \setminus \{0, 1\}$. Later on, other proofs were devised, relying only on more elementary facts; see, for instance, [160] or [88]. Many of these proofs use the *Schottky theorem* (proved by Schottky [366, 367] in 1904), that states that for any holomorphic function $f\colon \mathbb{D} \to \mathbb{C} \setminus \{0, 1\}$ we have

$$\log|f(z)| < \left(\pi + \log^+|f(0)|\right) \frac{1 + |z|}{1 - |z|},$$

where $\log^+$ is the positive part of the logarithm (this form of Schottky theorem is due to Hayman [181]; cf. also [16]). A proof of the Schottky theorem in our spirit is described in [237]: it relies on an explicit metric of negative Gaussian curvature on $\mathbb{C} \setminus \{0, 1\}$ (constructed by Grauert and Rieckziegel in 1965 [174]), and on the Ahlfors lemma we shall see in Theorem 1.10.7 (see [16, 18]). Actually, it is possible to prove that a weaker qualitative version of the Schottky theorem is equivalent to the normality of $\mathrm{Hol}(\mathbb{D}, \mathbb{C} \setminus \{z_0, z_1\})$; see, e. g., [210, Theorem 1.2.6]. For more on the history of the Schottky theorem, and for a generalization due to Miranda [293], see [88, Chapter XII].

Our proof of the Montel theorem is adapted from a more general argument due to Wu [422].

Usually, the name Montel theorem is ascribed to another result, previously proved in 1907 by Montel himself [296] (but cf. also [380] and [241]): every sequence of holomorphic functions from a Riemann surface X into a *bounded* domain $D \subset\subset \mathbb{C}$ has a subsequence converging in $\mathrm{Hol}(X, \mathbb{C})$; in other words, $\mathrm{Hol}(X, D)$ is relatively compact in $\mathrm{Hol}(X, \mathbb{C})$. This is now an immediate consequence of our

Montel theorem: it suffices to imbed D into a larger bounded domain $D_1 \supset\supset D$ and then to invoke the normality of $\mathrm{Hol}(X, D_1)$.

A domain $D \subset \widehat{X}$ is *tautly imbedded* in a Riemann surface $\widehat{X}$ if $\mathrm{Hol}(Y, D)$ is relatively compact in $\mathrm{Hol}(Y, \widehat{X})$ for all Riemann surfaces Y; so Proposition 1.7.15 (whose proof is taken from [12]) says that a hyperbolic domain $D \subset \widehat{X}$ is tautly imbedded in X. A related notion is the following: we say that a domain $D \subset\subset \widehat{X}$ relatively compact in a Riemann surface $\widehat{X}$ is *hyperbolically imbedded* in $\widehat{X}$ if for every $\tau \in \partial D$ and every neighborhood $U \subseteq \widehat{X}$ of τ there is a neighborhood $V \subset \widehat{X}$ of τ with $\overline{V} \subset U$ such that $\omega_D(\overline{V} \cap D, D \setminus U) > 0$. Kiernan [231] in 1973 has proved that a relatively compact domain $D \subset \widehat{X}$ is hyperbolically embedded in $\widehat{X}$ if and only if is tautly embedded in $\widehat{X}$ if and only if the following geometrical assertion holds: if $\{z_\nu\}$ and $\{w_\nu\}$ are sequences in D such that $z_\nu \to \sigma \in \overline{D}$ and $w_\nu \to \tau \in \overline{D}$ so that $\omega_D(z_\nu, w_\nu) \to 0$ as $\nu \to +\infty$ then $\sigma = \tau$; cf. also Proposition 1.7.27.

Theorem 1.7.19 is due to Vitali [404, 405] (see also Stieltjes [380], Porter [342], and Carathéodory and Landau [101]). For more on the history of the Montel and Vitali theorems, see [88, pp. 251–252].

Corollary 1.7.21 is the first step toward the proof of the fact that the automorphism group of a complex manifold equipped with a distance contracted by holomorphic maps is a finite-dimensional Lie group (see H. Cartan [106], Wu [422], and Kobayashi [237]).

The big Picard theorem originally has been proved by Picard [320] in 1879; it has subsequently been strengthened by Julia in 1924 [217]. Our proof is modeled on Huber [201] and Kobayashi [237]. For generalizations see, e. g., [278, 238, 257, 239] and references therein.

The proof of Proposition 1.7.27 is adapted from [287, Theorem 3.4].

1.8 Boundary behavior of the universal covering map

In this section, we want to study more accurately the boundary behavior of the universal covering map of a particular kind of multiply connected hyperbolic domains.

Definition 1.8.1. A (always noncompact) domain D of a compact Riemann surface $\widehat{X}$ is *of regular type* if

(a) every connected component of ∂D is either a Jordan curve (i. e., a closed simple continuous curve) or an isolated point;
(b) every connected component Σ of ∂D has a neighborhood V in $\widehat{X}$ such that $V \cap \partial D = \Sigma$;
(c) if Σ is an isolated point, we can take V to be simply connected; if Σ is a Jordan curve, we require that V is doubly connected and is such that $V \setminus \Sigma$ has exactly two connected components, both simply connected, one contained in D, and the other one contained in $\widehat{X} \setminus \overline{D}$.

If Σ is a connected component of ∂D, we shall say that Σ is a *point component* if it is an isolated point and that it is a *Jordan component* otherwise. Furthermore, to every component Σ of ∂D we associate the element $[\sigma_\Sigma]$ in $\pi_1(D)$ represented by any simple loop σ_Σ in $D \cap V$ separating Σ from $\partial D \setminus \Sigma$ and leaving Σ on its left side; two such loops are homotopic in $D \cap V$. We say that Σ is *irrelevant* if σ_Σ is null-homotopic in D; it is *relevant* otherwise. Finally, if $\pi_D: \widetilde{X} \to D$ is the universal covering map of D and $\Gamma_D \subset \mathrm{Aut}(\widetilde{X})$ is the automorphism group of the covering, we set $\gamma_\Sigma = \mu[\sigma_\Sigma] \in \Gamma_D$, where $\mu: \pi_1(D) \to \Gamma_D$ is the isomorphism introduced in Proposition 1.6.9 (see also Remark 1.6.10).

In a simply connected domain of regular type (e. g., the complement in $\widehat{\mathbb{C}}$ of a point or of a topological disk) all connected components of the boundary are clearly irrelevant. This is actually the only case where a connected component of the boundary of a domain of regular type can be irrelevant. To prove this assertion we need a couple of topological properties of Riemann surfaces that can be found in [142].

Lemma 1.8.2. *Let $D \subset \widehat{X}$ be a multiply connected domain of regular type. Then every connected component of ∂D is relevant.*

Proof. Let Σ be a connected component of ∂D. By definition, we have a simple loop σ_Σ inside D which is homotopic to Σ in $\overline{D}$ (more precisely, it is homotopic to Σ in the closure of $V \cap D$, where V is the neighborhood of Σ given in the definition). More precisely (see [142, Theorem 2.1]), σ_Σ is isotopic to Σ, and thus we can find a doubly connected domain $W \subset D \cap V$ whose boundary consists exactly of Σ and of the image of σ.

Assume, by contradiction, that Σ is irrelevant, so that σ_Σ is null-homotopic in D. In particular (see [142, Theorem 1.7]), σ_Σ must bound a topological disk $D' \subset D$. But then the domain $\tilde{D}$, obtained as union of D' with the image of σ_Σ and with W, is simply connected, contained in D, and such that $\partial\tilde{D} = \Sigma \subseteq \partial D$. Being D connected, this is possible only if $D' = D$; hence D should be simply connected, contradiction. □

The topology of domains of regular type is not too wild.

Lemma 1.8.3. *Let $D \subset \widehat{X}$ be a domain of regular type. Then ∂D has a finite number of connected components and $\pi_1(D)$ is finitely generated.*

Proof. Assume, by contradiction, that $\{\Sigma_\nu\}$ is an infinite sequence of distinct connected components of ∂D. Take $z_\nu \in \Sigma_\nu$; up to a subsequence, we can assume that $\{z_\nu\}$ converges to a point $w_0 \in \partial D$. But then the connected component of ∂D containing w_0 cannot be separated from the other components, contradiction.

To prove the second statement, we need a preliminary remark. If Σ is a connected component of ∂D, we can obtain a new (topological) orientable surface X_1 by adding Σ to D if Σ is a point component, or by gluing to D a topological disk along Σ (this can be done thanks to condition (c) in Definition 1.8.1). In both cases, X_1 can be obtained as the union of D and a simply connected open set; therefore, an application of the Seifert–Van Kampen theorem (see, e. g., [260, Theorem 10.1 and Corollary 10.8]) shows that $\pi_1(D)$ can be generated by adding to $\pi_1(X_1)$ at most one generator. Since the fundamental group of a (topological) orientable compact surface is finitely generated, arguing by induction on the number of connected components of ∂D we then get that $\pi_1(D)$ is finitely generated. □

Hyperbolic domains of regular type form a large class of Riemann surfaces, which are sufficiently well behaved for our needs and at the same time not too much specific.

Our aim in this section is to study the relationship between Σ and γ_Σ, where Σ is a connected component of the boundary of a domain of regular type. The results we shall obtain will be used in Section 3.3.

Let us first see what happens in two examples.

Example 1.8.4. Take $D = A(r,1) \subset \mathbb{C}$, with $0 < r < 1$, and denote by Σ_0 the inner boundary and by Σ_1 the outer boundary. Let $\pi\colon \mathbb{H}^+ \to A(r,1)$ be the universal covering map given by (1.49), that is

$$\pi(w) = \exp\left(2\pi i \frac{\log w}{\log a}\right),$$

where $a = 2\pi/\log r^{-1}$. Given $w_0 \in \mathbb{H}^+$, a representative of $[\sigma_{\Sigma_0}]$ issuing from $\pi(w_0)$ is the loop $\sigma_{\Sigma_0}(t) = \pi(w_0)e^{2\pi i t}$. The lifting of σ_{Σ_0} starting from w_0 is $\tilde{\sigma}_{\Sigma_0}(t) = a^t w_0$, and thus $\gamma_{\Sigma_0}(w_0) = aw_0$.

Analogously, one obtains $\gamma_{\Sigma_1}(w_0) = a^{-1}w_0$, and thus γ_{Σ_0} and γ_{Σ_1} are hyperbolic automorphisms of $\mathbb{H}^+$ with fixed points 0 and ∞. Notice furthermore that π extends continuously to $\mathbb{R}^* = \partial\mathbb{H}^+ \setminus \{0,\infty\}$.

Example 1.8.5. Take $D = \mathbb{D}^*$, and set $\Sigma_0 = \{0\}$ and $\Sigma_1 = \partial\mathbb{D}$. Let $\pi\colon \mathbb{H}^+ \to \mathbb{D}^*$ be the universal covering map given by (1.48), that is $\pi(w) = \exp(2\pi i w)$. Given $w_0 \in \mathbb{H}^+$, a representative of $[\sigma_{\Sigma_0}]$ issuing from $\pi(w_0)$ is still the loop $\sigma_{\Sigma_0}(t) = \pi(w_0)e^{2\pi i t}$, but this time the lifting of σ_{Σ_0} starting from w_0 is $\tilde{\sigma}_{\Sigma_0}(t) = w_0 + t$, and thus $\gamma_{\Sigma_0}(w_0) = w_0 + 1$.

Analogously, one obtains $\gamma_{\Sigma_1}(w_0) = w_0 - 1$, and thus γ_{Σ_0} and γ_{Σ_1} are parabolic automorphisms of $\mathbb{D}$ with fixed point ∞. Notice furthermore that π extends continuously to $\mathbb{R} = \partial\mathbb{H}^+ \setminus \{\infty\}$ and that $\pi(w)$ tends to 0 as w tends to infinity inside a vertical strip $\{w \in \mathbb{H}^+ \mid |\operatorname{Re} w| \le M\}$.

Let Σ be a connected component of the boundary of a multiply connected hyperbolic domain $D \subset \widehat{X}$ of regular type. Choose a simple loop $\sigma\colon [0,1] \to D$ in D separating Σ from $\partial D \setminus \Sigma$, leaving Σ on the left and close enough to Σ so that the domain $D_1 \subset D$ bounded by σ and Σ is doubly connected. We can choose σ so that it is isotopic to Σ (see the first part of the proof of Lemma 1.8.2); thus we have a continuous map $H\colon [0,1] \times [0,1] \to \overline{D_1}$ satisfying the following properties:

(a) $H(0,\cdot) \equiv \sigma$ and $H(1,\cdot) \equiv \Sigma$;
(b) $H(\cdot,0) \equiv H(\cdot,1)$;
(c) H sends $(0,1) \times [0,1)$ injectively into D_1.

Let $\tilde{\sigma}\colon [0,1] \to \mathbb{D}$ be a lifting of σ with respect to the universal covering map $\pi_D\colon \mathbb{D} \to D$; we know (Remark 1.6.10) that $\gamma_\Sigma(\tilde{\sigma}(0)) = \tilde{\sigma}(1)$. Since Σ is relevant (Lemma 1.8.2), $\gamma_\Sigma \ne \mathrm{id}_{\mathbb{D}}$; let $\tau_1, \tau_2 \in \partial\mathbb{D}$ be the fixed points of γ_Σ (possibly $\tau_1 = \tau_2$). We can enclose the compact image of $\tilde{\sigma}$ in a lens L bounded by two arcs of circumference connecting τ_1 and τ_2. By Lemma 1.4.17, L is invariant under γ_Σ; hence the image $\overline{\sigma}$ of $\tilde{\sigma}([0,1])$ under the action of the cyclic group Γ_Σ generated by γ_Σ is a Jordan arc contained in L and connecting τ_1 and τ_2 (by Proposition 1.4.10); see Figures 1.3 and 1.4.

Now let $\widetilde{H}\colon [0,1) \times [0,1] \to \mathbb{D}$ be a homotopy lifting H with respect to π_D such that $\widetilde{H}(0,\cdot) = \tilde{\sigma}$; using the action of Γ_Σ we can extend $\widetilde{H}$ to a homotopy $\overline{H}\colon [0,1)\times\mathbb{R} \to \mathbb{D}$ such that $\overline{H}(\{0\}\times\mathbb{R}) = \overline{\sigma}$, $\lim_{t\to-\infty} \overline{H}(\cdot,t) \equiv \tau_1$ and $\lim_{t\to+\infty} \overline{H}(\cdot,t) \equiv \tau_2$. Put $\Delta_1 = \overline{H}((0,1)\times\mathbb{R})$;

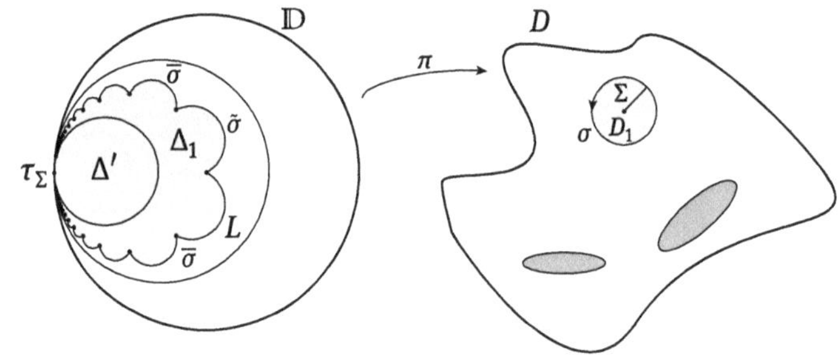

Figure 1.3: The point component case.

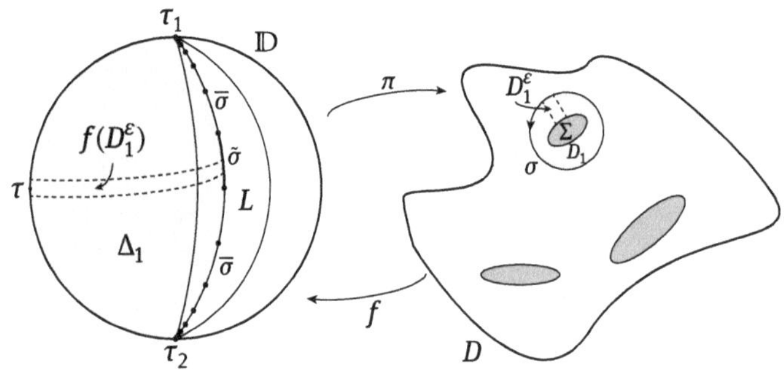

Figure 1.4: The Jordan component case.

we get $\pi_D(\Delta_1) = H((0,1)\times[0,1]) = D_1$. Moreover, we have $\partial\Delta_1 \cap \mathbb{D} = \overline{\sigma}$. Indeed, we clearly have $\overline{\sigma} \subseteq \partial\Delta_1 \cap \mathbb{D}$. To prove the converse inclusion, take $z \in \partial\Delta_1 \cap \mathbb{D}$ and a sequence $\{(s_k, t_k)\} \subset (0,1) \times \mathbb{R}$ such that $\overline{H}(s_k, t_k) \to z$. Since $z \in \mathbb{D}$, the sequence $\{|t_k|\}$ cannot diverge to $+\infty$; therefore, up to a subsequence we can assume that $\{(s_k, t_k)\}$ converges to some $(s_0, t_0) \in [0,1] \times \mathbb{R}$ such that $\overline{H}(s_0, t_0) = z$. In particular,

$$\pi_D(z) = \pi_D(\overline{H}(s_0, t_0)) = H(s_0, \{t_0\}) \in D,$$

where $\{t\}$ is the fractional part of t. In particular, $s_0 \neq 1$; being z in the boundary of Δ_1 it follows that $s_0 = 0$, and so $z \in \overline{\sigma}$, as claimed. Summing up, Δ_1 is an open connected set contained in $\mathbb{D} \setminus \overline{\sigma}$ and with $\overline{\sigma}$ as boundary; therefore, it is a connected component of $\mathbb{D} \setminus \overline{\sigma}$.

Definition 1.8.6. The open set Δ_1 just constructed is bounded by $\overline{\sigma}$ and by a closed subarc (possibly reduced to a point) of $\partial\mathbb{D}$; the open arc (possibly void) will be denoted by C_Σ and called the *principal arc* associated to Σ. If $\overline{\Delta_1} \cap \partial\mathbb{D}$ contains only one point, i. e., if $C_\Sigma = \emptyset$ (it can happen only if γ_Σ is parabolic), that point will be denoted by τ_Σ and called the *principal point* associated to Σ. Any open arc $\gamma(C_\Sigma)$ and any point $\gamma(\tau_\Sigma)$, with $\gamma \in \Gamma_D$, will be said *associated* to Σ.

Now that we have introduced the terminology we shall need, we can begin proving theorems. First of all, we study the situation for a point component.

Theorem 1.8.7. *Let $\Sigma = \{a\}$ be a point component of the boundary of a multiply connected hyperbolic domain $D \subset \widehat{X}$ of regular type. Denote by $\pi_D\colon \mathbb{D} \to D$ the universal covering map of D and let $\Gamma_D \subset \operatorname{Aut}(\mathbb{D})$ be the automorphism group of π_D. Then:*
(i) *γ_Σ is parabolic;*
(ii) *C_Σ is empty;*
(iii) *if $\tau \in \partial\mathbb{D}$ is associated to Σ, then $\pi_D(z)$ tends to a as $z \to \tau$ inside any angle centered about the radius ending at τ.*

Proof. We retain the notation introduced so far. Assume, by contradiction, that C_Σ is not empty, and let $\{z_\nu\} \subset \mathbb{D}$ be any sequence converging to a point of C_Σ. Then $z_\nu \in \Delta_1$ eventually, and so every limit point of $\{\pi_D(z_\nu)\}$ must belong to $\overline{D_1} \cap \partial D = \Sigma$; therefore, $\pi(z_\nu) \to a$ as $\nu \to \infty$. In other words, we have shown that $\pi_D(z)$ tends to a as z tends to C_Σ; the Fatou uniqueness theorem (see Corollary A.2.2) then implies that π_D is constant, impossible.

Thus C_Σ is empty; hence γ_Σ is parabolic and we have proved (i) and (ii). To prove (iii), we can assume $\tau = \tau_\Sigma$ because the statement is invariant under the action of Γ_D.

If we take any Euclidean disk Δ' internally tangent to $\partial\mathbb{D}$ in τ_Σ and containing (or not containing) the image of $\tilde{\sigma}$, then Δ' still contains (or does not contain) the image of $\overline{\sigma}$, by Lemma 1.4.17. In particular, we can find such a Δ' contained in Δ_1. Now let $\{z_\nu\}$ be a sequence converging to τ_Σ inside an angle centered about the radius ending at τ_Σ. Then $z_\nu \in \Delta' \subset \Delta_1$ eventually, and again we find that $\pi_D(z_\nu)$ must converge to a. □

Remark 1.8.8. Using a terminology we shall introduce in the next chapter, we have proved that π_D has *nontangential limit a* at τ_Σ; see Proposition 2.2.4.

Mutatis mutandis, an analogous argument works for Jordan components.

Theorem 1.8.9. *Let Σ be a Jordan component of the boundary of a multiply connected hyperbolic domain $D \subset \widehat{X}$ of regular type. Denote by $\pi_D\colon \mathbb{D} \to D$ the universal covering map of D and let $\Gamma_D \subset \operatorname{Aut}(\mathbb{D})$ be the automorphism group of π_D. Then:*
(i) *C_Σ is not empty;*
(ii) *if $C \subset \partial\mathbb{D}$ is an open arc associated to Σ, then π_D extends continuously to C, the image of C through this extension is exactly Σ and Γ_D is properly discontinuous at every point of C.*

Proof. We still retain the notation introduced so far. Fix $1/2 < \varepsilon < 1$, and set

$$D_1^\varepsilon = H((0,1) \times (1-\varepsilon, \varepsilon)).$$

The function $f = \widetilde{H} \circ H^{-1}: D_1^\varepsilon \to \Delta_1$ is holomorphic, being an inverse of π_D on D_1^ε. If C_Σ were empty, and thus γ_Σ parabolic, the same argument used in the proof of Theorem 1.8.7 would show that $f(z)$ would tend to the unique fixed point of γ_Σ as z goes to $\Sigma \cap \overline{D_1^\varepsilon}$, and this is impossible, again by the Fatou uniqueness theorem Corollary A.2.2 (that can be applied to D_1^ε thanks to Theorem 1.6.32).

To prove (ii), we can assume $C = C_\Sigma$. Fix a point $\tau \in C_\Sigma$; then we can choose the homotopy $\widetilde{H}$ and $\varepsilon > 0$ in such a way that τ is in the boundary of $f(D_1^\varepsilon)$; see Figure 1.4. Then f is a biholomorphism between D_1^ε and $f(D_1^\varepsilon)$, with inverse π_D, and both D_1^ε and $f(D_1^\varepsilon)$ are simply connected domains bounded by Jordan curves (because H is an isotopy). Therefore, by Theorem 1.6.32, π_D extends continuously to a neighborhood U in $\overline{\mathbb{D}}$ of τ, with $\pi(U \cap \partial\mathbb{D}) = D_1^\varepsilon \cap \Sigma$. Furthermore, π_D is then locally injective at τ; hence Γ_D must be properly discontinuous at τ. Since $\tau \in C_\Sigma$ is generic, it follows that π extends continuously to C_Σ and $\pi_D(C_\Sigma) = \Sigma$. □

It would be nice if the automorphism associated to a Jordan component were hyperbolic. Unfortunately, this is not true, as we have seen in Example 1.8.5; however, this is the only exception.

Proposition 1.8.10. *Let $D \subset \widehat{X}$ be a multiply connected hyperbolic domain of regular type, not biholomorphic to $\mathbb{D}^*$. Then a connected component Σ of ∂D is Jordan if and only if γ_Σ is hyperbolic.*

Proof. If γ_Σ is hyperbolic, C_Σ is not empty, and thus Σ must be Jordan, by Theorem 1.8.7. Conversely, let Σ be a Jordan component, and assume, by contradiction, that γ_Σ is parabolic. Then, since C_Σ must be not empty, $C_\Sigma = \partial\mathbb{D} \setminus \{\tau_\Sigma\}$, where τ_Σ is the unique fixed point of γ_Σ.

Let $\gamma \in \Gamma_D$ be different from the identity. Clearly, Γ_D cannot be properly discontinuous at a fixed point of γ; hence, by Theorem 1.8.9(ii), γ must be parabolic with fixed-point τ_Σ. But then, by Proposition 1.4.12, Γ_D is Abelian and composed only by parabolic elements; therefore, D is biholomorphic to $\mathbb{D}^*$, by Theorem 1.6.33. □

We end this section with a result that we shall need in Section 3.3. The proof depends on the classical *Fatou theorem*.

Theorem 1.8.11 (Fatou, 1906). *Let $f: \mathbb{D} \to \mathbb{C}$ be a bounded holomorphic function. Then the limit*

$$f^*(e^{i\theta}) = \lim_{r \to 1^-} f(re^{i\theta})$$

exists for almost every $e^{i\theta} \in \mathbb{S}^1$.

For a proof see, e. g., [360, Theorem 11.21].

Proposition 1.8.12. *Let $D \subset\subset \mathbb{C}$ be a bounded domain and let $\pi_D: \mathbb{D} \to D$ be the universal covering map. Take $\tau \in \partial D$ such that there exists a simply connected domain $\Omega \subset D$*

with $\tau \in \partial\Omega$ and such that $\partial\Omega$ is a Jordan curve. Assume moreover that $\partial\Omega \cap \partial D$ is a Jordan arc containing τ in its relative interior. Then there exists $\sigma \in \mathbb{D}$ such that π_D extends continuously to $\mathbb{D} \cup \{\sigma\}$ and $\pi_D(\sigma) = \tau$.

Proof. Since Ω is simply connected, we can find $\widetilde{\Omega} \subset \mathbb{D}$ such that $\pi_{\mathbb{D}}|_{\widetilde{\Omega}}$ is a biholomorphism between $\widetilde{\Omega}$ and Ω; let $g\colon \Omega \to \widetilde{\Omega}$ be its inverse. Notice that g extends continuously to $\overline{\Omega} \cap D$.

Since $\partial\Omega$ is a Jordan curve, we know (Theorem 1.6.32) that the Riemann biholomorphism $\pi_\Omega\colon \mathbb{D} \to \Omega$ extends to a homeomorphisms of the closures. In particular, we can apply Theorem 1.8.11 to $g \circ \pi_\Omega$, and thus up to shrinking Ω we can assume that g restricted to the Jordan arc $\gamma = \partial\Omega \cap D$ has a well-defined limit in $\partial\mathbb{D}$ at both extremities of γ. It follows that the boundary of $\widetilde{\Omega}$ is again a Jordan curve, given by the union of a Jordan arc $g(\gamma) \subset \mathbb{D}$ with an arc in $\partial\mathbb{D}$ bounded by the extremities of $g(\gamma)$. But then we can apply again Theorem 1.6.32 to show that g extends to a homeomorphisms of the closures, and taking $\sigma = g(\tau)$ we have the assertion. □

Notes to Section 1.8

This section is mostly adapted from Julia [220] and it gives a short account of the works of Schottky [364, 365], Picard [321], Kœbe [242, 243], de la Vallée-Poussin [400, 401], and Julia [219] himself on the uniformization theory of multiply connected domains.

Koebe [243] has shown that every domain $D \subset \widehat{\mathbb{C}}$ of finite connectivity (i. e., such that $\widehat{\mathbb{C}} \setminus D$ has a finite number of connected components) is biholomorphic to a plane domain of regular type. For a modern proof of this statement, and for conditions ensuring the extension of the biholomorphism to a homeomorphism of the closures, see, e. g., [125, Chapter 15]. More generally, any noncompact Riemann surface X with finitely generated first homology group $H_1(X;\mathbb{Z})$ can be realized as a domain of regular type; see [383].

For further information on these matters, consult [169].

The Fatou Theorem 1.8.11 has been proved by Fatou [145] in 1906. In 1927, Littlewood [269] has shown that if $\gamma_0\colon [0,1) \to \mathbb{D}$ is such that $\gamma(t) \to 1$ as $t \to 1^-$ tangentially to $\partial\mathbb{D}$ then there is a bounded holomorphic function f defined on $\mathbb{D}$ such that f does not admit a limit along $e^{i\theta}\gamma_0$ for almost all $\theta \in [0,2\pi]$. In 1957, Lohwater and Piranias [270] improved Littlewood's result showing the existence of a Blaschke product not admitting a limit along $e^{i\theta}\gamma_0$ for all $\theta \in [0,2\pi]$.

1.9 The Poincaré metric

In Section 1.7, we have shown how to define the Poincaré distance on all hyperbolic Riemann surfaces and we used it to deduce a number of important consequences, including Theorem 1.7.6, a far-reaching generalization of (1.6). In this section, we shall do the same for (1.7), introducing the Poincaré metric, an infinitesimal version of the Poincaré distance. To do so, we need to recall a few standard definitions. *Warning:* this section requires some basic knowledge of differential geometry to be fully understood.

Definition 1.9.1. Let X be a Riemann surface, and let TX denote its (complex) tangent bundle. A *pseudo-Hermitian metric* on X is an upper semicontinuous function $h\colon TX \to \mathbb{R}^+$ such that $h(p;\lambda v) = |\lambda| h(p;v)$ for all $p \in X$, $v \in T_pX$ and $\lambda \in \mathbb{C}$. Clearly, this implies that $h(p; O_p) = 0$ for all $p \in X$, where O_p is the origin of T_pX. If moreover $h(p; v) = 0$ if and only if $v = O_p$ we say that h is a *Hermitian metric*. Finally, if h is of class C^k outside the zero section of TX we say that h is a C^k *(pseudo-)Hermitian metric*.

A local chart (U, z) determines for all $p \in U$ a nonzero tangent vector $\frac{\partial}{\partial z}\big|_p \in T_pX$, that we shall denote for simplicity by $\partial_z|_p$. Every $v \in T_pX$ can then be written as $v = \lambda\partial_z|_p$ for a suitable $\lambda \in \mathbb{C}$; as a consequence, if h is a pseudo-Hermitian metric on X we have $h(p; v) = h(p; \partial_z|_p)|\lambda|$. The function $\rho_h: U \to \mathbb{R}^+$ given by $\rho_h(p) = h(p; \partial_z|_p)$ is the *density* of h in the chart (U, z); we shall write $h = \rho_h|dz|$. Clearly, h is a Hermitian metric if and only if its densities are everywhere positive, and is of class C^k if and only if its densities are of class C^k.

Remark 1.9.2. Let $h: TX \to \mathbb{R}^+$ be a pseudo-Hermitian metric on a Riemann surface X, and let ρ_h and $\tilde{\rho}_h$ the densities with respect to the local charts (U, z) and $(\widetilde{U}, \tilde{z})$. Denoting by $\frac{\partial z}{\partial \tilde{z}}$ the derivative of the change of coordinates, on $U \cap \widetilde{U}$ we have $\partial_{\tilde{z}} = \frac{\partial z}{\partial \tilde{z}}\partial_z$, and hence

$$\tilde{\rho}_h = \left|\frac{\partial z}{\partial \tilde{z}}\right| \rho_h. \tag{1.53}$$

Definition 1.9.3. Let h_1 and h_2 be pseudo-Hermitian metrics on a Riemann surface X. We shall say that h_1 is *dominated* by h_2, and we write $h_1 \leq h_2$, if $h_1(p; v) \leq h_2(p; v)$ for all $p \in X$ and $v \in T_pX$.

If h_2 is a Hermitian metric, we can define the *quotient* $h_1/h_2: X \to \mathbb{R}^+$ by setting

$$\frac{h_1}{h_2}(p) = \frac{\rho_{h_1}(p)}{\rho_{h_2}(p)},$$

where ρ_{h_1} and ρ_{h_2} are the densities of h_1 and h_2 with respect to the same local chart at p; by (1.53), this quotient does not depend on the local chart chosen to compute it. Clearly, $h_1 \leq h_2$ if and only if $\frac{h_1}{h_2} \leq 1$.

The first (and, for us, main) example of Hermitian metric is given by the Poincaré metric.

Definition 1.9.4. The *Poincaré metric* on the unit disk $\mathbb{D}$ is the Hermitian metric $\kappa_{\mathbb{D}}: T\mathbb{D} \to \mathbb{R}^+$ given by

$$\kappa_{\mathbb{D}}(z; v) = \frac{|v|}{1 - |z|^2} \tag{1.54}$$

for all $z \in \mathbb{D}$ and $v \in \mathbb{C}$; we are using the canonical identification of $T\mathbb{D}$ with $\mathbb{D} \times \mathbb{C}$. The density $\rho_{\mathbb{D}}$ of the Poincaré metric is the *Poincaré density* and it is given by

$$\rho_{\mathbb{D}}(z) = \frac{1}{1 - |z|^2}.$$

To correctly express the Schwarz–Pick lemma using the Poincaré metric we need one more definition.

Definition 1.9.5. Let $f: X \to Y$ be a holomorphic function between two Riemann surfaces, and let $h: TY \to \mathbb{R}^+$ be a pseudo-Hermitian metric on Y. The *pullback metric* is the pseudo-Hermitian metric f^*h on X defined by

$$(f^*h)(p; v) = h(f(p); df_p(v))$$

for all $p \in X$ and $v \in T_pX$. Clearly, if h is Hermitian then f^*h is Hermitian, and if h is of class C^k then f^*h is of class C^k.

Remark 1.9.6. Let $f: X \to Y$ be a holomorphic function between two Riemann surfaces, and let $h: TY \to \mathbb{R}^+$ be a pseudo-Hermitian metric on Y. Given $p \in X$, choose a local chart (U, z) of X at p and a local chart (V, w) of Y at $f(p)$ such that $f(U) \subseteq V$; denote by ρ_h the density of h with respect to (V, w) and by $f^*\rho_h$ the density of f^*h with respect to (U, z). Then it is easy to see that

$$f^*\rho_h = |f'|(\rho_h \circ f), \tag{1.55}$$

where f' represents the differential of f in the local coordinates, i. e., $df_p(\partial_z|_p) = f'(p)\partial_w|_{f(p)}$ for all $p \in U$. In particular, if h is Hermitian then f^*h is Hermitian except in the critical points of f (where df, and hence f', vanishes); analogously, if h is of class C^k then f^*h is of class C^k except possibly at the critical points of f.

Then we can express (1.7) in the following way.

Theorem 1.9.7 (Schwarz–Pick lemma, 1915). *Let $f: \mathbb{D} \to \mathbb{D}$ be holomorphic. Then*

$$f^*\kappa_{\mathbb{D}} \leq \kappa_{\mathbb{D}}. \tag{1.56}$$

Moreover, equality in (1.56) at one point (i. e., $f^*\kappa_{\mathbb{D}}(z_0; v) = \kappa_{\mathbb{D}}(z_0; v)$ for some $z_0 \in \mathbb{D}$ and $v \neq 0$) occurs if and only if $f \in \mathrm{Aut}(\mathbb{D})$ if and only if $f^*\kappa_{\mathbb{D}} \equiv \kappa_{\mathbb{D}}$.

Proof. It follows immediately from Corollary 1.1.16. □

Definition 1.9.8. Let X and Y be Riemann surfaces endowed with the pseudo-Hermitian metrics h_X and h_Y, respectively. A holomorphic function $f: X_1 \to X_2$ is a (holomorphic) *local isometry* if $f^*h_Y \equiv h_X$; it is a (holomorphic) *isometry* if it also is a biholomorphism. When an automorphism γ of a Riemann surface X is an isometry of a pseudo-Hermitian metric h on X, we shall say that h is *invariant* under γ.

Remark 1.9.9. Using the notation introduced in Remark 1.9.6, it is easy to see that $f: X \to Y$ is a local isometry for h_X and h_Y if and only if

$$|f'|(\rho_{h_Y} \circ f) = \rho_{h_X}.$$

An immediate consequence of Theorem 1.9.7 is that the automorphisms of $\mathbb{D}$ are isometries for the Poincaré metric. Actually, this condition characterizes the Poincaré metric up to a nonnegative multiple.

Proposition 1.9.10. *Let h be a pseudo-Hermitian metric on $\mathbb{D}$ such that $\gamma^* h \equiv h$ for all $\gamma \in \mathrm{Aut}(\mathbb{D})$. Then h is a nonnegative multiple of the Poincaré metric.*

Proof. Put $c = \rho_h(0) \geq 0$, where ρ_h is the density of h. Given $z \in \mathbb{D}$, let $\gamma \in \mathrm{Aut}(\mathbb{D})$ be such that $\gamma(0) = z$. Since both h and $\kappa_{\mathbb{D}}$ are invariant under γ, we have

$$\rho_h(z)|\gamma'(0)| = \rho_h(0) = c\rho_{\mathbb{D}}(0) = c\rho_{\mathbb{D}}(z)|\gamma'(0)|$$

because $\rho_{\mathbb{D}}(0) = 1$, and hence $\rho_h \equiv c\rho_D$, as claimed. □

The Poincaré distance and the Poincaré metric share more than just the name; having one of the two, it is possible to recover the other one.

Definition 1.9.11. Let $h\colon TX \to \mathbb{R}^+$ be a pseudo-Hermitian metric on a Riemann surface X. Given a piecewise C^1 curve $\sigma\colon [a, b] \to X$, the *length* $\ell_h(\sigma)$ of σ with respect to h is given by

$$\ell_h(\sigma) = \int_\sigma h = \int_a^b h(\sigma(t); \sigma'(t))\, dt.$$

We shall denote by $\ell_{\mathbb{D}}$ the length associated to the Poincaré metric on $\mathbb{D}$.

The *pseudo-distance* $\omega_h\colon X \times X \to \mathbb{R}^+$ *induced* by a pseudo-Hermitian metric h is defined by

$$\omega_h(z, w) = \inf_\sigma \{\ell_h(\sigma)\},$$

for all z, $w \in X$, where the infimum is with respect to all the piecewise C^1 curves $\sigma\colon [a, b] \to X$ with $\sigma(a) = z$ and $\sigma(b) = w$.

Remark 1.9.12. It is easy to check that ω_h is indeed a pseudo-distance. The symmetry and the fact that $\omega_h(z, w) \geq 0$ are obvious. To prove the triangular inequality, it suffices to notice that if $\sigma_1\colon [a_1, b_1] \to X$ is a piecewise C^1 curve from z_1 to z_2 and $\sigma_2\colon [a_2, b_2] \to X$ is a piecewise C^1 curve from z_2 to z_3 then $\sigma_1 \star \sigma_2\colon [a_1, b_1 + b_2 - a_2] \to X$ given by

$$\sigma_1 \star \sigma_2(t) = \begin{cases} \sigma_1(t) & \text{if } t \in [a_1, b_1], \\ \sigma_2(t + a_2 - b_1) & \text{if } t \in [b_1, b_1 + b_2 - a_2] \end{cases}$$

is a piecewise C^1 curve from z_1 to z_3 such that $\ell_h(\sigma_1 \star \sigma_2) = \ell_h(\sigma_1) + \ell_h(\sigma_2)$. It follows that $\omega_h(z_1, z_3) \leq \ell_h(\sigma_1) + \ell_h(\sigma_2)$ and taking the infimum with respect to σ_1 and σ_2 we obtain the desired triangular inequality. It is a basic fact of differential geometry (that we shall not need) that if h is a Hermitian metric then ω_h is an actual distance, that is $\omega_h(z,w) > 0$ when $z \neq w$.

Remark 1.9.13. If $h\colon [a',b'] \to [a,b]$ is a C^1 diffeomorphism and $\sigma\colon [a,b] \to X$ is a piecewise C^1 curve, it is easy to check that $\ell_h(\sigma \circ h) = \ell_h(\sigma)$. The curve $\sigma \circ h$ is usually called a *reparametrization* of the curve σ.

Then we have the following.

Proposition 1.9.14. (i) *The Poincaré distance is the distance induced by the Poincaré metric. In particular, if $\sigma\colon \mathbb{R} \to \mathbb{D}$ is a geodesic for the Poincaré distance, then for all $t_1 \leq t_2$ we have*

$$\ell_{\mathbb{D}}(\sigma|_{[t_1,t_2]}) = \omega(\sigma(t_1), \sigma(t_2)) = |t_1 - t_2|.$$

(ii) *If $\sigma\colon (-\varepsilon,\varepsilon) \to \mathbb{D}$ is a C^1 curve, then*

$$\lim_{t\to 0} \frac{\omega(\sigma(0),\sigma(t))}{|t|} = \kappa_{\mathbb{D}}(\sigma(0);\sigma'(0)).$$

Proof. (i) Since the automorphisms of $\mathbb{D}$ are isometries for both the Poincaré metric and the Poincaré distance it suffices to compute the induced distance from the origin to $x_0 \in (0,1)$ and to show that it is equal to the length of the unique geodesic connecting 0 to x_0.

Let $\sigma\colon [a,b] \to \mathbb{D}$ be a piecewise C^1 curve with $\sigma(a) = 0$ and $\sigma(b) = x_0$. Its length $\ell_{\mathbb{D}}(\sigma)$ satisfies

$$\ell_{\mathbb{D}}(\sigma) = \int_a^b \frac{|\sigma'(t)|}{1 - |\sigma(t)|^2}\,dt \geq \int_a^b \frac{\operatorname{Re}\sigma'(t)}{1 - (\operatorname{Re}\sigma(t))^2}\,dt = \frac{1}{2}\log\frac{1+x_0}{1-x_0} = \omega(0,x_0). \tag{1.57}$$

Considering the geodesic $\sigma_0\colon [0,\tanh^{-1}(x_0)] \to \mathbb{D}$ given by $\sigma_0(t) = \tanh(t)$, we obtain equality in (1.57) and the assertion follows.

(ii) Again up to an automorphism, we can assume $\sigma(0) = 0$. Then we have

$$\lim_{t\to 0} \frac{\omega(\sigma(0),\sigma(t))}{|t|} = \lim_{t\to 0}\left[\frac{\omega(0,\sigma(t))}{|\sigma(t)|}\left|\frac{\sigma(t)}{t}\right|\right] = |\sigma'(0)| = \kappa_{\mathbb{D}}(0;\sigma'(0)),$$

because the derivative at 0 of $\tanh^{-1}$ is 1, and we are done. □

Our next aim is to define the Poincaré metric on any hyperbolic Riemann surface. To do so, we shall push forward the Poincaré metric of the disk.

Definition 1.9.15. Let $f\colon X \to Y$ be a holomorphic function between two Riemann surfaces, and let $h\colon TX \to \mathbb{R}^+$ be a pseudo-Hermitian metric on X. Assume that f is surjective and a local biholomorphism (a typical example is a covering map); assume moreover that $f(p_1) = f(p_2)$ and $df_{p_1}(v_1) = df_{p_2}(v_2)$ implies $h(p_1;v_1) = h(p_2;v_2)$. Then the *push-forward metric* is the pseudo-Hermitian metric f_*h on Y defined by $f_*h(q;w) = h(p;v)$, where $p \in X$ and $v \in T_pX$ are such that $f(p) = q$ and $df_p(v) = w$.

Remark 1.9.16. Let $f: X \to Y$ be a holomorphic function between two Riemann surfaces, and let $h: TX \to \mathbb{R}^+$ be a pseudo-Hermitian metric on X. Assume that f and h satisfy the hypotheses needed to define the push-forward metric f_*h. Given $q \in Y$, choose $p \in X$ such that $f(p) = q$, a local chart (U, z) of X at p and a local chart (V, w) of Y at $f(p)$ such that $f(U) \subseteq V$; denote by ρ_h the density of h with respect to (U, z) and by $f_*\rho_h$ the density of f_*h with respect to (V, w). Then it is easy to see that

$$(f_*\rho_h) \circ f = \frac{1}{|f'|}\rho_h, \tag{1.58}$$

where f' does not vanish because f is a local biholomorphism. In particular, if h is Hermitian then f_*h is Hermitian and if h is of class C^k then f_*h is of class C^k.

Remark 1.9.17. By definition, we have

$$f^*(f_*h) = h \quad \text{and} \quad f_*(f^*h) = h$$

as soon as the left side of the equation is defined. This is also apparent comparing (1.55) with (1.58).

Analogously, it is easy to check that if $f: X \to Y$ and $g: Y \to Z$ are holomorphic then $(g \circ f)^*h = f^*(g^*h)$ for any pseudo-Hermitian metric h on Z.

Let X be a hyperbolic Riemann surface and $\pi_X: \mathbb{D} \to X$ a holomorphic universal covering map. Assume that $z_1, z_2 \in \mathbb{D}$ and $v_1, v_2 \in \mathbb{C}$ are such that $d(\pi_X)_{z_1}(v_1) = d(\pi_X)_{z_2}(v_2)$. In particular, this means that $\pi_X(z_1) = \pi_X(z_2)$, and thus (see Proposition 1.6.9) there exists an automorphism γ of the covering such that $z_2 = \gamma(z_1)$. Let $\hat{v}_1 \in \mathbb{C}$ be the unique element such that $d\gamma_{z_1}(\hat{v}_1) = v_2$; then

$$\begin{aligned} d(\pi_X)_{z_1}(\hat{v}_1) &= d(\pi_X \circ \gamma)_{z_1}(\hat{v}_1) = d(\pi_X)_{\gamma(z_1)}(d\gamma_{z_1}(\hat{v}_1)) = d(\pi_X)_{z_2}(v_2) \\ &= d(\pi_X)_{z_1}(v_1), \end{aligned}$$

and thus $\hat{v}_1 = v_1$ because π_X is a local biholomorphism. Now, the automorphisms of the covering are automorphisms of $\mathbb{D}$, and hence isometries of the Poincaré metric (by Theorem 1.9.7); in particular,

$$\kappa_{\mathbb{D}}(z_2; v_2) = \kappa_{\mathbb{D}}(\gamma(z_1); d\gamma_{z_1}(v_1)) = \kappa_{\mathbb{D}}(z_1; v_1).$$

In other words, the conditions required in Definition 1.9.15 are fulfilled and the push-forward metric $(\pi_X)_*\kappa_{\mathbb{D}}$ is defined.

Definition 1.9.18. Let X be a hyperbolic Riemann surface. The *Poincaré metric* κ_X on X is the push-forward metric $\kappa_X = (\pi_X)_*\kappa_{\mathbb{D}}$, where $\pi_X\colon \mathbb{D} \to X$ is a holomorphic universal covering map. Clearly, κ_X is a C^∞-Hermitian metric. We shall write ℓ_X instead of ℓ_{κ_X}.

Remark 1.9.19. If $\pi_X, \tilde{\pi}_X\colon \mathbb{D} \to X$ are two holomorphic universal covering maps of the hyperbolic Riemann surface X, then there is $\gamma \in \operatorname{Aut}(\mathbb{D})$ such that $\tilde{\pi}_X = \pi_X \circ \gamma$ (see Lemma 1.6.6). The fact that automorphisms of $\mathbb{D}$ are isometries of the Poincaré metric immediately implies that $(\tilde{\pi}_X)_*\kappa_{\mathbb{D}} = (\pi_X)_*\kappa_{\mathbb{D}}$, and thus the Poincaré metric of X does not depend on the chosen universal covering map.

Example 1.9.20 (The Poincaré metric of the upper half-plane). Let $\Psi\colon \mathbb{D} \to \mathbb{H}^+$ be the Cayley transform. Given $w \in \mathbb{H}^+$ and $v \in \mathbb{C}$ the Poincaré metric of $\mathbb{H}^+$ is

$$\kappa_{\mathbb{H}^+}(w; v) = \kappa_{\mathbb{D}}(\Psi^{-1}(w); (\Psi^{-1})'(w)v) = \frac{|v|}{2\operatorname{Im} w}.$$

Example 1.9.21 (The Poincaré metric of the pointed disk). Take $z \in \mathbb{D}^*$ and $v \in \mathbb{C}$. By definition, we have $\kappa_{\mathbb{D}^*}(z; v) = \kappa_{\mathbb{H}^+}(\tilde{z}; \tilde{v})$, where $\tilde{z} \in \mathbb{H}^+$ and $\tilde{v} \in \mathbb{C}$ are such that $\pi_{\mathbb{D}^*}(\tilde{z}) = z$ and $d(\pi_{\mathbb{D}^*})_{\tilde{z}}(\tilde{v}) = v$; here, $\pi_{\mathbb{D}^*}\colon \mathbb{H}^+ \to \mathbb{D}^*$ is the universal covering map given by (1.48). A quick computation then yields

$$\kappa_{\mathbb{D}^*}(z; v) = \frac{|v|}{2|z|\,|\log|z||}.$$

Example 1.9.22 (The Poincaré metric of an annulus). Given $0 < r < 1$, take $z \in A(r, 1)$ and $v \in \mathbb{C}$. By definition, we have $\kappa_{A(r,1)}(z; v) = \kappa_{\mathbb{H}^+}(\tilde{z}; \tilde{v})$, where $\tilde{z} \in \mathbb{H}^+$ and $\tilde{v} \in \mathbb{C}$ are such that $\pi_{A(r,1)}(\tilde{z}) = z$ and $d(\pi_{A(r,1)})_{\tilde{z}}(\tilde{v}) = v$; here, $\pi_{A(r,1)}\colon \mathbb{H}^+ \to A(r, 1)$ is the universal covering map given by (1.49). A computation then yields

$$\kappa_{A(r,1)}(z; v) = \frac{\pi}{|\log r|} \frac{|v|}{2|z| \sin\left(\pi \frac{\log|z|}{\log r}\right)}.$$

The Poincaré metrics of hyperbolic Riemann surfaces enjoy a *Schwarz–Pick lemma* completely analogous to Theorem 1.7.6 and generalizing Theorem 1.9.7.

Theorem 1.9.23. *Let X and Y be two hyperbolic Riemann surfaces, and $f\colon X \to Y$ a holomorphic function. Then $f^*\kappa_Y \le \kappa_X$, i. e.,*

$$\kappa_Y(f(z); df_z(v)) \le \kappa_X(z; v)$$

for all $z \in X$ and $v \in T_zX$. Furthermore, equality for some $z \in X$ and $v \ne O_z$ occurs if and only if f is a covering map if and only if $f^*\kappa_Y \equiv \kappa_X$.

Proof. Lift f to a holomorphic function $\tilde{f}\colon \mathbb{D} \to \mathbb{D}$ such that $\pi_Y \circ \tilde{f} = f \circ \pi_X$. Then

$$\begin{aligned}(\pi_X^* \circ f^*)\kappa_Y = (\tilde{f}^* \circ \pi_Y^*)\kappa_Y &= \tilde{f}^*(\pi_Y^*(\pi_Y)_*\kappa_{\mathbb{D}}) \\ &= \tilde{f}^*\kappa_{\mathbb{D}} \le \kappa_{\mathbb{D}} = \pi_X^*(\pi_X)_*\kappa_{\mathbb{D}} = \pi_X^*\kappa_X,\end{aligned}$$

where we repeatedly used Remark 1.9.17. Hence applying $(\pi_X)_*$ to both sides, we get $f^*\kappa_Y \le \kappa_X$, as claimed.

Assume now that $\kappa_Y(f(z); df_z(v)) = \kappa_X(z; v)$ for some $z \in X$ and some nonzero $v \in T_zX$. Choose $\tilde{z} \in \mathbb{D}$ and $\tilde{v} \in \mathbb{C}$ such that $\pi_X(\tilde{z}) = z$ and $d(\pi_X)_{\tilde{z}}(\tilde{v}) = v$. Then

$$\begin{aligned}\kappa_{\mathbb{D}}(\tilde{f}(\tilde{z}); d(\tilde{f})_{\tilde{z}}(\tilde{v})) &= \kappa_Y(\pi_Y \circ \tilde{f}(\tilde{z}); d(\pi_Y)_{\tilde{f}(z)}(d\tilde{f}_{\tilde{z}}(\tilde{v})))\\ &= \kappa_Y(f \circ \pi_X(\tilde{z}); d(f \circ \pi_X)_{\tilde{z}}(\tilde{v}))\\ &= \kappa_Y(f(z); df_z(v)) = \kappa_X(z; v) = \kappa_{\mathbb{D}}(\tilde{z}; \tilde{v}).\end{aligned}$$

Theorem 1.9.7 implies that $\tilde{f} \in \mathrm{Aut}(\mathbb{D})$, and from $\pi_Y \circ \tilde{f} = f \circ \pi_X$ we infer that f is a covering map, because $\pi_Y \circ \tilde{f}$ and π_X are.

Conversely, assume that f is a covering map. Then $\tilde{f}$ is a covering map, and thus an automorphism of $\mathbb{D}$. Choose $\tilde{z} \in \mathbb{D}$ and $\tilde{v} \in \mathbb{C}$ such that $\pi_X(\tilde{z}) = z$ and $d(\pi_X)_{\tilde{z}}(\tilde{v}) = v$. Then

$$\begin{aligned}\kappa_Y(f(z); df_z(v)) &= \kappa_Y(f(\pi_X(\tilde{z})); df_{\pi_X(\tilde{z})}(d(\pi_X)_{\tilde{z}}(\tilde{v})))\\ &= \kappa_Y(\pi_Y(\tilde{f}(\tilde{z})); d(\pi_Y)_{\tilde{f}(\tilde{z})}(d\tilde{f}_{\tilde{z}}(\tilde{v})))\\ &= \kappa_{\mathbb{D}}(\tilde{f}(\tilde{z}); d\tilde{f}_{\tilde{z}}(\tilde{v})) = \kappa_{\mathbb{D}}(\tilde{z}; \tilde{v}) = \kappa_X(z; v)\end{aligned}$$

and we are done. □

Corollary 1.9.24. *Let $\Omega \subset X$ be a proper subdomain of a hyperbolic Riemann surface X. Then $\kappa_X(z; v) < \kappa_\Omega(z; v)$ for all $z \in \Omega$ and $v \in T_zX$.*

Proof. Let $\iota: \Omega \to X$ be the inclusion. Theorem 1.9.23 applied to ι shows that $\kappa_X(z; v) \le \kappa_\Omega(z; v)$, with equality at one point if and only if ι is a covering map; since $\Omega \ne X$ the inclusion cannot be a covering map, and the assertion follows. □

In $\mathbb{D}$, we have seen that we can recover the Poincaré distance as the distance induced by the Poincaré metric. This holds in general.

Proposition 1.9.25. *Let X be a hyperbolic Riemann surface. Then the Poincaré distance of X is the distance induced by the Poincaré metric.*

Proof. Let $\sigma: [a, b] \to X$ be a piecewise C^1 curve and let $\tilde{\sigma}: [a, b] \to \mathbb{D}$ be a lifting of σ with respect to a universal covering map $\pi_X: \mathbb{D} \to X$. Then $\sigma = \pi_X \circ \tilde{\sigma}$ and

$$\begin{aligned}\ell_X(\sigma) = \int_a^b \kappa_X(\sigma(t); \sigma'(t))\,dt &= \int_a^b \kappa_X(\pi_X(\tilde{\sigma}(t)); d(\pi_X)_{\tilde{\sigma}(t)}(\tilde{\sigma}'(t)))\,dt\\ &= \int_a^b \kappa_{\mathbb{D}}(\tilde{\sigma}(t); \tilde{\sigma}'(t))\,dt = \ell_{\mathbb{D}}(\tilde{\sigma}).\end{aligned} \tag{1.59}$$

Fix $\tilde{z} \in \mathbb{D}$ such that $\pi_X(\tilde{z}) = z$. If $\sigma\colon [a,b] \to X$ is a piecewise C^1 curve connecting z to w, denote by $\tilde{\sigma}\colon [a,b] \to \mathbb{D}$ the unique lifting with $\tilde{\sigma}(a) = \tilde{z}$. Then

$$\omega_X(z,w) = \inf_{\tilde{w}\in\pi_X^{-1}(w)} \{\omega(\tilde{z},\tilde{w})\} \le \omega(\tilde{z},\tilde{\sigma}(b)) \le \ell_{\mathbb{D}}(\tilde{\sigma}) = \ell_X(\sigma),$$

where we used (1.51), Proposition 1.9.14(i) and (1.59). Taking the infimum over all admissible σ, we get

$$\omega_X(z,w) \le \inf_\sigma \{\ell_X(\sigma)\}.$$

Conversely, by Proposition 1.7.3(iv) we can find $\tilde{w} \in \mathbb{D}$ such that $\pi_X(\tilde{w}) = w$ and $\omega_X(z,w) = \omega(\tilde{z},\tilde{w})$. Let $\tilde{\sigma}_0\colon [0,\ell] \to \mathbb{D}$ be a geodesic from $\tilde{z}$ to $\tilde{w}$, so that $\ell_{\mathbb{D}}(\tilde{\sigma}_0) = \omega_{\mathbb{D}}(\tilde{z},\tilde{w})$. Then

$$\ell_X(\pi_X \circ \tilde{\sigma}_0) = \ell_{\mathbb{D}}(\tilde{\sigma}_0) = \omega_{\mathbb{D}}(\tilde{z},\tilde{w}) = \omega_X(z,w).$$

It follows that $\inf_\sigma\{\ell_X(\sigma)\} \le \omega_X(z,w)$ and we are done. □

Actually, the previous proof suggests that a curve realizing the Poincaré distance always exists.

Definition 1.9.26. Let X be a hyperbolic Riemann surface. A *minimal geodesic* connecting $z_0 \in X$ to $z_1 \in X$ is a C^∞ regular curve $\sigma\colon [a,b] \to X$ such that $\sigma(a) = z_0$, $\sigma(b) = z_1$ and $\ell_X(\sigma) = \omega_X(z_0,z_1)$. Moreover, if $z_0 \neq z_1$ we also require that $\sigma'(t) \neq O$ for every $t \in [0,1]$.

Remark 1.9.27. If $\sigma\colon [a_0,b_0] \to X$ is a minimal geodesic, then $\sigma|_{[a,b]}$ is a minimal geodesic for all $a_0 \le a \le b \le b_0$. Indeed, assume by contradiction that there is a curve σ_1 connecting $\sigma(a)$ and $\sigma(b)$ with $\ell_X(\sigma_1) < \ell_X(\sigma|_{[a,b]})$. Then we have

$$\begin{aligned}\omega_X(\sigma(a_0),\sigma(b_0)) &= \ell_X(\sigma) = \ell_X(\sigma|_{[a_0,a]}) + \ell_X(\sigma|_{[a,b]}) + \ell_X(\sigma|_{[b,b_0]})\\ &> \ell_X(\sigma|_{[a_0,a]}) + \ell_X(\sigma_1) + \ell_X(\sigma|_{[b,b_0]}) \ge \omega_X(\sigma(a_0),\sigma(b_0)),\end{aligned}$$

contradiction.

Remark 1.9.28. Remark 1.9.13 implies that a reparametrization of a minimal geodesic is still a minimal geodesic.

Then we have the following.

Proposition 1.9.29. *Let X be a hyperbolic Riemann surface. Then every pair of points $z_0, z_1 \in X$ can be connected by a minimal geodesic.*

Proof. If $X = \mathbb{D}$, then Proposition 1.9.14(i) shows that every geodesic for the Poincaré distance is also a minimal geodesic; thus in this case the assertion follows from Proposition 1.2.7(iv).

Let X now be any hyperbolic Riemann surface, and choose $\tilde{z}_0, \tilde{z}_1 \in \mathbb{D}$ such that $\pi_X(\tilde{z}_j) = z_j$ for $j = 0, 1$ and $\omega_X(z_0, z_1) = \omega(\tilde{z}_0, \tilde{z}_1)$, where $\pi_X\colon \mathbb{D} \to X$ is a universal covering map; this can be done thanks to Proposition 1.7.3(iv). Let $\tilde{\sigma}\colon [0,1] \to \mathbb{D}$ be a minimal geodesic connecting $\tilde{z}_0$ and $\tilde{z}_1$ and set $\sigma = \pi_X \circ \tilde{\sigma}$. Then

$$\omega_X(z_0, z_1) = \omega(\tilde{z}_0, \tilde{z}_1) = \ell_{\mathbb{D}}(\tilde{\sigma}) = \ell_X(\sigma),$$

by (1.59). In particular, σ is a minimal geodesic connecting z_0 to z_1, and we are done. □

We ought to clarify the relations between the notion of minimal geodesic and the notion of geodesic introduced in Definition 1.7.1.

Proposition 1.9.30. *Let X be a hyperbolic Riemann surface. Then:*
(i) *every not constant minimal geodesic is a reparametrization of a geodesic;*
(ii) *if $X = \mathbb{D}$, then every geodesic is also a minimal geodesic;*
(iii) *if $X \neq \mathbb{D}$, then there exist geodesics in X that are not a minimal geodesic.*

Proof. (i) Let $\sigma\colon [a,b] \to X$ be a (not constant) minimal geodesic in X. Remark 1.9.27 implies that

$$\ell_X(\sigma|_{[s,t]}) = \omega_X(\sigma(s), \sigma(t)) \tag{1.60}$$

for all $a \le s \le t \le b$.

Define now $h\colon [a,b] \to [0,\ell]$ by setting $h(t) = \ell_X(\sigma|_{[a,t]})$, where $\ell = \ell_X(\sigma)$. Since $\sigma'(t) \neq O$ for every $t \in [a,b]$, it is easy to check that h is a C^∞ diffeomorphism between $[a,b]$ and $[0,\ell]$; we claim that $\hat{\sigma} = \sigma \circ h^{-1}\colon [0,\ell] \to X$ is a geodesic. But indeed using (1.60) we get

$$\begin{aligned}\omega_X(\hat{\sigma}(s), \hat{\sigma}(t)) &= \omega_X(\sigma(h^{-1}(s)), \sigma(h^{-1}(t))) = \ell_X(\sigma_{[h^{-1}(s),h^{-1}(t)]})\\ &= \ell_X(\sigma_{[0,h^{-1}(t)]}) - \ell_X(\sigma_{[0,h^{-1}(s)]})\\ &= h(h^{-1}(t)) - h(h^{-1}(s)) = t - s\end{aligned}$$

for all $0 \le s \le t \le \ell$, as required.

(ii) It follows from Proposition 1.9.14(i).

(iii) Take $z_0, z_1 \in X$, and choose $\tilde{z}_0, \tilde{z}_1 \in \mathbb{D}$ with $\pi_X(\tilde{z}_j) = z_j$ for $j = 0, 1$, and $\omega(\tilde{z}_0, \tilde{z}_1) > \omega_X(z_0, z_1)$, where $\pi_X\colon \mathbb{D} \to X$ is a universal covering map; this can be done because the fiber $\pi_X^{-1}(z_1)$ is infinite and discrete in $\mathbb{D}$. Let $\tilde{\sigma}\colon [0,1] \to \mathbb{D}$ be a minimal geodesic connecting $\tilde{z}_0$ to $\tilde{z}_1$. Then $\sigma = \pi_X \circ \tilde{\sigma}$ is, by Proposition 1.7.3(vi), a geodesic in X; but

$$\ell_X(\sigma) = \ell_{\mathbb{D}}(\tilde{\sigma}) = \omega(\tilde{z}_0, \tilde{z}_1) > \omega_X(z_0, z_1),$$

and hence σ is not a minimal geodesic. □

We end this section with a technical result we shall need in Section 4.6.

Proposition 1.9.31. *Let $\{D_\nu\}$ be a sequence of domains in $\mathbb{C}$ such that $D_\nu \subseteq D_{\nu+1}$ for all $\nu \in \mathbb{N}$. Put $D_\infty = \bigcup_\nu D_\nu$.*
(i) *Assume that $D_\infty = \mathbb{D}$. Then $\kappa_{D_\nu} \to \kappa_{\mathbb{D}}$ uniformly on compact subsets of $\mathbb{D} \times \mathbb{C}$ and $\omega_{D_\nu} \to \omega$ uniformly on compact subsets of $\mathbb{D} \times \mathbb{D}$.*
(ii) *Assume that $D_\infty = \mathbb{C}$. Then $\kappa_{D_\nu} \to 0$ and $\omega_{D_\nu} \to 0$ uniformly on compact subsets of $\mathbb{C} \times \mathbb{C}$.*

Proof. Since $D_\nu \subseteq D_{\nu+1}$, we have $\kappa_{D_\nu} \geq \kappa_{D_{\nu+1}}$ for all $\nu \geq 0$; in particular, the sequence $\kappa_{D_\nu}(z; v)$ is converging for all $z \in D_\infty$ and $v \in \mathbb{C}$. Analogously, since $D_\nu \subseteq D_\infty$, we have $\kappa_{\mathbb{D}} \leq \kappa_{D_\nu}$ in case (i) and $0 \leq \kappa_{D_\nu}$ in case (ii).

Now fix $z_0 \in D_\infty$, and choose $r > 0$ such that $z_0 \in \overline{D(0,r)} \subset D_\infty$. Since the closed Euclidean disk $\overline{D(0,r)}$ is compact and contained in D_∞, we have $\overline{D(0,r)} \subset D_\nu$ for ν large enough. As a consequence, we have

$$\lim_{\nu \to +\infty} \kappa_{D_\nu}(z_0; v) \leq \kappa_{D(0,r)}(z_0; v) = \frac{r|v|}{r^2 - |z_0|^2}.$$

If $D_\infty = \mathbb{D}$, we can let r go to 1 obtaining $\lim_{\nu\to+\infty} \kappa_{D_\nu}(z_0; v) \leq \kappa_{\mathbb{D}}(z_0; v)$, and hence the equality. Analogously, if $D_\infty = \mathbb{C}$ we can let r go to $+\infty$, and we get $\kappa_{D_\nu}(z_0; v) \to 0$. All the limits are clearly uniform on compact subsets.

Now take $z_0, w \in D_\infty$. Up to an automorphism of D_∞, we can assume that $z_0 = 0$. Choose again $r > 0$ such that $w \in \overline{D(0,r)} \subset D_\infty$. Then, for ν large enough, setting $\omega_{\mathbb{C}} \equiv 0$ we have

$$\begin{aligned}\omega_{D_\infty}(0, w) &\leq \omega_{D_\nu}(0, w) \leq \int_0^1 \kappa_{D_\nu}(tw; w)\, dt \\ &\leq \int_0^1 \kappa_{D(0,r)}(tw; w)\, dt = r|w| \int_0^1 \frac{1}{r^2 - t^2|w|^2}\, dt = \frac{1}{2} \log \frac{1 + |w|/r}{1 - |w|/r}.\end{aligned}$$

Taking the limit as $\nu \to +\infty$ and then letting r go to 1 if $D_\infty = \mathbb{D}$ and to $+\infty$ if $D_\infty = \mathbb{C}$ we get the assertion. □

Notes to Section 1.9

The Poincaré metric was first investigated by Riemann [350] in 1854, as an example (in modern terminology) of metric with constant negative Gaussian curvature. The first one to use the Poincaré metric to study non-Euclidean geometry and to devise the disk model of the Lobačevski hyperbolic plane was Beltrami in 1868 [55, 56]. Only in 1882 Poincaré [331] began to deal with the metric that now bears his name, both in $\mathbb{D}$ and in $\mathbb{H}^+$, using it for his work on Fuchsian groups.

It is possible to prove that the group of all isometries of class C^1 for the Poincaré metric consists of all holomorphic and antiholomorphic automorphisms of $\mathbb{D}$; see, e. g., [210, Proposition 1.1.20].

Proposition 1.9.31 is a very particular case of a much more general result, due to Hejhal [192], essentially saying that the universal covering map of a hyperbolic domain in $\mathbb{C}$, suitably normalized, depends continuously on the domain with respect to the so-called Carathéodory kernel convergence; the original theorem by Carathéodory [93] assumed that all the domains were simply connected. Since the Poincaré metric of a hyperbolic domain is obtained as the push-forward of the Poincaré metric of $\mathbb{D}$, and the Poincaré distance is the integrated form of the Poincaré metric, from Hejhal's result it follows the continuous dependence on the domain of the Poincaré distance and metric.

1.10 The Ahlfors lemma

An important (possibly the most important) geometric quantity associated to a C^2 Hermitian metric is the curvature. The main goal of this section is to show that Theorem 1.9.23 can be interpreted as a statement concerning metrics with curvature bounded above by a negative constant.

Definition 1.10.1. Let $h: TX \to \mathbb{R}^+$ be a C^2 Hermitian metric. The *Gaussian curvature* (or simply *curvature*) $K_h: X \to \mathbb{R}$ of h is defined by

$$K_h(p) = -\left.\frac{\Delta \log \rho_h}{\rho_h^2}\right|_p ,$$

where ρ_h is the density of h with respect to a local chart (U, z) at p, and Δ is the Laplacian computed in local coordinates as

$$\Delta = 4\frac{\partial^2}{\partial \bar{z} \partial z}.$$

Using (1.53), it is easy to check that $K(p)$ does not depend on the local chart used for the computation.

Example 1.10.2. A direct computation shows that the Poincaré metric in $\mathbb{D}$ has constant negative curvature -4.

One of the reasons for the importance of the curvature is that it is invariant under pullback and push-forward.

Proposition 1.10.3. *Let $f: X \to Y$ be a local biholomorphism between Riemann surfaces. Then:*
(i) *If h is a C^2 Hermitian metric on Y, then $K_{f^*h} = K_h \circ f$.*
(ii) *If f is surjective and h is a C^2 Hermitian metric on X such that $f_* h$ is defined, then $K_{f_* h} \circ f = K_h$.*

Proof. We shall compute in local coordinates; so fix local charts (U, z) in X and (V, w) in Y such that $f(U) \subseteq V$. Using (1.55), we get

$$K_{f^*h} = -\frac{4}{|f'|^2(\rho_h \circ f)^2} \frac{\partial^2}{\partial \bar{z} \partial z} \log(|f'|(\rho_h \circ f))$$

$$= \frac{4}{(\rho_h \circ f)^4}\left|\frac{\partial \rho_h}{\partial w}\circ f\right|^2 - \frac{4}{(\rho_h \circ f)^3}\left(\frac{\partial^2 \rho_h}{\partial \overline{w}\partial w}\right)\circ f$$
$$= K_h \circ f,$$

and (i) is proved. To prove (ii), it suffices to recall that $h = f^*(f_* h)$ and to apply (i) to $f_* h$. □

In particular, then the Poincaré metric of any hyperbolic Riemann surface has constant negative curvature −4. It turns out that the Poincaré metric actually is the largest possible Hermitian metric with curvature bounded above by −4; to prove this (and more), we need a way to deal with general pseudo-Hermitian metrics, where we cannot compute the curvature directly.

Definition 1.10.4. Let $h: TX \to \mathbb{R}^+$ be a (in general only upper semicontinuous) pseudo-Hermitian metric on a Riemann surface X. Let $p \in X$ be a point where $h|_{T_pX} \not\equiv 0$. A *supporting metric* at p is a C^2 Hermitian metric h_0 defined in a neighborhood U of p such that $\frac{h}{h_0}(p) = 1$ and $\frac{h}{h_0} \geq 1$ on U; moreover, h_0 is a *hyperbolic supporting metric* if its curvature is bounded above by −4. We say that h is *ultrahyperbolic* if at each point where it does not vanish it has a hyperbolic supporting metric.

Remark 1.10.5. If h is a ultrahyperbolic metric on a Riemann surface Y and $f \in \mathrm{Hol}(X, Y)$, then f^*h is a ultrahyperbolic metric on X. Indeed, the condition on the existence of a hyperbolic supporting metric is required only where the metric does not vanish, that is outside the critical points of f, and there we can apply Proposition 1.10.3(i) to show that the pullback of a hyperbolic supporting metric is still a hyperbolic supporting metric.

In particular, metrics of the form $f^*\kappa_{\mathbb{D}}$ with $f \in \mathrm{Hol}(\mathbb{D}, \mathbb{D})$ are ultrahyperbolic. We can now state and prove the famous *Ahlfors lemma*, a far reaching generalization of Theorem 1.9.7.

Theorem 1.10.6 (Ahlfors, 1938; Heins, 1962). *Let $h: T\mathbb{D} \to \mathbb{R}^+$ be a ultrahyperbolic metric on $\mathbb{D}$. Then $h \leq \kappa_{\mathbb{D}}$. Furthermore, equality holds at one point if and only if $h \equiv \kappa_{\mathbb{D}}$.*

Proof. We start by proving that $h \leq \kappa_{\mathbb{D}}$. Let us first suppose that ρ_h is bounded above. In particular, this implies that $\frac{h}{\kappa_{\mathbb{D}}} = \frac{\rho_h}{\rho_{\mathbb{D}}}$ tends to 0 as $|z| \to 1$, because $\rho_{\mathbb{D}}(z) \to +\infty$ as $|z| \to 1$. As a consequence, by upper semicontinuity we can find a global maximum $z_0 \in \mathbb{D}$ of $\frac{\rho_h}{\rho_{\mathbb{D}}}$. If $\rho_h(z_0) = 0$, then $h \equiv 0$ and we are done. If instead $\rho_h(z_0) > 0$, let h_0 be a hyperbolic supporting metric for h defined in a neighborhood of z_0. In particular, $\rho_{h_0} \leq \rho_h$ nearby z_0 implies that z_0 is a local minimum of $\log\rho_{\mathbb{D}} - \log\rho_{h_0}$. The curvature conditions say that $\Delta\log\rho_{h_0} \geq 4\rho_{h_0}^2$ and $\Delta\log\rho_{\mathbb{D}} = 4\rho_{\mathbb{D}}$, where $\rho_{\mathbb{D}}$ is the density of the Poincaré metric; therefore,

$$\Delta(\log\rho_{\mathbb{D}} - \log\rho_{h_0}) \leq 4(\rho_{\mathbb{D}}^2 - \rho_{h_0}^2). \tag{1.61}$$

Since z_0 is a local minimum, we have $\Delta(\log\rho_{\mathbb{D}} - \log\rho_{h_0})(z_0) \geq 0$; then $\frac{\rho_{\mathbb{D}}}{\rho_{h_0}}(z_0) \geq 1$, and hence $\frac{\rho_h}{\rho_{\mathbb{D}}}(z_0) = \frac{\rho_{h_0}}{\rho_{\mathbb{D}}}(z_0) \leq 1$. Since z_0 is a global maximum for $\frac{\rho_h}{\rho_{\mathbb{D}}}$, it follows that $h \leq \kappa_{\mathbb{D}}$ everywhere, as claimed.

In the general case, given $0 < r < 1$ consider the metric h_r with density $\rho_{h_r}(z) = r\rho_h(rz)$. Clearly, h_r is ultrahyperbolic; moreover, the upper semicontinuity of ρ_h implies that ρ_{h_r} is bounded above. The previous argument gives $h_r \leq \kappa_{\mathbb{D}}$; letting $r \to 1$, we obtain $h \leq \kappa_{\mathbb{D}}$, as claimed.

Now assume that ρ_h is equal to $\rho_{\mathbb{D}}$ somewhere but, by contradiction, that $\rho_h \not\equiv \rho_{\mathbb{D}}$. Let $A = \{z \in \mathbb{D} \mid \rho_h(z) < \rho_{\mathbb{D}}(z)\}$; then $\emptyset \neq A \neq \mathbb{D}$ and A is open, by upper semicontinuity.

Take $z_0 \in \partial A \cap \mathbb{D}$; clearly, $\rho_h(z_0) = \rho_{\mathbb{D}}(z_0) > 0$. Let h_0 be a hyperbolic supporting metric for h at z_0. Since A is open, we can find $z_1 \in A$ and $0 < r_1 < r_2$ and $0 < r_3$ such that $D(z_1, r_1) \subset A$, $z_0 \in D(z_1, r_2) \subset\subset D(0, r_3)$ and h_0 is defined in a neighborhood of $\overline{D(z_1, r_2)}$.

Put $u_0 = \log(\rho_{h_0}/\rho_{\mathbb{D}})$. By construction, we have $u_0 \leq 0$ on $\overline{D(z_1, r_2)}$ and $u_0(z_0) = 0$. Furthermore,

$$\Delta u_0 \geq 4(\rho_{h_0}^2 - \rho_{\mathbb{D}}^2) = 4(\rho_{h_0} + \rho_{\mathbb{D}})(\rho_{h_0} - \rho_{\mathbb{D}}) \geq 8M(\rho_{h_0} - \rho_{\mathbb{D}}) \geq 8M^2 u_0$$

on $\overline{D(z_1, r_2)}$, where $M = (1 - r_3^2)^{-1}$ and we have used the inequalities

$$\rho_{h_0} + \rho_{\mathbb{D}} \leq \rho_h + \rho_{\mathbb{D}} \leq 2\rho_{\mathbb{D}} \leq \frac{2}{1 - r_3^2},$$

which is valid because $\overline{D(z_1, r_2)} \subset D(0, r_3)$, and

$$t - s \leq M \log \frac{t}{s},$$

which is valid as soon as $0 < s \leq t \leq M$.

Now let us introduce the auxiliary function

$$v_0(z) = \exp(-\alpha|z - z_1|^2) - \exp(-\alpha r_2^2),$$

where $\alpha > 0$ is chosen so that

$$\Delta v_0(z) - 8M^2 v_0(z) \geq e^{-\alpha|z - z_1|^2}(4\alpha^2|z - z_1|^2 - 4\alpha - 8M^2) > 0$$

when $r_1 \leq |z - z_1| \leq r_2$. Notice that $v_0 > 0$ on $D(z_1, r_2)$ and $v_0|_{\partial D(z_1, r_2)} \equiv 0$.

Finally, put $w_0 = u_0 + \varepsilon v_0$, where $\varepsilon > 0$ is chosen so that $w_0 \leq 0$ on $\overline{D(z_1, r_1)}$. Since $w_0|_{\partial D(z_1, r_2)} \equiv u_0|_{\partial D(z_1, r_2)} \leq 0$ and $w_0(z_0) = \varepsilon v_0(z_0) > 0$, the function w_0 attains a local positive maximum at a point $z_2 \in D(z_1, r_2) \setminus \overline{D(z_1, r_1)}$. But then $0 \geq \Delta w_0(z_2) > 8M^2 w_0(z_2) > 0$, contradiction, and we are done. □

Actually, we have an *Ahlfors lemma* on all hyperbolic Riemann surfaces.

Corollary 1.10.7. *Let $h: TX \to \mathbb{R}^+$ be a ultrahyperbolic metric on a hyperbolic Riemann surface X. Then $h \le \kappa_X$. Moreover, equality holds at one point if and only if $h \equiv \kappa_X$.*

Proof. Let $\pi_X: \mathbb{D} \to X$ be a universal covering map. Thanks to Proposition 1.10.3, $\pi_X^* h$ is a ultrahyperbolic metric on $\mathbb{D}$; therefore, Theorem 1.10.6 yields $\pi_X^* h \le \kappa_{\mathbb{D}}$. Applying $(\pi_X)_*$ to both sides, we get

$$h = (\pi_X)_* \pi_X^* h \le (\pi_X)_* \kappa_{\mathbb{D}} = \kappa_X,$$

as claimed. Finally, if equality holds at one point for h and κ_X it holds at one point for $\pi_X^* h$ and $\kappa_{\mathbb{D}}$; therefore, $\pi_X^* h \equiv \kappa_{\mathbb{D}}$ by Theorem 1.10.6, and hence $h \equiv \kappa_X$. □

We can now generalize Theorem 1.9.23 obtaining a *Ahlfors–Schwarz–Pick lemma.*

Corollary 1.10.8. *Let X, Y be hyperbolic Riemann surfaces, and $h: TY \to \mathbb{R}^+$ a ultrahyperbolic metric on Y. Then $f^* h \le \kappa_X$ for all $f \in \operatorname{Hol}(X, Y)$, with equality at one point if and only if $h \equiv \kappa_Y$ and f is a covering map.*

In particular, if $\Omega \subseteq X$ is a domain in X then $\kappa_X|_{T\Omega} \le \kappa_\Omega$, with equality at one point if and only if $\Omega = X$.

Proof. The fact that $f^* h \le \kappa_X$ follows from Theorem 1.10.7 because $f^* h$ is a ultrahyperbolic metric on X. If the equality holds at one point, again Theorem 1.10.7 yields $f^* h \equiv \kappa_X$, and hence $f^* \kappa_Y \equiv \kappa_X$, because $h \le \kappa_Y$ always by Theorem 1.10.7. We can then recall Theorem 1.9.23 to conclude that f is a covering map, and thus $h \equiv f_* f^* h \equiv f_* \kappa_X \equiv \kappa_Y$. Conversely, if $h \equiv \kappa_Y$ and f is a covering map we already know that $f^* h \equiv \kappa_X$.

Finally, the last assertions follow considering the inclusion $\iota: \Omega \hookrightarrow X$, because $\iota^* \kappa_X = \kappa_X|_{T\Omega}$, and ι is a covering map if and only if $\Omega = X$. □

We end this section showing that, as the name suggests, nontrivial ultrahyperbolic metrics can exist only on hyperbolic Riemann surfaces.

Corollary 1.10.9. *If there exists a ultrahyperbolic pseudo-Hermitian metric $h: TX \to \mathbb{R}^+$ with $h \not\equiv 0$ on a Riemann surface X then X is hyperbolic.*

Proof. Assume, by contradiction, that X is not hyperbolic, and let $\pi_X: \widetilde{X} \to X$ be a universal covering map with $\widetilde{X} = \mathbb{C}$ or $\widehat{\mathbb{C}}$. Then for any $r > 0$ the metric $\tilde{h}_r = (\pi_X^* h)|_{T\mathbb{D}_r}$ is ultrahyperbolic on the disk $\mathbb{D}_r$. Corollary 1.10.7 then implies

$$\rho_{h_r}(z) \le \rho_{\mathbb{D}_r}(z) = \frac{r}{r^2 - |z|^2}.$$

Letting $r \to +\infty$ we then get $\pi_X^* h \equiv 0$, and hence $h \equiv 0$, contradiction. □

Notes to Section 1.10

In 1938, Ahlfors [16] proved Theorem 1.10.6 and Corollary 1.10.7 (with the exception of the equality statements which are due to Heins [187]; see below), and used it to give another proof of Schottky theorem and to get a good lower bound for the Bloch constant. The Ahlfors lemma has been very influential, because it has indicated that contraction properties of holomorphic maps are related to bounds on the curvature. Yau [428], in 1973, has shown that in the statement of Ahlfors lemma one can replace $\mathbb{D}$ with a Riemann surface endowed with a complete Hermitian metric with curvature bounded *below* by a constant, and in 1978 [429] he has given a several variables version of these results for holomorphic maps between a complete Kähler manifold with Ricci curvature bounded below and a Kähler manifold with biholomorphic sectional curvature bounded above by a negative constant. In 1979, Chen, Cheng, and Look [112] proved similar statements using the holomorphic sectional curvature both in the domain and in the target manifolds. For other related results see, e. g., [394], [345], the surveys [359, 312], the books [18, 239], and references therein.

As mentioned above, the equality statement has been proved by Heins [187] in 1962 (but see also Jørgensen [214]). In this paper, Heins introduced a larger class of pseudo-Hermitian metric on a Riemann surface having negative curvature in a suitable weak sense, the S-K metrics. A upper semicontinuous pseudo-Hermitian metric $h\colon TX \to \mathbb{R}^+$ is a *S-K metric* if for any $p_0 \in X$ and local chart (U, z) with $z(p_0) = 0$ and $\rho_h(0) > 0$ we have

$$[\rho_h(0)]^2 \le \liminf_{r\to 0^+} \frac{1}{r^2}\left[\frac{1}{2\pi}\int_0^{2\pi} \log \rho_h(re^{i\theta})\,d\theta - \log \rho_h(0)\right].$$

If h is of class C^2, then this condition is exactly equivalent to having curvature bounded above by -4. Heins has proved that ultrahyperbolic metrics are S-K metrics and that S-K metrics enjoy an Ahlfors lemma completely equivalent to Theorem 1.10.7; moreover, he has given a number of applications of this notion, including an improvement of the estimate of the Bloch constant.

Our proof of the equality statement in adapted from [292]; for other proofs and more applications, see also [290, 292, 358, 113].

A generalization of Corollary 1.10.8 is contained in [291]: if $\Omega_1, \Omega_2 \subseteq X$ are hyperbolic subdomains of a Riemann surface X such that there is $p \in \Omega_1 \cap \Omega_2$ so that $\kappa_{\Omega_2} \le \kappa_{\Omega_1}$ in a neighborhood of p with equality at p, then $\Omega_1 = \Omega_2$.

Corollary 1.10.9 is taken from [187, Theorem 9.1].

1.11 Bloch domains

We have seen that the Poincaré distance in an hyperbolic Riemann surface X is necessarily complete; in particular, for every $z \in X$ and $r > 0$ the Poincaré ball $B_X(z, r)$ is relatively compact in X. On the other hand, if Ω is a subdomain of X it might well happen that there exists a radius $r > 0$ such that all Poincaré balls with respect to the Poincaré distance of X of radius r centered in a point of Ω intersect $\partial\Omega$. Such subdomains will be used in Chapter 3 in our discussion of random iteration; this section is devoted to study and characterize them.

Definition 1.11.1. If $\Omega \subset X$ is a domain in a hyperbolic Riemann surface X and $z \in \Omega$ put

$$R(z;\Omega,X) = \sup\{r > 0 \mid B_X(z,r) \subseteq \Omega\}$$

and

$$R(\Omega,X) = \sup_{z\in\Omega} R(z;\Omega,X) = \sup\{r > 0 \mid \exists z \in \Omega : B_X(z,r) \subseteq \Omega\}.$$

We say that Ω is a *Bloch domain* of X if $R(\Omega,X) < +\infty$.

It is not too difficult to find examples of Bloch domains and of domains that are not Bloch.

Example 1.11.2. Let $\Omega \subset\subset X$ be a relatively compact domain of a hyperbolic Riemann surface X, and fix $z_0 \in \Omega$. Since $\overline{\Omega}$ is compact, there exists $R > 0$ such that $\overline{\Omega} \subset B_X(z_0,R)$. Therefore, if z, w are in Ω we must have $\omega_X(z,w) < 2R$; hence it follows that $R(\Omega,X) \leq 2R$, and thus Ω is a Bloch domain of X. In particular, all domains in a compact hyperbolic Riemann surface are (trivially) Bloch domains.

Example 1.11.3. A subdomain of a Bloch domain clearly still is a Bloch domain.

Example 1.11.4. Let $X \subset \widehat{X}$ be a noncompact hyperbolic domain in a Riemann surface $\widehat{X}$ and $\Omega = X \setminus F$, where $F \subset X$ is a not empty finite set. Then Ω is not a Bloch domain in X. Indeed, if $z \in \Omega$ tends to a point in ∂X the ω_X distance from z to the points in F tends to infinity, because ω_X is complete. Therefore, for every $R > 0$ we can find $z \in \Omega$ such that $R(z;\Omega,X) > R$, and so $R(\Omega,X) = +\infty$, as claimed.

Example 1.11.5. Let X be a noncompact hyperbolic Riemann surface. Given $r > 0$, let $\{z_\nu\} \subset X$ be a sequence with no accumulation points in X and intersecting all Poincaré balls of radius r in X. Set $\Omega = X \setminus \{z_\nu\}$; then $R(\Omega,X) \leq r$, and thus Ω is a Bloch domain of X.

One can construct such a sequence as follows. Fix a point $p_0 \in X$, and for $n \in \mathbb{N}^*$ set $B_n = B_X(p_0,n)$. The closed ball $\overline{B_1}$ is compact; therefore, it can be covered by a finite number of Poincaré balls of radius r. Next, the set $\overline{B_2} \setminus B_1$ is compact, and thus it can be covered by a finite number of Poincaré balls of radius r. Analogously, for every $n \geq 2$ we can cover $\overline{B_{n+1}} \setminus B_n$ by a finite number of Poincaré balls of radius r. Let $\{z_\nu\}$ be the sequence consisting of the centers of these balls. Clearly, this sequence has no accumulation points in X; furthermore, every point of X is contained in a Poincaré ball of radius r and center in $\{z_\nu\}$, which is another way of saying that $\{z_\nu\}$ intersects every Poincaré ball of radius r, as claimed.

Example 1.11.6. Given $r > 0$ and $\tau \in \partial\mathbb{D}$, let

$$\Omega = \bigcup_{t\in(-1,1)} B_{\mathbb{D}}(t\tau,r);$$

we would like to prove that Ω is a Bloch domain of $\mathbb{D}$.

Clearly, up to a rotation we can assume that $\tau = 1$. First of all, we claim that $\omega(t, i\alpha) \geq \omega(0, i\alpha)$ for all $\alpha, t \in (-1, 1)$. Indeed, using (1.5) we get

$$\omega(0, i\alpha) \leq \omega(t, i\alpha) \iff 1 - |\alpha|^2 \geq 1 - \left|\frac{t - i\alpha}{1 + i\alpha t}\right|^2 \iff |1 + i\alpha t|^2 \geq 1 - t^2,$$

and the latter inequality is obvious. In particular, if $\omega(0, i\alpha) \geq r$ then $i\alpha \notin \Omega$.

Take now any $z_0 \in \Omega$. By definition, there exists $t_0 \in (-1, 1)$ such that $z_0 \in B_{\mathbb{D}}(t_0, r)$. Put

$$\gamma_{t_0}(z) = \frac{z + t_0}{1 + t_0 z}.$$

Then $\gamma_{t_0} \in \operatorname{Aut}(\mathbb{D})$ is such that $\gamma_{t_0}(0) = t_0$; in particular, $B_{\mathbb{D}}(t_0, r) = \gamma_{t_0}(B_{\mathbb{D}}(0, r))$. Furthermore, $\gamma_{t_0}((-1, 1)) = (-1, 1)$, and hence $\gamma_{t_0}(\Omega) = \Omega$. Fix $\alpha \in (0, 1)$ such that $\omega(0, i\alpha) = r$; in particular, $i\alpha \notin \Omega$, and hence $\gamma_{t_0}(i\alpha) \notin \Omega$. But

$$\begin{aligned}\omega(z_0, \gamma_{t_0}(i\alpha)) &\leq \omega(z_0, t_0) + \omega(t_0, \gamma_{t_0}(i\alpha)) \\ &= \omega(z_0, t_0) + \omega(\gamma_{t_0}(0), \gamma_{t_0}(i\alpha)) = \omega(z_0, t_0) + \omega(0, i\alpha) < 2r,\end{aligned}$$

and thus the Poincaré ball $B_{\mathbb{D}}(z_0, 2r)$ is not contained in Ω. Since z_0 is a generic point of Ω we have shown that $R(\Omega, \mathbb{D}) \leq 2r$, and we are done.

Remark 1.11.7. In the next chapter (Proposition 2.2.7), we shall see that Ω is comparable to Stolz regions in $\mathbb{D}$, regions that will play an important role in the study of the boundary behavior of holomorphic functions.

It turns out that to study Bloch domains of a generic hyperbolic Riemann surface it is sufficient to understand Bloch domains of $\mathbb{D}$. Indeed, we have the following result.

Proposition 1.11.8. *Let $\Omega \subset X$ be a domain in a hyperbolic Riemann surface X and fix $p \in \Omega$. Let $\pi_X \colon \mathbb{D} \to X$ be a universal covering map such that $\pi_X(0) = p$ and let $\Sigma \subset \mathbb{D}$ be the connected component of $\pi_X^{-1}(\Omega)$ containing the origin. Then*

$$R(\pi_X(\tilde{z}); \Omega, X) = R(\tilde{z}; \Sigma, \mathbb{D}) \tag{1.62}$$

for all $\tilde{z} \in \Sigma$, and hence $R(\Omega, X) = R(\Sigma, \mathbb{D})$. In particular, Ω is a Bloch domain of X if and only if Σ is a Bloch domain of $\mathbb{D}$.

Proof. We have seen in Proposition 1.7.3(ii) that

$$\pi_X(B_{\mathbb{D}}(\tilde{z}, r)) = B_X(\pi_X(\tilde{z}), r) \tag{1.63}$$

for all $\tilde{z} \in \mathbb{D}$ and $r > 0$. In particular, this immediately says that $B_{\mathbb{D}}(\tilde{z}, r) \subseteq \Sigma$ implies $B_X(\pi_X(\tilde{z}), r) \subseteq \Omega$, and thus $R(\pi_X(\tilde{z}); \Omega, X) \geq R(\tilde{z}; \Sigma, \mathbb{D})$.

Conversely, take $r > 0$ such that $B_X(\pi_X(\tilde{z}), r) \subseteq \Omega$. From (1.63), we get $B_{\mathbb{D}}(\tilde{z}, r) \subseteq \pi_X^{-1}(\Omega)$; since $B_{\mathbb{D}}(\tilde{z}, r)$ is connected and $\tilde{z} \in \Sigma$ we infer $B_{\mathbb{D}}(\tilde{z}, r) \subseteq \Sigma$. But then $R(\tilde{z}; \Sigma, \mathbb{D}) \geq R(\pi_X(\tilde{z}); \Omega, X)$, and (1.62) is proved. Since $\pi_X(\Sigma) = \Omega$, we immediately get $R(\Omega, X) = R(\Sigma, \mathbb{D})$ and we are done. □

The main goal of this section is to give a different characterization of Bloch domains. We start with the following.

Definition 1.11.9. Let $\Omega \subseteq X$ be a domain in a hyperbolic Riemann surface X. For $z \in \Omega$, put

$$\mu(z; \Omega, X) = \frac{\kappa_X}{\kappa_\Omega}(z)$$

and

$$\mu(\Omega, X) = \sup_{z \in \Omega} \mu(z; \Omega, X);$$

notice that Corollary 1.10.8 yields $\mu(\Omega, X) \leq 1$.

Our aim is to prove that Ω is a Bloch domain of X if and only if $\mu(\Omega, X) < 1$. First of all, we will show that we can assume $X = \mathbb{D}$. Indeed, we have the following analogous of Proposition 1.11.8.

Proposition 1.11.10. *Let $\Omega \subset X$ be a domain in a hyperbolic Riemann surface X and fix $p \in \Omega$. Let $\pi_X \colon \mathbb{D} \to X$ be a universal covering map such that $\pi_X(0) = p$ and let $\Sigma \subset \mathbb{D}$ be the connected component of $\pi_X^{-1}(\Omega)$ containing the origin. Then*

$$\mu(\pi_X(\tilde{z}); \Omega, X) = \mu(\tilde{z}; \Sigma, \mathbb{D}) \tag{1.64}$$

for all $\tilde{z} \in \Sigma$, and hence $\mu(\Omega, X) = \mu(\Sigma, \mathbb{D})$.

Proof. Respectively, by definition and by Theorem 1.9.23 we know that

$$\kappa_X\big(\pi_X(\tilde{z}); d(\pi_X)_{\tilde{z}}(\tilde{v})\big) = \kappa_{\mathbb{D}}(\tilde{z}; \tilde{v})$$

and

$$\kappa_\Omega\big(\pi_X(\tilde{z}); d(\pi_X)_{\tilde{z}}(\tilde{v})\big) = \kappa_\Sigma(\tilde{z}; \tilde{v})$$

for all $\tilde{z} \in \Sigma$ and $\tilde{v} \in \mathbb{C}$. Thus (1.64) is clear and $\mu(\Omega, X) = \mu(\Sigma, \mathbb{D})$ follows again because $\pi_X(\Sigma) = \Omega$. □

We shall need two technical lemmas.

Lemma 1.11.11. (i) *The function $A\colon \mathbb{R} \to \mathbb{R}$ given by*

$$A(t) = t \exp\left(\frac{1-t^2}{1+t^2}\right)$$

is increasing in $(0,1)$ and satisfies $1 > A(t) > t$ there.

(ii) *Given $T \in (0,1)$, let $\psi_T\colon (0, A(T)) \to \mathbb{R}$ be defined by*

$$\psi_T(x) = \frac{2x \log(A(T)/x)}{1-x^2}. \tag{1.65}$$

Then ψ_T is increasing in $(0,T]$, decreasing on $[T, A(T))$ and satisfies

$$\psi_T(T) = \frac{2T}{1+T^2}. \tag{1.66}$$

In particular, taking $T = \tanh R$ we get $\psi_T(\tanh R) = \tanh(2R)$.

Proof. Since

$$A'(t) = \left(\frac{1-t^2}{1+t^2}\right)^2 \exp\left(\frac{1-t^2}{1+t^2}\right)$$

is strictly positive in $(0,1)$ and $\exp(\frac{1-t^2}{1+t^2}) > 1$ in that interval, (i) follows because $A(1) = 1$.

Next,

$$\psi_T'(x) = \frac{2(1+x^2)}{(1-x^2)^2}\left[\log\frac{A(T)}{x} - \frac{1-x^2}{1+x^2}\right].$$

Therefore, $\psi_T'(x) > 0$ if and only if $A(x) < A(T)$, and thus, being A increasing, if and only if $x \in (0,T)$.

Finally, the equality (1.66) is trivial and the last equality follows from the formula

$$\frac{2(\tanh R)}{1+(\tanh R)^2} = \tanh(2R),$$

which in turns is a consequence of the addition formula (1.9). □

Lemma 1.11.12. *Given $p \in \mathbb{D}$ and $R > 0$ let $B^* = B_{\mathbb{D}}(p,R) \setminus \{p\}$ be the pointed Poincaré ball centered at p and of radius R. Then*

$$\kappa_{B^*}(z; v) = \frac{1}{\psi_T(\tanh \omega(z,p))}\kappa_{\mathbb{D}}(z; v)$$

for all $z \in B^*$ and $v \in \mathbb{C}$, where ψ_T is the function defined in (1.65) and $T \in (0,1)$ is the unique solution of the equation $A(t) = \tanh R$.

Proof. Let us first consider the case $p = 0$. Then $B_0^* = B_{\mathbb{D}}(0, R) \setminus \{0\}$ is a pointed Euclidean disk $D(0, \tanh R) \setminus \{0\}$ centered at the origin of radius $\tanh R$. Let $\gamma_0 \colon B_0^* \to \mathbb{D}^*$ be the biholomorphism given by $\gamma_0(z) = z/\tanh R$. Recalling Example 1.9.21, we get

$$\begin{aligned}\kappa_{B_0^*}(z; v) &= \kappa_{\mathbb{D}^*}\left(\frac{z}{\tanh R}; \frac{v}{\tanh R}\right) = \frac{|v|}{2|z|\left|\log \frac{|z|}{\tanh R}\right|} \\ &= \frac{1 - |z|^2}{2|z|\left|\log \frac{|z|}{\tanh R}\right|}\kappa_{\mathbb{D}}(z; v) = \frac{1}{\psi_T(|z|)}\kappa_{\mathbb{D}}(z; v) \\ &= \frac{1}{\psi_T(\tanh \omega(0, z))}\kappa_{\mathbb{D}}(z; v),\end{aligned}$$

where $T \in (0, 1)$ is the unique solution of the equation $A(t) = \tanh R$.

For the general case, the automorphism $\gamma_p(z) = (z-p)/(1-\bar{p}z)$ is a biholomorphism between B^* and B_0^*; therefore,

$$\begin{aligned}\kappa_{B^*}(z; v) = \kappa_{B_0^*}(\gamma_p(z); \gamma_p'(z)v) &= \frac{1}{\psi_T(\tanh \omega(0, \gamma_p(z)))}\kappa_{\mathbb{D}}(\gamma_p(z); \gamma_p'(z)v) \\ &= \frac{1}{\psi_T(\tanh \omega(p, z))}\kappa_{\mathbb{D}}(z; v)\end{aligned}$$

because γ_p is an isometry of $\kappa_{\mathbb{D}}$ with $\gamma_p(p) = 0$ and we are done. □

We are now ready to prove the main result of this section.

Theorem 1.11.13. *Let $\Omega \subset X$ be a domain in a hyperbolic Riemann surface X. Then*

$$\tanh R(\Omega, X) \le \mu(\Omega, X) \le \tanh(2R(\Omega, X)). \tag{1.67}$$

In particular, Ω is a Bloch domain of X if and only if $\mu(\Omega, X) < 1$.

Proof. By Propositions 1.11.8 and 1.11.10, we can assume that $X = \mathbb{D}$ and $\Omega \subseteq \mathbb{D}$.

First of all, we would like to prove that

$$\tanh R(z; \Omega, \mathbb{D}) \le \mu(z; \Omega, \mathbb{D}) \tag{1.68}$$

for all $z \in \Omega$. Notice that, since both $R(z; \Omega, \mathbb{D})$ and $\mu(z; \Omega, \mathbb{D})$ are invariant under automorphisms of $\mathbb{D}$, it suffices to prove (1.68) for $z = 0$.

Choose $r > 0$ so that $B = B_{\mathbb{D}}(0, r) \subseteq \Omega$; in particular, $\kappa_B(0; v) \ge \kappa_\Omega(0; v)$ for every $v \in \mathbb{C}$. Now, B is an Euclidean disk centered at the origin of radius $t = \tanh r$; therefore, the map $\phi(z) = z/t$ is a biholomorphism between B and $\mathbb{D}$. As a consequence, $\kappa_B(0; v) = \kappa_{\mathbb{D}}(0; v/t) = \frac{1}{t}\kappa_{\mathbb{D}}(0; v)$. Putting everything together, we get

$$\mu(0; \Omega, \mathbb{D}) = \frac{\kappa_{\mathbb{D}}(0; v)}{\kappa_\Omega(0; v)} = (\tanh r)\frac{\kappa_B(0; v)}{\kappa_\Omega(0; v)} \ge \tanh r.$$

Taking the supremum over all admissible r we obtain (1.68). The left inequality in (1.67) then follows taking the supremum over $z \in \Omega$.

If $R(\Omega, \mathbb{D}) = +\infty$, then the right inequality in (1.67) is trivial; so we can assume that $R(\Omega, \mathbb{D}) < +\infty$. Take $z \in \Omega$, and for brevity set $\tau(z) = \tanh R(z; \Omega, \mathbb{D})$ and $T_0 = \tanh R(\Omega, \mathbb{D})$. We claim that

$$\mu(z; \Omega, \mathbb{D}) \leq \psi_{T_0}(\tau(z)), \tag{1.69}$$

where ψ_{T_0} is the function defined in Lemma 1.11.11.

Before proving the claim, let us show why (1.69) yields the right inequality in (1.67). Indeed, since $\tau(z) \leq T_0$ and ψ_{T_0} is increasing in $(0, T_0]$ we have $\psi_{T_0}(\tau(z)) \leq \psi_{T_0}(T_0) = \tanh(2R(\Omega, \mathbb{D}))$, again by Lemma 1.11.11. Taking the supremum over $z \in \Omega$, we then get the assertion.

So, we are left to prove (1.69), which is equivalent to

$$\frac{1}{\psi_{T_0}(\tau(z))} \kappa_{\mathbb{D}}(z; v) \leq \kappa_{\Omega}(z; v) \tag{1.70}$$

for all $z \in \Omega$ and $v \in \mathbb{C}$.

Given $T \in (T_0, 1)$ put

$$h_T(z; v) = \frac{1}{\psi_T(\tau(z))} \kappa_{\mathbb{D}}(z; v)$$

for all $z \in \Omega$ and $v \in \mathbb{C}$. If we prove that h_T is ultrahyperbolic on Ω, then Theorem 1.10.7 yields $h_T \leq \kappa_\Omega$ and letting $T \to T_0^+$ we get (1.70).

Fix $z_0 \in \Omega$, and choose $p \in \mathbb{D} \cap \partial\Omega$ such that $\omega(z_0, p) = R(z_0; \Omega, \mathbb{D})$; we clearly have $T > T_0 \geq \tanh \omega(z_0, p)$. If $z \in \Omega$ is such that

$$\omega(z_0, z) < \tanh^{-1}(T) - \omega(z_0, p)$$

we have

$$R(z; \Omega, \mathbb{D}) \leq \omega(z, p) \leq \omega(z, z_0) + \omega(z_0, p) < \tanh^{-1}(T),$$

and hence

$$\tau(z) \leq \tanh \omega(z, p) < T$$

with equality on the left when $z = z_0$. Since ψ_T is increasing in $(0, T]$, we get

$$h_T(z; v) \geq \frac{1}{\psi_T(\tanh \omega(z, p))} \kappa_{\mathbb{D}}(z; v) \tag{1.71}$$

for all $v \in \mathbb{C}$ and $z \in \Omega$ such that $\omega(z_0, z) < \tanh^{-1}(T) - \omega(z_0, p)$, with equality for $z = z_0$. But the right-hand side of (1.71) is, by Lemma 1.11.12, the Poincarè metric of the pointed

Poincaré disk $B_{\mathbb{D}}(p,R) \setminus \{p\}$ with $R = \tanh^{-1} A(T)$, and thus it has constant negative curvature -4. This shows that h_T is ultrahyperbolic and the proof is complete. □

Notes to Section 1.11

The content of this section comes from [52].

2 Boundary Schwarz lemmas

In the first chapter we learned to appreciate the importance of the Schwarz lemma. Unfortunately, its original form has a shortcoming: it cannot be directly applied to get information about boundary behaviors. Julia first and Wolff shortly later overcame this difficulty, proving the lemmas known under their names, which are the main topic of this chapter.

Their idea is simple. The Schwarz–Pick lemma says that a holomorphic function $f \in \mathrm{Hol}(\mathbb{D}, \mathbb{D})$ sends Poincaré balls into Poincaré balls. Then to get information about a boundary point $\tau \in \partial\mathbb{D}$ one can choose a sequence of Poincaré balls with centers converging to τ and constant Euclidean radius, apply the Schwarz–Pick lemma to each one of them, and take the limit. In particular, it turns out that the right geometrical objects to consider are the *horocycles*: Euclidean disks internally tangent to a point of $\partial\mathbb{D}$. In fact, the Julia and Wolff lemmas say that, under suitable hypotheses, a holomorphic function $f \in \mathrm{Hol}(\mathbb{D}, \mathbb{D})$ sends horocycles into horocycles in a very controlled way.

In this chapter, we shall also discuss some applications of the Julia and Wolff lemmas, leaving the main one—dynamics—to the next chapters. The first application concerns the behavior of the angular derivative. Let $f \in \mathrm{Hol}(\mathbb{D}, \mathbb{D})$, take $\sigma \in \partial\mathbb{D}$, and assume for the sake of simplicity, that $f(z) \to \tau \in \partial\mathbb{D}$ as $z \to \sigma$. Then we would like to know something about the behavior of the derivative f' near σ. A natural approach is to study the incremental ratio $(f(z) - \tau)/(z - \sigma)$, and the Julia lemma turns out to be the ideal tool for this investigation. We shall see that a suitable limit, the *nontangential limit*, of the incremental ratio at σ exists, possibly equal to infinity; moreover, if it is finite, it coincides with the nontangential limit of the derivative at σ. This will also allow us to give a criterion for the existence of the nontangential limit of f' at a boundary point.

We shall also present the Lindelöf principle: under suitable assumptions, the existence of the limit along a curve ending at a boundary point implies the existence of the nontangential limit at that point.

The second application of the Julia and Wolff lemmas concerns the structure of the automorphism group of hyperbolic Riemann surfaces. We shall show that a function $f \in \mathrm{Hol}(\mathbb{D}, \mathbb{D})$ can commute with a hyperbolic automorphism γ of $\mathbb{D}$ if and only if f itself is a hyperbolic automorphism with the same fixed points as γ—a first extension of Proposition 1.4.12. From this, we shall infer several properties of $\mathrm{Aut}(X)$; for instance, we shall prove that the automorphism group of a compact hyperbolic Riemann surface is finite.

The last section of this chapter is devoted to a much more recent boundary Schwarz lemma. One way to state the equality case of the Schwarz lemma is to say that if $f \in \mathrm{Hol}(\mathbb{D}, \mathbb{D})$ is such that $f(z) - z = O(z^2)$ then $f \equiv \mathrm{id}_{\mathbb{D}}$. A boundary version of this statement could be that if $f \in \mathrm{Hol}(\mathbb{D}, \mathbb{D})$ is such that $f(z) - z = O((z - \tau)^k)$ as $z \to \tau \in \partial\mathbb{D}$ then $f \equiv \mathrm{id}_{\mathbb{D}}$. We shall show that such a statement, due to Burns and Krantz, indeed holds for $k = 3$ and that this value of k is optimal.

https://doi.org/10.1515/9783110601978-002

More precisely, Section 2.1 will be devoted to the proof of the Julia lemma, both in $\mathbb{D}$ and in $\mathbb{H}^+$. In particular, we shall officially define the horocycles and we shall introduce the boundary dilation of a function at a boundary point, a concept that will be fundamental in the sequel. In Section 2.2, we shall introduce the Stolz regions and the notion of nontangential limit. In Section 2.3, we shall prove the Julia–Wolff–Carathéodory theorem that gives conditions for the existence of the nontangential limit both of the incremental ratio and of the derivative at a given point in the boundary of $\mathbb{D}$ in terms of the boundary dilation. We shall also obtain a few inequalities involving the nontangential limit of the derivative at boundary points.

Section 2.4 is devoted to the Lindelöf theorem and to a couple of applications we shall need later on. In Section 2.5, we shall prove the Wolff lemma. This result will be crucial for the study of the holomorphic dynamics, but in this chapter we shall use it to study in detail the structure of the automorphism group of hyperbolic Riemann surfaces in Section 2.6. Finally, in Section 2.7, we shall discuss the Burns–Krantz theorem mentioned above together with some very recent generalizations.

2.1 The Julia lemma

In this section, we shall introduce the horocycles, a sort of boundary Poincaré balls. Using them we shall state, prove, and discuss the Julia lemma and, in the next sections, its consequences concerning angular derivatives.

Fix $\tau \in \partial\mathbb{D}$. From a geometrical point of view, the limit of Poincaré balls of constant Euclidean radius and center z for $z \to \tau$ is an Euclidean disk tangent to the boundary of $\mathbb{D}$ in τ. We are thus led to the following definition.

Definition 2.1.1. The *horocycle* $E(\tau, R) \subset \mathbb{D}$ *of center* $\tau \in \partial\mathbb{D}$ and *radius* $R > 0$ is the Euclidean disk of radius $R/(R+1)$ tangent to $\partial\mathbb{D}$ in τ; see Figure 2.1. Analytically, it is easy to see that the horocycle $E(\tau, R)$ is given by

$$E(\tau, R) = \left\{ z \in \mathbb{D} \;\middle|\; \frac{|\tau - z|^2}{1 - |z|^2} < R \right\}. \tag{2.1}$$

Remark 2.1.2. A formula that sometimes will be useful and that relates the horocycles with the generalized Cayley transform introduced in Definition 1.3.5 is the following:

$$\frac{1 - |z|^2}{|\tau - z|^2} = \operatorname{Re}\left(\frac{\tau + z}{\tau - z}\right) = \operatorname{Im} \Psi_\tau(z), \tag{2.2}$$

valid for all $\tau \in \partial\mathbb{D}$ and $z \in \mathbb{D}$.

A horocycle $E(\tau, R)$ is the limit of Poincaré balls in the sense made precise by the following proposition.

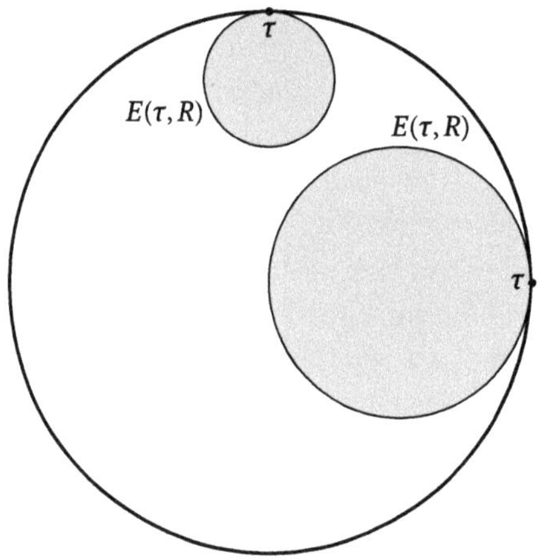

Figure 2.1: Horocycles in $\mathbb{D}$.

Proposition 2.1.3. *Let $\{B_\nu = B_{\mathbb{D}}(z_\nu, R_\nu)\}$ be a sequence of Poincaré balls in $\mathbb{D}$ such that $z_\nu \to \tau \in \partial\mathbb{D}$ and*

$$\lim_{\nu\to\infty} \frac{1-|z_\nu|}{1-\tanh R_\nu} = R \neq 0, \infty. \tag{2.3}$$

Then:
(i) *if $z \in B_\nu$ for infinitely many ν, then $z \in \overline{E(\tau, R)}$;*
(ii) *if $z \in E(\tau, R)$, then $z \in B_\nu$ for all sufficiently large ν.*

Proof. We observe that, thanks to (1.5), having $z \in B_\nu$ is equivalent to having

$$\frac{|1-\overline{z_\nu}z|^2}{1-|z|^2} < \frac{1-|z_\nu|^2}{1-(\tanh R_\nu)^2} = \frac{1+|z_\nu|}{1+\tanh R_\nu} \cdot \frac{1-|z_\nu|}{1-\tanh R_\nu}. \tag{2.4}$$

If $z \in B_\nu$ for infinitely many ν, passing to a subsequence we may send $\nu \to \infty$ in (2.4), and using (2.3) we obtain $z \in \overline{E(\tau, R)}$.

Conversely, if $z \in E(\tau, R)$ then

$$\lim_{\nu\to\infty} \frac{|1-\overline{z_\nu}z|^2}{1-|z|^2} = \frac{|\tau - z|^2}{1-|z|^2} < R = \lim_{\nu\to\infty} \frac{1-|z_\nu|^2}{1-(\tanh R_\nu)^2},$$

and (2.4) must hold for all sufficiently large ν. □

The geometrical meaning of (2.3) is that the Euclidean radius of $B_{\mathbb{D}}(z_\nu, R_\nu)$ tends to the Euclidean radius of $E(\tau, R)$; cf. (1.11).

There is another way to describe the horocycles that shows how they can be defined in completely intrinsic terms by using the Poincaré distance.

Proposition 2.1.4. *Let $\tau \in \partial\mathbb{D}$ and $R > 0$. Then*

$$E(\tau, R) = \left\{ z \in \mathbb{D} \;\middle|\; \lim_{w\to\tau} [\omega(z, w) - \omega(0, w)] < \frac{1}{2}\log R \right\}. \tag{2.5}$$

Proof. For any $z \in \mathbb{D}$, let $\gamma_z \in \operatorname{Aut}(\mathbb{D})$ be such that $\gamma_z(z) = 0$. We have

$$\begin{aligned}\omega(z, w) - \omega(0, w) = \omega(0, \gamma_z(w)) - \omega(0, w) &= \frac{1}{2}\log\left(\frac{1 + |\gamma_z(w)|}{1 - |\gamma_z(w)|} \cdot \frac{1 - |w|}{1 + |w|}\right) \\ &= \frac{1}{2}\log\left(\frac{1 - |w|^2}{1 - |\gamma_z(w)|^2}\right) + \log\left(\frac{1 + |\gamma_z(w)|}{1 + |w|}\right).\end{aligned}$$

Now (1.5) yields

$$\frac{1 - |w|^2}{1 - |\gamma_z(w)|^2} = \frac{|1 - \overline{w}z|^2}{1 - |z|^2};$$

therefore, recalling that $|\gamma_z(w)| \to 1$ as $w \to \tau$, we get

$$\lim_{w\to\tau} [\omega(z, w) - \omega(0, w)] = \frac{1}{2}\log\frac{|\tau - z|^2}{1 - |z|^2} \tag{2.6}$$

and we are done. □

Actually, we have already met the horocycles, though in disguise, in Lemma 1.4.17, where we proved that a parabolic automorphism of $\mathbb{D}$ with fixed-point $\tau \in \partial\mathbb{D}$ sends any horocycle centered at τ into itself. More generally, automorphisms of $\mathbb{D}$ send horocycles in horocycles.

Proposition 2.1.5. *Let $\gamma \in \operatorname{Aut}(\mathbb{D})$, and choose $\sigma \in \partial\mathbb{D}$ and $R > 0$. Put $z_0 = \gamma^{-1}(0)$, and $\beta = (1 - |z_0|^2)/(|\sigma - z_0|^2)$. Then $\gamma(E(\sigma, R)) = E(\gamma(\sigma), \beta R)$.*

Proof. For every $z \in \mathbb{D}$, we have

$$|\gamma(\sigma) - \gamma(z)|^2 = \frac{|\sigma - z|^2(1 - |z_0|^2)^2}{|1 - \overline{z_0}z|^2|1 - \overline{z_0}\sigma|^2}.$$

Hence (1.5) yields

$$\frac{|\gamma(\sigma) - \gamma(z)|^2}{1 - |\gamma(z)|^2} = \frac{|\sigma - z|^2}{1 - |z|^2} \cdot \frac{1 - |z_0|^2}{|\sigma - z_0|^2}, \tag{2.7}$$

and the assertion follows. □

Corollary 2.1.6. *Let $\gamma \in \operatorname{Aut}(\mathbb{D})$ be a parabolic automorphism with fixed-point $\tau \in \partial\mathbb{D}$. Then $\tau - \gamma(0)$ is orthogonal to $\gamma(0)$.*

Proof. Indeed, we already know (Lemma 1.4.17) that $\gamma^{-1}(E(\tau, R)) = E(\tau, R)$ for every $R > 0$. Then Proposition 2.1.5 applied to γ^{-1} yields $1 - |\gamma(0)|^2 = |\tau - \gamma(0)|^2$, i. e., $|\gamma(0)|^2 = \operatorname{Re}(\overline{\tau}\gamma(0))$, and the assertion follows. □

We have seen that can be often useful to transfer the problem under consideration back and forth from $\mathbb{D}$ to the upper half-plane $\mathbb{H}^+$; thus we shall need the description of horocycles in $\mathbb{H}^+$.

Definition 2.1.7. The *horocycle* $E(a, R) \subset \mathbb{H}^+$ of *center* $a \in \partial\mathbb{H}^+ = \mathbb{R} \cup \{\infty\}$ and *radius* $R > 0$ is given by

$$E(a, R) = \left\{ w \in \mathbb{H}^+ \;\middle|\; \lim_{z \to a} [\omega_{\mathbb{H}^+}(w, z) - \omega_{\mathbb{H}^+}(i, z)] < \frac{1}{2} \log R \right\}.$$

From the invariance property of the Poincaré distance, it follows immediately that $\Psi(E(\tau, R)) = E(\Psi(\tau), R)$, where $\Psi\colon \mathbb{D} \to \mathbb{H}^+$ is the Cayley transform extended to $\partial\mathbb{D}$ by setting $\Psi(1) = \infty$. As a consequence, we can easily obtain an analytic description of the horocycles in the upper half-plane.

Lemma 2.1.8. *Given $a \in \partial\mathbb{H}^+$ and $R > 0$, we have*

$$E(a, R) = \left\{ w \in \mathbb{H}^+ \;\middle|\; \frac{1 + a^2}{|w - a|^2} \operatorname{Im} w > \frac{1}{R} \right\} \tag{2.8}$$

if $a \in \mathbb{R}$ and

$$E(\infty, R) = \left\{ w \in \mathbb{H}^+ \;\middle|\; \operatorname{Im} w > \frac{1}{R} \right\}, \tag{2.9}$$

if $a = \infty$.

Proof. By construction, $w \in E(a, R)$ if and only if $\Psi^{-1}(w) \in E(\Psi^{-1}(a), R)$, i. e., if and only if

$$\frac{|\Psi^{-1}(a) - \Psi^{-1}(w)|^2}{1 - |\Psi^{-1}(w)|^2} < R.$$

If $a = \infty$, we have $\Psi^{-1}(a) = 1$ and so

$$\frac{|\Psi^{-1}(a) - \Psi^{-1}(w)|^2}{1 - |\Psi^{-1}(w)|^2} = \frac{1}{\operatorname{Im} w};$$

if instead $a \in \mathbb{R}$ we have

$$\frac{|\Psi^{-1}(a) - \Psi^{-1}(w)|^2}{1 - |\Psi^{-1}(w)|^2} = \frac{|a - w|^2}{(1 + a^2) \operatorname{Im} w}$$

and we are done. □

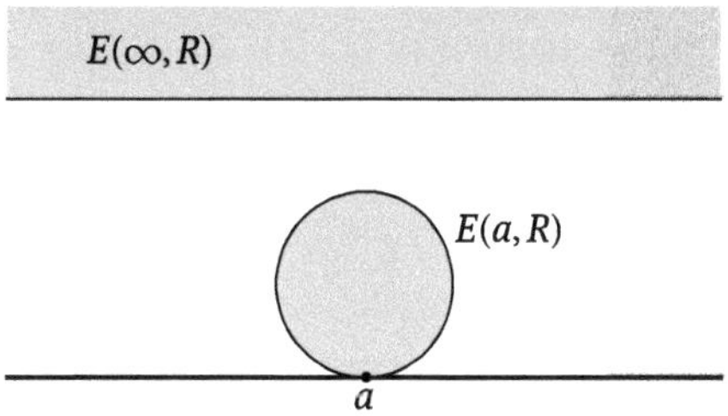

Figure 2.2: Horocycles in $\mathbb{H}^+$.

Geometrically, the horocycles in $\mathbb{H}^+$ are Euclidean disks tangent to the real axis or half-planes parallel to the real axis; see Figure 2.2.

Now we proceed toward a first boundary version of the Schwarz lemma. If we denote by $\mathcal{B}_z$ the family of Poincaré balls centered at $z \in \mathbb{D}$, the Schwarz–Pick lemma says that a function $f \in \mathrm{Hol}(\mathbb{D},\mathbb{D})$ maps each element of $\mathcal{B}_z$ into an element of $\mathcal{B}_{f(z)}$. In a similar way, we can associate to each $\sigma \in \partial\mathbb{D}$ the family $\mathcal{E}_\sigma$ of horocycles centered at σ; then a boundary Schwarz lemma is a statement saying that, under suitable hypotheses, a function $f \in \mathrm{Hol}(\mathbb{D},\mathbb{D})$ maps each element of $\mathcal{E}_\sigma$ into an element of $\mathcal{E}_\tau$ for a suitable $\tau \in \partial\mathbb{D}$, possibly also when f does not extend continuously to $\partial\mathbb{D}$.

Let $f\colon \mathbb{D} \to \mathbb{D}$ be holomorphic. If there does not exist a sequence $\{z_\nu\} \subset \mathbb{D}$ converging toward the boundary such that $|f(z_\nu)| \to 1$, this means that $f(\mathbb{D})$ is relatively compact in $\mathbb{D}$; hence, by the Ritt theorem (see Corollary 1.1.15), f has a fixed point in $\mathbb{D}$, and we can apply the standard Schwarz lemma. So, in our quest for a boundary Schwarz lemma it is natural to assume that there is a sequence $\{z_\nu\} \subset \mathbb{D}$ such that $|z_\nu| \to 1$ and $|f(z_\nu)| \to 1$. Proposition 2.1.3 then suggests that we need some information about the behavior of $(1 - |f(z_\nu)|)/(1 - |z_\nu|)$. One side is provided by the following.

Lemma 2.1.9. *Let $f\colon \mathbb{D} \to \mathbb{D}$ be holomorphic. Then*

$$\frac{1 - |f(z)|}{1 - |z|} \geq \frac{1 - |f(0)|}{1 + |f(0)|} > 0 \tag{2.10}$$

for every $z \in \mathbb{D}$. Moreover, equality in (2.10) holds at one point $z_0 \neq 0$ (and hence everywhere) if and only if $f(z) = e^{i\theta} z$ for a suitable $\theta \in \mathbb{R}$.

Proof. By the Schwarz–Pick lemma and the triangular inequality, for every $z \in \mathbb{D}$ we have

$$\omega(0, f(z)) \leq \omega(0, f(0)) + \omega(f(0), f(z)) \leq \omega(0, f(0)) + \omega(0, z),$$

i. e.,

$$\frac{1 + |f(z)|}{1 - |f(z)|} \leq \frac{1 + |f(0)|}{1 - |f(0)|} \cdot \frac{1 + |z|}{1 - |z|}. \tag{2.11}$$

Let $a_0 = (|f(0)| + |z|)/(1 + |f(0)||z|)$. Then the right-hand side of (2.11) is equal to $(1 + a_0)/(1 - a_0)$, and thus we get $|f(z)| \le a_0$, i. e.,

$$1 - |f(z)| \ge 1 - a_0 = (1 - |z|)\,\frac{1 - |f(0)|}{1 + |f(0)||z|} \ge (1 - |z|)\,\frac{1 - |f(0)|}{1 + |f(0)|}$$

as required. Moreover, we have equality at one point $z_0 \ne 0$ if and only if $f(0) = 0$ and $|f(z_0)| = |z_0|$, and thus, by the Schwarz lemma, if and only if $f(z) = e^{i\theta}z$ for some $\theta \in \mathbb{R}$. □

It is interesting to notice that (2.10) reduces to (1.1) when $f(0) = 0$; in other words, Lemma 2.1.9 is just another incarnation of the Schwarz lemma.

Now we can state and prove the *Julia lemma*.

Theorem 2.1.10 (Julia lemma, 1920). *Let $f\colon \mathbb{D} \to \mathbb{D}$ be a holomorphic function and take $\sigma \in \partial\mathbb{D}$ such that*

$$\liminf_{z\to\sigma} \frac{1 - |f(z)|}{1 - |z|} = \alpha < \infty. \tag{2.12}$$

Then there exists a unique $\tau \in \partial\mathbb{D}$ such that

$$\frac{|\tau - f(z)|^2}{1 - |f(z)|^2} \le \alpha\frac{|\sigma - z|^2}{1 - |z|^2} \tag{2.13}$$

for every $z \in \mathbb{D}$, i. e.,

$$f(E(\sigma, R)) \subseteq E(\tau, \alpha R) \tag{2.14}$$

for every $R > 0$. Moreover, equality in (2.13) *holds at one point (and hence everywhere) if and only if $f \in \mathrm{Aut}(\mathbb{D})$.*

Proof. The Schwarz–Pick lemma yields

$$\left|\frac{f(z) - f(w)}{1 - \overline{f(w)}f(z)}\right| \le \left|\frac{z - w}{1 - \overline{w}z}\right|$$

for all $z, w \in \mathbb{D}$. Taking the square, subtracting 1 and recalling (1.5) we get

$$\frac{|1 - \overline{f(w)}f(z)|^2}{1 - |f(z)|^2} \le \frac{1 - |f(w)|^2}{1 - |w|^2} \cdot \frac{|1 - \overline{w}z|^2}{1 - |z|^2}. \tag{2.15}$$

Now choose a sequence $\{z_\nu\} \subset \mathbb{D}$ such that $z_\nu \to \sigma \in \partial\mathbb{D}$ and

$$\lim_{\nu\to\infty} \frac{1 - |f(z_\nu)|}{1 - |z_\nu|} = \alpha;$$

moreover up to a subsequence we can assume that $\{f(z_\nu)\}$ converges to some $\tau \in \partial\mathbb{D}$ as $\nu \to \infty$. Then setting $w = z_\nu$ in (2.15) and taking the limit as $\nu \to \infty$ we obtain (2.13).

The point τ is unique: if (2.14) holds for two distinct points $\tau_1, \tau_2 \in \partial\mathbb{D}$, then we get a contradiction taking R so small that $E(\tau_1, \alpha R) \cap E(\tau_2, \alpha R) = \emptyset$.

The proof of the last statement is only a bit more involved. If $f \in \mathrm{Aut}(\mathbb{D})$, then using (2.7) it is easy to show that (2.13) is an equality for all $z \in \mathbb{D}$ when $\tau = f(\sigma)$.

For the converse, using (2.2) we can rewrite (2.13) as

$$\mathrm{Re}\left(\frac{1}{\alpha}\frac{\sigma + z}{\sigma - z} - \frac{\tau + f(z)}{\tau - f(z)}\right) \le 0.$$

If the equality holds at some point in $\mathbb{D}$, the maximum principle for harmonic functions yields

$$\frac{\tau + f(z)}{\tau - f(z)} = \frac{1}{\alpha}\frac{\sigma + z}{\sigma - z} + ib$$

for some $b \in \mathbb{R}$, i. e.,

$$f(z) = \sigma_0 \frac{z - a}{1 - \bar{a}z}, \tag{2.16}$$

where

$$\sigma_0 = \tau\bar{\sigma}\frac{1 + \alpha - ib\alpha}{1 + \alpha + ib\alpha} \in \partial\mathbb{D}$$

and

$$a = \sigma\frac{\alpha - ib\alpha - 1}{\alpha - ib\alpha + 1} \in \mathbb{D};$$

thus $f \in \mathrm{Aut}(\mathbb{D})$, as claimed □

In particular, if the liminf in (2.12) is finite (it is positive by Lemma 2.1.9), then it is possible to associate a point $\tau \in \partial\mathbb{D}$ to f and σ in a very definite way. To better understand what is going on, let us look at the situation from a slightly different point of view.

Definition 2.1.11. Given $f\colon \mathbb{D} \to \mathbb{D}$ holomorphic and $\sigma, \tau \in \partial\mathbb{D}$, set

$$\beta_f(\sigma, \tau) = \sup_{z\in\mathbb{D}}\left\{\frac{|\tau - f(z)|^2}{1 - |f(z)|^2} \Big/ \frac{|\sigma - z|^2}{1 - |z|^2}\right\} \in (0, +\infty]. \tag{2.17}$$

The *boundary dilation* $\beta_f(\sigma)$ of f at σ is

$$\beta_f(\sigma) = \inf_{\tau\in\partial\mathbb{D}} \beta_f(\sigma, \tau). \tag{2.18}$$

The boundary dilation encodes the behaviour of horocycles under the action of a holomorphic function.

Lemma 2.1.12. *Let $f \in \mathrm{Hol}(\mathbb{D},\mathbb{D})$ and $\sigma,\tau \in \partial\mathbb{D}$. Then*

$$f(E(\sigma,R)) \subseteq E(\tau,\beta_f(\sigma,\tau)R) \tag{2.19}$$

for every $R > 0$. Furthermore, there exists at most one $\tau \in \partial\mathbb{D}$ such that $\beta_f(\sigma,\tau) < +\infty$; in particular, if $\beta_f(\sigma) < +\infty$ then there exists a unique $\tau \in \partial\mathbb{D}$ such that $\beta_f(\sigma) = \beta_f(\sigma,\tau)$.

Proof. Formula (2.19) is an immediate consequence of the definitions. If there exist distinct $\tau_1,\tau_2 \in \partial\mathbb{D}$ such that $\beta_f(\sigma,\tau_1),\beta_f(\sigma,\tau_2) < +\infty$, then we get a contradiction choosing $R > 0$ so small that $E(\tau_1,\beta_f(\sigma,\tau_1)R) \cap E(\tau_2,\beta_f(\sigma,\tau_2)R) = \emptyset$. □

Remark 2.1.13. It might happen that $\beta_f(\sigma) = +\infty$; an easy example is given by $f(z) = \frac{1}{2}z$, when we have $\beta_f(\sigma) = +\infty$ for all $\sigma \in \partial\mathbb{D}$ because $|f(z)| < \frac{1}{2}$ for all $z \in \mathbb{D}$. A less trivial example, having $|f(z)| \to 1$ as $z \to \sigma$, will be given in Example 2.1.18.

On the other hand, if $f \in \mathrm{Hol}(\mathbb{D},\mathbb{D})$ extends continuously to a point $\sigma \in \partial\mathbb{D}$ then it is clear that $\beta_f(\sigma,\tau) = +\infty$ if $\tau \neq f(\sigma)$, and hence $\beta_f(\sigma) = \beta_f(\sigma,f(\sigma))$.

Example 2.1.14. If $\gamma \in \mathrm{Aut}(\mathbb{D})$ and $\sigma \in \partial\mathbb{D}$ then using the previous remark and (2.7), it is easy to check that

$$\beta_\gamma(\sigma) = \frac{1 - |\gamma^{-1}(0)|^2}{|\sigma - \gamma^{-1}(0)|^2}.$$

The boundary dilation measure how fast $f(z)$ tends to the boundary of $\partial\mathbb{D}$ as $z \to \sigma$. Indeed, the boundary dilation is a fancy way to express the liminf (2.12).

Proposition 2.1.15. *Take $f \in \mathrm{Hol}(\mathbb{D},\mathbb{D})$ and $\sigma \in \partial\mathbb{D}$. Then*

$$\liminf_{z\to\sigma} \frac{1 - |f(z)|}{1 - |z|} = \beta_f(\sigma).$$

Proof. For the sake of brevity, set $\beta = \beta_f(\sigma)$ and

$$\alpha = \liminf_{z\to\sigma} \frac{1 - |f(z)|}{1 - |z|}.$$

Theorem 2.1.10 tells that $\beta \le \alpha$; so $\beta = +\infty$ implies $\alpha = +\infty$ and it remains to prove that $\alpha \le \beta$ when β is finite.

For every $\nu \in \mathbb{N}^*$ set $z_\nu = \frac{\nu-1}{\nu+1}\sigma$. Clearly, $z_\nu \in \mathbb{D}$ and $z_\nu \to \sigma$ as $\nu \to +\infty$; moreover, a quick computation shows that $z_\nu \in \partial E(\sigma,1/\nu)$. Since $\beta < +\infty$, by Lemma 2.1.12 there is a unique $\tau \in \partial\mathbb{D}$ such that $f(z_\nu) \in \overline{E(\tau,\beta/\nu)}$. Now, the Euclidean diameter of $E(\tau,R)$ is $2R/(1+R)$; therefore,

$$|\tau - f(z_\nu)| \le \frac{2\beta}{\beta + \nu}.$$

Since $1 - |z_\nu| = 2/(\nu + 1)$, it follows that

$$\alpha \le \limsup_{\nu\to\infty} \frac{1 - |f(z_\nu)|}{1 - |z_\nu|} \le \lim_{\nu\to\infty} \beta \cdot \frac{\nu + 1}{\nu + \beta} = \beta$$

and we are done. □

In particular, the boundary dilation is finite if and only if z does not approach the boundary too much faster than $f(z)$. To give an example of a function f with infinite boundary dilation, we transfer the stage to $\mathbb{H}^+$.

Definition 2.1.16. Given $F \in \mathrm{Hol}(\mathbb{H}^+, \mathbb{H}^+)$ and $a, b \in \partial\mathbb{H}^+ = \mathbb{R} \cup \{\infty\}$, set

$$\beta_F^{\mathbb{H}^+}(a, b) = \inf_{w\in\mathbb{H}^+} \left\{ \frac{1}{1 + a^2} \frac{|w - a|^2}{\operatorname{Im} w} \Big/ \frac{1}{1 + b^2} \frac{|F(w) - b|^2}{\operatorname{Im} F(w)} \right\} \in [0, +\infty),$$

where we put $|w - \infty|^2/(1 + \infty^2) = 1$ for all $w \in \mathbb{H}^+$. The *boundary dilation* $\beta_F^{\mathbb{H}^+}(a)$ of F at a is then given by

$$\beta_F^{\mathbb{H}^+}(a) = \sup_{b\in\partial\mathbb{H}^+} \beta_F^{\mathbb{H}^+}(a, b).$$

Remark 2.1.17. The formulas used in the proof of Lemma 2.1.8 show that the boundary dilation of $F \in \mathrm{Hol}(\mathbb{H}^+, \mathbb{H}^+)$ is related to the boundary dilation of $f = \Psi^{-1} \circ F \circ \Psi \in \mathrm{Hol}(\mathbb{D}, \mathbb{D})$, where $\Psi\colon \mathbb{D} \to \mathbb{H}^+$ is the Cayley transform, by

$$\beta_F^{\mathbb{H}^+}(a, b) = \frac{1}{\beta_f(\Psi^{-1}(a), \Psi^{-1}(b))} \tag{2.20}$$

and

$$\beta_F^{\mathbb{H}^+}(a) = \frac{1}{\beta_f(\Psi^{-1}(a))}. \tag{2.21}$$

In particular, $\beta_f(\sigma) = +\infty$ if and only if $\beta_F^{\mathbb{H}^+}(\Psi(\sigma)) = 0$; moreover, if $\beta_F^{\mathbb{H}^+}(a) > 0$ then there exists a unique $b \in \partial\mathbb{H}^+$ such that $\beta_F^{\mathbb{H}^+}(a) = \beta_F^{\mathbb{H}^+}(a, b)$.

Example 2.1.18. Let $F \in \mathrm{Hol}(\mathbb{H}^+, \mathbb{H}^+)$ be given by $F(w) = \log w$, where log is the principal branch of the logarithm in $\mathbb{H}^+$. Then consideration of the points $w_\nu = i\nu$ with $\nu \in \mathbb{N}^*$ shows that

$$\beta_F^{\mathbb{H}^+}(\infty, b) \le \inf_{\nu\in\mathbb{N}^*} \left\{ \frac{1}{\nu} \Big/ \frac{1}{1 + b^2} \frac{|\log \nu + i\frac{\pi}{2} - b|^2}{\pi/2} \right\} = 0$$

for all $b \in \partial\mathbb{H}^+$, and thus $\beta_F^{\mathbb{H}^+}(\infty) = 0$. In particular, by (2.20) we get $\beta_f(1) = +\infty$, where $f = \Psi^{-1} \circ F \circ \Psi \in \mathrm{Hol}(\mathbb{D}, \mathbb{D})$.

Using the Cayley transform, we get the following version of the Julia lemma for the upper half-plane

Theorem 2.1.19. *Let $F\colon \mathbb{H}^+ \to \mathbb{H}^+$ be a holomorphic function and take $a \in \partial\mathbb{H}^+$. Then*

$$\beta_F^{\mathbb{H}^+}(a) = \limsup_{w\to a} \frac{|F(w)+i|^2}{\operatorname{Im} F(w)} \frac{\operatorname{Im} w}{|w+i|^2}. \tag{2.22}$$

In particular, if $\beta_F^{\mathbb{H}^+}(a) > 0$ then the unique $b \in \partial\mathbb{H}^+$ so that $\beta_F^{\mathbb{H}^+}(a) = \beta_F^{\mathbb{H}^+}(a,b)$ is such that

$$F(E(a,R)) \subseteq E\left(b, \frac{1}{\beta_F^{\mathbb{H}^+}(a)} R\right) \tag{2.23}$$

for every $R > 0$, i. e.,

$$\frac{1+b^2}{|F(w)-b|^2} \operatorname{Im} F(w) \ge \beta_F^{\mathbb{H}^+}(a) \frac{1+a^2}{|w-a|^2} \operatorname{Im} w \tag{2.24}$$

for every $w \in \mathbb{H}^+$, where we put $(1+\infty^2)/|w-\infty|^2 = 1$. In particular, if $\beta = \beta_F^{\mathbb{H}^+}(\infty,\infty) > 0$ we have

$$\operatorname{Im} F(w) \ge \beta \operatorname{Im} w$$

for all $w \in \mathbb{H}^+$. Furthermore, equality in (2.24) holds at one point (and hence everywhere) if and only if $F \in \operatorname{Aut}(\mathbb{H}^+)$.

Proof. Put $f = \Psi^{-1} \circ F \circ \Psi \in \operatorname{Hol}(\mathbb{D},\mathbb{D})$ and $\sigma = \Psi^{-1}(a)$. Then (2.21) and Proposition 2.1.15 yield

$$\beta_F^{\mathbb{H}^+}(a) = \frac{1}{\beta_f(\sigma)} = \limsup_{w\to a} \frac{1-|\Psi^{-1}(w)|}{1-|\Psi^{-1}(F(w))|}. \tag{2.25}$$

Now,

$$\begin{aligned} \frac{1-|\Psi^{-1}(w)|}{1-|\Psi^{-1}(F(w))|} &= \frac{1-|\Psi^{-1}(w)|^2}{1-|\Psi^{-1}(F(w))|^2} \cdot \frac{1+|\Psi^{-1}(F(w))|}{1+|\Psi^{-1}(w)|} \\ &= \frac{|F(w)+i|^2}{\operatorname{Im} F(w)} \frac{\operatorname{Im} w}{|w+i|^2} \frac{1+|\Psi^{-1}(F(w))|}{1+|\Psi^{-1}(w)|}. \end{aligned}$$

Choose a sequence $\{w_\nu\} \subset \mathbb{H}^+$ converging to a and realizing the lim sup in (2.25). Up to a subsequence, we can also assume that $|\Psi^{-1}(F(w_\nu))| \to \delta \in [0,1]$. Since $|\Psi^{-1}(w_\nu)| \to 1$, if $\delta < 1$ then automatically $\beta_F^{\mathbb{H}^+}(a) = 0$. If $\delta = 1$, the right-most fraction in the last formula, computed in $w = w_\nu$, tends to 1 as $\nu \to +\infty$, and hence in all cases we get (2.22).

The rest of the statement follows immediately from Lemma 2.1.8 and from the usual Julia lemma in the disk Theorem 2.1.10. □

We now prove a few more formulas, that we shall need later on, for the computation of the boundary dilation.

Lemma 2.1.20. *Take $f \in \mathrm{Hol}(\mathbb{D}, \mathbb{D})$ and $\sigma \in \partial\mathbb{D}$. Then*

$$\beta_f(\sigma) = \lim_{r\to 1^-} \frac{1 - |f(r\sigma)|}{1-r} = \lim_{r\to 1^-} \frac{1 - |f(r\sigma)|^2}{1-r^2}. \tag{2.26}$$

Proof. By Proposition 2.1.15, we have

$$\beta_f(\sigma) \le \liminf_{r\to 1^-} \frac{1 - |f(r\sigma)|}{1-r};$$

in particular the first equality in (2.26) is proven when $\beta_f(\sigma) = +\infty$. Assume then that $\beta_f(\sigma) < +\infty$. An easy computation shows that $r\sigma \in \partial E(\sigma, \frac{1-r}{1+r})$; therefore, Theorem 2.1.10 yields $f(r\sigma) \in \overline{E(\tau, \beta_f(\sigma)\frac{1-r}{1+r})}$ for a suitable $\tau \in \partial\mathbb{D}$. Since $E(\tau, R)$ is an Euclidean disk of radius $R/(R+1)$ internally tangent to $\partial\mathbb{D}$ in τ it follows that

$$1 - |f(r\sigma)| \le |\tau - f(r\sigma)| \le 2\frac{\beta_f(\sigma)\frac{1-r}{1+r}}{1+\beta_f(\sigma)\frac{1-r}{1+r}}.$$

Therefore,

$$\limsup_{r\to 1^-} \frac{1 - |f(r\sigma)|}{1-r} \le \limsup_{r\to 1^-} \frac{2\beta_f(\sigma)}{1+r+\beta_f(\sigma)(1-r)} = \beta_f(\sigma)$$

and the first equality is proved.

To prove the second equality in (2.26), first of all notice that

$$\frac{2}{1+r}\frac{1 - |f(r\sigma)|}{1-r} \ge \frac{1 + |f(r\sigma)|}{1+r}\frac{1 - |f(r\sigma)|}{1-r} = \frac{1 - |f(r\sigma)|^2}{1-r^2} \ge \frac{1}{1+r}\frac{1 - |f(r\sigma)|}{1-r}. \tag{2.27}$$

From this, the second equality immediately follows when $\beta_f(\sigma) = +\infty$. If $\beta_f(\sigma) < +\infty$ then $|f(r\sigma)| \to 1$ as $r \to 1^-$, and thus the assertion follows again from (2.27). □

Proposition 2.1.21. *Take $f \in \mathrm{Hol}(\mathbb{D}, \mathbb{D})$ and $\sigma \in \partial\mathbb{D}$. Then*

$$\frac{1}{2}\log \beta_f(\sigma) = \liminf_{z\to\sigma}[\omega(0, z) - \omega(0, f(z))]. \tag{2.28}$$

Proof. First of all,

$$\begin{aligned}\omega(0,z) - \omega(0, f(z)) &= \frac{1}{2}\log\frac{1+|z|}{1+|f(z)|} + \frac{1}{2}\log\frac{1-|f(z)|}{1-|z|}\\ &\ge \frac{1}{2}\log\frac{1}{2} + \frac{1}{2}\log\frac{1-|f(z)|}{1-|z|}\end{aligned} \tag{2.29}$$

and, therefore,

$$\liminf_{z\to\sigma}[\omega(0,z)-\omega(0,f(z))] \geq \frac{1}{2}\log\frac{1}{2}+\frac{1}{2}\log\beta_f(\sigma),$$

where we used Proposition 2.1.15. In particular, if $\beta_f(\sigma) = +\infty$ the assertion follows; assume then that $\beta_f(\sigma) < +\infty$. Choose a sequence $z_k \to \sigma$ such that $(1-|f(z_k)|)/(1-|z_k|) \to \beta_f(\sigma)$. Then $|f(z_k)| \to 1$, and thus

$$\begin{aligned}\liminf_{z\to\sigma}[\omega(0,z)-\omega(0,f(z))] &\leq \liminf_{k\to+\infty}[\omega(0,z_k)-\omega(0,f(z_k))]\\ &= \frac{1}{2}\log\lim_{k\to+\infty}\frac{1+|z_k|}{1+|f(z_k)|}+\frac{1}{2}\log\lim_{k\to+\infty}\frac{1-|f(z_k)|}{1-|z_k|}\\ &= \frac{1}{2}\log\beta_f(\sigma).\end{aligned}$$

In particular, the right-hand side of (2.28) is finite. If $\{z_k\}$ is instead a sequence converging to σ realizing the lim inf in (2.28), by (2.29) we must have $|f(z_k)| \to 1$, and hence

$$\begin{aligned}\liminf_{z\to\sigma}[\omega(0,z)-\omega(0,f(z))] &= \lim_{k\to+\infty}[\omega(0,z_k)-\omega(0,f(z_k))]\\ &= \frac{1}{2}\log\lim_{k\to+\infty}\frac{1+|z_k|}{1+|f(z_k)|}+\frac{1}{2}\log\lim_{k\to+\infty}\frac{1-|f(z_k)|}{1-|z_k|}\\ &\geq \frac{1}{2}\log\beta_f(\sigma),\end{aligned}$$

again by Proposition 2.1.15, and we are done. □

Using this formula, we can give a version of the Julia lemma valid in hyperbolic domains in compact Riemann surfaces.

Proposition 2.1.22. *Let $D \subset \widehat{X}$ be a hyperbolic domain in a compact Riemann surface $\widehat{X}$. Let $f \in \mathrm{Hol}(D,D)$ and $\sigma \in \partial D$, and for a fixed $z_0 \in D$, set*

$$\frac{1}{2}\log\beta_f(\sigma) = \liminf_{w\to\sigma}[\omega_D(z_0,w)-\omega_D(z_0,f(w))]. \tag{2.30}$$

Assume that $\beta_f(\sigma) < +\infty$. Then there is a $\tau \in \partial D$ such that

$$\liminf_{w\to\tau}[\omega_D(f(z),w)-\omega_D(z_0,w)] \leq \limsup_{w\to\sigma}[\omega_D(z,w)-\omega_D(z_0,w)]+\frac{1}{2}\log\beta_f(\sigma). \tag{2.31}$$

Proof. Choose a sequence $\{w_\nu\} \subset D$ converging to σ and realizing the lim inf in (2.30); up to a subsequence, we can also assume that $f(w_\nu) \to \tau \in \overline{D}$. If $\tau \in D$, then $\beta_f(\sigma) = +\infty$, because ω_D is complete; so necessarily $\tau \in \partial D$. Then we have

$$\begin{aligned}\liminf_{w\to\tau}[\omega(f(z),w)-\omega(z_0,w)] &\leq \liminf_{\nu\to+\infty}[\omega(f(z),f(w_\nu))-\omega(z_0,f(w_\nu))]\\ &= \liminf_{\nu\to+\infty}[\omega(f(z),f(w_\nu))-\omega(z_0,w_\nu)]+\liminf_{\nu\to+\infty}[\omega(z_0,w_\nu)-\omega(z_0,f(w_\nu))]\end{aligned}$$

$$\leq \limsup_{\nu\to+\infty}[\omega(z,w_\nu)-\omega(z_0,w_\nu)] + \lim_{\nu\to\infty}[\omega(z_0,w_\nu)-\omega(z_0,f(w_\nu))]$$
$$\leq \limsup_{w\to\sigma}[\omega(z,w)-\omega(z_0,w)] + \frac{1}{2}\log\beta_f(\sigma)$$

and we are done. □

Recalling Propositions 2.1.4 and 2.1.21, we see that for $D = \mathbb{D}$ we have recovered the usual Julia lemma. Actually, in Section 3.3 we shall see that if $D \subset \widehat{X}$ is a hyperbolic domain of regular type then the lim inf and lim sup in (2.31) are actually limits and they can be used to define horocycles as in Proposition 2.1.4 obtaining a Julia lemma for hyperbolic domains of regular type. Thus using the Poincaré distance it is possible to express and prove the Julia lemma in intrinsic metric terms, without using extrinsic Euclidean constructions.

We end this section with an appealing measure-theoretic interpretation of the boundary dilation obtained using the Herglotz representation formula (A.3).

Let $f \in \mathrm{Hol}(\mathbb{D},\mathbb{D})$ and $\sigma,\tau \in \partial\mathbb{D}$. Put $F = (\tau+f)/(\tau-f)$; by construction,

$$\operatorname{Re}F(z) = \operatorname{Re}\frac{\tau+f(z)}{\tau-f(z)} = \frac{1-|f(z)|^2}{|\tau-f(z)|^2} \geq \frac{1}{\beta_f(\sigma,\tau)}\frac{1-|z|^2}{|\sigma-z|^2} \geq 0$$

for all $z \in \mathbb{D}$. Then Theorem A.3.3 associates to F a positive Borel measure μ_τ on $\partial\mathbb{D}$ such that

$$F(z) = \int_{\partial\mathbb{D}} \frac{\zeta+z}{\zeta-z}\,d\mu_\tau(\zeta) + i\operatorname{Im}F(0). \tag{2.32}$$

Let $c_{\sigma,\tau} \geq 0$ denote $\mu_\tau(\{\sigma\})$, i. e., the mass of μ_τ concentrated at the point σ, and put $\mu_0 = \mu - c_{\sigma,\tau}\delta_\tau$, where δ_τ is the Dirac measure concentrated in τ. Then (2.32) becomes

$$\frac{\tau+f(z)}{\tau-f(z)} = c_{\sigma,\tau}\frac{\sigma+z}{\sigma-z} + \int_{\partial\mathbb{D}} \frac{\zeta+z}{\zeta-z}\,d\mu_0(\zeta) + i\operatorname{Im}F(0). \tag{2.33}$$

Taking the real part of this equation, we obtain

$$\frac{1-|f(z)|^2}{|\tau-f(z)|^2} = c_{\sigma,\tau}\frac{1-|z|^2}{|\sigma-z|^2} + \int_{\partial\mathbb{D}} \frac{1-|z|^2}{|\zeta-z|^2}\,d\mu_0(\zeta), \tag{2.34}$$

that we can rewrite as

$$\frac{1-|f(z)|^2}{|\tau-f(z)|^2} \Big/ \frac{1-|z|^2}{|\sigma-z|^2} = c_{\sigma,\tau} + \int_{\partial\mathbb{D}} \frac{|\sigma-z|^2}{|\zeta-z|^2}\,d\mu_0(\zeta).$$

Taking the infimum for $z \in \mathbb{D}$ of the left-hand side we get $1/\beta_f(\sigma,\tau)$. We claim that the infimum for $z \in \mathbb{D}$ of the integral in the right-hand side is 0. Indeed, fix $\varepsilon > 0$ and choose $\delta > 0$ so small that the μ_0-measure of the arc C of amplitude δ around σ is less than ε; this can be done because, by construction, $\mu_0(\{\sigma\}) = 0$. Taking $z = r\sigma$ with $r \in (0,1)$, we get

$$\begin{aligned}\int_{\partial\mathbb{D}} \frac{|\sigma - z|^2}{|\zeta - z|^2}\, d\mu_0(\zeta) &= (1-r)^2 \int_{\partial\mathbb{D}} \frac{1}{|\zeta - r\sigma|^2}\, d\mu_0(\zeta) \\ &= (1-r)^2 \int_C \frac{1}{|\zeta - r\sigma|^2}\, d\mu_0(\zeta) + (1-r)^2 \int_{\partial\mathbb{D}\setminus C} \frac{1}{|\zeta - r\sigma|^2}\, d\mu_0(\zeta) \\ &= I_0(r) + I_1(r).\end{aligned}$$

Now, $|\zeta - r\sigma| \geq 1 - r$; therefore, $I_0(r) \leq \mu_0(C) < \varepsilon$ for all $r \in (0,1)$. On the other hand, there exists $\eta > 0$ such that $|\zeta - r\sigma| > \eta$ for all $\zeta \in \partial\mathbb{D} \setminus C$ and all $r \in (0,1)$; therefore,

$$I_1(r) \leq \frac{(1-r)^2}{\eta^2} \mu_0(\partial\mathbb{D})$$

and hence $I_1(r)$ tends to 0 as $r \to 1^-$. Putting all together, we find that

$$\inf_{z\in\mathbb{D}} \int_{\partial\mathbb{D}} \frac{|\sigma - z|^2}{|\zeta - z|^2}\, d\mu_0 = 0,$$

and this proves that

$$\beta_f(\sigma, \tau) = \mu_\tau(\{\sigma\})^{-1},$$

that is the interpretation we announced.

***Notes to Section* 2.1**

Horocycles were born with the non-Euclidean geometry; in particular, with the Poincaré model of hyperbolic geometry. In Euclidean geometry, a circumference can be defined as a trajectory orthogonal to a pencil of straight lines issuing from a given point. If we take as center of the pencil the point at infinity (i. e., if we have a pencil of lines parallel to a given one), we obtain another straight line. In hyperbolic geometry, a trajectory orthogonal to a pencil of hyperbolic lines (i. e., geodesics for the Poincaré metric) is called *cycle*. If the pencil consists of geodesics issuing from an interior point, we obtain a Poincaré circle with the same center as the pencil. If we instead take a trajectory orthogonal to a pencil of hyperbolic lines passing through a given point in the boundary (i. e., a pencil of lines parallel to a given one), we obtain exactly a horocycle (Figure 2.3).

It is interesting to notice that in hyperbolic geometry there is a third kind of cycle. If we take a pencil of hyperbolic lines orthogonal to a given one, an associated orthogonal trajectory is called a *hypercycle*. In Euclidean geometry, horocycles and hypercycles are one and the same thing; in hyperbolic geometry, they are distinct. In fact, geometrically, a hypercycle is a circular arc connecting two distinct points of $\partial\mathbb{D}$ (Figure 2.4).

A classical exposition of the geometric theory of cycles in hyperbolic geometry is [218]; more recent treatments can be found in [160], [159], and [229].

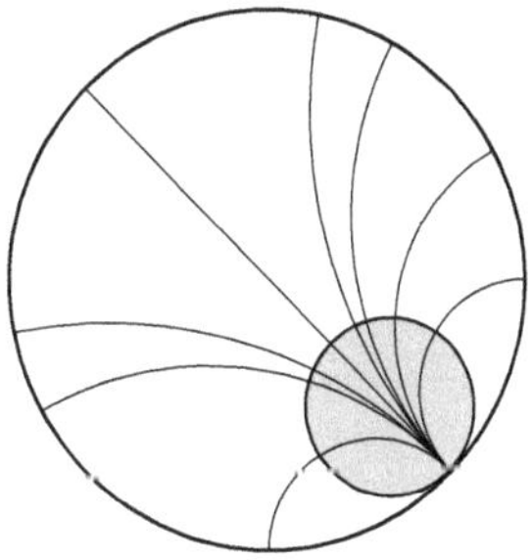

Figure 2.3: A horocycle orthogonal to a pencil of hyperbolic lines passing through a given point in the boundary.

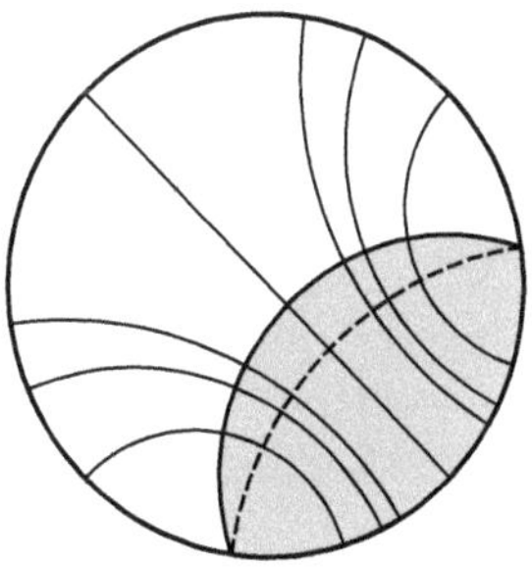

Figure 2.4: A hypercycle orthogonal to a pencil of hyperbolic lines orthogonal to a given hyperbolic line (dashed).

A remark about notation. We used E to denote horocycles (instead of, e. g., H) because the multidimensional horocycles in the unit ball of $\mathbb{C}^n$, defined with formula (2.5) replacing the Poincaré distance with the Bergmann (or Kobayashi) distance of the ball, are ellipsoids.

Proposition 2.1.3 is classical; it can be found for instance in [218]. On the other hand, Proposition 2.1.4 is the one-variable version of the lesser known characterization given by Yang [427] (see also [307]) of horospheres in the n-dimensional ball.

In connection with Proposition 2.1.5, it should be mentioned that every pair of horocycles is congruent under $\mathrm{Aut}(\mathbb{D})$, as it is easily verified.

A consequence of Lemma 2.1.9 is that

$$\beta_f(\sigma) \geq \frac{1-|f(0)|}{1+|f(0)|}.$$

This inequality can be improved. For instance, in [313] (see also Corollary 2.3.11) it is shown that

$$\beta_f(\sigma) \geq \frac{2}{1+|f^h(0)|}\frac{|\tau - f(0)|^2}{1-|f(0)|^2} \geq \frac{2}{1+|f^h(0)|}\frac{1-|f(0)|}{1+|f(0)|},$$

where $\tau \in \partial\mathbb{D}$ is the unique point such that $\beta_f(\sigma) = \beta_f(\sigma, \tau)$ and f^h is the hyperbolic derivative introduced in Definition 1.5.1. A more precise statement is contained in [9, 155] (see also [286]) where it is proved that if $f \in \mathrm{Hol}(\mathbb{D}, \mathbb{D}) \setminus \mathrm{Aut}(\mathbb{D})$ then

$$\beta_f(\sigma) \geq \left[\frac{1-|z_1|^2}{|\sigma - z_1|^2} + \frac{1-|z_2|^2}{|\sigma - z_2|^2}\frac{|f^*(\sigma,z_1) - f^*(z_2,z_1)|^2}{1-|f^*(z_2,z_1)|^2}\right]\frac{|\tau - f(z_1)|^2}{1-|f(z_1)|^2}$$

for all $z_1, z_2 \in \mathbb{D}$, with equality in one pair of points (and hence everywhere) if and only if f is a Blaschke product of degree 2. In this formula, f^* is the hyperbolic difference quotient (see Definition 1.5.1) and

$$f^*(\sigma,z) = \overline{\tau}\sigma\frac{\tau - f(z)}{\overline{\tau} - \overline{f(z)}}\frac{\overline{\sigma} - \overline{z}}{\sigma - z} \in \partial\mathbb{D}.$$

The papers [9] and [155] contain even stronger versions of this inequality involving higher order derivatives.

The Julia lemma (Theorem 2.1.10) was first proved by Julia [216] in 1920; in [218], there is an analogous result for hypercycles. In the literature, the Julia lemma is often ancillary to Theorem 2.3.2, and sometimes it is this latter result which is called "Julia lemma."

[63] contains two results related to the Julia lemma. Let $f \in \mathrm{Hol}(\mathbb{D},\mathbb{D})$ and $\sigma, \tau \in \partial\mathbb{D}$. We define $\Psi_f, \Phi_f \colon \mathbb{R}^+ \to \mathbb{R}^+$ by setting

$$\Psi_f(r) = \inf\{R > 0 \mid f(E(\sigma,r)) \subseteq E(\tau,R)\} \quad \text{and} \quad \Phi_f(r) = \frac{D_\tau(f(E(\sigma,r)))}{2 - D_\tau(f(E(\sigma,r)))},$$

where $D_\tau(A) = \sup\{|\tau - z| \mid z \in A\}$; notice that $D_\tau(E(\tau,R))$ is the Euclidean diameter d of $E(\tau,R)$, and hence $\frac{d}{2-d} = R$. Then the functions $r \mapsto \frac{1}{r}\Psi_f(r)$ and $r \mapsto \frac{1}{r}\Phi_f(r)$ are decreasing and converge to $\beta_f(\sigma,\tau)$ as $r \to 0^+$. Moreover, the function $r \mapsto \frac{1}{r}\Phi_f(r)$ is strictly decreasing unless $f \in \mathrm{Aut}(\mathbb{D})$, and in that case it is constant. Finally, either $\Psi_f \equiv +\infty$ or $\Psi_f(r) < +\infty$ for all $r > 0$.

Propositions 2.1.21 and 2.1.22 come from [4]. This intrinsic metric approach plays a fundamental rôle in the study of holomorphic dynamics in several complex variables; see, e. g., [8] and references therein.

The measure-theoretic interpretation of the boundary dilation is due to R. Nevanlinna [305].

A multipoint version of Julia lemma, along the lines of the multipoint Schwarz–Pick lemma discussed in Section 1.5, is contained in [9]. Take $f \in \mathrm{Hol}(\mathbb{D},\mathbb{D}) \setminus \mathrm{Aut}(\mathbb{D})$ and $\sigma \in \partial\mathbb{D}$ such that $\beta_f(\sigma) < +\infty$, and let $\tau \in \partial\mathbb{D}$ be the unique point such that $\beta_f(\sigma) = \beta_f(\sigma,\tau)$. Then the boundary dilation of $f^*(\cdot,w)$ at $\sigma \in \partial\mathbb{D}$ is given by

$$\beta_{f^*(\cdot,w)}(\sigma) = \beta_f(\sigma)\frac{1-|f(w)|^2}{|\tau - f(w)|^2} - \frac{1-|w|^2}{|\sigma - w|^2}$$

and we have

$$\frac{|f^*(\sigma,w) - f^*(z,w)|^2}{1-|f^*(z,w)|^2} \leq \beta_{f^*(\cdot,w)}(\sigma)\frac{|\sigma - z|^2}{1-|z|^2}$$

for all $z, w \in \mathbb{D}$, with equality for some $z, w \in \mathbb{D}$ if and only if there is equality everywhere if and only if f is a Blaschke product of degree 2. In the same paper is explained how to obtain versions of the Julia lemma involving more than 2 points.

See also [48, 285] for other kinds of multipoint Julia lemmas, and [68] for a good account of the history of the Julia lemma.

2.2 Stolz regions and nontangential limits

Let $f \in \mathrm{Hol}(\mathbb{D}, \mathbb{D})$ and $\sigma \in \partial\mathbb{D}$ be such that $\beta = \beta_f(\sigma)$ is finite, and let τ be the unique point of $\partial\mathbb{D}$ such that $\beta_f(\sigma, \tau) = \beta$ (Lemma 2.1.12). Now choose a sequence $\{z_\nu\} \subset \mathbb{D}$ converging to σ so that $(1-|f(z_\nu)|)/(1-|z_\nu|)$ admits a finite limit. Therefore, $|f(z_\nu)| \to 1$ as $\nu \to \infty$ and so every limit point of the sequence $\{f(z_\nu)\}$ must belong to $\partial\mathbb{D}$. Let $\tau_1 \in \partial\mathbb{D}$ be one of these limit points; the same argument used to prove the Julia lemma then shows that $\beta_f(\sigma, \tau_1)$ is finite. But this means that $\tau_1 = \tau$, and so $f(z_\nu) \to \tau$ as $\nu \to +\infty$.

The conclusion of this argument is that $\beta_f(\sigma) < +\infty$ should imply that f admits a limit in some sense when z tends to σ. Actually, much more is true: even f' admits the limit at σ and this limit can be computed starting from $\beta_f(\sigma)$. This is the content of the Julia–Wolff–Carathéodory theorem that we shall describe in the next section. But first in this section we shall introduce the kind of limit we are interested in.

Definition 2.2.1. Given $\tau \in \partial\mathbb{D}$ and $M > 0$, the *Stolz region* $K(\tau, M)$ *of vertex* τ and *amplitude* M is given by

$$K(\tau, M) = \left\{ z \in \mathbb{D} \;\middle|\; \frac{|\tau - z|}{1 - |z|} < M \right\}. \tag{2.35}$$

Note that $K(\tau, M) = \emptyset$ if $M \leq 1$, because $|\tau - z| \geq 1 - |z|$.

We can define Stolz regions also using the Poincaré distance.

Lemma 2.2.2. *Given* $\tau \in \partial\mathbb{D}$ *and* $M > 1$, *we have*

$$K(\tau, M) = \left\{ z \in \mathbb{D} \;\middle|\; \lim_{w \to \tau} [\omega(z, w) - \omega(0, w)] + \omega(0, z) < \log M \right\}.$$

In particular, for every $M > 1$ *we have*

$$K(\tau, M) \subset E(\tau, M^2). \tag{2.36}$$

Proof. It follows immediately from (2.6) and from the equality

$$\frac{|\tau - z|^2}{1 - |z|^2} \frac{1 + |z|}{1 - |z|} = \left(\frac{|\tau - z|}{1 - |z|} \right)^2. \qquad \square$$

Geometrically, $K(\tau, M)$ is a sort of angle with vertex at τ. Graphically, this is shown in Figure 2.5; to prove it analytically, let us introduce other another approach region at a point $\tau \in \partial\mathbb{D}$, defined in Euclidean terms.

Definition 2.2.3. The *angular region* $A(\tau, \alpha)$ *of vertex* $\tau \in \partial\mathbb{D}$ and *amplitude* $\alpha \in (0, \pi/2)$ is the intersection of $\mathbb{D}$ with an angle of vertex τ and opening 2α symmetric with respect to the radius ending at τ; see Figure 2.6. Analytically,

$$A(\tau, \alpha) = \{ z \in \mathbb{D} \mid |\mathrm{Im}(\bar{\tau} z)| < (\tan \alpha)(1 - \mathrm{Re}(\bar{\tau} z)) \}.$$

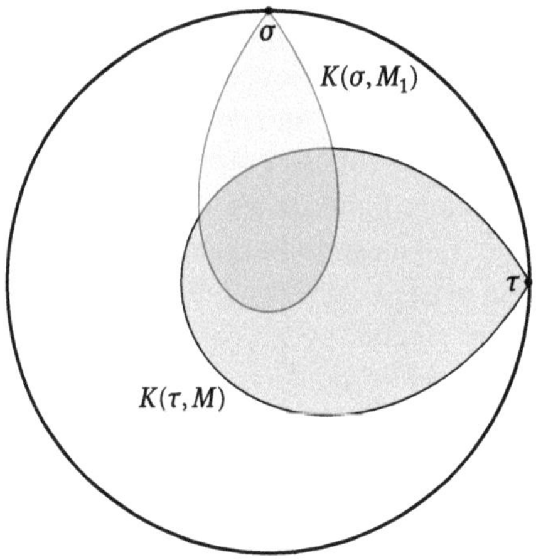

Figure 2.5: Stolz regions in $\mathbb{D}$.

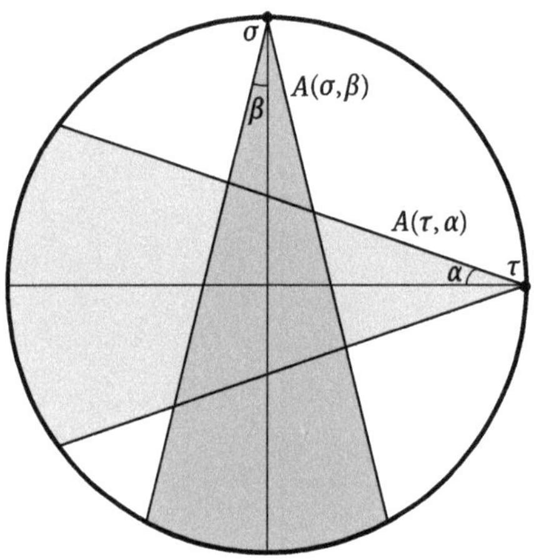

Figure 2.6: Angular regions in $\mathbb{D}$.

Stolz regions and angular regions are comparable nearby the vertex.

Proposition 2.2.4. *For every $M > 1$, there exist $\alpha_1, \alpha_2 \in (0, \pi/2)$ and $\varepsilon > 0$ such that*

$$A(\tau, \alpha_1) \cap D(\tau, \varepsilon) \subset K(\tau, M) \subset A(\tau, \alpha_2) \tag{2.37}$$

for all $\tau \in \partial\mathbb{D}$. More precisely, $\alpha_2 = \arctan\sqrt{M^2 - 1}$ and we can take α_1 arbitrarily close to α_2 by choosing ε small enough.

Proof. Up to a rotation, we can assume that $\tau = 1$. If $z \in K(1, M)$, we have

$$(1 - \operatorname{Re} z)^2 + |\operatorname{Im} z|^2 = |1 - z|^2 \le M^2(1 - |z|)^2 \le M^2(1 - \operatorname{Re} z)^2, \tag{2.38}$$

and hence $|\operatorname{Im} z| \le \sqrt{M^2-1}(1-\operatorname{Re} z)$, i. e., $K(1,M) \subset A(1, \arctan\sqrt{M^2-1})$.

Conversely, assume that $z \in A(1,\alpha)$ and put $c = \tan\alpha$. Then

$$|1-z|^2 \le (1+c^2)(1-\operatorname{Re} z)^2$$

and

$$1-|z|^2 \ge 1-(\operatorname{Re} z)^2 - c^2(1-\operatorname{Re} z)^2 = (1-\operatorname{Re} z)(1+\operatorname{Re} z - c^2(1-\operatorname{Re} z));$$

hence

$$\frac{|1-z|}{1-|z|} \le \frac{1+|z|}{1+\operatorname{Re} z - c^2(1-\operatorname{Re} z)}\sqrt{1+c^2}. \tag{2.39}$$

Now, given $0<\delta$ put

$$\varepsilon_\delta = \frac{\delta}{c+c^2(1+\delta)};$$

notice that $\varepsilon_\delta < 1$ always when $c \ge 1$, and also when $\delta < c/(1-c)$ if $c < 1$. If $z \in A(1,\alpha) \cap D(1,\varepsilon_\delta)$, we have $1+|z| < 1+\operatorname{Re} z + c(1-\operatorname{Re} z)$ and $1-\operatorname{Re} z < \varepsilon_\delta$; therefore,

$$\begin{aligned}\frac{1+|z|}{1+\operatorname{Re} z - c^2(1-\operatorname{Re} z)} &< \frac{1+\operatorname{Re} z + c(1-\operatorname{Re} z)}{1+\operatorname{Re} z - c^2(1-\operatorname{Re} z)}\\ &< \frac{1+\operatorname{Re} z + c\varepsilon_\delta}{1+\operatorname{Re} z - c^2\varepsilon_\delta} = 1+\frac{(c^2+c)\varepsilon_\delta}{1+\operatorname{Re} z - c^2\varepsilon_\delta}\\ &< 1+\frac{(c^2+c)\varepsilon_\delta}{1-c^2\varepsilon_\delta} = 1+\delta,\end{aligned}$$

where the last equality follows from the choice of ε_δ. Therefore, we have proved that

$$A(1,\alpha) \cap D(1,\varepsilon_\delta) \subset K\big(1, (1+\delta)\sqrt{1+(\tan\alpha)^2}\big),$$

as required. □

Actually there is a third kind of approach region we shall need to use later on, defined this time by using the hyperbolic geometry induced by the Poincaré distance.

Definition 2.2.5. The *hyperbolic approach region* $\Omega_{R,\tau}$ *of vertex* $\tau \in \partial\mathbb{D}$ and *amplitude* $R>0$ is given by

$$\Omega_{R,\tau} = \bigcup_{t\in[0,1)} B_{\mathbb{D}}(t\tau, R);$$

see Figure 2.7. In other words, $\Omega_{R,\tau}$ is the set of points of $\mathbb{D}$ whose Poincaré distance from the radial geodesic starting at 0 and ending at τ is less then R.

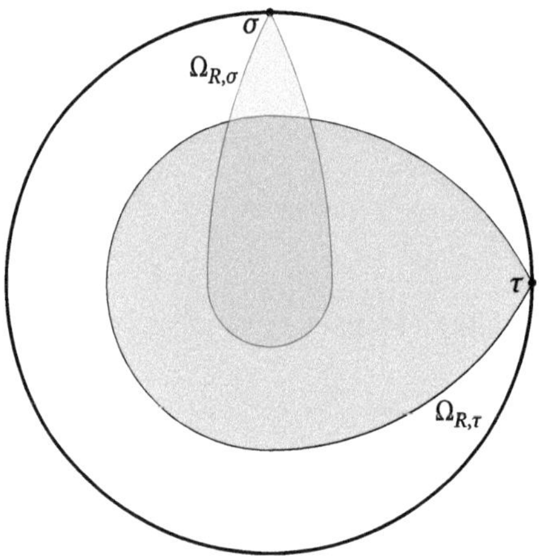

Figure 2.7: Hyperbolic approach regions in $\mathbb{D}$.

To show that Stolz regions and hyperbolic approach regions are comparable, it will be useful to transfer the stage to $\mathbb{H}^+$. For the sake of simplicity, in $\mathbb{H}^+$ we define only Stolz regions with vertex at infinity.

Definition 2.2.6. Given $\varepsilon \in (0,1)$ the *Stolz region* $K(\infty,\varepsilon) \subset \mathbb{H}^+$ of *vertex* ∞ and *amplitude* ε is given by

$$K(\infty,\varepsilon) = \Psi(K(1,1/\varepsilon)) = \left\{ w \in \mathbb{H}^+ \;\middle|\; \frac{|w+i| - |w-i|}{2} > \varepsilon \right\},$$

where $\Psi\colon \mathbb{D} \to \mathbb{H}^+$ is the Cayley transform; see Figure 2.8.

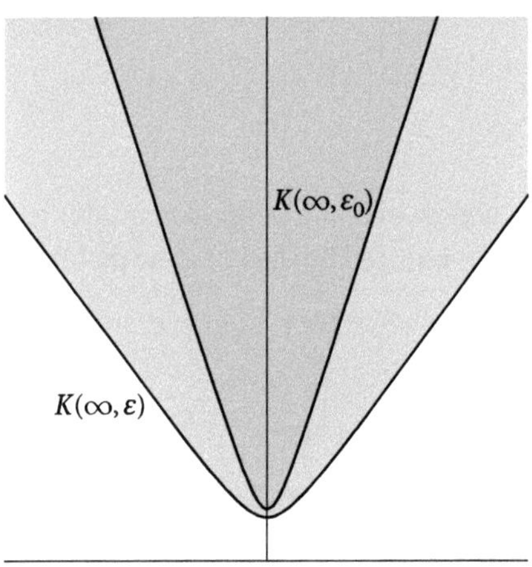

Figure 2.8: Stolz regions at infinity in $\mathbb{H}^+$.

We can give a more explicit expression of $K(\infty,\varepsilon)$; as a consequence, we shall be able to compare Stolz regions and hyperbolic approach regions.

Proposition 2.2.7. (i) *Given* $\varepsilon \in (0,1)$, *we have*

$$K(\infty,\varepsilon) = \{w \in \mathbb{H}^+ \mid (\operatorname{Im} w)^2 > \varepsilon^2(|w|^2 + 1 - \varepsilon^2)\}.$$

In particular, we have

$$\left\{w \in \mathbb{H}^+ \;\middle|\; \operatorname{Im} w > \delta\varepsilon|w|,\ |w|^2 > \frac{1-\varepsilon^2}{\delta^2 - 1}\right\} \subset K(\infty,\varepsilon) \subset \{w \in \mathbb{H}^+ \mid \operatorname{Im} w > \varepsilon|w|\}$$

for every $\delta > 1$.

(ii) *Given* $R > 0$ *put* $\rho = \tanh R$ *and choose* $\alpha \in [0,\pi/2)$ *such that* $\tan\frac{\alpha}{2} = \rho$. *Then*

$$\Omega_{R,\tau} = B_{\mathbb{D}}(0,R) \cup \Psi_\tau^{-1}(V(i,\alpha)),$$

where Ψ_τ *is the generalized Cayley transform* (1.15) *and*

$$\begin{aligned} V(i,\alpha) &= \{w = ire^{i\theta} \in \mathbb{H}^+ \mid r > 1, |\theta| < \alpha\} \\ &= \{w \in \mathbb{H}^+ \mid |w| > 1, |\operatorname{Re} w| < (\tan\alpha)\operatorname{Im} w\}. \end{aligned}$$

(iii) *For every* $M > 1$, *we can find* $0 < R_1 < R_2$ *and* $\varepsilon > 0$ *such that*

$$\Omega_{R_1,\tau} \cap D(\tau,\varepsilon) \subset K(\tau,M) \subset \Omega_{R_2,\tau} \tag{2.40}$$

for every $\tau \in \partial\mathbb{D}$. *More precisely,* $R_2 = \tanh^{-1}(\tan\frac{\alpha}{2})$ *with* $\tan\alpha = \sqrt{M^2-1}$ *and we can take* R_1 *arbitrarily close to* R_2 *by choosing* ε *small enough.*

Proof. (i) Squaring $|w+i| > 2\varepsilon + |w-i|$ and simplifying we get that $w \in K(\infty,\varepsilon)$ if and only if $\operatorname{Im} w - \varepsilon^2 > \varepsilon|w-i|$. Again squaring and simplifying, we get that $w \in K(\infty,\varepsilon)$ if and only if $(\operatorname{Im} w)^2 + \varepsilon^4 > \varepsilon^2(|w|^2+1)$, and we are done.

(ii) Applying the generalized Cayley transform Ψ_τ, we are reduced to show that

$$\Omega^{\mathbb{H}^+}_{R,\infty} = B_{\mathbb{H}^+}(i,R) \cup V(i,\alpha),$$

where

$$\Omega^{\mathbb{H}^+}_{R,\infty} = \bigcup_{t\ge 1} B_{\mathbb{H}^+}(it,R).$$

If $w = ire^{i\theta} \in \mathbb{H}^+$ and $t > 0$, we have

$$\omega_{\mathbb{H}^+}(w,it) = \tanh^{-1}\left|\frac{re^{i\theta}-t}{re^{i\theta}+t}\right| = \tanh^{-1}\left|\frac{r^2+t^2-2rt\cos\theta}{r^2+t^2+2rt\cos\theta}\right|^{1/2}.$$

In particular, the point of $i\mathbb{R}^+$ closest to w is ir and we have

$$\omega_{\mathbb{H}^+}(ir, ire^{i\theta}) = \tanh^{-1}\left|\frac{1-\cos\theta}{1+\cos\theta}\right|^{1/2} = \tanh^{-1}\left|\tan\frac{\theta}{2}\right|. \tag{2.41}$$

Using (2.41), we immediately see that we have chosen α so that $\omega_{\mathbb{H}^+}(i, ie^{i\alpha}) = R$. This implies that if $w = ire^{i\theta}$ with $|\theta| < \alpha$ we have $\omega_{\mathbb{H}^+}(ir, w) < R$; since $B_{\mathbb{H}^+}(i, R) \subset \Omega_{R,\infty}^{\mathbb{H}^+}$ by definition we have proved that $B_{\mathbb{H}^+}(i, R) \cup V(i, \alpha) \subseteq \Omega_{R,\infty}^{\mathbb{H}^+}$.

Conversely, take $w = ire^{i\theta} \in \Omega_{R,\infty}^{\mathbb{H}^+}$. This means that there exists $t \geq 1$ such that $w \in B_{\mathbb{H}^+}(it, R)$. We have already noticed that the point of $i\mathbb{R}^+$ closest to w is ir; in particular, $\omega_{\mathbb{H}^+}(w, ir) < R$. Then (2.41) immediately shows that $|\theta| < \alpha$, and hence if $r > 1$ we have $w \in V(i, \alpha)$. If instead $r \leq 1$, there still exists $t \geq 1$ such that $\omega_{\mathbb{H}^+}(w, it) < R$. But then

$$\omega_{\mathbb{H}^+}(w, i) = \tanh^{-1}\left|\frac{r^2+1-2r\cos\theta}{r^2+1+2r\cos\theta}\right|^{1/2} \leq \tanh^{-1}\left|\frac{r^2+t^2-2rt\cos\theta}{r^2+t^2+2rt\cos\theta}\right|^{1/2} < R$$

and we are done.

(iii) Without loss of generality, we can assume $\tau = 1$. As usual, we transfer the stage to $\mathbb{H}^+$. Choose $\alpha \in [0, \pi/2)$ as in the statement and fix $0 \leq \alpha_1 < \alpha$. If $w \in V(i, \alpha_1)$ we have $|w| < \sqrt{1+(\tan\alpha_1)^2}\operatorname{Im} w$; therefore, taking $\delta = \sqrt{\frac{1+(\tan\alpha)^2}{1+(\tan\alpha_1)^2}} > 1$ and using (i) we see that $w \in K(\infty, M^{-1})$ if $|w|$ is large enough. But if $z \in \mathbb{D}$ is such that $|z-1| < \varepsilon$ and $\operatorname{Re} z > 0$ setting $w = \Psi(z)$, we have

$$|w|^2 = \frac{|z+1|^2}{|z-1|^2} > \frac{1+|z|^2+2\operatorname{Re} z}{\varepsilon^2} > \frac{1}{\varepsilon^2};$$

thus choosing ε small enough we get $|w|$ as large as we need. In particular, take $R_1 = \tanh^{-1}(\tan\frac{\alpha_1}{2})$; then $w \notin B_{\mathbb{H}^+}(i, R_1)$ if $|w|$ is large enough, and thus we can find $\varepsilon > 0$ so that the first containment in (2.40) is satisfied.

Conversely, given $M > 1$ choose $\alpha \in [0, \pi/2)$ such that $\tan\alpha = \sqrt{M^2-1}$ and put $\rho = \tan\frac{\alpha}{2}$ and $R = \tanh^{-1}\rho$. Notice that the trigonometric formula expressing $\tan\alpha$ in terms of $\tan(\alpha/2)$ yields

$$M^2 - 1 = (\tan\alpha)^2 = \frac{4\rho^2}{(1-\rho^2)^2}. \tag{2.42}$$

Now take $z \in K(1, M) \cap \{\operatorname{Re} z > 0\}$ and put $w = \Psi(z)$. It is easy to check that $\operatorname{Re} z > 0$ implies $|w| > 1$. Moreover, by (i), $z \in K(1, M)$ implies $M \operatorname{Im} w > |w|$. Therefore, $(M^2 - 1)(\operatorname{Im} w)^2 > |\operatorname{Re} w|^2$; recalling that $M^2 - 1 = (\tan\alpha)^2$, this means that $\Psi(z) \in V(i, \alpha)$, and hence $K(1, M) \cap \{\operatorname{Re} z > 0\} \subset \Omega_{R,1}$ by (ii).

To complete the proof, it suffices to show that $K(1, M) \cap \{\operatorname{Re} z \le 0\} \subset B_{\mathbb{D}}(0, R)$. Now, if $\operatorname{Re} z \le 0$ we have $1 + |z|^2 \le |1 - z|^2$; therefore, $z \in K(1, M) \cap \{\operatorname{Re} z \le 0\}$ implies

$$\frac{1 + |z|^2}{(1 - |z|)^2} \le \frac{|1 - z|^2}{(1 - |z|)^2} < M^2.$$

This yields $(M^2 - 1)|z|^2 - 2M^2|z| + M^2 - 1 > 0$, and hence

$$|z| < \frac{M^2}{M^2 - 1} - \sqrt{\frac{M^4}{(M^2 - 1)^2} - 1} < \rho,$$

where the last inequality follows expressing M^2 in terms of ρ using (2.42). Therefore, $K(1, M) \cap \{\operatorname{Re} z < 0\} \subset B_{\mathbb{D}}(0, \mathbb{R}) \subset \Omega_{R,1}$ and we are done. □

In particular, Stolz regions are Bloch domains, thanks to Examples 1.11.3 and 1.11.6.

Before using Stolz regions to define the kind of limit we are interested in, we note two results we shall need later. The first one is an estimate of the distance between the boundaries of two Stolz regions.

Lemma 2.2.8. *Fix $\sigma \in \partial\mathbb{D}$ and $1 < M_1 < M_2$. Then for each $z \in K(\sigma, M_1)$ we have*

$$\inf_{w \in \partial K(\sigma, M_2)} \{\omega(z, w)\} \ge \frac{1}{2} \log \frac{M_2}{M_1}$$

and

$$d(z, \partial K(\sigma, M_2)) \ge \frac{M_2 - M_1}{M_2 + M_1}(1 - |z|) \ge \frac{1}{M_1}\frac{M_2 - M_1}{M_2 + M_1}|\sigma - z|.$$

Proof. Fix $z \in K(\sigma, M_1)$ and let $1 < M < M_1$ be so that $z \in \partial K(\sigma, M)$. Take $w \in \partial K(\sigma, M_2)$; then Lemma 2.2.2 yields

$$\begin{aligned}
\log M_2 &= \lim_{\zeta \to \sigma} [\omega(w, \zeta) - \omega(0, \zeta)] + \omega(0, w) \\
&\le 2\omega(z, w) + \lim_{\zeta \to \sigma} [\omega(z, \zeta) - \omega(0, \zeta)] + \omega(0, z) \\
&= 2\omega(z, w) + \log M < 2\omega(z, w) + \log M_1,
\end{aligned}$$

i. e., $\omega(z, w) \ge \frac{1}{2} \log \frac{M_2}{M_1}$ and the first inequality is proved.

In particular, if $R = \frac{1}{2} \log \frac{M_2}{M_1}$ we have that $B_{\mathbb{D}}(z, R) \subset K(\sigma, M_2)$, and thus the Euclidean distance between z and $\partial K(\sigma, M_2)$ is at least equal to the Euclidean distance from z to $\partial B_{\mathbb{D}}(z, R)$. Recalling Proposition 1.2.7, we then get

$$d(z, \partial K(\sigma, M_2)) \ge \frac{r(1 - |z|^2)}{1 - r^2|z|^2} - |z|\left(1 - \frac{1 - r^2}{1 - r^2|z|^2}\right) = \frac{r(1 - |z|^2)}{1 + r|z|},$$

where $r = \tanh R = \frac{M_2 - M_1}{M_2 + M_1} \in (0,1)$. Recalling that $z \in K(\sigma, M_1)$, we finally get

$$d(z, \partial K(\sigma, M_2)) \geq r(1 - |z|) > \frac{r}{M_1}|\sigma - z|$$

and we are done. □

Next, we can state and prove a Julia lemma for Stolz regions.

Proposition 2.2.9. *Let $f \in \mathrm{Hol}(\mathbb{D}, \mathbb{D})$ and $\sigma \in \partial\mathbb{D}$ be such that $\beta_f(\sigma) < +\infty$. Then there exists $\tau \in \partial\mathbb{D}$ such that*

$$f(K(\sigma, M)) \subseteq K(\tau, C\sqrt{\beta_f(\sigma)}M)$$

for all $M > 1$, where $C = \exp \omega(0, f(0))$.

Proof. Let $\tau \in \partial\mathbb{D}$ be given by the Julia lemma (Theorem 2.1.10). Then

$$\begin{aligned}\frac{1 - |f(z)|}{1 + |f(z)|}\left(\frac{|\tau - f(z)|}{1 - |f(z)|}\right)^2 &= \frac{|\tau - f(z)|^2}{1 - |f(z)|^2} \\ &\leq \beta_f(\sigma)\frac{|\sigma - z|^2}{1 - |z|^2} = \beta_f(\sigma)\left(\frac{|\sigma - z|}{1 - |z|}\right)^2 \frac{1 - |z|}{1 + |z|}.\end{aligned}$$

On the other hand, from

$$\omega(0, f(z)) \leq \omega(0, f(0)) + \omega(f(0), f(z)) \leq \omega(0, f(0)) + \omega(0, z)$$

we get

$$\frac{1 + |f(z)|}{1 - |f(z)|} \leq \frac{1 + |f(0)|}{1 - |f(0)|} \cdot \frac{1 + |z|}{1 - |z|}.$$

Hence

$$\frac{|\tau - f(z)|}{1 - |f(z)|} \leq \sqrt{\beta_f(\sigma)\frac{1 + |f(0)|}{1 - |f(0)|}}\frac{|\sigma - z|}{1 - |z|},$$

and we are done. □

Remark 2.2.10. Using (2.55), we shall see that

$$C\sqrt{\beta_f(\sigma)} = \sqrt{\beta_f(\sigma)\frac{1 + |f(0)|}{1 - |f(0)|}} \geq \sqrt{\frac{2}{1 + |f^h(0)|}} \geq 1.$$

We can at last define the kind of limits at the boundary we are interested in.

Definition 2.2.11. We say that a function $f\colon \mathbb{D} \to \widehat{\mathbb{C}}$ has *nontangential limit* (or *angular limit*) $c \in \widehat{\mathbb{C}}$ at $\sigma \in \partial\mathbb{D}$ if $f(z) \to c$ as z tends to σ within $K(\sigma, M)$ for all $M > 1$. When this happen we shall write

$$K\text{-}\lim_{z\to\tau} f(z) = c.$$

Sometimes when f has nontangential limit c at σ we shall also write $f(\sigma) = c$. Other symbols commonly used in the literature to denote the nontangential limit are nt-lim and $\angle$-lim.

We shall also say that a sequence $\{z_\nu\} \subset \mathbb{D}$ converges *nontangentially* to $\sigma \in \partial\mathbb{D}$ if $z_\nu \to \sigma$ and there is $M > 1$ such that $z_\nu \in K(\sigma, M)$ eventually.

Analogously, we shall say that a function $f\colon \mathbb{H}^+ \to \widehat{\mathbb{C}}$ has *nontangential limit* $L \in \mathbb{C}$ at infinity, and we shall write $L = K\text{-}\lim_{w\to\infty} f(w)$, if $f(w) \to L$ as $w \to \infty$ within $K(\infty, \varepsilon)$ for all $\varepsilon \in (0, 1)$. If this happen, we shall sometimes write $f(\infty) = L$. We shall also say that a sequence $\{w_\nu\} \subset \mathbb{H}^+$ converges *nontangentially* to ∞ if $w_\nu \to \infty$ and there is $\varepsilon \in (0, 1)$ such that $w_\nu \in K(\infty, \varepsilon)$ eventually.

In the rest of this book, we shall encounter many instances of holomorphic functions having nontangential limit at a point in the boundary. A typical example is given by the classical Fatou Theorem 1.8.11 that says that a bounded holomorphic function defined on $\mathbb{D}$ admits radial limit (and hence, by the Lindelöf Theorem 2.4.2, nontangential limit) at almost any point of $\partial\mathbb{D}$. Another example is given by the Julia lemma itself, that as we shall see (Lemma 2.2.14) can be restated to say that $\beta_f(\delta) < \infty$ implies that f has nontangential limit at σ. At this point, the inquisitive reader might ask whether it is really indispensable the use of the nontangential limit; maybe deeper arguments can provide the existence of unrestricted limits. This is not the case: there are examples of bounded holomorphic functions admitting a nontangential limit but not a unrestricted limit at a point in the boundary.

Example 2.2.12. Let $f\colon \mathbb{D} \to \mathbb{C}$ be given by

$$f(z) = \exp\left(-\frac{1}{1-z}\right).$$

If we write $z = 1 - re^{i\theta}$ with $r \in (0, 2)$ and $\theta \in (-\pi/2, \pi/2)$, we get

$$f(z) = \exp\left(\frac{\cos(\pi-\theta)}{r}\right)\exp\left(i\frac{\sin(\pi-\theta)}{r}\right);$$

in particular f is bounded by 1 because $\cos(\pi - \theta) < 0$ for all $\theta \in (-\pi/2, \pi/2)$.

If $z \in K(1, M)$, Proposition 2.2.4 implies that $|\theta| < \arctan\sqrt{M^2-1} < \pi/2$. Thus

$$\forall z \in K(1, M) \quad |f(z)| \le \exp\left(-\frac{|\cos(\pi - a_M)|}{r}\right)$$

where $a_M = \arctan\sqrt{M^2-1}$ and, therefore, f has nontangential limit 0 at 1, because $r \to 0$ as $z \to 1$.

On the other hand, for $c > 1/2$ let $\gamma_c(t) = 1 - t\exp(i(\frac{\pi}{2} - ct))$. Since

$$1 - |\gamma_c(t)|^2 = 2t\cos\left(\frac{\pi}{2} - ct\right) - t^2 = (2c-1)t^2 + O(t^4),$$

we have $\gamma_c(t) \in \mathbb{D}$ for t small enough; moreover, clearly $\gamma_c(t) \to 1$ as $t \to 0^+$. But

$$|f(\gamma_c(t))| = \exp\left(\frac{\cos(\frac{\pi}{2} + ct)}{t}\right) = \exp(-c + O(t^2)) \to e^{-c}$$

as $t \to 0^+$, and thus $f(z)$ cannot have unrestricted limit as $z \to 1$ outside of Stolz regions.

Remark 2.2.13. In Proposition 3.3.12, we shall see that the difference $\omega_D(z,\zeta) - \omega_D(z_0,\zeta)$ has a limit when ζ tends to a Jordan boundary point of a hyperbolic domain of regular type D, where $z_0 \in D$ is a fixed reference point. So, using the metric characterization given by Lemma 2.2.2, we can define Stolz regions and hence nontangential limits in hyperbolic domains of regular type even when the boundary is not smooth.

As promised, we can now use the Julia lemma to prove the existence of a nontangential limit at a point where the boundary dilation is finite, giving a rigorous expression for the heuristic argument described at the beginning of this section.

Lemma 2.2.14. *Let $f \in \mathrm{Hol}(\mathbb{D},\mathbb{D})$ and $\sigma \in \partial\mathbb{D}$ be such that $\beta_f(\sigma) < +\infty$. Then f has nontangential limit $\tau \in \partial\mathbb{D}$ at σ, where τ is the unique point in $\partial\mathbb{D}$ such that $\beta_f(\sigma) = \beta_f(\sigma,\tau)$.*

Proof. The existence and uniqueness of τ are proved in Lemma 2.1.12. Put $\beta = \beta_f(\sigma,\tau)$ and take $z \in K(\tau, M)$. By definition, we have

$$|\tau - f(z)|^2 < \frac{|\tau - f(z)|^2}{1 - |f(z)|^2} \le \beta\frac{|\sigma - z|^2}{1 - |z|^2} < \beta M\frac{|\sigma - z|}{1 + |z|}.$$

Therefore, if we let $z \to \sigma$ inside $K(\sigma, M)$ we get $f(z) \to \tau$, as required. □

In the next section, we shall see (in the Julia–Wolff–Carathéodory Theorem 2.3.2) that we can push this argument up to also get the existence of the nontangential limit of the derivative. For now, we shall end this section with a few technical results we shall need later.

In the sequel, we shall often be interested in determining when a sequence converges to a point of the boundary not only nontangentially but with a precise slope. The next lemma contains a few equivalent conditions making precise what we mean by “precise slope.” Notice that if $z \in \mathbb{D}$ then $\mathrm{Re}(1 - \overline{\sigma}z) > 0$ for any $\sigma \in \partial\mathbb{D}$, and hence $\arg(1 - \overline{\sigma}z)$ is a well-defined number in $(-\pi/2, \pi/2)$.

Lemma 2.2.15. *Let $\{z_\nu\} \subset \mathbb{D}$ be a sequence converging toward $\sigma = e^{i\phi} \in \partial\mathbb{D}$, where $\phi = \arg(\sigma) \in [0, 2\pi)$. Fix $\theta \in [-\pi/2, \pi/2]$. Then the following assertions are equivalent:*
(i) $\arg(\sigma - z_\nu) \to \arg(\sigma) + \theta \pmod{2\pi}$ *as* $\nu \to +\infty$;
(ii) $\arg(1 - \overline{\sigma} z_\nu) \to \theta$ *as* $\nu \to +\infty$;
(iii) $\frac{\sigma - z_\nu}{|\sigma - z_\nu|} \to \sigma e^{i\theta}$ *as* $\nu \to +\infty$;
(iv) $\frac{1-\overline{\sigma} z_\nu}{|1-\overline{\sigma} z_\nu|} \to e^{i\theta}$ *as* $\nu \to +\infty$;
(v) $\frac{\operatorname{Im}(\overline{\sigma} z_\nu)}{1-\operatorname{Re}(\overline{\sigma} z_\nu)} \to -\tan\theta$ *as* $\nu \to +\infty$.
Furthermore,

$$\frac{|\operatorname{Im}(\overline{\sigma} z_\nu)|}{1-\operatorname{Re}(\overline{\sigma} z_\nu)} \to |\tan\theta| \quad\Longleftrightarrow\quad \frac{|\sigma - z_\nu|}{1-|z_\nu|} \to \frac{1}{\cos\theta}. \tag{2.43}$$

In particular, $z_\nu \to \sigma$ radially (i. e., $\theta = 0$) if and only if $\frac{|\sigma - z_\nu|}{1-|z_\nu|} \to 1$, and $z_\nu \to \sigma$ tangentially (i. e., $\theta = \pm\pi/2$) if and only if $\frac{|\sigma - z_\nu|}{1-|z_\nu|} \to +\infty$.

Proof. The equivalence of (i) and (ii) and the equivalence of (iii) and (iv) are trivial. Since $1 - \overline{\sigma} z_\nu = |1 - \overline{\sigma} z_\nu| e^{i\arg(1-\overline{\sigma} z_\nu)}$, the equivalence of (ii) and (iv) immediately follows recalling that $\arg(1 - \overline{\sigma} z_\nu) \in (-\pi/2, \pi/2)$ always.

Now we have

$$\frac{1-\overline{\sigma} z_\nu}{|1-\overline{\sigma} z_\nu|} = \frac{1 - i\frac{\operatorname{Im}(\overline{\sigma} z_\nu)}{1-\operatorname{Re}(\overline{\sigma} z_\nu)}}{\sqrt{1 + \left(\frac{\operatorname{Im}(\overline{\sigma} z_\nu)}{1-\operatorname{Re}(\overline{\sigma} z_\nu)}\right)^2}};$$

therefore,

$$\frac{1-\overline{\sigma} z_\nu}{|1-\overline{\sigma} z_\nu|} \to e^{i\theta} \iff \begin{cases} \dfrac{1}{\sqrt{1+\left(\frac{\operatorname{Im}(\overline{\sigma} z_\nu)}{1-\operatorname{Re}(\overline{\sigma} z_\nu)}\right)^2}} \to \cos\theta, \\ -\dfrac{\frac{\operatorname{Im}(\overline{\sigma} z_\nu)}{1-\operatorname{Re}(\overline{\sigma} z_\nu)}}{\sqrt{1+\left(\frac{\operatorname{Im}(\overline{\sigma} z_\nu)}{1-\operatorname{Re}(\overline{\sigma} z_\nu)}\right)^2}} \to \sin\theta, \end{cases}$$

and the equivalence of (iv) and (v) follows.

Finally, we can write

$$\frac{|\sigma - z_\nu|}{1-|z_\nu|} = (1+|z_\nu|)\frac{\sqrt{1 + \left(\frac{\operatorname{Im}(\overline{\sigma} z_\nu)}{1-\operatorname{Re}(\overline{\sigma} z_\nu)}\right)^2}}{1 + \operatorname{Re}(\overline{\sigma} z_\nu) - \operatorname{Im}(\overline{\sigma} z_\nu)\frac{\operatorname{Im}(\overline{\sigma} z_\nu)}{1-\operatorname{Re}(\overline{\sigma} z_\nu)}},$$

and (2.43) follows. □

Remark 2.2.16. A useful formula that we shall need a few times when studying convergence to the boundary along a precise slope is

$$
\begin{aligned}
\frac{w-z}{1-\overline{z}w} &= -\frac{1}{\overline{z}}\left[\frac{1}{\overline{\sigma z}}\frac{\overline{\sigma}-\overline{z}}{|\overline{\sigma}-\overline{z}|}\frac{|\sigma-z|}{\sigma-z}+\frac{\sigma-w}{\sigma-z}\right]^{-1}\left[\frac{\sigma-w}{\sigma-z}-1\right] \\
&= -\frac{1}{\overline{z}}\left[\frac{1}{\sigma\overline{z}}\frac{1-\sigma\overline{z}}{1-\overline{\sigma}z}+\frac{\sigma-w}{\sigma-z}\right]^{-1}\left[\frac{\sigma-w}{\sigma-z}-1\right],
\end{aligned}
\tag{2.44}
$$

which is valid for all $z, w \in \mathbb{D}$ and $\sigma \in \partial\mathbb{D}$ and whose proof is just a computation.

We end this section improving Proposition 1.7.27.

Lemma 2.2.17. *Let $\{z_\nu\}, \{w_\nu\} \subset \mathbb{D}$ be two sequences such that*

$$C = \limsup_{\nu\to\infty} \omega(z_\nu, w_\nu) < +\infty.$$

(i) *If $z_\nu \to \sigma \in \partial\mathbb{D}$ nontangentially then $w_\nu \to \sigma$ nontangentially.*
(ii) *If furthermore $C = 0$, then*

$$\lim_{\nu\to+\infty} \frac{\sigma - w_\nu}{\sigma - z_\nu} = 1. \tag{2.45}$$

In particular, if moreover $(\sigma - z_\nu)/|\sigma - z_\nu| \to \tau \in \partial\mathbb{D}$ then $(\sigma - w_\nu)/|\sigma - w_\nu| \to \tau$.

Proof. (i) Proposition 1.7.27 implies that $w_\nu \to \sigma$. Assume that $z_\nu \to \sigma$ non-tangentially, and choose $M > 1$ so that $z_\nu \in K(\sigma, M)$ eventually. Since

$$
\begin{aligned}
&\omega(w_\nu, \zeta) - \omega(0, \zeta) + \omega(0, w_\nu) \\
&\qquad \le \omega(z_\nu, \zeta) - \omega(0, \zeta) + \omega(0, z_\nu) + 2\omega(z_\nu, w_\nu)
\end{aligned}
$$

for all $\zeta \in \mathbb{D}$, recalling Lemma 2.2.2 we see that

$$\lim_{\zeta\to\sigma}\left[\omega(w_\nu, \zeta) - \omega(0, \zeta)\right] + \omega(0, w_\nu) < \log M + 2\tilde{C},$$

where $\tilde{C} = \sup_\nu \omega(z_\nu, w_\nu) < +\infty$, and hence $w_\nu \in K(\sigma, \tilde{M})$ for a suitable $\tilde{M} > 1$; thus $w_\nu \to \sigma$ nontangentially, as claimed.

(ii) The formula (2.44) yields

$$\frac{w_\nu - z_\nu}{1 - \overline{z_\nu} w_\nu} = -\frac{1}{\overline{z_\nu}}\left[\frac{1}{\overline{\sigma z_\nu}}\frac{\overline{\sigma}-\overline{z_\nu}}{|\overline{\sigma}-\overline{z_\nu}|}\frac{|\sigma-z_\nu|}{\sigma-z_\nu}+\frac{\sigma-w_\nu}{\sigma-z_\nu}\right]^{-1}\left[\frac{\sigma-w_\nu}{\sigma-z_\nu}-1\right].$$

The assumption $C = 0$ implies that the left-hand side converges to 0 as $\nu \to +\infty$ and thus (2.45) follows. Finally,

$$\frac{\sigma - w_\nu}{|\sigma - w_\nu|} = \frac{\sigma - w_\nu}{\sigma - z_\nu}\frac{|\sigma - z_\nu|}{|\sigma - w_\nu|}\frac{\sigma - z_\nu}{|\sigma - z_\nu|} \to \tau$$

and we are done. □

***Notes to Section* 2.2**

Stolz regions where introduced by Stolz [381] in 1875, while generalizing a classical theorem of Abel about convergence of power series. Abel [13] in 1826 proved that if $f(z) = \sum_{k=0}^{\infty} a_k z^k$ is a power series convergent in $\mathbb{D} \cup \{1\}$ then

$$\lim_{r \to 1^-} f(r) = \sum_{k=0}^{\infty} a_k;$$

Stolz showed that $f(z)$ actually admits limit when $z \to 1$ inside $K(1, M)$ for all $M > 1$. For a study of the existence of limits along tangential directions, see [354] and references therein.

In the literature, nontangential limits are defined sometimes using Stolz regions and sometimes using angular regions, more rarely using hyperbolic approach regions; Propositions 2.2.4 and 2.2.7 show that all these definitions are equivalent. Another kind of approach region at $\sigma \in \partial\mathbb{D}$ occasionally used is obtained by taking the interior of the convex hull of $\{\sigma\} \cup B_{\mathbb{D}}(0, R)$; these regions are clearly comparable to angular regions.

Following [361], we used the letter K for Stolz regions and nontangential limits because their multidimensional analogues have been introduced by Korányi in [248].

Proposition 2.2.9 is in [4], but possibly it was known before.

2.3 The Julia–Wolff–Carathéodory theorem

We have seen (Lemma 2.2.14) that a holomorphic self-map of $\mathbb{D}$ having finite boundary dilation at a point $\sigma \in \partial\mathbb{D}$ has nontangential limit $\tau \in \partial\mathbb{D}$ at σ. As anticipated before, actually much more is true: also the derivative has a finite nontangential limit. To prove the existence of the limit of the derivative, it is natural to study the behavior of the incremental ratio; the link between the two in our context is given by the following result.

Proposition 2.3.1. *Let $f \in \mathrm{Hol}(\mathbb{D}, \mathbb{C})$ and $\sigma \in \partial\mathbb{D}$ be given. Then the following statements are equivalent:*

(i) *there exist $L \in \mathbb{C}$ and $\alpha \in \mathbb{C}$ such that*

$$K\text{-}\lim_{z \to \sigma} f(z) = L \quad \textit{and} \quad K\text{-}\lim_{z \to \sigma} \frac{L - f(z)}{\sigma - z} = \alpha;$$

(ii) *there exists $\alpha \in \mathbb{C}$ such that*

$$K\text{-}\lim_{z \to \sigma} f'(z) = \alpha.$$

Proof. Assume that (i) holds. To compute the nontangential limit of the derivative, fix $M > 1$ and take $z \in K(1, M)$. Given $R > 0$, notice that Lemma 2.2.8 yields $B_{\mathbb{D}}(z, R) \subset K(\sigma, e^{2R} M)$. Put $r = \tanh R$ and set

$$\rho_z = \frac{r}{1 + r|z|}(1 - |z|^2).$$

A quick computation using Proposition 1.2.7 shows that $D_z = D(z, \rho_z)$ is contained in $B_{\mathbb{D}}(z, R)$, and hence in $K(\sigma, e^{2R}M)$. Now, the Cauchy formula yields

$$\begin{aligned}|f'(z) - \alpha| &= \frac{1}{2\pi}\left|\int_{\partial D_z} \frac{f(\zeta) - L - \alpha(\zeta - \sigma)}{(\zeta - z)^2}\, d\zeta\right| \\ &\leq \sup_{\zeta\in\partial D_z} \frac{|\sigma - \zeta|}{\rho_z} \sup_{\zeta\in\partial D_z}\left|\frac{L - f(\zeta)}{\sigma - \zeta} - \alpha\right|.\end{aligned}$$

By construction, we have $\partial D_z \subset \overline{K(1, e^{2R}M)}$; in particular, $|\sigma - \zeta| \leq e^{2R}M(1 - |\zeta|)$ for all $\zeta \in \partial D_z$. Moreover, since D_z is an Euclidean disk of center $|z|$ and radius ρ_z, we have

$$\sup_{\zeta\in\partial D_z}(1 - |\zeta|) = 1 - \inf_{\zeta\in\partial D_z} |\zeta| = 1 - (|z| - \rho_z) = \rho_z + (1 - |z|).$$

Summing up, we have

$$\sup_{\zeta\in\partial D_z} \frac{|\sigma - \zeta|}{\rho_z} \leq e^{2R}M\left(1 + \frac{1 - |z|}{\rho_z}\right) \leq e^{2R}M\left(1 + \frac{1 + r|z|}{r(1 + |z|)}\right) \leq e^{2R}M\frac{1 + 2r}{r}.$$

Finally, the same computation shows that $|\sigma - \zeta|$ is bounded uniformly on ∂D_z by a constant times $|\sigma - z|$; in particular, for every $\varepsilon > 0$ we can find $\delta > 0$ such that $|\sigma - z| < \delta$ implies $|\sigma - \zeta| < \varepsilon$ for all $\zeta \in \partial D_z$. Therefore, (i) implies that

$$\sup_{\zeta\in\partial D_z}\left|\frac{L - f(\zeta)}{\sigma - \zeta} - \alpha\right| \longrightarrow 0$$

as $z \to \sigma$ in $K(\sigma, M)$, and we are done.

Conversely, assume that (ii) holds. For all $z, w \in \mathbb{D}$, we have

$$f(w) = f(z) + \int_z^w f'(\zeta)\, d\zeta = f(z) + (w - z)\int_0^1 f'(z + t(w - z))\, dt.$$

Now, for every $z \in \mathbb{D}$ if $w \to \sigma$ nontangentially then there is a Stolz region K with vertex σ containing all segments $\{z + t(w - z) \mid t \in [0, 1]\}$. In particular, K also contains the segment $\{z + t(\sigma - z) \mid t \in [0, 1)\}$ and, by (ii), $f'(z + t(\sigma - z)) \to \alpha$ as $t \to 1^-$. We can then apply the Lebesgue dominated convergence theorem obtaining

$$K\text{-}\lim_{w\to\sigma} f(w) = f(z) + (\sigma - z)\int_0^1 f'(z + t(\sigma - z))\, dt.$$

Denoting by L this nontangential limit, since if $z \in K(\sigma, M)$ then $z + t(\sigma - z) \in K(\sigma, M)$ for all $t \in [0, 1)$, we can apply again the Lebesgue dominated convergence theorem to obtain

$$K\text{-}\lim_{z\to\sigma}\frac{L-f(z)}{\sigma-z}=K\text{-}\lim_{z\to\sigma}\int_0^1 f'(z+t(\sigma-z))\,dt=\alpha$$

and the proof is complete. □

We can now show that when the boundary dilation is finite then both the incremental ratio and the derivative have a finite nontangential limit. This is the content of the *Julia–Wolff–Carathéodory theorem.*

Theorem 2.3.2 (Julia–Wolff–Carathéodory, 1926). *Let $f\in \mathrm{Hol}(\mathbb{D},\mathbb{D})$ and take $\tau,\sigma\in\partial\mathbb{D}$. Then*

$$K\text{-}\lim_{z\to\sigma}\frac{\tau-f(z)}{\sigma-z}=\tau\bar{\sigma}\beta_f(\sigma,\tau). \tag{2.46}$$

If this nontangential limit is finite then f has nontangential limit τ at σ and

$$K\text{-}\lim_{z\to\sigma}f'(z)=\tau\bar{\sigma}\beta_f(\sigma). \tag{2.47}$$

In particular, if $\tau=\sigma$ then the nontangential limit of f' at σ is a positive real number.

Proof. If $z\in K(\sigma,M)$, then

$$\left|\frac{\tau-f(z)}{\sigma-z}\right|\geq\frac{1}{M}\frac{1-|f(z)|}{1-|z|}; \tag{2.48}$$

therefore, (2.46) is proved if $\beta_f(\sigma)=+\infty$, by Proposition 2.1.15. Assume then $\beta=\beta_f(\sigma)$ finite, and let (Lemma 2.1.12) τ_0 be the unique point of $\partial\mathbb{D}$ such that $\beta_f(\sigma,\tau_0)=\beta$; in particular we know (Lemma 2.2.14) that f has nontangential limit τ_0 at σ. In particular, if $\tau\neq\tau_0$, i. e., if $\beta_f(\sigma,\tau)=+\infty$, then both the left-hand side and the right-hand side of (2.46) are infinite; so we are left with proving (2.46) when $\beta_f(\sigma,\tau)<+\infty$, i. e., for $\tau=\tau_0$.

By definition, we have

$$\frac{1-|z|^2}{|\sigma-z|^2}=\operatorname{Re}\frac{\sigma+z}{\sigma-z}\leq\beta\operatorname{Re}\frac{\tau_0+f(z)}{\tau_0-f(z)}=\beta\frac{1-|f(z)|^2}{|\tau_0-f(z)|^2},$$

with equality at one point (and hence everywhere) if and only if $f\in\mathrm{Aut}(\mathbb{D})$. Therefore, we can write

$$\beta\frac{\tau_0+f(z)}{\tau_0-f(z)}-\frac{\sigma+z}{\sigma-z}=\frac{\tau_0+F(z)}{\tau_0-F(z)} \tag{2.49}$$

for a suitable $F\colon\mathbb{D}\to\mathbb{C}$ holomorphic with $|F|\leq 1$. Now $|F|=1$ at one point if and only if $f\in\mathrm{Aut}(\mathbb{D})$; in this case, (2.46) and (2.47) are easily verified using Example 2.1.14. So, we can assume $F\in\mathrm{Hol}(\mathbb{D},\mathbb{D})$.

Now, we have

$$\frac{1}{\beta_F(\sigma,\tau_0)} = \beta \cdot \inf_{z\in\mathbb{D}} \left\{ \operatorname{Re} \frac{\tau_0 + f(z)}{\tau_0 - f(z)} \Big/ \operatorname{Re} \frac{\sigma + z}{\sigma - z} \right\} - 1 = 0. \tag{2.50}$$

So, $\beta_F(\sigma,\tau_0) = +\infty$ and, since we have already proved that (2.46) holds in this case, we get that $(\sigma - z)/(\tau_0 - F(z))$ has nontangential limit 0 at σ. Moreover, (2.49) yields

$$\beta \frac{\sigma - z}{\tau_0 - f(z)} = \frac{\sigma + z}{\tau_0 + f(z)} + \frac{\tau_0 + F(z)}{\tau_0 + f(z)} \frac{\sigma - z}{\tau_0 - F(z)};$$

it follows that $(\tau_0 - f(z))/(\sigma - z)$ has nontangential limit $\overline{\sigma}\tau_0\beta$ at σ.

When $\beta_f(\sigma) < +\infty$, to get (2.47) from (2.46) it suffices now to apply Proposition 2.3.1. □

Definition 2.3.3. Let $f \in \operatorname{Hol}(\mathbb{D},\mathbb{D})$ and $\sigma \in \partial\mathbb{D}$. Assume that f has nontangential limit $\tau \in \partial\mathbb{D}$ at σ. The nontangential limit in (2.46) is the *angular derivative* of f at σ and is denoted by $f'(\sigma)$.

The Julia–Wolff–Carathéodory theorem has, of course, an upper half-plane version. For the sake of simplicity, we shall prove it at infinity only.

Corollary 2.3.4. *Let $F \in \operatorname{Hol}(\mathbb{H}^+,\mathbb{H}^+)$ and assume that $\beta_F^{\mathbb{H}^+}(\infty,\infty) > 0$. Then we have*

$$K\text{-}\lim_{w\to\infty} \frac{F(w)}{w} = K\text{-}\lim_{w\to\infty} F'(w) = \beta_F^{\mathbb{H}^+}(\infty,\infty) < +\infty.$$

Proof. Let $\Psi\colon \mathbb{D} \to \mathbb{H}^+$ be the Cayley transform and set $f = \Psi^{-1} \circ F \circ \Psi$. Then it is easy to check that

$$\frac{F(w)}{w} = \frac{1 - \Psi^{-1}(w)}{1 - f(\Psi^{-1}(w))} \cdot \frac{1 + f(\Psi^{-1}(w))}{1 + \Psi^{-1}(w)}.$$

Recalling (2.20) we see that $\beta_f(1,1) < +\infty$ and thus f has nontangential limit 1 at 1. Moreover, by construction, w tends nontangentially to ∞ if and only if $\Psi^{-1}(w)$ tends nontangentially to 1. Therefore we can apply Theorem 2.3.2 to get

$$K\text{-}\lim_{w\to\infty} \frac{F(w)}{w} = \frac{1}{\beta_f(1,1)} = \beta_F^{\mathbb{H}^+}(\infty,\infty).$$

Next, it is easy to check that

$$F'(w) = \left(\frac{1 - \Psi^{-1}(w)}{1 - f(\Psi^{-1}(w))} \right)^2 f'(\Psi^{-1}(w)),$$

and hence

$$K\text{-}\lim_{w\to\infty} F'(w) = \frac{1}{\beta_f(1,1)} = \beta_F^{\mathbb{H}^+}(\infty,\infty),$$

as required. □

Definition 2.3.5. Let $F \in \mathrm{Hol}(\mathbb{H}^+, \mathbb{H}^+)$ be such that $\beta_F^{\mathbb{H}^+}(\infty, \infty) > 0$. Then the *angular derivative* $F'(\infty)$ of F at ∞ is given by

$$F'(\infty) = K\text{-}\lim_{w \to \infty} \frac{F(w)}{w} = \beta_F(\infty, \infty).$$

When $\beta_F(\infty, \infty) = 0$, we set by convention $F'(\infty) = 0$; in such a way we have $F'(\infty) = \beta_F^{\mathbb{H}^+}(\infty, \infty)$ always.

In particular, the angular derivative at infinity of a holomorphic self-map of $\mathbb{H}^+$ is always finite—and we shall need this fact.

Coming back to the disk, the Julia–Wolff–Carathéodory theorem implies that when the angular derivative is finite then it coincides with the nontangential limit of the derivative. When the angular derivative is infinite, in general we cannot infer anything about the behavior of f' at σ.

Example 2.3.6. Given $\lambda \in \mathbb{C}$ set $f_\lambda(z) = \frac{\lambda}{k_\lambda} z^{k_\lambda}$, where $k_\lambda \in \mathbb{N}^*$ is the smallest integer greater than $|\lambda|$. Then $f_\lambda \in \mathrm{Hol}(\mathbb{D}, \mathbb{D})$ is such that $\beta_{f_\lambda}(1) = +\infty$ (because $|f_\lambda(1)| < 1$) and $f_\lambda'(1) = \lambda$.

In this example, we obtained $\beta_f(1) = +\infty$ by taking a function f whose image was strictly contained in $\mathbb{D}$. If we rule out this possibility, the link between the angular derivative and the usual derivative is much tighter.

Proposition 2.3.7. *Let $f \in \mathrm{Hol}(\mathbb{D}, \mathbb{D})$ and $\sigma \in \partial\mathbb{D}$ be such that*

$$\limsup_{r \to 1} |f(r\sigma)| = 1. \tag{2.51}$$

Then

$$\beta_f(\sigma) = \limsup_{r \to 1} |f'(r\sigma)|. \tag{2.52}$$

Proof. If the limsup in (2.52) is infinite, then $\{|f'(r\sigma)|\}$ cannot be bounded as $r \to 1$, and thus, by Theorem 2.3.2, $\beta_f(\sigma) = +\infty$. So assume it is finite, and thus $|f'(r\sigma)| < M$ for all $r \in (0, 1)$ and a suitable $M < +\infty$; again by Theorem 2.3.2 it suffices to show that $\beta_f(\sigma)$ is also finite.

Now for all $r_1, r_2 \in (0, 1)$ we have

$$|f(r_2\sigma) - f(r_1\sigma)| = \left| \int_{r_1}^{r_2} f'(r\sigma)\, dr \right| \le M|r_2 - r_1|. \tag{2.53}$$

By (2.51), there are $\tau \in \partial\mathbb{D}$ and a sequence $\{r_\nu\} \subset (0, 1)$ converging to 1 such that $f(r_\nu\sigma) \to \tau$ as $\nu \to +\infty$. Therefore, putting $r_2 = r_\nu$ in (2.53) and passing to the limit, for any $r \in (0, 1)$ we get

$$|\tau - f(r\sigma)| \le M(1-r).$$

Hence

$$\beta_f(\sigma) = \liminf_{z\to\sigma} \frac{1-|f(z)|}{1-|z|} \le \liminf_{r\to 1} \frac{1-|f(r\sigma)|}{1-r} \le \liminf_{r\to 1} \frac{|\tau - f(r\sigma)|}{1-r} \le M,$$

and we are done. □

Proposition 2.3.7 gives a more practical way to compute $\beta_f(\sigma)$ than Proposition 2.1.15; furthermore, it is the last step toward the following result, which summarizes the relationships between the boundary dilation and the angular derivative.

Corollary 2.3.8. *Let $f \in \mathrm{Hol}(\mathbb{D},\mathbb{D})$ and $\sigma \in \partial\mathbb{D}$. Then the following assertions are equivalent:*
(i) *we have*

$$\limsup_{r\to 1} |f(r\sigma)| = 1 \quad \textit{and} \quad \limsup_{r\to 1} |f'(r\sigma)| < +\infty;$$

(ii) *the boundary dilation $\beta_f(\sigma)$ is finite;*
(iii) *f has nontangential limit $\tau \in \partial\mathbb{D}$ at σ and the angular derivative $f'(\sigma)$ is finite;*
(iv) $\limsup_{r\to 1} |f(r\sigma)| = 1$ *and f' has finite nontangential limit at σ.*

Furthermore, if any of these assertions holds then $f'(\sigma) = \tau\bar{\sigma}\beta_f(\sigma)$.

Proof. Proposition 2.3.7 shows that (i) implies (ii). By Theorem 2.3.2, we have that (ii) implies (iii). Proposition 2.3.1 yields that (iii) implies (iv). Finally, (iv) clearly implies (i), and the last assertion follows by invoking again Theorem 2.3.2. □

Remark 2.3.9. The condition on the lim sup of $|f(r\sigma)|$ in (iv) is necessary. Indeed, take $f(z) = \frac{1}{2}z$. Then f and f' clearly have nontangential (actually unrestricted) limit at 1 but $\beta_f(1) = +\infty$ because $|f(1)| = \frac{1}{2} < 1$.

As an application, we get another formula involving the boundary dilation that we shall need later on.

Corollary 2.3.10. *Take $f \in \mathrm{Hol}(\mathbb{D},\mathbb{D})$ and $\sigma \in \partial\mathbb{D}$. Let $\{z_\nu\} \subset \mathbb{D}$ be a sequence converging to σ such that $(\sigma - z_\nu)/|\sigma - z_\nu|$ converges to $\tau \in \partial\mathbb{D}$, and write $\tau = \sigma e^{i\theta}$. Then*

$$\lim_{\nu\to+\infty} \omega(z_\nu, f(z_\nu)) = \frac{1}{2}\log \frac{|\beta_f(\sigma,\sigma) + e^{-2i\theta}| + |\beta_f(\sigma,\sigma) - 1|}{|\beta_f(\sigma,\sigma) + e^{-2i\theta}| - |\beta_f(\sigma,\sigma) - 1|}.$$

In particular,

$$\lim_{r\to 1^-} \omega(r\sigma, f(r\sigma)) = \frac{1}{2}\log \beta_f(\sigma,\sigma).$$

Moreover, if $\beta_f(\sigma,\sigma) < +\infty$ then $f(\sigma) = \sigma$ and

$$\lim_{r\to1^-}\omega(r\sigma,f(r\sigma))=\frac{1}{2}\log f'(\sigma). \tag{2.54}$$

Proof. Applying (2.44) with $w=f(z)$ and recalling (2.46), we get

$$\frac{f(z)-z}{1-\overline{z}f(z)}\longrightarrow-\sigma\frac{\beta_f(\sigma,\sigma)-1}{\beta_f(\sigma,\sigma)+\sigma^2\overline{\tau}^2},$$

and the first formula follows immediately because $\sigma^2\overline{\tau}^2=e^{-2i\theta}$. The second formula is a consequence of the first because in that case $\theta=0$. The final assertion follows from Lemma 2.2.14 and Theorem 2.3.2. □

So, the Julia lemma gives us an effective way to deal with the boundary behavior of the derivative of a function $f\in\mathrm{Hol}(\mathbb{D},\mathbb{D})$. As an application, we shall now prove two statements giving bounds on the angular derivative, which are somehow akin to the bound on the derivative at a fixed point given by the Schwarz lemma.

The first result can be thought of as a boundary version of (1.34). In the statement, f^h is the hyperbolic derivative of f introduced in Definition 1.5.1.

Corollary 2.3.11. *Let $f\in\mathrm{Hol}(\mathbb{D},\mathbb{D})$ and $\sigma\in\partial\mathbb{D}$ be such that the angular derivative $f'(\sigma)$ exists and it is finite. Then*

$$|f'(\sigma)|\geq\frac{2}{1+|f^h(0)|}\frac{|\tau-f(0)|^2}{1-|f(0)|^2}\geq\frac{2}{1+|f^h(0)|}\frac{1-|f(0)|}{1+|f(0)|}. \tag{2.55}$$

In particular, if $f(0)=0$ we get

$$|f'(\sigma)|\geq\frac{2}{1+|f'(0)|}\geq1; \tag{2.56}$$

moreover, $|f'(\sigma)|=1$ if and only if f is a rotation.

Proof. Assume first $f\in\mathrm{Aut}(\mathbb{D})$, and write $f(z)=e^{i\theta}\gamma_a(z)$ for suitable $\theta\in\mathbb{R}$ and $a\in\mathbb{D}$, where $\gamma_a(z)=(z-a)/(1-\overline{a}z)$ as usual. Then $|f^h(0)|=1$ and

$$|f'(\sigma)|=\frac{1-|a|^2}{|1-\overline{a}\sigma|^2}\geq\frac{1-|a|^2}{(1+|a|)^2}=\frac{1-|f(0)|}{1+|f(0)|}$$

as desired.

Take now $f\notin\mathrm{Aut}(\mathbb{D})$ and let us first suppose $f(0)=0$. Then (1.34) yields

$$\frac{1-|f(z)|}{1-|z|}\geq\frac{1+|z|}{1+|f'(0)||z|};$$

taking the lim inf as $z\to\sigma$ Propositions 2.1.15 and 2.3.7 yield

$$|f'(\sigma)|\geq\frac{2}{1+|f'(0)|},$$

and (2.56) is proved.

In the general case, put $a = f(0)$ and $F = \gamma_a \circ f$. Since $F(0) = 0$ and, by Corollary 2.3.8, f has nontangential limit $\tau \in \partial\mathbb{D}$ as $z \to \sigma$, we can apply the previous argument to F obtaining

$$|F'(\sigma)| \geq \frac{2}{1 + |F'(0)|}.$$

But

$$F'(z) = \gamma_a'(f(z))f'(z) = \frac{1 - |a|^2}{(1 - \overline{a}f(z))^2} f'(z);$$

in particular, $F'(0) = f^h(0)$ and

$$F'(\sigma) = \frac{1 - |f(0)|^2}{(1 - \overline{f(0)}\tau)^2} f'(\sigma).$$

Hence

$$\begin{aligned} |f'(\sigma)| &\geq \frac{2}{1 + |f^h(0)|} \frac{|1 - \overline{f(0)}\tau|^2}{1 - |f(0)|^2} = \frac{2}{1 + |f^h(0)|} \frac{|\tau - f(0)|^2}{1 - |f(0)|^2} \\ &\geq \frac{2}{1 + |f^h(0)|} \frac{(1 - |f(0)|)^2}{1 - |f(0)|^2} = \frac{2}{1 + |f^h(0)|} \frac{1 - |f(0)|}{1 + |f(0)|} \end{aligned}$$

as claimed.

The final statement follows immediately recalling that, by the usual Schwarz lemma, when $f(0) = 0$ then $|f'(0)| \leq 1$ with equality if and only if f is a rotation. □

Example 2.3.12. The inequality (2.55) is sharp. Indeed, given $0 \leq a < 1$ let $f_a \in \mathrm{Hol}(\mathbb{D}, \mathbb{D})$ be given by

$$f_a(z) = z \frac{z + a}{1 + az}.$$

Then $f_a(0) = 0$, $f_a^h(0) = f_a'(0) = a$, $f_a(1) = 1$ and $f_a'(1) = 2/(1 + a)$, so in this case equality holds in (2.55) with $\sigma = 1$.

The next bound on the angular derivative is possibly more intriguing.

Theorem 2.3.13. *Let $f \in \mathrm{Hol}(\mathbb{D}, \mathbb{D})$. Assume there are $\sigma_1, \sigma_2, \tau_1, \tau_2 \in \partial\mathbb{D}$ with $\sigma_1 \neq \sigma_2$ such that $f(r\sigma_1) \to \tau_1$ and $f(r\sigma_2) \to \tau_2$ as $r \to 1$. Write $\sigma_2 = e^{i\phi}\sigma_1$ and $\tau_2 = e^{i\psi}\tau_1$ for suitable $\phi, \psi \in [-\pi, \pi)$. Then*

$$\beta_f(\sigma_1)\beta_f(\sigma_2) \geq \left[\frac{\sin(\psi/2)}{\sin(\phi/2)}\right]^2,$$

with equality if and only if $f \in \mathrm{Aut}(\mathbb{D})$*. In particular, if both* $\beta_f(\sigma_1)$ *and* $\beta_f(\sigma_2)$ *are finite then*

$$|f'(\sigma_1)||f'(\sigma_2)| \geq \left[\frac{\sin(\psi/2)}{\sin(\phi/2)}\right]^2.$$

Proof. If either of $\beta_f(\sigma_j)$ is infinite, or if $\tau_1 = \tau_2$, then there is nothing to prove. So, assume that both $\beta_f(\sigma_j)$ are finite and that $\psi \neq 0$. Let $\gamma_1, \gamma_2 \in \mathrm{Aut}(\mathbb{D})$ be given by

$$\gamma_j(z) = \alpha_j \frac{z - a_j}{1 - \overline{a_j} z}$$

with

$$\alpha_1 = \sigma_1 \frac{1 + ie^{i\phi/2}}{1 - ie^{-i\phi/2}}, \quad a_1 = \frac{ie^{i\phi/2} - 1}{ie^{i\phi/2} + 1}$$

and

$$\alpha_2 = \tau_1 \frac{1 + ie^{i\psi/2}}{1 - ie^{-i\psi/2}}, \quad a_2 = \frac{ie^{i\psi/2} - 1}{ie^{i\psi/2} + 1}.$$

Then it is not too difficult to check that $\gamma_1(1) = \sigma_1$, $\gamma_1(-1) = \sigma_2$, $\gamma_2(1) = \tau_1$, $\gamma_2(-1) = \tau_2$, and that

$$\gamma_1'(1) \cdot \gamma_1'(-1) = -\sigma_1\sigma_2[\sin(\phi/2)]^2,$$
$$\gamma_2'(1) \cdot \gamma_2'(-1) = -\tau_1\tau_2[\sin(\psi/2)]^2.$$

Let $g = \gamma_2^{-1} \circ f \circ \gamma_1$. Then (since, by Theorem 2.3.2, f has nontangential limit τ_j at σ_j for $j = 1, 2$) g has nontangential limit 1 at 1 and nontangential limit −1 at −1; moreover, (2.47) yields

$$\beta_g(1)\beta_g(-1) = \frac{\gamma_1'(1)\gamma_1'(-1)}{\gamma_2'(1)\gamma_2'(-1)} f'(\sigma_1)f'(\sigma_2) = \left[\frac{\sin(\phi/2)}{\sin(\psi/2)}\right]^2 \beta_f(\sigma_1)\beta_f(\sigma_2).$$

Therefore, it suffices to show that $\beta_g(1)\beta_g(-1) \geq 1$, with equality if and only if g is a hyperbolic automorphism of $\mathbb{D}$ fixing 1 and −1.

By definition of boundary dilation, we have

$$g(0) \in \overline{E(1, \beta_g(1))} \cap \overline{E(-1, \beta_g(-1))}.$$

Now, it is easy to check that $\overline{E(1, R_1)} \cap \overline{E(-1, R_2)} \neq \emptyset$ if and only if $R_1 R_2 \geq 1$, and $R_1 R_2 = 1$ if and only if the intersection is just one point. Therefore, $\beta_g(1)\beta_g(-1) \geq 1$; furthermore, $\beta_g(1)\beta_g(-1) = 1$ implies that $g(0) \in \partial E(1, \beta_g(1))$, and so $g \in \mathrm{Aut}(\mathbb{D})$ by the equality statement in the Julia lemma. □

We can use this result to control the angular derivative at boundary fixed points.

Definition 2.3.14. Let $f \in \mathrm{Hol}(\mathbb{D}, \mathbb{D})$, and $\sigma \in \partial\mathbb{D}$. Assume that f has nontangential limit $f(\sigma) \in \overline{\mathbb{D}}$ at σ. We say that σ is a *boundary contact point* if $|f(\sigma)| = 1$ and that it is a *boundary fixed point* if $f(\sigma) = \sigma$. Notice that we are not assuming that f extend continuously at σ; we are just assuming that f has a nontangential limit at σ.

Remark 2.3.15. Let $\sigma \in \partial\mathbb{D}$ be a boundary fixed point of $f \in \mathrm{Hol}(\mathbb{D}, \mathbb{D})$ such that $\beta_f(\sigma) < +\infty$. Then Lemma 2.2.14 and Theorem 2.3.2 imply that $f'(\sigma)$ exists and it is equal to the positive real number $\beta_f(\sigma) > 0$. With a slight abuse of notation, we shall write $f'(\sigma) = +\infty$ when $\beta_f(\sigma) = +\infty$, so that $f'(\sigma) = \beta_f(\sigma)$ for all boundary fixed points.

Corollary 2.3.16. *Let $f \in \mathrm{Hol}(\mathbb{D}, \mathbb{D})$. Assume that f has two distinct boundary fixed points, $\sigma_1, \sigma_2 \in \partial\mathbb{D}$. Then*

$$f'(\sigma_1)f'(\sigma_2) \geq 1,$$

with equality if and only if f is a hyperbolic automorphism of $\mathbb{D}$. Furthermore:
(i) *if $f \not\equiv \mathrm{id}_{\mathbb{D}}$ has also a fixed point in $\mathbb{D}$, then $f'(\sigma) > 1$ for any boundary fixed point $\sigma \in \partial\mathbb{D}$;*
(ii) *if f has no fixed points in $\mathbb{D}$, then there is at most one boundary fixed point $\sigma \in \partial\mathbb{D}$ such that $f'(\sigma) \leq 1$.*

Proof. The first assertion follows from Theorem 2.3.13 applied with $\tau_j = \sigma_j$ for $j = 1, 2$, because an automorphism with two fixed points in the boundary is necessarily hyperbolic.

To prove (i), up to conjugating with an automorphism we can assume that $f(0) = 0$; then Corollary 2.3.11 implies $f'(\sigma) > 1$, because f, having a fixed point in $\partial\mathbb{D}$, cannot be a rotation.

Finally, (ii) follows immediately if there is a boundary fixed point σ with $f'(\sigma) < 1$. If we had two distinct boundary fixed points $\sigma_1, \sigma_2 \in \partial\mathbb{D}$ with $f'(\sigma_1) = f'(\sigma_2) = 1$, then f should be a hyperbolic automorphism and Lemma 1.4.9 would give a contradiction. □

Remark 2.3.17. In Corollary 2.5.5, we shall see that a $f \in \mathrm{Hol}(\mathbb{D}, \mathbb{D})$ without fixed points in $\mathbb{D}$ always has a (necessarily unique) boundary fixed point $\sigma \in \partial\mathbb{D}$ such that $0 < f'(\sigma) \leq 1$.

Notes to Section 2.3

Our proof of Proposition 2.3.1 is mostly adapted from [255] and [80]; see also Corollary 2.4.11 for a generalization of this result.

The Julia–Wolff–Carathéodory Theorem 2.3.2 has been proved in several forms by several people; the most important versions are due to Wolff [417] in 1926 and to Carathéodory [97] and Landau and Valiron [255] in 1929. The presence of Julia's name is due to the essential role played by the Julia lemma in Wolff's and Carathéodory's proofs. A measure theoretic proof of the Julia–Wolff–Carathéodory theorem, based on arguments similar to the one we described at the end of Section 2.1, has been given by R. Nevanlinna [305] (cf. [15, 18]). Our proof is a combination of the proofs in [97, 255].

Consult [99, 395, 337, 317] for other applications and generalizations of the Julia–Wolff–Carathéodory theorem.

An immediate consequence of (2.47) is that $\sigma\overline{\tau}f'(\sigma)$ is a positive real number. This is known as the *Jack lemma*, because it has been rediscovered by Jack in [206]; see also [155] and references therein.

The discussion about the angular derivative following Theorem 2.3.2 is essentially taken from [97].

Corollary 2.3.11 is a variation on a result stated by Osserman [313] but was probably known before; our proof is inspired by [48]. Ünkelbach [396, 397] proved a refined version of (2.56): if $f(0) = 0$, then

$$|f'(\sigma)| \geq 1 + \frac{|1 - f'(0)|^2}{1 - |f'(0)|^2}.$$

Mercer [285] has removed the condition $f(0) = 0$, obtaining

$$|f'(\sigma)| \geq 2\frac{|f(\sigma) - f(0)|^2}{1 - |f(0)|^2}\,\frac{1 - \operatorname{Re}\left(\sigma f(\sigma)\frac{\overline{f(\sigma)} - \overline{f(0)}}{f(\sigma) - f(0)} f^h(0)\right)}{1 - |f^h(0)|^2};$$

it should be remarked that Mercer stated it only for the case $\sigma = f(\sigma) = 1$. This estimate generalizes previous results contained in [245, 158, 348]. A different proof can be found in [9], where it is also shown that, when f is not an automorphism, equality holds if and only if f is a Blaschke product of degree 2.

Herzig [195] has refined Ünkelbach's estimate in another direction showing that if $f(z) = a_k z^k + O(z^{k+1})$ with $k \geq 1$ and $a_k \neq 0$ then

$$|f'(\sigma)| \geq k + \frac{|1 - a_k|^2}{1 - |a_k|^2}.$$

Similarly, Dubinin [138] has shown that if $f(0) = 0$ then

$$|f'(\sigma)| \geq 1 + \frac{2(1 - |f'(0)|)^2}{1 - |f'(0)|^2 + |f''(0)|/2}.$$

An elementary proof of this inequality is in [286]; see also [259]. Actually, all these estimates are particular cases of a much more powerful estimate obtained independently in [155] and [9], expressed in terms of the iterated hyperbolic difference quotients introduced in [39].

Corollary 2.3.10 comes from [374, p. 159] and [118].

Theorem 2.3.13 seems to have appeared for the first time (with different proofs) in [265] and in [53], but possibly it was known before; indeed a similar result is in [421]. For a related result, see [262].

Cowen and Pommerenke [130] have given a more precise version of Corollary 2.3.16: if $f \in \operatorname{Hol}(\mathbb{D}, \mathbb{D})$ is such that $f(1) = 1$ and $f(-1) = -1$, then

$$f'(1)f'(-1) \geq 1 + \sup_{-1<x<1}\left\{\frac{4(\operatorname{Im} f(x))^2}{(1 - |f(x)|^2)^2}\right\}.$$

Furthermore, [130] generalized Corollary 2.3.16 to more than two boundary fixed points, as we shall see in Section 4.8.

2.4 The Lindelöf theorem

We have seen that the existence of a nontangential limit can often be forced by some kind of assumption about radial behavior. This is a particular instance of a more general phenomenon: under suitable conditions, the existence of the limit along a curve ending at a boundary point of $\mathbb{D}$ implies the existence of the nontangential limit at that point. The main technical part of the proof of the results of this kind we shall see is contained in the following proposition.

Proposition 2.4.1. *Let* $\Sigma = \{z \in \mathbb{C} \mid |\operatorname{Re} z| < 1\}$. *Take* $F \in \operatorname{Hol}(\Sigma, \mathbb{C})$ *bounded on* $\{z \in \Sigma \mid \operatorname{Im} z \geq c_0\}$ *for a suitable* $c_0 \in \mathbb{R}$. *Assume that there exists a continuous curve* $\tilde{\gamma}\colon [0,1) \to \Sigma$ *with* $\lim_{t\to 1^-} \operatorname{Im} \tilde{\gamma}(t) = +\infty$ *such that*

$$\lim_{t\to 1^-} F(\tilde{\gamma}(t)) = L \in \mathbb{C}$$

exists. Then for all $0 < \delta < 1$ *we have*

$$\lim_{\operatorname{Im} z\to +\infty} F(z) = L$$

uniformly in $\{z \in \Sigma \mid |\operatorname{Re} z| \leq 1 - \delta\}$.

Proof. First of all, without loss of generality we can assume that $L = 0$ and that F is bounded by 1 in $\{z \in \Sigma \mid \operatorname{Im} z \geq c_0\}$. Up to replacing $F(z)$ by $\overline{F(\overline{z} + 2ic_0)}$ in $\{z \in \Sigma \mid \operatorname{Im} z \leq c_0\}$, we can also assume that F is bounded by 1 everywhere, i. e., $F \in \operatorname{Hol}(\Sigma, \mathbb{D})$.

Fix $\varepsilon \in (0,1)$, and take $y_0 > \operatorname{Im} \tilde{\gamma}(0)$ so large that $|F(\tilde{\gamma}(t))| < \varepsilon$ whenever $\operatorname{Im} \tilde{\gamma}(t) \geq y_0$. We claim that if $\operatorname{Im} z \geq y_0$ and $|\operatorname{Re} z| \leq 1 - \delta$ then

$$|F(z)| \leq \varepsilon^{\delta/4}; \tag{2.57}$$

the proposition will obviously follow from this.

Fix $z_0 \in \Sigma$ with $\operatorname{Im} z_0 \geq y_0$ and $|\operatorname{Re} z_0| \leq 1 - \delta$, and choose $t_0 < 1$ so that $\operatorname{Im} \tilde{\gamma}(t_0) = \operatorname{Im} z_0$ and $\operatorname{Im} \tilde{\gamma}(t) > \operatorname{Im} z_0$ for all $t \in (t_0, 1)$. Let E be the image of $[t_0, 1)$ under $\tilde{\gamma}$, and let $\overline{E}$ be the reflection of E with respect to the axis $\{\operatorname{Im} z = \operatorname{Im} z_0\}$. In particular, $E \cup \overline{E}$ intersects this axis only in $\tilde{\gamma}(t_0)$.

If $z_0 \in E$, (2.57) is clear; so assume $z_0 \notin E$. Notice that we can find a curve in Σ starting from z_0 with imaginary part going to $+\infty$ and not crossing $E \cup \overline{E}$; therefore, z_0 is contained in a connected component Σ_0 of $\Sigma \setminus (E \cup \overline{E})$ containing points with imaginary part arbitrarily high. Moreover, $\partial\Sigma_0 \subset \partial\Sigma \cup (E \cup \overline{E})$ and $\partial\Sigma_0$ cannot intersect both $\{\operatorname{Re} z = 1\}$ and $\{\operatorname{Re} z = -1\}$.

Assume for the moment $\operatorname{Re} z_0 > \operatorname{Re} \tilde{\gamma}(t_0)$, so that $\partial\Sigma_0 \supset \{\operatorname{Re} z = 1\}$. For every $\eta > 0$, define $G_\eta \in \operatorname{Hol}(\Sigma, \mathbb{C})$ by

$$G_\eta(z) = \frac{F(z)\overline{F(\overline{z} + 2i \operatorname{Im} z_0)}\varepsilon^{(1+z)/2}}{1 + \eta(1+z)}. \tag{2.58}$$

When $z \in \Sigma$, we have $|F| < 1$, $|\overline{F}| < 1$,

$$|\varepsilon^{(1+z)/2}| = \varepsilon^{(1+\operatorname{Re} z)/2} < 1$$

and

$$\left|\frac{1}{1+\eta(1+z)}\right| = \frac{1}{\sqrt{\eta^2|\operatorname{Im} z|^2 + |1+\eta(1+\operatorname{Re} z)|^2}} \le \frac{1}{1+\eta(1+\operatorname{Re} z)} < 1;$$

therefore, $|G_\eta(z)| < 1$ for all $z \in \Sigma$. Moreover, we have $|F(z)| < \varepsilon$ on E and $|F(\overline{z} + 2i \operatorname{Im} z_0)| < \varepsilon$ on $\overline{E}$; hence $|G_\eta(z)| < \varepsilon$ on $E \cup \overline{E}$. Furthermore,

$$\limsup_{\operatorname{Re} z \to 1} \left|\frac{\varepsilon^{(1+z)/2}}{1+\eta(1+z)}\right| = \frac{\varepsilon}{1+2\eta} < \varepsilon \quad \text{and} \quad \limsup_{\operatorname{Im} z \to \pm\infty} \left|\frac{\varepsilon^{(1+z)/2}}{1+\eta(1+z)}\right| = 0;$$

therefore,

$$\limsup_{z \to \partial\Sigma_0} |G_\eta(z)| < \varepsilon.$$

Then we can apply the maximum principle to infer that $|G_\eta(z)| < \varepsilon$ on Σ_0 and, in particular, in z_0. Letting $\eta \to 0$ we finally get

$$|F(z_0)|^2 \le \varepsilon \cdot \varepsilon^{-(1+\operatorname{Re} z)/2} = \varepsilon^{(1-\operatorname{Re} z)/2} < \varepsilon^{\delta/2},$$

and thus (2.57) is proved if $\operatorname{Re} z_0 > \operatorname{Re} \tilde{\gamma}(t_0)$. If $\operatorname{Re} z_0 < \operatorname{Re} \tilde{\gamma}(t_0)$, replacing $1+z$ by $1-z$ in (2.58) we are led to the same conclusion, and the proof is completed. □

We can now prove the main result of this section, the *Lindelöf theorem*.

Theorem 2.4.2 (Lindelöf, 1915). *Let $\gamma: [0,1) \to \mathbb{D}$ be a continuous curve such that*

$$\lim_{t \to 1} \gamma(t) = \sigma \in \partial\mathbb{D}.$$

If $f: \mathbb{D} \to \mathbb{C}$ is a bounded holomorphic function such that the limit

$$\lim_{t \to 1} f(\gamma(t)) = L \in \mathbb{C}$$

exists then f has nontangential limit L at σ.

Proof. Without loss of generality, we can suppose $\sigma = 1$. We begin by transferring everything to $\mathbb{H}^+$, via the Cayley transform Ψ. Then we have a bounded holomorphic function $f_1 = f \circ \Psi^{-1}: \mathbb{H}^+ \to \mathbb{C}$ and a curve $\gamma_1 = \Psi \circ \gamma: [0,1) \to \mathbb{H}^+$ such that $\gamma_1(t) \to \infty$ and $f_1(\gamma_1(t)) \to L$ as $t \to 1$. We claim that $f_1(w) \to L$ as $w \to \infty$ within $K_\varepsilon = \{w \in \mathbb{H}^+ \mid \operatorname{Im} w > \varepsilon|w|\}$ for all $\varepsilon \in (0,1)$; this will give the assertion thanks to Proposition 2.2.7(i).

Let $\Sigma = \{z \in \mathbb{C} \mid |\operatorname{Re} z| < 1\}$ be an infinite vertical strip and define $\Phi\colon \mathbb{H}^+ \to \Sigma$ by

$$\Phi(w) = \frac{2i}{\pi} \log w + 1,$$

where log denotes the principal branch of the logarithm, so that $\log(i) = i\frac{\pi}{2}$. Then Φ is a biholomorphism between $\mathbb{H}^+$ and Σ, with inverse $\Phi^{-1}(z) = \exp(-i\frac{\pi}{2}(z-1))$, such that $\Phi(i) = 0$, $\operatorname{Im}\Phi(w) \to +\infty$ as $w \to \infty$, and $\operatorname{Im}\Phi(w) \to -\infty$ as $w \to 0$. Furthermore, it is easy to check that

$$\Phi(K_\varepsilon) = \left\{ z \in \mathbb{C} \;\middle|\; |\operatorname{Re} z| < \frac{2}{\pi}\arccos(\varepsilon) \right\}.$$

Setting $F = f_1 \circ \Phi^{-1}$ and $\tilde{\gamma} = \Phi \circ \gamma_1$, to conclude the proof it suffices to apply Proposition 2.4.1 to F and $\tilde{\gamma}$, because $\operatorname{Im}\tilde{\gamma}(t) \to +\infty$ and $F(\tilde{\gamma}(t)) \to L$ as $t \to 1$. □

Actually, a slightly more general result is true: to get the existence of the nontangential limit it suffices to assume that f is bounded on Stolz regions. Let us fix the terminology.

Definition 2.4.3. Let $\sigma \in \partial\mathbb{D}$. A *$\sigma$-curve* is a continuous curve $\gamma\colon [0,1) \to \mathbb{D}$ such that $\gamma(t) \to \sigma$ as $t \to 1^-$. A σ-curve is *nontangential* if there exist $t_0 \in [0,1)$ and $M > 1$ such that $\gamma(t) \in K(\sigma, M)$ for all $t_0 \le t < 1$.

Definition 2.4.4. Given $\tau \in \partial\mathbb{D}$, we say that a function $f\colon \mathbb{D} \to \mathbb{C}$ is *K-bounded* at τ if for every $M > 1$ there exists $C_M > 0$ such that $|f(z)| \le C_M$ for all $z \in K(\tau, M)$.

Remark 2.4.5. Clearly, a bounded function is K-bounded, but the converse might not be true. For instance, the function

$$f(z) = \frac{1}{z+1}$$

is obviously unbounded on $\mathbb{D}$ but it is K-bounded at 1 for the simple reason that $-1 \notin \overline{K(1,M)}$ for all $M > 1$. However, the function f is bounded in a neighborhood of 1, and so we might still apply the Lindelöf theorem to f. To get an example of a function which is K-bounded but not bounded in any neighborhood of 1, let us move to $\mathbb{H}^+$.

Let $F \in \operatorname{Hol}(\mathbb{H}^+, \mathbb{C})$ be given by

$$F(w) = \frac{1}{1 - \exp(-i\log w)};$$

we claim that F is K-bounded at ∞ but it is not bounded in any neighborhood of ∞.

Fix $\varepsilon > 0$; we must prove that F is bounded in $K(\infty, \varepsilon)$. By Proposition 2.2.7(i), it suffices to show that F is bounded in the set $K_\varepsilon = \{w \in \mathbb{H}^+ \mid \operatorname{Im} w > \varepsilon|w|\}$. Since ε is fixed, we can find $\eta > 0$ such that $\eta < \arg(w) < \pi - \eta$ for all $w \in K_\varepsilon$. In particular, $|\exp(-i\log w)| > e^\eta > 1$, and thus

$$|F(w)| = \frac{1}{|1 - \exp(-i \log w)|} \leq \frac{1}{e^{\eta} - 1}$$

uniformly on $K(\infty, \varepsilon)$. So F is K-bounded; to prove that it is unbounded in a neighborhood of ∞, let $\gamma: [0, +\infty) \to \mathbb{H}^+$ be given by $\gamma(t) = t + i\sqrt{1+t}$. We clearly have $|\gamma(t)| \to +\infty$ and $\arg(\gamma(t)) \to 0$ as $t \to +\infty$. For every $k \in \mathbb{N}$, choose $t_k \geq 0$ such that $|\gamma(t_k)| = \exp(2k\pi)$. Then

$$\begin{aligned}\exp(-i \log \gamma(t_k)) &= \exp(\arg(\gamma(t_k)))[\cos(-\log|\gamma(t_k)|) + i \sin(-\log|\gamma(t_k)|)] \\ &= \exp(\arg(\gamma(t_k))) \to 1\end{aligned}$$

as $k \to +\infty$. Therefore, $|F(\gamma(t_k))| \to +\infty$ and F is unbounded in a neighborhood of ∞, as claimed.

We now generalize the Lindelöf theorem to K-bounded functions.

Proposition 2.4.6. *Let $f \in \mathrm{Hol}(\mathbb{D}, \mathbb{C})$ be K-bounded at $\sigma \in \partial\mathbb{D}$ and such that the limit*

$$\lim_{t \to 1} f(\gamma(t)) = L \in \mathbb{C}$$

exists for a nontangential σ-curve γ. Then f has nontangential limit L at σ.

Proof. By definition, there exists $M_0 > 1$ and $t_0 \in [0, 1)$ such that $\gamma(t) \in K(\sigma, M)$ for all $M \geq M_0$ and $t_0 \leq t < 1$. Fix $M_1 > M_0$; we must prove that $f(z) \to L$ as $z \to \sigma$ inside $K(\sigma, M_1)$.

Without loss of generality, we can assume $\sigma = 1$. Let $\Phi \circ \Psi: \mathbb{D} \to \Sigma$ be the biholomorphism between $\mathbb{D}$ and $\Sigma = \{z \in \mathbb{C} \mid |\operatorname{Re} z| < 1\}$ used in the proof of Theorem 2.4.2. Given $M_1' > M_1$, by Proposition 2.2.7 we can find $0 < \delta_1' < \delta_1 < 1$ and $c_0 > 0$ such that

$$\begin{aligned}\Phi \circ \Psi(K(\sigma, M_1)) \cap \{z \in \Sigma \mid \operatorname{Im} z \geq c_0\} &\subset \{z \in \Sigma \mid |\operatorname{Re} z| < 1 - \delta_1,\ \operatorname{Im} z \geq c_0\} \\ &\subset \{z \in \Sigma \mid |\operatorname{Re} z| < 1 - \delta_1',\ \operatorname{Im} z \geq c_0\} \\ &\subset \Phi \circ \Psi(K(\sigma, M_1')).\end{aligned} \tag{2.59}$$

Since f is K-bounded, we have that $F = f \circ (\Phi \circ \Psi)^{-1}$ is bounded in

$$\{z \in \Sigma \mid |\operatorname{Re} z| < 1 - \delta_1',\ \operatorname{Im} z \geq c_0\}.$$

Moreover, the image of $\tilde{\gamma} = \Phi \circ \Psi \circ \gamma|_{[t_0,1)}$ is contained in $\Sigma_1 = \{z \in \Sigma \mid |\operatorname{Re} z| < 1 - \delta_1'\}$. Since Σ_1 is linearly biholomorphic to Σ, we are in the hypotheses of Proposition 2.4.1, and the assertion follows. □

As a first application of this generalization of the Lindelöf theorem, we prove that the usual chain rule holds for angular derivatives, too.

Proposition 2.4.7. *Let $f, g \in \mathrm{Hol}(\mathbb{D}, \mathbb{D})$, and let $\sigma \in \partial\mathbb{D}$ be such that $f(\sigma) = \tau \in \partial\mathbb{D}$ and $g(\tau) = \eta \in \partial\mathbb{D}$. Then*

$$K\text{-}\lim_{z\to\sigma}\frac{g(f(z))-\eta}{z-\sigma}=K\text{-}\lim_{z\to\sigma}(g\circ f)'(z)=g'(\tau)f'(\sigma).$$

Proof. First of all, assume $\beta_f(\sigma)$ and $\beta_g(\tau)$ finite. We claim that the curve $\rho(t)=f(t\sigma)$ goes to τ nontangentially as $t\to 1$. Indeed, $\rho'(1)$ is tangent to $\partial\mathbb{D}$ in τ if and only if $\operatorname{Re}(\rho'(1)\overline{\tau})=0$; but

$$\operatorname{Re}(\rho'(1)\overline{\tau})=\operatorname{Re}(f'(\sigma)\sigma\overline{\tau})=\beta_f(\sigma)>0$$

by Theorem 2.3.2 and the claim is proved.

As a consequence, $g(\rho(t))\to\eta$ as $t\to 1$; but $g(\rho(t))=(g\circ f)(t\sigma)$, and then Theorem 2.4.2 implies that $g\circ f(z)\to\eta$ as $z\to\sigma$ nontangentially.

Furthermore, since g' has finite nontangential limit at τ, it is K-bounded. Moreover, f sends Stolz regions into Stolz regions, by Proposition 2.2.9; so $g'\circ f$ is K-bounded, too. Since $(g'\circ f)(t\sigma)=g'(\rho(t))\to g'(\tau)$ as $t\to 1^-$ because ρ is a non-tangential τ-curve, Proposition 2.4.6 implies that $g'\circ f$ has nontangential limit $g'(\tau)$ at σ.

The usual chain rule then gives

$$K\text{-}\lim_{z\to\sigma}(g\circ f)'(z)=\left[K\text{-}\lim_{z\to\sigma}g'(f(z))\right]\cdot\left[K\text{-}\lim_{z\to\sigma}f'(\sigma)\right]=g'(\tau)f'(\sigma),$$

and the assertion follows from Theorem 2.3.2 and Proposition 2.3.1.

Now assume either $g'(\tau)$ or $f'(\sigma)$ infinite. Since

$$\frac{1-|g(f(z))|}{1-|z|}=\frac{1-|g(f(z))|}{1-|f(z)|}\cdot\frac{1-|f(z)|}{1-|z|}$$

and both factors in the right-hand side have strictly positive infimum (by Lemma 2.1.9), if one of them goes to infinity the product does. Hence

$$\liminf_{z\to\sigma}\frac{1-|g(f(z))|}{1-|z|}=+\infty,$$

and the assertion follows from Proposition 2.1.15. □

We end this section with a definition and a few technical results we shall need later.

Definition 2.4.8. A domain $\Omega\subseteq\mathbb{D}$ has an *inner tangent* at $\sigma\in\mathbb{D}$ if for every $M>1$ there exists $\varepsilon>0$ such that $K(\sigma,M)\cap D(\sigma,\varepsilon)\subset\Omega$. In particular, any sequence in $\mathbb{D}$ converging nontangentially to σ is eventually contained in Ω.

Proposition 2.4.9. *Let $\Omega\subseteq\mathbb{D}$ be a domain having an inner tangent at $\sigma\in\partial\mathbb{D}$. Let $g\in\operatorname{Hol}(\Omega,\mathbb{C})$ be such that $\operatorname{Re}g$ has a finite nontangential limit at σ. Then*

$$K\text{-}\lim_{z\to\sigma}(\sigma-z)g'(z)=0.$$

Proof. Fix $M > 1$; then we can find $M_1 > M$ and $\varepsilon > 0$ so that $h = \operatorname{Re} g$ may be extended continuously to the closure of $\Omega_{M_1} = K(\sigma, M_1) \cap D(\sigma, \varepsilon) \subset \Omega$. Let $\Phi\colon \mathbb{D} \to \Omega_{M_1}$ a biholomorphism; since $\partial\Omega_{M_1}$ is a Jordan curve we can assume (Theorem 1.6.32) that Φ extends to a homeomorphism of $\overline{\mathbb{D}}$ with $\overline{\Omega_{M_1}}$ so that $\Phi(1) = \sigma$. We claim that

$$|\Phi^h(w)| \geq \frac{M_1 - M}{M_1 + M} \tag{2.60}$$

for all $w \in \Phi^{-1}(\Omega_M)$, where Φ^h is the hyperbolic derivative of Φ (see Definition 1.5.1) and $\Omega_M = K(\sigma, M) \cap D(\sigma, \varepsilon)$. Indeed, given $w \in \Phi^{-1}(\Omega_M)$, put $r(w) = \inf_{z \in \partial\Omega_{M_1}} \omega(\Phi(w), z)$ and set

$$B = \{z \in \mathbb{D} \mid \omega(z, \Phi(w)) < r(w)\} \subset \Omega_{M_1}.$$

The map $\Psi\colon \mathbb{D} \to B$ given by

$$\Psi(z) = \frac{(\tanh r(w))z + \Phi(w)}{1 + \overline{\Phi(w)}(\tanh r(w))z}$$

is a biholomorphism between $\mathbb{D}$ and B with $\Psi(0) = \Phi(w)$. Therefore, the Schwarz–Pick lemma Corollary 1.1.16 applied to $\Phi^{-1} \circ \Psi$ and to $z = 0$ yields

$$1 \geq \frac{|(\Phi^{-1} \circ \Psi)'(0)|}{1 - |w|^2} = \frac{(\tanh r(w))(1 - |\Phi(w)|^2)}{|\Phi'(w)|(1 - |w|^2)},$$

and hence recalling Lemma 2.2.8 we get

$$|\Phi^h(w)| \geq \tanh r(w) \geq \frac{M_1 - M}{M_1 + M} > 0$$

for all $w \in \Phi^{-1}(\Omega_M)$, as claimed.

Put $F = g \circ \Phi - \ell$, where $\ell \in \mathbb{R}$ is the nontangential limit of $h = \operatorname{Re} g$. Then $H = \operatorname{Re} F = h \circ \Phi - \ell$ is continuous on $\overline{\mathbb{D}}$, harmonic in $\mathbb{D}$, and $H(1) = 0$. A classical estimate on harmonic functions (see Proposition A.3.1) implies that

$$(1 - |w|^2)|F'(w)| \leq 2U(w)$$

for all $w \in \mathbb{D}$, where $U\colon \overline{\mathbb{D}} \to \mathbb{R}$ is the harmonic extension of $|H|_{\partial\mathbb{D}}|$. Since $H(1) = 0$ and $U \in C^0(\overline{\mathbb{D}})$, we have

$$\lim_{w \to 1} (1 - |w|^2)|F'(w)| = 0.$$

Now,

$$(1 - |\Phi(w)|^2)|g'(\Phi(w))| = \frac{1}{|\Phi^h(w)|}(1 - |w|^2)|F'(w)|.$$

Therefore, thanks to (2.60), we have

$$\lim_{\substack{z\to\sigma\\ z\in\Omega_M}} (1-|z|^2)|g'(z)| = 0,$$

and the assertion follows because in Ω_M the quotient $|\sigma - z|/(1-|z|)$ is bounded. □

Corollary 2.4.10. *Let $\Omega \subseteq \mathbb{D}$ be a simply connected domain having an inner tangent at $\sigma \in \partial\mathbb{D}$. Take $f \in \mathrm{Hol}(\Omega, \mathbb{C})$ and assume that there is $\tau \in \mathbb{C}\setminus f(\Omega)$ such that*

$$K\text{-}\lim_{z\to\sigma} \arg\frac{\tau - f(z)}{\sigma - z} = \theta \in [0, 2\pi). \tag{2.61}$$

Then

$$K\text{-}\lim_{z\to\sigma} \arg f'(z) = \theta. \tag{2.62}$$

Proof. Since Ω is simply connected and $\tau \notin f(\Omega)$, we can define $g \in \mathrm{Hol}(\Omega, \mathbb{C})$ by setting

$$g(z) = -i\log\frac{\tau - f(z)}{\sigma - z};$$

in particular, $\arg\frac{\tau - f(z)}{\sigma - z} = \operatorname{Re} g$ is well-defined in Ω. An easy computation shows that

$$g'(z) = -i\left[\frac{1}{\sigma - z} - \frac{f'(z)}{\tau - f(z)}\right].$$

By (2.61), we can apply Proposition 2.4.9 to g, obtaining that $(\sigma - z)g'(z) \to 0$ as $z \to \sigma$ nontangentially, i. e.,

$$K\text{-}\lim_{z\to\sigma} f'(z)\frac{\sigma - z}{\tau - f(z)} = 1.$$

Taking the real part of $-i$ times the logarithm of this last equality, we get (2.62) and we are done. □

Similarly, we get the following.

Corollary 2.4.11. *Let $\Omega \subseteq \mathbb{D}$ be a simply connected domain having an inner tangent at $\sigma \in \partial\mathbb{D}$. For $f \in \mathrm{Hol}(\Omega, \mathbb{C})$, the following assertions are equivalent:*

(i) *there exist $\tau \in \mathbb{C}\setminus f(\Omega)$ and $\beta \in \mathbb{C}^*$ such that*

$$K\text{-}\lim_{z\to\sigma} \frac{\tau - f(z)}{\sigma - z} = \beta \in \mathbb{C}^*; \tag{2.63}$$

(ii) *there exists $\beta \in \mathbb{C}^*$ such that*

$$K\text{-}\lim_{z\to\sigma} f'(z) = \beta. \tag{2.64}$$

Proof. Assume that (i) holds. Since Ω is simply connected and $\tau \notin f(\Omega)$, we can again define $g \in \mathrm{Hol}(\Omega, \mathbb{C})$ by setting

$$g(z) = -i \log \frac{\tau - f(z)}{\sigma - z};$$

in particular, $\operatorname{Re} g = \arg \frac{\tau - f(z)}{\sigma - z}$ and

$$(\sigma - z)g'(z) = -i\left[1 - f'(z)\frac{\sigma - z}{\tau - f(z)}\right].$$

By (2.63), we can apply Proposition 2.4.9 to g, obtaining that $(\sigma - z)g'(z) \to 0$ as $z \to \sigma$ nontangentially, i. e.,

$$K\text{-}\lim_{z \to \sigma} f'(z)\frac{\sigma - z}{\tau - f(z)} = 1.$$

Then (ii) follows recalling again (2.63).

Conversely, if (ii) holds we can prove (i) arguing as in the second part of the proof of Proposition 2.3.1. □

***Notes to Section* 2.4**

Theorem 2.4.2 was originally proved by Montel [297] in 1912 using normal family techniques and assuming γ to be radial. The general statement (with the idea of using maximum modulus arguments) has been obtained in 1915 by Lindelöf [268]; our proof is taken from [361]. Actually, Lindelöf proved Theorem 2.4.2 for functions $f \in \mathrm{Hol}(\mathbb{D}, D)$, where $D \subset \widehat{\mathbb{C}}$ is a hyperbolic domain; moreover, previously [267] he also proved that if $f \in \mathrm{Hol}(\mathbb{D}, D)$ is bounded along a σ-curve, with $\sigma \in \partial\mathbb{D}$, then it is K-bounded at σ. See [88, Sections V.5 and XII.5] for more comments, history, and related results. Much later, Lehto and Virtanen [261] in 1957 have proved that Theorem 2.4.2 holds for *normal functions*, i. e., for functions $f \in \mathrm{Hol}(\mathbb{D}, \widehat{\mathbb{C}})$ such that $\{f \circ \gamma \mid \gamma \in \mathrm{Aut}(\mathbb{D})\}$ is a normal family.

Arguing as in Example 2.2.12, it is possible to prove that the function $f(z) = \exp(-(1 - z)^{-2})$ has radial limit 0 at 1 but it does not admit nontangential limit at 1; in particular, it is not K-bounded.

Proposition 2.4.9 and Corollary 2.4.10 are due to Yamashita [424]. Actually, Yamashita has proved that the conditions (2.61) and (2.62) are equivalent, and that they imply that f has nontangential limit τ at σ.

Arosio and Bracci in [22] have proved (in $\mathbb{C}^n$ but it works in one variable too) a discrete version of the Lindelöf theorem: let $f \in \mathrm{Hol}(\mathbb{D}, \mathbb{D})$ and let $\{z_\nu\} \subset \mathbb{D}$ be a sequence converging to $\sigma \in \partial\mathbb{D}$. Assume there exists $C > 0$ such that $\omega(z_\nu, z_{\nu+1}) < C$ for all $\nu \in \mathbb{N}$. Then if the sequence $\{f(z_\nu)\}$ converges to $\tau \in \partial\mathbb{D}$ it follows that f has nontangential limit τ at σ.

2.5 The Wolff lemma

We now discuss the Wolff lemma, the second boundary version of the Schwarz lemma; in the next section, we shall see some of its consequences, mainly on the structure of the automorphism group of hyperbolic Riemann surfaces, even though for us the main applications will be in the study of the dynamics of holomorphic self-maps, as we shall see in the next chapter.

The original Schwarz lemma said something about functions $f \in \mathrm{Hol}(\mathbb{D},\mathbb{D})$ with a fixed point in $\mathbb{D}$. We now assume instead as hypothesis that f has *no* fixed points in $\mathbb{D}$. It turns out that then it exists a point $\tau \in \partial\mathbb{D}$ such that f sends every horocycle centered in τ into itself, exactly as a function with a fixed point $z_0 \in \mathbb{D}$ sends every Poincaré ball centered in z_0 into itself. This is the content of the *Wolff lemma.*

Theorem 2.5.1 (Wolff lemma, 1926). *Let $f \in \mathrm{Hol}(\mathbb{D},\mathbb{D})$ be without fixed points. Then there is a unique $\tau \in \partial\mathbb{D}$ such that for all $z \in \mathbb{D}$,*

$$\frac{|\tau - f(z)|^2}{1 - |f(z)|^2} \le \frac{|\tau - z|^2}{1 - |z|^2}, \tag{2.65}$$

i. e.,

$$f(E(\tau,R)) \subseteq E(\tau,R) \tag{2.66}$$

for all $R > 0$. Moreover, the equality in (2.65) *holds at one point (and hence everywhere) if and only if f is a parabolic automorphism of $\mathbb{D}$ leaving τ fixed.*

Proof. To prove the uniqueness of τ, assume that (2.66) holds for two distinct points $\tau, \tau_1 \in \partial\mathbb{D}$. Then we can construct two horocycles, one centered at τ and the other centered at τ_1, tangent to each other at a point of $\mathbb{D}$. By (2.66), this point would be a fixed point of f, contradiction.

For the existence, pick a sequence $\{r_\nu\} \subset (0,1)$ with $r_\nu \to 1$, and set $f_\nu = r_\nu f$. Then $f_\nu(\mathbb{D})$ is relatively compact in $\mathbb{D}$; by the Ritt theorem (Corollary 1.1.15), each f_ν has a unique fixed-point $w_\nu \in \mathbb{D}$. Up to a subsequence, we can assume $w_\nu \to \tau \in \overline{\mathbb{D}}$. If τ were in $\mathbb{D}$, we would have

$$f(\tau) = \lim_{\nu\to\infty} f_\nu(w_\nu) = \lim_{\nu\to\infty} w_\nu = \tau,$$

which is impossible; therefore, $\tau \in \partial\mathbb{D}$.

Now, by the Schwarz–Pick lemma

$$1 - \left|\frac{f_\nu(z) - w_\nu}{1 - \overline{w_\nu} f_\nu(z)}\right|^2 \ge 1 - \left|\frac{z - w_\nu}{1 - \overline{w_\nu} z}\right|^2,$$

or, equivalently,

$$\frac{|1 - \overline{w_\nu} f_\nu(z)|^2}{1 - |f_\nu(z)|^2} \le \frac{|1 - \overline{w_\nu} z|^2}{1 - |z|^2}.$$

Taking the limit as $\nu \to \infty$ we get (2.65), as claimed.

Finally, assume the equality holds for some $z \in \mathbb{D}$. Then, exactly as in Theorem 2.1.10, we see that f must be an automorphism of $\mathbb{D}$ of the form (2.16) with $\alpha = 1$

and $\sigma = \tau$. Recalling Remark 1.4.7 we deduce that f is a parabolic automorphism of $\mathbb{D}$ leaving τ fixed. The converse follows from Lemma 1.4.17. □

Remark 2.5.2. A function $f \in \mathrm{Hol}(\mathbb{D}, \mathbb{D})$, $f \not\equiv \mathrm{id}_{\mathbb{D}}$, satisfying (2.66) for a point $\tau \in \partial\mathbb{D}$ cannot have fixed points. Indeed, assume by contradiction that it has a fixed point; up to composing with an automorphism, we can assume that the fixed point is the origin. Fix $R < 1$, and let z_0 be the point of $\partial E(\tau, R)$ closest to 0. Then $f(z_0) \in \overline{E(\tau, R)}$; on the other hand, by the Schwarz lemma, $|f(z_0)| \leq |z_0|$. This implies that $|f(z_0)| = |z_0|$, and again by the Schwarz lemma, f should be a rotation, and thus $f \equiv \mathrm{id}_{\mathbb{D}}$, by (2.66), contradiction.

Definition 2.5.3. Let $f \in \mathrm{Hol}(\mathbb{D}, \mathbb{D})$, with $f \not\equiv \mathrm{id}_{\mathbb{D}}$. The *Wolff point* $\tau_f \in \overline{\mathbb{D}}$ is:
- the unique fixed point of f if f has a fixed point in $\mathbb{D}$;
- the unique point in $\partial\mathbb{D}$ given by the Wolff lemma Theorem 2.5.1 if f has no fixed points in $\mathbb{D}$.

Using the Wolff point, we can give a unified version of the Schwarz and Wolff lemmas.

Corollary 2.5.4. *Let $f \in \mathrm{Hol}(\mathbb{D}, \mathbb{D})$, with $f \not\equiv \mathrm{id}_{\mathbb{D}}$, and let $\tau \in \overline{\mathbb{D}}$ be its Wolff point. Then*

$$\frac{|1 - \overline{\tau} f(z)|^2}{1 - |f(z)|^2} \leq \frac{|1 - \overline{\tau} z|^2}{1 - |z|^2} \tag{2.67}$$

for every $z \in \mathbb{D}$.

Proof. If f is fixed points free, (2.67) is exactly (2.65). On the other hand, if f has a fixed point in $\mathbb{D}$ (which is, by definition, τ), then (2.67) follows from (1.6) setting $w = \tau$. □

Entering the Julia–Wolff–Carathéodory theorem into play, we can see even better why the Wolff lemma should be considered a boundary Schwarz lemma.

Corollary 2.5.5. *Let $f \in \mathrm{Hol}(\mathbb{D}, \mathbb{D})$ be without fixed points. Then*

$$K\text{-}\lim_{z \to \tau_f} f(z) = \tau_f$$

and the derivative f' has nontangential limit $f'(\tau_f) = \beta_f(\tau_f)$ at τ_f with

$$\frac{1 - |f(0)|}{1 + |f(0)|} \leq f'(\tau_f) \leq 1.$$

Moreover, τ_f is the unique point $\tau \in \partial\mathbb{D}$ such that

$$K\text{-}\lim_{z \to \tau} f(z) = \tau \quad \text{and} \quad K\text{-}\lim_{z \to \tau} |f'(z)| \leq 1.$$

Proof. The assertions follow from Theorem 2.5.1, Lemma 2.2.14, Theorem 2.3.2, Lemma 2.1.9, and Proposition 2.3.7. □

Definition 2.5.6. Let $f \in \mathrm{Hol}(\mathbb{D}, \mathbb{D})$ be without fixed points and let $\tau \in \partial\mathbb{D}$ be its Wolff point. We shall say that f is *hyperbolic* if $0 < f'(\tau) < 1$ and that it is *parabolic* if $f'(\tau) = 1$. If $f \in \mathrm{Hol}(\mathbb{D}, \mathbb{D})$ has a fixed point in $\mathbb{D}$, we shall sometimes say that f is *elliptic*.

It is very natural to conjecture that we might complete the statement of Corollary 2.5.5 with a sentence like “Furthermore, $f'(\tau_f) = 1$ if and only if f is a parabolic automorphism of $\mathbb{D}$ leaving τ_f fixed.” However, this is false, as the following example shows.

Example 2.5.7. Let $f\colon \mathbb{D} \to \mathbb{C}$ be given by

$$f(z) = \frac{1 + 3z^2}{3 + z^2}.$$

It is easy to check that $f(\mathbb{D}) \subseteq \mathbb{D}$, that f has no fixed points in $\mathbb{D}$, that $\lim_{z\to 1} f(z) = 1$ and that $\lim_{z\to 1} f'(z) = 1$; therefore, Corollary 2.5.5 yields $\tau_f = 1$ and $\beta_f(1) = 1$ even though $f \notin \mathrm{Aut}(\mathbb{D})$. We shall come back to this example in Section 2.7.

We end this section with the Wolff lemma in $\mathbb{H}^+$.

Proposition 2.5.8. *Let $F \in \mathrm{Hol}(\mathbb{H}^+, \mathbb{H}^+)$ be without fixed points. Then either there exists a unique point $a \in \mathbb{R}$ such that*

$$\frac{\operatorname{Im} w}{|w - a|^2} \le \frac{\operatorname{Im} F(w)}{|F(w) - a|^2} \tag{2.68}$$

for all $w \in \mathbb{H}^+$, i. e., $F(E(a, R)) \subseteq E(a, R)$ for all $R > 0$, or

$$\operatorname{Im} w \le \operatorname{Im} F(w) \tag{2.69}$$

for all $w \in \mathbb{H}^+$, i. e., $F(E(\infty, R)) \subseteq E(\infty, R)$ for all $R > 0$. Equality in (2.68) or in (2.69) holds at one point, and hence everywhere, if and only if F is a parabolic automorphism of $\mathbb{H}^+$ with fixed point $a \in \mathbb{R}$, respectively ∞. Finally, if (2.69) holds then

$$F'(\infty) \ge 1,$$

where $F'(\infty)$ is the angular derivative of F at infinity.

Proof. It is just a translation (using (2.8), (2.9), and the Cayley transform) of Theorem 2.5.1 and Corollary 2.5.5. □

Notes to Section 2.5

The Wolff lemma was partially stated by Julia [216] in 1920, but its true birth occurred in 1926 in Wolff's paper [416] on dynamics. It is less known than its akin due to Julia, but it is the cornerstone of complex dynamics on hyperbolic domains. For other applications of Wolff lemma see, e. g., [188, 189].

2.6 The automorphism group of hyperbolic Riemann surfaces

The main applications of the Wolff lemma are in dynamics, as we shall see in the next chapter. Here, we shall instead describe a different application of the Wolff and Julia lemmas, with remarkable consequences for the study of Riemann surfaces.

In Proposition 1.4.12, we saw that two automorphisms of $\mathbb{D}$ commute if and only if they have the same fixed points. We shall now prove a first extension of that result.

Theorem 2.6.1. *Let $\gamma \in \mathrm{Aut}(\mathbb{D})$ be hyperbolic, and $f \in \mathrm{Hol}(\mathbb{D}, \mathbb{D})$ be such that*

$$f \circ \gamma = \gamma \circ f. \tag{2.70}$$

Then either f is a hyperbolic automorphism of $\mathbb{D}$ with the same fixed points as γ or $f \equiv \mathrm{id}_{\mathbb{D}}$.

Proof. Assume $f \not\equiv \mathrm{id}_{\mathbb{D}}$; in particular, f cannot have more than one fixed point (Corollary 1.1.14). If f has a fixed point $z_0 \in \mathbb{D}$, then by (2.70)

$$f(\gamma(z_0)) = \gamma(f(z_0)) = \gamma(z_0),$$

i. e., $\gamma(z_0) = z_0$, impossible. Hence f is fixed point free and we can apply the Wolff lemma to get a point $\tau \in \partial\mathbb{D}$ satisfying (2.66). However, $\gamma(\tau)$ still satisfies (2.66), by (2.70) and Proposition 2.1.5; therefore, the uniqueness part of Wolff lemma implies $\gamma(\tau) = \tau$.

Now transfer everything to $\mathbb{H}^+$ in such a way that τ goes into ∞ and the fixed point set of γ goes into $\{0, \infty\}$. Then $\gamma(z) = \lambda z$ for some positive $\lambda \neq 1$ and, up to replacing γ by γ^{-1}, we can assume $\lambda > 1$. So, f satisfies

$$f(\lambda w) = \lambda f(w) \tag{2.71}$$

for all $w \in \mathbb{H}^+$. By Corollary 2.3.4 and Proposition 2.5.8, there exists $\beta \geq 1$ such that, for any $w \in \mathbb{H}^+$,

$$\lim_{k\to\infty} \frac{f(\lambda^k w)}{\lambda^k w} = \beta.$$

Then (2.71) implies $f(w) = \beta w$ and f is a hyperbolic automorphism with the same fixed points as γ, as claimed. □

Using this theorem, we shall now do a deeper study of the structure of the automorphism group of a hyperbolic Riemann surface. The first new fact is the following.

Theorem 2.6.2. *Let X be a hyperbolic Riemann surface with non-Abelian fundamental group. Then id_X is isolated in $\mathrm{Hol}(X, X)$. In particular, $\mathrm{Aut}(X)$ is discrete.*

Proof. Let $\pi_X\colon \mathbb{D} \to X$ be the universal covering map and realize as usual (see Proposition 1.6.9) the fundamental group of X as the group Γ_X of automorphisms of the covering, properly discontinuous on $\mathbb{D}$ (Proposition 1.6.12).

Assume, by contradiction, that there exists a sequence $\{f_\nu\} \subset \mathrm{Hol}(X, X)$ converging to id_X. Let $\tilde{f}_\nu \in \mathrm{Hol}(\mathbb{D}, \mathbb{D})$ be a lifting of f_ν; we claim that we can choose $\tilde{f}_\nu$ so that $\tilde{f}_\nu \to \mathrm{id}_{\mathbb{D}}$ in $\mathrm{Hol}(\mathbb{D}, \mathbb{D})$. Indeed, choose $z_0 \in X$ and fix $\tilde{z}_0 \in \pi_X^{-1}(z_0)$. Since $f_\nu(z_0) \to z_0$, we can choose $\tilde{f}_\nu$ so that $\tilde{f}_\nu(\tilde{z}_0) \to \tilde{z}_0$ as $\nu \to +\infty$. In particular, $\{\tilde{f}_\nu\}$ has no compactly divergent subsequences. Let $\tilde{f}$ be a limit point of $\{\tilde{f}_\nu\}$. Clearly, $\tilde{f}(\tilde{z}_0) = \tilde{z}_0$ and $\pi_X \circ \tilde{f} = \pi_X$; therefore, $\tilde{f} \equiv \mathrm{id}_{\mathbb{D}}$ in a neighborhood of $\tilde{z}_0$, and hence everywhere. So, $\mathrm{id}_{\mathbb{D}}$ is the unique limit point of $\{\tilde{f}_\nu\}$ and Theorem 1.7.12 and Lemma 1.7.18 yield $\tilde{f}_\nu \to \mathrm{id}_{\mathbb{D}}$ as $\nu \to \infty$.

Now, for all $\gamma \in \Gamma_X$ and $\nu \in \mathbb{N}$ we have

$$\pi_X \circ \tilde{f}_\nu \circ \gamma = f_\nu \circ \pi_X \circ \gamma = f_\nu \circ \pi_X;$$

therefore, $\tilde{f}_\nu \circ \gamma$ is another lifting of f_ν, and thus (Proposition 1.6.14) there is $\alpha_\nu(\gamma) \in \Gamma_X$ such that

$$\tilde{f}_\nu \circ \gamma = \alpha_\nu(\gamma) \circ \tilde{f}_\nu. \tag{2.72}$$

Since Γ_X is non-Abelian, by Corollary 1.4.15 and Proposition 1.4.12, it must contain at least one hyperbolic automorphism γ_1 and another element γ_2 with fixed-point set different from the fixed point set of γ_1. Now, (2.72) implies $\alpha_\nu(\gamma) \to \gamma$ as $\nu \to +\infty$ because $\tilde{f}_\nu \to \mathrm{id}_{\mathbb{D}}$. Since Γ_X is properly discontinuous (and thus discrete by Lemma 1.6.22), we must have $\alpha_\nu(\gamma) = \gamma$ for sufficiently large ν. Then Theorem 2.6.1 applied with $\gamma = \gamma_1$ and $f = \tilde{f}_\nu$ for large enough ν shows that $\tilde{f}_\nu$ is a hyperbolic automorphism with the same fixed point set as γ_1. Finally, a second application of Theorem 2.6.1 to $\gamma = \tilde{f}_\nu$ and $f = \gamma_2$ implies that γ_2 has the same fixed point set as γ_1, contradiction. □

Theorem 2.6.2 has several interesting consequences. We begin with the classical *Klein–Poincaré theorem.*

Corollary 2.6.3 (Klein–Poincaré theorem, 1885). *Let X be a Riemann surface. Then X is hyperbolic with non-Abelian fundamental group if and only if* $\mathrm{Aut}(X)$ *is properly discontinuous on X.*

Proof. Assume that X is hyperbolic with a non-Abelian fundamental group. Then Theorem 2.6.2 shows that $\mathrm{Aut}(X)$ is discrete, and thus properly discontinuous on X, by Proposition 1.7.22.

Conversely, if X is not hyperbolic or it has an Abelian fundamental group then $\mathrm{Aut}(X)$ cannot be properly discontinuous on X because it is not discrete; see Propositions 1.1.5, 1.6.19, and 1.6.38. □

We also get a bound on the cardinality of $\mathrm{Aut}(X)$.

Corollary 2.6.4. *Let X be a hyperbolic Riemann surface with non-Abelian fundamental group. Then* $\mathrm{Aut}(X)$ *is countable.*

Proof. Assume, by contradiction, $\mathrm{Aut}(X)$ uncountable. Fix $z_0 \in X$, and define a function $\mu\colon \mathrm{Aut}(X) \to \mathbb{R}^+$ by

$$\mu(\gamma) = \omega_X(z_0, \gamma(z_0)).$$

Since $\mathrm{Aut}(X)$ is uncountable, we can find a sequence $\{\gamma_\nu\}$ of distinct elements of $\mathrm{Aut}(X)$ such that $\{\mu(\gamma_\nu)\}$ is bounded in $\mathbb{R}^+$. In particular, then, $\{\gamma_\nu\}$ cannot have compactly diverging subsequences; so, by the Montel theorem and Corollary 1.7.21, up to a subsequence we can suppose that $\{\gamma_\nu\}$ converges to an element of $\mathrm{Aut}(X)$, contradicting Theorem 2.6.2. □

Corollary 2.6.5. *Let X be a compact hyperbolic Riemann surface. Then* $\mathrm{Aut}(X)$ *is finite.*

Proof. By Theorem 1.6.33, X has a non-Abelian fundamental group; thus (Theorem 2.6.2) $\mathrm{Aut}(X)$ is discrete. On the other hand, by Theorem 1.7.12 and Corollary 1.7.21 $\mathrm{Aut}(X)$ is compact (because X is compact), and hence it must be finite. □

Theorem 2.6.2 does not apply to doubly connected domains, because they have an Abelian fundamental group by Theorem 1.6.33. Later on we shall need some sort of substitute, that we now describe.

Proposition 2.6.6. *Let* $D \subset \mathbb{C}$ *be a doubly connected hyperbolic domain not biholomorphic to* $\mathbb{D}^*$. *Denote by* $\pi_D\colon \mathbb{H}^+ \to D$ *a universal covering map and by* Γ_D *the automorphism group of this covering. Take* $f \in \mathrm{Hol}(D, D)$, $f \not\equiv \mathrm{id}_D$. *Then the following statements are equivalent:*
(i) $f \notin \mathrm{Aut}(D)$;
(ii) $f_*(\pi_1(D))$ *is trivial;*
(iii) *there exists* $\hat{f} \in \mathrm{Hol}(D, \mathbb{H}^+)$ *such that* $f = \pi \circ \hat{f}$;
(iv) *a lifting (and hence any lifting)* $\tilde{f} \in \mathrm{Hol}(\mathbb{H}^+, \mathbb{H}^+)$ *is automorphic under* Γ_D.
Furthermore, if $\tilde{f} \in \mathrm{Hol}(\mathbb{H}^+, \mathbb{H}^+)$ *is a lifting of* $f \in \mathrm{Aut}(D)$ *then* $\tilde{f}$ *is conjugated either to* $\gamma_0(w) = cw$ *or to* $\gamma_0(w) = -c/w$ *for some* $c > 0$.

Proof. We already know (Proposition 1.6.17) that (ii) $\Longleftrightarrow$ (iii) $\Longleftrightarrow$ (iv), and it is clear that (ii) $\Longrightarrow$ (i). We shall now show that (i) $\Longrightarrow$ (iv). We know (Theorem 1.6.33) that Γ_D is cyclic generated by a hyperbolic automorphism of $\mathbb{H}^+$; without loss of generality, we can assume that Γ_D is generated by $\gamma(z) = \lambda z$ for a suitable $\lambda > 1$. Take $f \in \mathrm{Hol}(D, D)$, and let $\tilde{f} \in \mathrm{Hol}(\mathbb{H}^+, \mathbb{H}^+)$ be a lifting. Since $\tilde{f} \circ \gamma$ is another lifting of f, by Proposition 1.6.14 there exists $n \in \mathbb{Z}$ such that

$$\tilde{f}(\lambda w) = \lambda^n \tilde{f}(w)$$

for all $w \in \mathbb{H}^+$; our aim is to prove that if $f \notin \mathrm{Aut}(D)$ then $n = 0$.

We know, by Corollary 2.3.4, that the angular derivative $\beta < +\infty$ of $\tilde{f}$ at ∞ is given by

$$\beta = \lim_{y \to +\infty} \frac{\tilde{f}(iy)}{iy}.$$

Therefore, for any $y_0 > 0$ we have

$$\beta = \lim_{k \to \infty} \frac{\tilde{f}(\lambda^k iy_0)}{\lambda^k iy_0} = \frac{\tilde{f}(iy_0)}{iy_0} \lim_{k \to \infty} \lambda^{(n-1)k},$$

and hence $n \leq 1$.

Let $g = -1/\tilde{f}$. Then $g \in \mathrm{Hol}(\mathbb{H}^+, \mathbb{H}^+)$ and $g(\lambda w) = \lambda^{-n} g(w)$; arguing as before we then find $n \geq -1$. Summing up, the only possibilities left are $n = -1, 0, 1$.

If $n = 1$, then, by Theorem 2.6.1, $\tilde{f}(w) = cw$; in particular $\tilde{f} \in \mathrm{Aut}(\mathbb{H}^+)$, and thus $f \in \mathrm{Aut}(D)$. If $n = -1$, we can apply Theorem 2.6.1 to $g = -1/\tilde{f}$ to get $g(w) = c^{-1}w$. Then $\tilde{f}(w) = -c/w$; in particular, $\tilde{f} \in \mathrm{Aut}(\mathbb{H}^+)$ and again $f \in \mathrm{Aut}(D)$. So if $f \notin \mathrm{Aut}(D)$ necessarily $n = 0$ and we are done. □

From this, we infer the following.

Corollary 2.6.7. *Let X be a hyperbolic Riemann surface not biholomorphic to $\mathbb{D}$ or $\mathbb{D}^*$. Then* $\mathrm{Aut}(X)$ *is open and closed in* $\mathrm{Hol}(X, X)$.

Proof. The closure is Corollary 1.7.21.

Suppose that X is not doubly connected. If, by contradiction, $\mathrm{Aut}(X)$ is not open in $\mathrm{Hol}(X, X)$ there exists a sequence $\{f_\nu\} \subset \mathrm{Hol}(X, X) \setminus \mathrm{Aut}(X)$ converging to $\gamma \in \mathrm{Aut}(X)$. But then $\gamma^{-1} \circ f_\nu \to \mathrm{id}_X$ and this is impossible by Theorem 2.6.2.

Finally, if X is doubly connected (and different from $\mathbb{D}^*$) then the set $\mathrm{Hol}(X, X) \setminus \mathrm{Aut}(X)$ is closed in $\mathrm{Hol}(X, X)$ by Proposition 2.6.6 because a limit of automorphic functions is automorphic and we are done. □

Remark 2.6.8. Corollary 2.6.7 does not hold for $\mathbb{D}^*$. Take $f_\nu(z) = (1 - 1/\nu)z$; then $f_\nu \in \mathrm{Hol}(\mathbb{D}^*, \mathbb{D}^*)$ for all ν, each f_ν is not surjective and yet $f_\nu \to \mathrm{id}_{\mathbb{D}^*}$.

We saw that the automorphism group of a compact hyperbolic Riemann surface is finite. This is still true in another important case.

Theorem 2.6.9. *Let $D \subset \widehat{X}$ be a multiply connected domain of regular type. Assume that D is not doubly connected. Then* $\mathrm{Aut}(D)$ *is finite.*

Proof. Suppose first that ∂D has no Jordan components; then ∂D is a finite set and $D = \widehat{X} \setminus \partial D$. By the big Picard theorem 1.7.25, every automorphism of D extends to an automorphism of $\widehat{X}$ sending ∂D into itself. There are three cases to consider:

(a) $\widehat{X}$ hyperbolic. Then $\mathrm{Aut}(\widehat{X})$ is finite (by Corollary 2.6.5), and thus, *a fortiori*, $\mathrm{Aut}(D)$ is finite.

(b) $\widehat{X}$ is a torus. In this case, given two points $z_0, z_1 \in \widehat{X}$ there is only a finite number of automorphisms of $\widehat{X}$ sending z_0 in z_1 (see Proposition 1.6.38); it follows that $\mathrm{Aut}(D)$ must be finite.

(c) $\widehat{X} = \widehat{\mathbb{C}}$. In this case, ∂D contains at least 3 points, because D is not doubly connected. Now, an automorphism of $\widehat{\mathbb{C}}$ is completely determined by its action on 3 points (Corollary 1.6.20); in particular, it is completely determined by its action on ∂D. In this way, $\mathrm{Aut}(D)$ is identified with a subgroup of the permutation group on a finite set, and hence is finite.

So, assume ∂D has at least one Jordan component; in particular, $\overline{D} \neq \widehat{X}$, and D is hyperbolic. Let P denote the set of point components of ∂D. Then, again by the big Picard Theorem 1.7.25, every automorphism of D extends holomorphically to an automorphism of $D \cup P$ sending P onto itself, and $D \cup P$ is still a domain of regular type. If $D \cup P$ is doubly connected, it is easily checked (see Corollary 1.6.36 and Proposition 1.6.38) that the invariance of P singles out a finite subgroup of $\mathrm{Aut}(D \cup P)$. Hence we are reduced to the case $P = \emptyset$ and D not doubly connected.

Assume, by contradiction, $\mathrm{Aut}(D)$ infinite, and let $\{\gamma_\nu\}$ be an infinite sequence of distinct elements of $\mathrm{Aut}(D)$. By Theorem 2.6.2, $\mathrm{Aut}(D)$ is discrete; hence, by Proposition 1.7.20 and Montel Theorem 1.7.12, we can assume that $\{\gamma_\nu\}$ converges to a constant $\sigma \in \partial D$.

Let $\pi_D : \mathbb{D} \to D$ be a universal covering map of D and realize its fundamental group as the automorphism group Γ_D of the covering. We associate to each γ_ν an automorphism $\tilde{\gamma}_\nu$ of $\mathbb{D}$ such that $\gamma_\nu \circ \pi_D = \pi_D \circ \tilde{\gamma}_\nu$. Moreover, by Theorem 1.8.9, we can find a point $\tau \in \partial\mathbb{D}$ such that $\tilde{\gamma}_\nu \to \tau$ as $\nu \to \infty$, and such that Γ_D is properly discontinuous at τ.

Let $\tilde{\gamma} \in \Gamma_D$ be different from $\mathrm{id}_{\mathbb{D}}$; then, by Proposition 1.6.14, for any $\nu \in \mathbb{N}$ there is $\alpha_\nu(\tilde{\gamma}) \in \Gamma_D$ such that $\tilde{\gamma}_\nu \circ \tilde{\gamma} = \alpha_\nu(\tilde{\gamma}) \circ \tilde{\gamma}_\nu$. Let V be an open neighborhood of τ in $\overline{\mathbb{D}}$ and K a compact subset of V; since $\tilde{\gamma}_\nu \to \tau$, we have $\tilde{\gamma}_\nu(K) \subset V$ for any large enough ν. But we also have $\alpha_\nu(\tilde{\gamma}) \circ \tilde{\gamma}_\nu \to \tau$; hence $\alpha_\nu(\tilde{\gamma})(V) \cap V \neq \emptyset$ for any large enough ν. Since Γ_D is properly discontinuous at τ, this implies that $\alpha_\nu(\tilde{\gamma}) \equiv \mathrm{id}_{\mathbb{D}}$ for large enough ν. But then $\tilde{\gamma}_\nu \circ \tilde{\gamma} = \tilde{\gamma}_\nu$ implies $\tilde{\gamma} \equiv \mathrm{id}_{\mathbb{D}}$, contradiction. □

We conclude this section describing what happens in Theorem 2.6.1 if we assume γ elliptic or parabolic.

Proposition 2.6.10. *Let $\gamma \not\equiv \mathrm{id}_{\mathbb{D}}$ be an elliptic automorphism of $\mathbb{D}$ with fixed point $z_0 \in \mathbb{D}$ and choose $\gamma_1 \in \mathrm{Aut}(\mathbb{D})$ such that $\gamma_1(0) = z_0$. Let $f \in \mathrm{Hol}(\mathbb{D}, \mathbb{D})$ be such that $f \circ \gamma = \gamma \circ f$. Then either:*
(i) *$(\gamma_1^{-1} \circ f \circ \gamma_1)(z) = az$ for some $a \in \overline{\mathbb{D}}$, or*
(ii) *$(\gamma_1^{-1} \circ f \circ \gamma_1)(z) = zg(z^n)$ for some $n \in \mathbb{N}$ and $g \in \mathrm{Hol}(\mathbb{D}, \mathbb{D})$. This latter possibility can occur only if $\gamma_1^{-1} \circ \gamma \circ \gamma_1(z) = e^{2\pi i k/n}$ for some $k \in \mathbb{Z}$.*

Proof. We can assume, conjugating by γ_1 if necessary, $z_0 = 0$; so $\gamma(z) = e^{2\pi i\theta}z$ for a suitable $\theta \in \mathbb{R}$. If f commutes with γ, it is obvious that $f(0) = 0$ and that f' is automorphic under the group generated by γ. If $\theta \notin \mathbb{Q}$, this implies that $f'(\tau z) = f'(z)$ for infinitely many distinct $\tau \in \partial\mathbb{D}$; hence f' is constant and we are in case (i).

If $\theta = k/n \in \mathbb{Q}\setminus\mathbb{Z}$, let $h(z) = f(z)/z$; by the Schwarz lemma, if we are not in case (i) then $h \in \mathrm{Hol}(\mathbb{D}, \mathbb{D})$. Now, h is automorphic under the group generated by γ. Expanding h in Taylor series centred at 0 we deduce that $h^{(j)}(0) = 0$ if n does not divide j. This is equivalent to saying that $h(z) = g(z^n)$ for a suitable $g \in \mathrm{Hol}(\mathbb{D}, \mathbb{D})$ and we are done. □

Finally, for the parabolic case we transfer everything on $\mathbb{H}^+$, and obtain the following.

Proposition 2.6.11. *Let $\gamma \in \mathrm{Aut}(\mathbb{H}^+)$ be parabolic with fixed point $a \in \partial\mathbb{H}^+$ and choose $\gamma_1 \in \mathrm{Aut}(\mathbb{H}^+)$ such that $\gamma_1(\infty) = a$. Let $f \in \mathrm{Hol}(\mathbb{H}^+, \mathbb{H}^+)$ be such that $f \circ \gamma = \gamma \circ f$. Then either:*
(i) *$(\gamma_1^{-1} \circ f \circ \gamma_1)(w) = w + c$ for some $c \in \mathbb{R}$, or*
(ii) *$(\gamma_1^{-1} \circ f \circ \gamma_1)(w) = w + g(\exp(2\pi i w/b))$ for some $g \in \mathrm{Hol}(\mathbb{D}^*, \mathbb{H}^+)$, where $b \in \mathbb{R}^*$ is such that $(\gamma_1^{-1} \circ \gamma \circ \gamma_1)(w) = w + b$ for all $w \in \mathbb{H}^+$.*

Proof. We can assume, conjugating by γ_1 if necessary, that $a = \infty$; so $\gamma(w) = w + b$ for a suitable $b \in \mathbb{R}^*$. If f has a fixed point $w_0 \in \mathbb{H}^+$, then we have

$$f(w_0 + b) = f(w_0) + b = w_0 + b,$$

and thus $f \equiv \mathrm{id}_{\mathbb{H}^+}$, by Corollary 1.1.14. So, assume that f has no fixed points; arguing as in the proof of Theorem 2.6.1, we see that the Wolff point of f is exactly ∞. By the Wolff lemma, $\mathrm{Im} f(w) \geq \mathrm{Im}\, w$ for all $w \in \mathbb{H}^+$, with equality if and only if $f(w) = w + c$ for some $c \in \mathbb{R}$. Excluding this case, the function $h(w) = f(w) - w$ sends $\mathbb{H}^+$ into itself and is automorphic under the group generated by γ. Hence, by Proposition 1.6.17, Theorem 1.6.33, and (1.48), we can write $h(w) = g(\exp(2\pi i w/b))$ for a suitable $g \in \mathrm{Hol}(\mathbb{D}^*, \mathbb{H}^+)$ and we are done. □

Notes to Section 2.6

Theorems 2.6.1 and 2.6.2, as well as the approach we followed in the rest of Section 2.5, are due to Heins [183]. In particular, Theorem 2.6.2 is considered by Heins to be a generalization of the *Aumann–Carathéodory Starrheitssatz* that we shall prove later (see Corollary 3.1.11) applicable also to self-maps without fixed points.

Corollaries 2.6.3 and 2.6.4 has been proved by Poincaré [334] in 1885, following Klein's ideas. With regard to Corollary 2.6.4, it should be mentioned that every countable group can be realized as automorphism group of a Riemann surface; see [175].

Corollary 2.6.5 has been proved by Schwarz [370] in 1879 and Klein (see [334]). It is complemented by the renowned *Hurwitz theorem* ([204]; cf. also [87]): if X is a compact Riemann surface of genus $g \geq 2$ then the order of $\mathrm{Aut}(X)$ is at most $84(g - 1)$. For a modern proof see, e. g., [294, Theorem 3.9]. Furthermore, later on (Corollary 3.3.3) we shall show that under these hypotheses $\mathrm{Hol}(X, X)$ reduces to the union of $\mathrm{Aut}(X)$ with the set of constant maps.

Proposition 2.6.6 is due to Heins [184].

Corollary 2.6.7 has been proved by H. Cartan [105] in 1932 for multiply connected domains D bounded by a finite number of disjoint Jordan curves. More precisely, he proved that $f \in \mathrm{Hol}(D, D)$ is an automorphism if and only if the induced homomorphism $f_*\colon \pi_1(D) \to \pi_1(D)$ is not nilpotent; we shall recover this result in Corollary 3.3.18.

Theorem 2.6.9 is due to Kœbe [243]; our proof is taken from [183].

For more information on the automorphism group of plane domains see, e. g., [185, 233, 50, 51] and references therein. In particular, Heins [185] has given sharp bounds on the cardinality of the automorphism group of a plane domain in terms of the number of connected components of the complement. It is also known (see, e. g., [50]) that almost any multiply connected plane domain has no nontrivial automorphisms.

2.7 The Burns–Krantz theorem

In Example 2.5.7, we noticed that we can find a function $f \in \mathrm{Hol}(\mathbb{D},\mathbb{D}) \setminus \mathrm{Aut}(\mathbb{D})$ such that $f(1) = 1$ and $f'(1) = 1$. This in sharp contrast with the uniqueness part of the Schwarz lemma, which says that if $f \in \mathrm{Hol}(\mathbb{D},\mathbb{D})$ is such that $f(0) = 0$ and $f'(0) = 1$ then $f \equiv \mathrm{id}_{\mathbb{D}}$. This section is devoted to finding a satisfying boundary version of the uniqueness part of the Schwarz lemma.

A possible reformulation of the uniqueness part of the Schwarz lemma is the following: if $f \in \mathrm{Hol}(\mathbb{D},\mathbb{D})$ is such that $f(z) = z + o(|z|)$, then $f \equiv \mathrm{id}_{\mathbb{D}}$. To get a boundary version of this statement, one might look for $c > 0$ such that if $f \in \mathrm{Hol}(\mathbb{D},\mathbb{D})$ is such that $f(z) = z + o(|\sigma - z|^c)$ as $z \to \sigma \in \partial\mathbb{D}$ then $f \equiv \mathrm{id}_{\mathbb{D}}$. Example 2.5.7 (see also Example 2.7.5 below) shows that c must be necessarily at least 3; indeed, we shall prove (Theorem 2.7.4) that the best value of c is exactly 3. To do so, we shall start from the (more invariant) uniqueness part of the Schwarz–Pick lemma Corollary 1.1.16 that can be stated as follows: if $f \in \mathrm{Hol}(\mathbb{D},\mathbb{D})$ is such that

$$|f^h(z)| = |f'(z)|\frac{1-|z|^2}{1-|f(z)|^2} = 1 + o(1)$$

as $z \to z_0 \in \mathbb{D}$ then $f \in \mathrm{Aut}(\mathbb{D})$. In Theorem 2.7.2, we shall show that if $f \in \mathrm{Hol}(\mathbb{D},\mathbb{D})$ is such that

$$|f^h(z)| = 1 + o(|\sigma - z|^2)$$

as $z \to \sigma \in \partial\mathbb{D}$ then $f \in \mathrm{Aut}(\mathbb{D})$, and the exponent 2 is optimal. From this, it will not be too difficult to deduce Theorem 2.7.4.

The proof of Theorem 2.7.2 will depend on a consequence of Corollary 1.5.9.

Lemma 2.7.1. *Let $f \in \mathrm{Hol}(\mathbb{D},\mathbb{D}) \setminus \mathrm{Aut}(\mathbb{D})$. Then*

$$|f^h(z)| \le \frac{|f^h(0)| + \frac{2|z|}{1+|z|^2}}{1 + |f^h(0)|\frac{2|z|}{1+|z|^2}} \tag{2.73}$$

for all $z \in \mathbb{D}$, with equality for some $z \neq 0$ if and only if f is a Blaschke product of degree 2 and $z = -tc$ for some $t > 0$, where c is the critical point of f.

Proof. If we apply (1.28) with $z = v$ and $u = w = 0$, we get

$$\omega(0, f^h(z)) \le \omega(0, f^h(0)) + 2\omega(0, z),$$

i. e.,

$$\frac{1 + |f^h(z)|}{1 - |f^h(z)|} \le \frac{1 + |f^h(0)|}{1 - |f^h(0)|}\left(\frac{1 + |z|}{1 - |z|}\right)^2,$$

which in turns is equivalent to (2.73). If we have equality for some $z \neq 0$, by Corollary 1.5.9, f must be a Blaschke product of degree 2 such that $R_c(z)$, $R_c(0)$, 0, and z lie in this order on a geodesic issuing from the critical point c of f. Since the geodesics passing through 0 are rays, this means that c and z are on the same diameter but on opposite sides with respect to 0, i. e., $z = -tc$ for some $t > 0$. □

With this, we can now prove the main result of this section.

Theorem 2.7.2. *Let $f \in \mathrm{Hol}(\mathbb{D}, \mathbb{D})$ be such that*

$$|f^h(z_n)| = 1 + o\big((1 - |z_n|)^2\big) \tag{2.74}$$

for some sequence $\{z_n\} \subset \mathbb{D}$ with $|z_n| \to 1$. Then $f \in \mathrm{Aut}(\mathbb{D})$, and hence $|f^h| \equiv 1$.

Proof. Assume, by contradiction, that $f \notin \mathrm{Aut}(\mathbb{D})$. Then $|f^h(0)| < 1$, by Lemma 1.5.2, and (2.73) holds. In particular,

$$1 - |f^h(z)| \ge 1 - \frac{|f^h(0)| + \frac{2|z|}{1+|z|^2}}{1 + |f^h(0)|\frac{2|z|}{1+|z|^2}} = \frac{(1 - |f^h(0)|)(1 - |z|)^2}{1 + 2|f^h(0)||z| + |z|^2}.$$

Therefore,

$$\liminf_{n\to+\infty} \frac{1 - |f^h(z_n)|}{(1 - |z_n|)^2} \ge \lim_{n\to+\infty} \frac{1 - |f^h(0)|}{1 + 2|f^h(0)||z_n| + |z_n|^2} = \frac{1 - |f^h(0)|}{2(1 + |f^h(0)|)} > 0,$$

contradicting (2.74) and we are done. □

The exponent 2 in (2.74) is sharp, as the following example shows.

Example 2.7.3. Take $f_0(z) = z^2$. An easy computation (or Remark 1.5.8) yields

$$f_0^h(z) = \frac{2z}{1 + |z|^2}.$$

Therefore,

$$|f_0^h(z)| = 1 - \frac{1}{2}(1 - |z|)^2 + O\big((1 - |z|)^3\big),$$

and thus in (2.74) we can neither lower the exponent nor replace $o((1-|z|)^2)$ by $O((1-|z|)^2)$.

However, this example is somewhat unsatisfactory because $f(z)=z^2$ has a fixed point in $\mathbb{D}$. Let us instead consider $f(z)=(1+3z^2)/(3+z^2)$. We have already seen in Example 2.5.7 that $f\in \mathrm{Hol}(\mathbb{D},\mathbb{D})\backslash \mathrm{Aut}(\mathbb{D})$ and that f has no fixed points in $\mathbb{D}$. Moreover, a quick computation shows that we again have

$$|f^h(z)|=1-\frac{1}{2}(1-|z|)^2+O((1-|z|)^3)$$

and so (2.74) is optimal for fixed-points free maps, too.

As a consequence of Theorem 2.7.2, we can get the *Burns–Krantz theorem.*

Theorem 2.7.4 (Burns–Krantz, 1994). *Let* $f\in \mathrm{Hol}(\mathbb{D},\mathbb{D})$ *be such that*

$$f(z)=z+o(|\sigma-z|^3) \tag{2.75}$$

as $z\to\sigma$ *nontangentially for some* $\sigma\in\partial D$. *Then* $f\equiv \mathrm{id}_{\mathbb{D}}$.

Proof. Without loss of generality, we can assume that $\sigma=1$. We first show that (2.75) implies that $|f^h(z)|=1+o((1-|z|)^2)$ as $z\to 1$ nontangentially.

Fix $1<M<M'$ and for $z\in K(1,M)$ let $C(z)$ be the circumference with center z and radius $r(z)=d(z,\partial K(1,M'))$. Notice that, by Lemma 2.2.8, there exists $c\in(0,1)$ so that

$$c|1-z|\le r(z)<|1-z|; \tag{2.76}$$

furthermore,

$$f'(z)=\frac{1}{2\pi i}\int_{C(z)}\frac{\zeta+(f(\zeta)-\zeta)}{(\zeta-z)^2}\,d\zeta=1+\frac{1}{2\pi i}\int_{C(z)}\frac{f(\zeta)-\zeta}{(\zeta-z)^2}\,d\zeta.$$

Now fix $\varepsilon>0$. By (2.75), we can find $\delta>0$ such that $|f(\zeta)-\zeta|<\varepsilon|1-\zeta|^3$ as soon as $|\zeta-1|<\delta$ and $\zeta\in K(1,M')$. Now if $z\in K(1,M)$ and $\zeta\in C(z)$ we have

$$|\zeta-1|\le|\zeta-z|+|z-1|=r(z)+|z-1|<2|z-1|;$$

therefore, if $z\in K(1,M)$ is such that $|z-1|<\delta/2$ recalling (2.76) we get

$$\begin{aligned}\left|\frac{1}{2\pi i}\int_{C(z)}\frac{f(\zeta)-\zeta}{(\zeta-z)^2}\,d\zeta\right|&\le\frac{\varepsilon}{2\pi}\int_{C(z)}\frac{|1-\zeta|^3}{|\zeta-z|^2}|d\zeta|\le\frac{\varepsilon}{r(z)}\max_{\zeta\in C(z)}|1-\zeta|^3\\&\le\varepsilon\left(1+\frac{|1-z|}{r(z)}\right)^3 r(z)^2\le\varepsilon\left(1+\frac{|1-z|}{r(z)}\right)^3|1-z|^2\\&\le\varepsilon\left(1+\frac{1}{c}\right)^3|1-z|^2.\end{aligned}$$

So, we have proved that

$$f'(z) = 1 + o(|1 - z|^2)$$

as $z \to 1$ nontangentially. Furthermore, we have

$$\frac{1 - |f(z)|^2}{1 - |z|^2} - 1 = \frac{|z|^2 - |f(z)|^2}{1 - |z|^2} = \frac{o(|1 - z|^3)}{1 - |z|^2} = o(|1 - z|^2)$$

as $z \to 1$ nontangentially, because in $K(z, M)$ we have $1 - |z| = O(|1 - z|)$. Putting all this together, we get

$$|f^h(z)| = |f'(z)|\frac{1 - |z|^2}{1 - |f(z)|^2} = 1 + o(|1 - z|^2)$$

as $z \to 1$ nontangentially, as claimed.

Applying Theorem 2.7.2, we therefore infer $f \in \operatorname{Aut}(\mathbb{D})$. But (2.75) also implies $f(1) = 1$, $f'(1) = 1$ and $f''(1) = 0$, and it is easy to check that the only automorphism of $\mathbb{D}$ satisfying these conditions is $\mathrm{id}_{\mathbb{D}}$. □

The function already used in Examples 2.5.7 and 2.7.3 shows that the exponent 3 is optimal.

Example 2.7.5. Let $f \in \operatorname{Hol}(\mathbb{D}, \mathbb{D})$ be given by

$$f(z) = \frac{1 + 3z^2}{3 + z^2}.$$

We already noticed in Example 2.5.7 that $f(1) = 1$ and $f'(1) = 1$. A quick computation shows that

$$f(z) = z + \frac{1}{4}(1 - z)^3 + O(|1 - z|^4),$$

and thus we can neither lower the exponent nor replace $o(|1-z|^3)$ by $O(|1-z|^3)$ in (2.75).

Notes to Section 2.7

Lemma 2.7.1 is originally due to Goluzin [168]; our proof is inspired by [48].

Theorem 2.7.2 has been obtained by Bracci, Kraus, and Roth [82] in 2020; it is actually a particular case of a much more general statement expressed in terms of Hermitian pseudometrics on $\mathbb{D}$. Bracci, Kraus, and Roth have also proved boundary versions of the equality case in the Ahlfors lemma and in the Nehari theorem (see the notes to Section 1.1); they can also deal with sequences of Hermitian pseudometrics. Their approach shows that all these theorems actually are comparison theorems for Hermitian pseudometrics with appropriate curvature bounds.

Theorem 2.7.4 has been proved by Burns and Krantz [91] in 1994. The original proof uses ideas similar to the ones we introduced at the end of Section 2.1; our proof is taken from [82]. Other proofs can be found, e. g., in [388, 68] and [111]. This latter paper discusses comparisons of the map f with Blaschke products, in the spirit of Nehari theorem; [82] contains cleaner results of this kind. Other generalizations of Theorem 2.7.4 can be found in [389, 85, 376]. The papers [250] and [141] include surveys on various versions of the Burns–Krantz theorem up to, respectively, 2010 and 2014.

3 Discrete dynamics on Riemann surfaces

In this chapter, we begin to deal with the main argument of this book: holomorphic dynamics. As anticipated in the Introduction (and in the title of the book), we shall mainly deal with hyperbolic Riemann surfaces, where the whole strength of the Montel theorem is available. The idea is that if X is a hyperbolic Riemann surface and $f \in \mathrm{Hol}(X, X)$ then the sequence of iterates of f is a normal family and so its behavior cannot be chaotic. For this reason, holomorphic dynamics on hyperbolic Riemann surfaces is completely different from holomorphic dynamics on elliptic Riemann surfaces (i. e., $\widehat{\mathbb{C}}$) or parabolic Riemann surfaces (e. g., $\mathbb{C}$), where a large part of the theory is devoted to studying the chaotic part of the dynamics, concentrated on the so-called Julia set. On hyperbolic Riemann surfaces, the Julia set is empty: indeed, we shall be able to prove that (with a few exceptions completely classified in the case of automorphisms) the sequence of iterates of a holomorphic self-map of a hyperbolic Riemann surface either is compactly divergent or converges, uniformly on compact sets, to a constant.

This is the best result of this kind for a generic hyperbolic Riemann surface. But if $D \subset \widehat{X}$ is a hyperbolic domain then we can say something more. In this case, in fact, $\mathrm{Hol}(D, D)$ is contained in $\mathrm{Hol}(D, \widehat{X})$, a space without compactly divergent sequences; therefore, the sequence of iterates of a function $f \in \mathrm{Hol}(D, D)$ is relatively compact in $\mathrm{Hol}(D, \widehat{X})$ and so it always has converging subsequences, converging possibly to a point of ∂D.

This observation (already somewhat anticipated in Proposition 1.7.20) leads to the core of this chapter: the *Heins theorem*, stating that if $D \subset \widehat{X}$ is a hyperbolic domain of regular type and $f \in \mathrm{Hol}(D, D)$ is not an automorphism then the sequence of iterates of f converges, uniformly on compact sets, to a constant $\tau \in \overline{D}$, the *Wolff point* of f.

The proof of the Heins theorem will occupy the first three sections of this chapter. In Section 3.1, we shall study in detail the dynamics of a holomorphic function with a fixed point, using yet another generalization of the Schwarz lemma, the *Radó–Cartan–Carathéodory Theorem* 3.1.10. In Section 3.2, we shall prove the *Wolff–Denjoy theorem* (that is the Heins theorem in $\mathbb{D}$), which is the model along which the whole theory has developed. We shall come back to the dynamics in $\mathbb{D}$ in the next chapter. Finally, in Section 3.3 we shall show how to recover in hyperbolic domains the tools used to prove the Wolff–Denjoy theorem (namely, the horocycles) and with them we shall get the Heins theorem in its full generality.

Having established this, the rest of the chapter is devoted to the study of deeper and subtler dynamical properties in hyperbolic Riemann surfaces. For instance, in Section 3.4 we shall prove that the Wolff point depends continuously on the function. Then in Section 3.5 we shall introduce the notion of *model* of a holomorphic self-map. Roughly speaking, a model of a self-map $f \in \mathrm{Hol}(X, X)$ is an automorphism F of a (usually) simply connected Riemann surface that, in a very precise sense, captures and reproduces the dynamics of the original map f nearby its Wolff point. For instance, if f

https://doi.org/10.1515/9783110601978-003

has a fixed point $z_0 \in X$ with $0 < |f'(z_0)| < 1$ then the model will be the automorphism $F(w) = f'(z_0)w$ of $\mathbb{C}$; if instead the Wolff point is in the boundary, then the model will be a hyperbolic or parabolic automorphism of $\mathbb{D}$ (or sometimes a translation in $\mathbb{C}$). Here we shall discuss an abstract approach to the existence of models on hyperbolic Riemann surfaces; we shall come back to this topic in much more detail in the next chapter.

Finally, the last two sections are devoted to *random dynamics*. Discrete dynamics is concerned with the asymptotic behavior of the sequence of functions obtained by composing a given self-map with itself; random dynamics is instead concerned with the asymptotic behavior of sequences obtained by composing self-maps picked (for instance, randomly) inside a given family of self-maps. We shall present fairly complete results in two cases: when the self-maps are picked from a family of the form $\mathrm{Hol}(X, \Omega)$ where Ω is a Bloch domain in the hyperbolic Riemann surface X (Section 3.6), and when the self-maps are close enough to a given self-map (Section 3.7).

3.1 The fixed-point case

The aim of this first section is to study the dynamics of a function with a fixed point. If $f \in \mathrm{Hol}(\mathbb{D}, \mathbb{D})$ is such that $f(z_0) = z_0$ for some $z_0 \in \mathbb{D}$, then Corollary 1.1.16 yields $|f'(z_0)| \le 1$, with equality if and only if $f \in \mathrm{Aut}(\mathbb{D})$. It turns out that this is true for any hyperbolic Riemann surface (Theorem 3.1.10) and this fact will be a main tool in this section.

Before starting, let us introduce the basic terminology of dynamics.

Definition 3.1.1. Let $f: X \to X$ be a self-map of a set X. For $k \in \mathbb{N}$, we define the map $f^k: X \to X$ by induction as $f^0 = \mathrm{id}_X$ and $f^k = f \circ f^{k-1}$ for $k \ge 1$. The map f^k is the kth *iterate* of f. If f is invertible and $k \in \mathbb{N}$, we set $f^{-k} = (f^{-1})^k$.

We say that f is *periodic* if there exists $k \ge 1$ such that $f^k \equiv \mathrm{id}_X$. The least such k is the *period* of f.

If X is a topological space and f is continuous, a *limit point* of the sequence of iterates $\{f^k\}$ is the limit in $C^0(X, X)$ of a subsequence $\{f^{k_\nu}\}$ of iterates, taken with respect to the compact-open topology. We shall denote by $\Gamma(f)$ the set of limit points of $\{f^k\}$. Notice that if X is a Riemann surface and f is holomorphic then $\Gamma(f) \subset \mathrm{Hol}(X, X)$, because $\mathrm{Hol}(X, X)$ is closed in $C^0(X, X)$.

If the topological space X is a subspace of a larger topological space $\widehat{X}$ (for instance, if $X = D \subset \widehat{X}$ is a hyperbolic domain), a map $g: X \to \widehat{X}$ obtained as limit of a subsequence of iterates of f will still be called a *limit point* of $\{f^k\}$ in $C^0(X, \widehat{X})$.

Definition 3.1.2. Let $f: X \to X$ be a self-map of a set X. The (*forward*) *orbit* of a point $x_0 \in X$ is the set $O^+(x_0) = \{f^k(x_0)\}_{k \in \mathbb{N}}$.

Setting $x_k = f^k(x_0)$, we clearly have $f(x_k) = x_{k+1}$ for all $k \in \mathbb{N}$. A *backward orbit* of $x_0 \in X$ is instead a sequence $\{x_k\}_{k\in\mathbb{N}}$ such that $f(x_k) = x_{k-1}$ for all $k \geq 1$. If f is invertible, each x_0 has a unique backward orbit, given by its forward orbit under f^{-1}; but if f is not invertible a point x_0 can have many backward orbits or none at all.

A point x_0 is *periodic* if there exists $k \geq 1$ such that $f^k(x_0) = x_0$. The least such k is the *period* of x_0. We shall say that x_0 is *preperiodic* if there exist $h > k \geq 1$ such that $f^h(x_0) = f^k(x_0)$. Clearly, x_0 is preperiodic if and only if $O^+(x_0)$ is a finite set.

The core business of discrete dynamics is the study of the asymptotic behavior of orbits and of sequences of iterates. In this chapter, we shall study in detail what happens for holomorphic self-map of hyperbolic Riemann surfaces; the study of the dynamics on nonhyperbolic Riemann surfaces (and in particular on $\widehat{\mathbb{C}}$ and $\mathbb{C}$) is a wonderful subject but it needs techniques and it gives results completely different from what we shall see here. A very good introduction to holomorphic dynamics on nonhyperbolic Riemann surfaces is [287].

We start our work with the following easy lemma.

Lemma 3.1.3. *Let X be a Riemann surface, and $f \in \mathrm{Hol}(X,X)$. Then id_X can be a limit point of $\{f^k\}$ only if $f \in \mathrm{Aut}(X)$.*

Proof. Obviously, f should be one-to-one. Furthermore, let $\{f^{k_\nu}\}$ be a subsequence of iterates converging to id_X and take $z_0 \in X$. By one of the Hurwitz theorems (see Corollary A.1.3), we must have $z_0 \in f^{k_\nu}(X) \subseteq f(X)$ for all ν large enough and so f must also be surjective. □

Definition 3.1.4. A continuous self-map $f: X \to X$ of a topological space X is *pseudoperiodic* if it is not periodic and id_X is a limit point of $\{f^k\}$.

We can now prove the following fundamental characterization of limit points of sequence of iterates of holomorphic self-maps of a hyperbolic Riemann surface.

Theorem 3.1.5. *Let X be a hyperbolic Riemann surface, and take $f \in \mathrm{Hol}(X,X)$. Let $h \in \mathrm{Hol}(X,X)$ be a limit point of the sequence $\{f^k\}$. Then either:*
(i) *h is a constant $z_0 \in X$, or*
(ii) *h is an automorphism of X. In this case, $f \in \mathrm{Aut}(X)$, too.*

Proof. If h is constant, there is nothing to prove. Suppose h is not constant, write $h = \lim_{\nu\to\infty} f^{k_\nu}$, and set $m_\nu = k_{\nu+1} - k_\nu$; without loss of generality, we can assume that $m_\nu \to +\infty$ as $\nu \to +\infty$. By the Montel Theorem 1.7.12, up to a subsequence we can also assume that $\{f^{m_\nu}\}$ either converges to a holomorphic map $g \in \mathrm{Hol}(X,X)$ or is compactly divergent.

Now, for any $z \in X$ we have

$$\lim_{\nu\to\infty} f^{m_\nu}(f^{k_\nu}(z)) = \lim_{\nu\to\infty} f^{k_{\nu+1}}(z) = h(z);$$

therefore, $\{f^{m_\nu}\}$ cannot be compactly divergent and hence it converges to a map g such that $g \circ h = h$. It follows that g is the identity on the open subset $h(X)$ of X; hence $g \equiv \mathrm{id}_X$ and, by Lemma 3.1.3, f is an automorphism.

It remains to show that h itself is an automorphism, but this follows immediately from Corollary 1.7.21. □

For hyperbolic domains, we can take into account the existence of a boundary.

Corollary 3.1.6. *Let $D \subset \widehat{X}$ be a hyperbolic domain and take $f \in \mathrm{Hol}(D, D)$. Let $h\colon D \to \widehat{X}$ be a limit point in $\mathrm{Hol}(D, \widehat{X})$ of the sequence $\{f^k\}$. Then either:*
(i) *h is a constant $z_0 \in \overline{D}$, or*
(ii) *h is an automorphism of D. In this case, $f \in \mathrm{Aut}(D)$, too.*

Proof. Clearly, $h(D) \subseteq \overline{D}$, and thus if h is constant we are done. If h is not constant, then $h(D) \subseteq D$ by a Hurwitz theorem (Corollary A.1.3). In particular, $h \in \mathrm{Hol}(D, D)$ and the assertion follows from Theorem 3.1.5. □

Remark 3.1.7. If $D \subset \widehat{X}$ is a hyperbolic domain and Y is another Riemann surface (e. g., $Y = D$), then every sequence $\{f_k\} \subset \mathrm{Hol}(Y, D)$ admits a limit point in $\mathrm{Hol}(Y, \widehat{X})$ by Proposition 1.7.15.

To state the result we mentioned at the beginning of this section, we ought to define what we mean by derivative at a fixed point of a function defined on a Riemann surface.

Definition 3.1.8. Let X be a Riemann surface, and let $f \in \mathrm{Hol}(X, X)$. If $z_0 \in X$ is a fixed point of f, then the differential df_{z_0} sends the complex tangent space $T_{z_0}X$ at X in z_0 into itself; hence df_{z_0} acts on $T_{z_0}X$ by multiplication by a complex number, that we shall denote by $f'(z_0)$ and call the *derivative* (or *multiplier*) of f at the fixed point z_0. Clearly, if X actually is a plane domain then $f'(z_0)$ is the usual derivative of f at z_0.

Remark 3.1.9. Assume we have a biholomorphism $\phi\colon X \to Y$ between two Riemann surfaces and a function $f \in \mathrm{Hol}(X, X)$ with a fixed-point $x_0 \in X$. Then $y_0 = \phi(x_0)$ is a fixed point of $g = \phi \circ f \circ \phi^{-1} \in \mathrm{Hol}(Y, Y)$; moreover, $dg_{y_0} = d\phi_{x_0} \circ df_{x_0} \circ (d\phi_{x_0})^{-1}$. It follows that

$$
\begin{aligned}
g'(y_0)v = dg_{y_0}(v) &= d\phi_{x_0} \circ df_{x_0} \circ (d\phi_{x_0})^{-1}(v) = f'(x_0) d\phi_{x_0}((d\phi_{x_0})^{-1}(v)) \\
&= f'(x_0)v
\end{aligned}
$$

for all $v \in T_{y_0}Y$, i. e., $g'(y_0) = f'(x_0)$. In other words, the multiplier is invariant under biholomorphisms.

And now we can prove the *Radó–Cartan–Carathéodory theorem.*

Theorem 3.1.10 (Radó, 1924). *Let $f \in \mathrm{Hol}(X, X)$ be a holomorphic self-map of a hyperbolic Riemann surface X. Assume that f has a fixed point $z_0 \in X$. Then:*

(i) $|f'(z_0)| \leq 1$;
(ii) $f'(z_0) = 1$ *if and only if* $f \equiv \mathrm{id}_X$;
(iii) $|f'(z_0)| = 1$ *if and only if* $f \in \mathrm{Aut}(X)$.

Proof. Let $\pi_X\colon \mathbb{D} \to X$ be the universal covering map of X. Since π_X is a local isometry for the Poincaré distances (Proposition 1.7.3), there is an open ball B for ω_X centered at z_0 and biholomorphic to $\mathbb{D}$. Since, by Theorem 1.7.6, $f(B) \subseteq B$ and the multiplier is invariant under biholomorphisms (Remark 3.1.9), the Schwarz–Pick lemma Corollary 1.1.16 immediately yields (i) and (ii). In particular, if $f \in \mathrm{Aut}(X)$ then (i) applied to f and f^{-1} yields $|f'(z_0)| = 1$.

Finally, assume $|f'(z_0)| = 1$. By Theorem 1.7.12, the sequence $\{f^k\}$ has a subsequence $\{f^{k_\nu}\}$ converging to a holomorphic function $h \in \mathrm{Hol}(X, X)$. Obviously, $|h'(z_0)| = 1$; hence h is not constant and, by Theorem 3.1.5, f is an automorphism. □

On the unit disk $\mathbb{D}$, we can find functions $f \in \mathrm{Hol}(\mathbb{D}, \mathbb{D})$ with $f(0) = 0$ that are not automorphisms but have anyway $|f'(0)|$ arbitrarily close to 1. Surprisingly, this is not true in multiply connected hyperbolic Riemann surfaces, as shown in the *Aumann–Carathéodory Starrheitssatz.*

Corollary 3.1.11 (Aumann–Carathéodory Starrheitssatz, 1934). *Let X be a multiply connected hyperbolic Riemann surface. Then for every $z_0 \in X$, we have*

$$\sup\{|f'(z_0)| \mid f \in \mathrm{Hol}(X, X),\ f \notin \mathrm{Aut}(X) \text{ and } f(z_0) = z_0\} < 1. \tag{3.1}$$

Proof. First of all, assume that X is not biholomorphic to $\mathbb{D}^*$. Suppose, by contradiction, that there are $z_0 \in X$ and a sequence $\{f_\nu\} \subset \mathrm{Hol}(X, X) \setminus \mathrm{Aut}(X)$ such that $f_\nu(z_0) = z_0$ for all $\nu \in \mathbb{N}$ and $|f_\nu'(z_0)| \to 1$. By Theorem 1.7.12, up to a subsequence we can assume that $\{f_\nu\}$ tends toward a holomorphic function $g\colon X \to X$. Obviously, $g(z_0) = z_0$ and $|g'(z_0)| = 1$; therefore, by Theorem 3.1.10, $g \in \mathrm{Aut}(X)$. Hence we have constructed a sequence of nonautomorphisms converging toward an automorphism, but by Corollary 2.6.7 this is impossible.

The argument for $X = \mathbb{D}^*$ is slightly different. Assume, by contradiction, that there exist $z_0 \in \mathbb{D}^*$ and a sequence $\{f_\nu\} \subset \mathrm{Hol}(\mathbb{D}^*, \mathbb{D}^*) \setminus \mathrm{Aut}(\mathbb{D}^*)$ such that $f_\nu(z_0) = z_0$ for all $\nu \in \mathbb{N}$ and $|f_\nu'(z_0)| \to 1$; as in the general case, we can assume that $f_\nu \to g \in \mathrm{Hol}(\mathbb{D}^*, \mathbb{D}^*)$. Clearly, $g(z_0) = z_0$ and $|g'(z_0)| = 1$; so, again by Theorem 3.1.10, $g \in \mathrm{Aut}(\mathbb{D}^*)$ and, since g has a fixed point, $g \equiv \mathrm{id}_{\mathbb{D}^*}$, by Proposition 1.6.38.

To conclude, remark that every $f \in \mathrm{Hol}(\mathbb{D}^*, \mathbb{D}^*)$ is the restriction of a holomorphic function $\tilde{f} \in \mathrm{Hol}(\mathbb{D}, \mathbb{D})$ such that $\tilde{f}(\mathbb{D}^*) \subseteq \mathbb{D}^*$. So, we have constructed a sequence $\{\tilde{f}_\nu\} \subset \mathrm{Hol}(\mathbb{D}, \mathbb{D})$ converging (by the Vitali Theorem 1.7.19) to $\mathrm{id}_{\mathbb{D}}$ such that $\tilde{f}_\nu(z_0) = z_0$ and $\tilde{f}_\nu(\mathbb{D}^*) \subseteq \mathbb{D}^*$ for all $\nu \in \mathbb{N}$. By one of the Hurwitz theorems (Corollary A.1.3), we must have $\tilde{f}_\nu(0) = 0$ for all sufficiently large ν and this implies (Corollary 1.1.14) $\tilde{f}_\nu \equiv \mathrm{id}_{\mathbb{D}}$ eventually, contradiction. □

Using Theorem 3.1.10, we can completely describe the structure of the isotropy group of a point in a hyperbolic Riemann surface.

Corollary 3.1.12. *Let X be a hyperbolic Riemann surface. Then $\mathrm{Aut}_{z_0}(X)$ is either finite cyclic for all $z_0 \in X$ (and X is multiply connected) or it is isomorphic to $\mathbb{S}^1$ for all $z_0 \in X$ (and X is simply connected).*

Proof. Fix $z_0 \in X$, and define $D\colon \mathrm{Aut}_{z_0}(X) \to \mathbb{S}^1$ by $D(\gamma) = \gamma'(z_0)$. By Theorem 3.1.10, D is a continuous injective homomorphism of $\mathrm{Aut}_{z_0}(X)$ into $\mathbb{S}^1$; since (Corollary 1.7.21) $\mathrm{Aut}_{z_0}(X)$ is compact, D is an isomorphism of topological groups between $\mathrm{Aut}_{z_0}(X)$ and a closed subgroup of $\mathbb{S}^1$.

If X is simply connected, we already know that $\mathrm{Aut}_{z_0}(X)$ is isomorphic to $\mathbb{S}^1$ for all $z_0 \in X$. If X is multiply connected with non-Abelian fundamental group, $\mathrm{Aut}_{z_0}(X)$ is discrete for all $z_0 \in X$ (Theorem 2.6.2), and hence finite cyclic (Proposition 1.4.13). Finally, if X is doubly connected Proposition 1.6.38 shows that $\mathrm{Aut}_{z_0}(X)$ is either cyclic of order 2 or trivial for all $z_0 \in X$. □

Now we are ready to describe the dynamics of a function with a fixed point.

Theorem 3.1.13. *Let $f \in \mathrm{Hol}(X,X)$ be a holomorphic self-map of a hyperbolic Riemann surface X. Assume that f has a fixed point $z_0 \in X$. Then either:*
(i) *$|f'(z_0)| < 1$ and the sequence $\{f^k\}$ converges to z_0, or*
(ii) *f is a periodic automorphism, or*
(iii) *f is a pseudoperiodic automorphism. This latter possibility can occur only if X is simply connected.*

Proof. By Theorem 3.1.10, $|f'(z_0)| \le 1$. If $|f'(z_0)| = 1$, f is an automorphism, and the assertion follows from Corollary 3.1.12 and Proposition 1.4.13.

Finally, assume $|f'(z_0)| < 1$. Since f has a fixed point, $\{f^k\}$ cannot have compactly divergent subsequences. Let h_1, h_2 be two limit points of $\{f^k\}$. Since $f \notin \mathrm{Aut}(X)$, by Theorem 3.1.5 both h_1 and h_2 are constant; but z_0 should be a fixed point for both h_1 and h_2, and so $h_1 \equiv h_2 \equiv z_0$. In other words, z_0 is the unique limit point of $\{f^k\}$, and thus $f^k \to z_0$ by Lemma 1.7.18. □

Definition 3.1.14. Let X be a Riemann surface, and $f \in \mathrm{Hol}(X,X)$. A fixed-point $z_0 \in X$ of f is called *attracting* if $|f'(z_0)| < 1$ and *superattracting* if $f'(z_0) = 0$.

As a consequence, self-coverings of a hyperbolic Riemann surface with a fixed point are necessarily automorphisms.

Corollary 3.1.15. *Let X be a hyperbolic Riemann surface and $f \in \mathrm{Hol}(X,X)$ a self-covering of X. Assume that f has a fixed-point $z_0 \in X$. Then f is a periodic or pseudoperiodic automorphisms of X and z_0 is not attracting.*

Proof. Let $\pi_X\colon \mathbb{D} \to X$ be a universal covering map. Given $\tilde{z}_0 \in \pi_X^{-1}(z_0)$, we can choose a lifting $\tilde{f}\colon \mathbb{D} \to \mathbb{D}$ of f such that $\tilde{f}(\tilde{z}_0) = \tilde{z}_0$. Since (see, e. g., [260, Lemma 12.1]) $\tilde{f}$ is a self-covering of $\mathbb{D}$, it must be an automorphism; therefore, $|\tilde{f}'(\tilde{z}_0)| = 1$, by the Schwarz–Pick lemma. This clearly implies that $|f'(z_0)| = 1$ and the assertion follows from Theorem 3.1.13. □

We can also generalize Corollary 1.1.14.

Corollary 3.1.16. *Let X be a hyperbolic Riemann surface and assume there exists a function $f \in \mathrm{Hol}(X, X)$ with two fixed points. Then $f \equiv \mathrm{id}_X$ or X is multiply connected and f is a periodic automorphism of X.*

Proof. If $f \not\equiv \mathrm{id}_X$, then by Corollary 1.1.14, X must be multiply connected. If f were not an automorphism, the sequence $\{f^k\}$ would have to converge to each of the distinct fixed points (by Theorem 3.1.13), impossible. So, $f \in \mathrm{Aut}(X)$ and the assertion follows from Corollary 3.1.12. □

Example 3.1.17. Let $D \subset \mathbb{C}$ be the doubly connected domain

$$D = \{z \in \mathbb{C} \mid 1/2 < |z| < 2\}.$$

Then the function $f \in \mathrm{Hol}(D, D)$ given by $f(z) = 1/z$ has two distinct fixed points, 1 and -1.

We end this section by adding other pieces to the *ménage à trois* between dynamics, commuting functions and common fixed points. We already saw in two different occasions (Proposition 1.4.12 and Theorem 2.6.1) that in some instances commuting functions have a common fixed point. Another easy instance of the existence of common fixed points for commuting functions is contained in the following.

Lemma 3.1.18. *Let X be a hyperbolic Riemann surface and $f, g \in \mathrm{Hol}(X, X) \setminus \{\mathrm{id}_X\}$ be such that $f \circ g = g \circ f$. Assume moreover that either X is simply connected or X is multiply connected and $f \notin \mathrm{Aut}(X)$. Then if $z_0 \in X$ is a fixed point of f it is also a fixed point of g.*

Proof. By Corollaries 1.1.14 and 3.1.16, we know that z_0 is the unique fixed point of f. Since $f(g(z_0)) = g(f(z_0)) = g(z_0)$, it then follows that $g(z_0) = z_0$ and we are done. □

Example 3.1.19. Let $D = \{z \in \mathbb{C} \mid 1/2 < |z| < 2\}$ and $f, g \in \mathrm{Aut}(D)$ be given by $f(z) = 1/z$ and $g(z) = -z$. Then $f \circ g = g \circ f$ but $\mathrm{Fix}(f) = \{1, -1\}$ and $\mathrm{Fix}(g) = \emptyset$.

Using dynamical arguments, we can prove the *Shields theorem*, giving the existence of common fixed points in a fairly general context.

Theorem 3.1.20 (Shields, 1964). *Let $D \subset \widehat{X}$ be a hyperbolic domain and $\mathcal{F} \subset \mathrm{Hol}(D, D)$ a family of holomorphic self-maps of D extending continuously to $\overline{D}$ and commuting with each other under composition. Assume that either:*
(i) *D is simply connected, or*
(ii) *D is multiply connected and $\mathcal{F}$ is not contained in $\mathrm{Aut}(D)$.*

Then $\mathcal{F}$ has a common fixed point, i. e., there exists $z_0 \in \overline{D}$ such that $f(z_0) = z_0$ for all $f \in \mathcal{F}$.

Proof. Take $f_0 \in \mathcal{F}$; if D is multiply connected, we also assume that $f_0 \notin \mathrm{Aut}(D)$.

If f_0 has a fixed point $z_0 \in D$, then the assertion follows from Lemma 3.1.18.

If f_0 has no fixed points in D, let $h \in \mathrm{Hol}(D, \widehat{X})$ be a limit point of $\{f_0^k\}$, which exists by Proposition 1.7.15. Corollary 3.1.6 (recalling that if D is simply connected and $f_0 \in \mathrm{Aut}(D)$ then the sequence $\{f_0^k\}$ is compactly divergent, by Proposition 1.4.10, and hence h cannot be an automorphism of D) says that $h \equiv z_0 \in \overline{D}$; we claim that z_0 is a common fixed point of $\mathcal{F}$. Indeed, let $\{f_0^{k_\nu}\}$ be a subsequence of iterates converging to z_0, and take $g \in \mathcal{F}$. Then

$$g(z_0) = \lim_{\nu\to\infty} g(f_0^{k_\nu}(z)) = \lim_{\nu\to\infty} f_0^{k_\nu}(g(z)) = z_0,$$

where z is any point of D, and we are done. □

Example 3.1.21. Example 3.1.19 shows that when D is multiply connected a commuting family of automorphisms of D might not have a common fixed point.

We shall return to this topic in Section 4.10.

Notes to Section 3.1

The first mathematician to deal with problems in holomorphic dynamics seems to have been Schröder. In his 1870 paper [368], he studied the local situation near a fixed point, essentially obtaining Theorem 3.1.13(i); another proof is in [244]. Theorem 3.1.5 is proved, in a slightly different form, in [392].

Theorem 3.1.10 has been proved by Radó [344] in 1924, but some special cases were known before. For instance, Bieberbach [66] in 1913 proved that the unique automorphism f of a hyperbolic domain D with a fixed-point $z_0 \in D$ such that $f'(z_0) > 0$ is the identity. This is important in uniformization theory of multiply connected domains. A different proof of Theorem 3.1.10 is in [290]. The names of Cartan and Carathéodory are associated to this theorem because H. Cartan [103, 104] in 1930 proved a version of this result for bounded domains in $\mathbb{C}^2$ and then Carathéodory [98] in 1932 extended Cartan's results to bounded domains in $\mathbb{C}^n$.

Corollary 3.1.11 was originally proved by Aumann and Carathéodory [26] in 1934; our proof is taken from [183]. Heins [182] and Hervé [194] studied the supremum in (3.1) and described the functions attaining it in doubly connected domains and, partially, in multiply connected domains; see also [289] for a simpler approach, and [290] for generalizations.

A direct proof of Corollary 3.1.16 is in [288]. Leschinger [263] has proved that a holomorphic self-map of a domain $D \subset \widehat{\mathbb{C}}$ with three fixed points is necessarily the identity; see [318] for a simpler proof. More generally, Minda [288] has proved that a holomorphic self-map of a hyperbolic domain $D \subset \widehat{X}$ of a compact Riemann surface of genus $g \geq 0$ with $2g + 3$ fixed points is the identity, and has given examples of nontrivial automophisms with $2g + 2$ fixed points. In [290] it is also shown that a periodic automorphism of a hyperbolic domain $D \subset \widehat{X}$ with a fixed point in D has period at most $4g + 2$ if $\widehat{X}$ has genus $g \geq 1$; if $\widehat{X}$ has genus 0 and ∂D has $k \geq 2$ connected components, then the period is at most k, and again these bounds are the best possible. Finally, if $D \subset \widehat{X}$ is a hyperbolic domain with k boundary components in a compact Riemann surface $\widehat{X}$ of genus g such that $g + k \geq 2$ then an automorphism of D extending continuously to ∂D with a fixed point in ∂D is necessarily the identity.

The results in [290] depends on a result contained in [280], interesting by itself: any hyperbolic domain $D \subset \widehat{X}$ is biholomorphic to a hyperbolic domain $\widetilde{D} \subset \widehat{X}_1$ such that every automorphism of $\widetilde{D}$ is the restriction of an automorphism of $\widehat{X}_1$.

Shields [375] proved Theorem 3.1.20 for $D = \mathbb{D}$ in 1964. The generalization to hyperbolic domains is in [12]. It should be remarked that there are examples of commuting continuous functions mapping the closed unit interval into itself without common fixed points (see, e. g., [74, 202]).

3.2 The Wolff–Denjoy theorem

In this short (but important) section, we shall study dynamics in $\mathbb{D}$, proving the fundamental *Wolff–Denjoy theorem.*

Theorem 3.2.1 (Wolff–Denjoy, 1926). *Let $f \in \mathrm{Hol}(\mathbb{D}, \mathbb{D})$ and assume f is neither an elliptic automorphism nor the identity. Then the sequence of iterates $\{f^k\}$ converges, uniformly on compact sets, to the Wolff point $\tau_f \in \overline{\mathbb{D}}$ of f.*

Proof. If f has a fixed point, the assertion follows from Theorem 3.1.13; so assume f has no fixed points.

If f is an automorphism, we already proved the assertion in Proposition 1.4.10. So, assume now that $f \notin \mathrm{Aut}(\mathbb{D})$, and let $h = \lim_{\nu\to\infty} f^{k_\nu}$ be a limit point of $\{f^k\}$ in $\mathrm{Hol}(\mathbb{D}, \mathbb{C})$; by Corollary 3.1.6, h is a constant $\tau \in \overline{\mathbb{D}}$. If τ were an interior point of $\mathbb{D}$, we would have

$$f(\tau) = \lim_{\nu\to\infty} f(f^{k_\nu}(\tau)) = \lim_{\nu\to\infty} f^{k_\nu}(f(\tau)) = \tau, \tag{3.2}$$

impossible; therefore, $\tau \in \partial\mathbb{D}$.

We claim that $\tau = \tau_f$. In fact, by the Wolff lemma (Theorem 2.5.1), for any $R > 0$ and $\nu \in \mathbb{N}$ we have

$$f^{k_\nu}(E(\tau_f, R)) \subseteq E(\tau_f, R);$$

hence

$$\{\tau\} = h(E(\tau_f, R)) \subseteq \overline{E(\tau_f, R)} \cap \partial\mathbb{D} = \{\tau_f\},$$

that is our claim. Therefore, τ_f is the unique limit point of $\{f^k\}$, and thanks to Lemma 1.7.18 we are done, because by the Montel theorem the sequence $\{f^k\}$ is relatively compact in $\mathrm{Hol}(\mathbb{D}, \mathbb{C})$. □

Therefore, beside the trivial cases of the identity and of elliptic automorphisms, a sequence of iterates of self-maps of $\mathbb{D}$ always converges, and the limit is always a constant function. If $f \in \mathrm{Hol}(\mathbb{D}, \mathbb{D})$ has a fixed point $z_0 \in \mathbb{D}$, then the result could be expected, because by the Schwarz lemma f contracts the Poincaré balls centered at z_0. If f has no fixed points, then again the result could be expected when f has angular derivative strictly less than 1 at its Wolff point τ, for in this case f contracts the horocycles centered at τ, by the Wolff lemma. The remarkable fact is that the theorem holds even if f has angular derivative 1 at τ_f, when the horocycles are not shrinked. Even more remarkable will be the generalization to hyperbolic domains of regular type, but we defer this to the next section.

If γ is an automorphism, we can say a bit more. If γ is elliptic we know that $\{\gamma^k\}$ does not converge. If γ is parabolic, then $\{\gamma^k\}$ converges to the unique fixed point in the boundary. If γ is hyperbolic, $\{\gamma^k\}$ still converges to a fixed point in the boundary, but which one? A first answer is given by Proposition 1.4.10: it is the unique fixed point $\tau \in \partial\mathbb{D}$ such that $|\gamma'(\tau)| < 1$. A more geometric answer is contained in the following.

Proposition 3.2.2. *Let $\gamma \in \mathrm{Aut}(\mathbb{D})$ be hyperbolic. Then $\{\gamma^k\}$ converges to the fixed point of γ farthest (for the Euclidean distance) from $\gamma^{-1}(0)$.*

Proof. Let $a = \gamma^{-1}(0) \neq 0$, and write

$$\gamma(z) = e^{i\theta}\frac{z-a}{1-\bar{a}z},$$

for a suitable $\theta \in \mathbb{R}$. Let $\tau_1, \tau_2 \in \partial\mathbb{D}$ be the distinct fixed points of γ, where τ_1 is its Wolff point and τ_2 is the other one. Now, for every $z \in \overline{\mathbb{D}}$ we have

$$\gamma'(z) = e^{i\theta}\frac{1-|a|^2}{(1-\bar{a}z)^2}.$$

We know that $|\gamma'(\tau_1)| < 1 < |\gamma'(\tau_2)|$, by Lemma 1.4.9 and Proposition 1.4.10; hence

$$|\tau_1 - a|^2 > 1 - |a|^2 > |\tau_2 - a|^2,$$

and the assertion follows from Theorem 3.2.1. □

Due to the importance of the Wolff–Denjoy theorem in this book, we shall now describe a second, totally independent proof.

Second proof of Theorem 3.2.1. Take $f \in \mathrm{Hol}(\mathbb{D},\mathbb{D})$, not an elliptic automorphism nor the identity. If f is a hyperbolic or parabolic automorphism, the assertion follows from Proposition 1.4.10.

Assume then $f \notin \mathrm{Aut}(\mathbb{D})$. Take two distinct points $z_1, z_2 \in \mathbb{D}$; then the sequence $\{\omega(f^k(z_1), f^k(z_2))\}$ is either eventually zero or strictly decreasing, by Theorem 1.2.3.

Let $h = \lim_{\nu\to\infty} f^{k_\nu}$ be a limit point of $\{f^k\}$ in $\mathrm{Hol}(\mathbb{D},\mathbb{C})$. Assume first that $h \in \mathrm{Hol}(\mathbb{D},\mathbb{D})$; we claim that h is constant. Indeed, take $z_1, z_2 \in \mathbb{D}$ such that $h(z_1) \neq h(z_2)$, if possible. Let $a \geq 0$ be the limit (that we know to exist) of the sequence $\{\omega(f^k(z_1), f^k(z_2))\}$; obviously, $a = \omega(h(z_1), h(z_2)) > 0$. But, since $f \notin \mathrm{Aut}(\mathbb{D})$, we have

$$\begin{aligned} a = \omega(h(z_1), h(z_2)) &> \omega(f(h(z_1)), f(h(z_2))) \\ &= \lim_{\nu\to\infty} \omega(f^{k_\nu+1}(z_1), f^{k_\nu+1}(z_2)) = a, \end{aligned}$$

contradiction. So h is constant, say $h \equiv \tau \in \mathbb{D}$. Now

$$\omega(f(\tau),\tau) = \lim_{\nu\to\infty} \omega(f(f^{k_\nu}(\tau)), f^{k_\nu}(\tau)) = \lim_{\nu\to\infty} \omega(f^{k_\nu}(f(\tau)), f^{k_\nu}(\tau)) = 0.$$

Therefore, τ is the unique fixed point of f; hence for all $z \in \mathbb{D}$,

$$\lim_{k\to\infty} \omega(\tau, f^k(z)) = \lim_{k\to\infty} \omega(f^k(\tau), f^k(z)) = \lim_{\nu\to\infty} \omega(f^{k_\nu}(\tau), f^{k_\nu}(z)) = 0,$$

because the sequence $\{\omega(f^k(\tau), f^k(z))\}$ is decreasing, and thus $f^k \to \tau$.

Finally, assume then that $\{f^k\}$ has no limit points in $\mathrm{Hol}(\mathbb{D},\mathbb{D})$; therefore, every limit point should be a constant in $\partial\mathbb{D}$ (Corollary 3.1.6). In particular, f cannot have fixed points.

We claim that for every $z_0 \in \mathbb{D}$ there is a subsequence $\{f^{k_\nu}\}$ such that

$$|f(f^{k_\nu}(z_0))| > |f^{k_\nu}(z_0)| \tag{3.3}$$

for all $\nu \in \mathbb{N}$. Indeed, if this were not true, there should exist $k_1 \in \mathbb{N}$ such that

$$|f(f^k(z_0))| \le |f^k(z_0)|$$

as soon as $k \ge k_1$. Then for all $k > k_1$ we would have

$$|f^k(z_0)| = |f(f^{k-1}(z_0))| \le |f^{k-1}(z_0)| \le \cdots \le |f^{k_1}(z_0)|,$$

and thus $\{f^k\}$ could not have limit points in $\partial\mathbb{D}$, contradiction.

So let $\{f^{k_\nu}\}$ be a subsequence satisfying (3.3) for a given $z_0 \in \mathbb{D}$; we may assume that $f^{k_\nu} \to \tau_0 \in \partial\mathbb{D}$. Set $z_\nu = f^{k_\nu}(z_0)$; then (3.3) implies

$$\liminf_{z\to\tau_0} \frac{1-|f(z)|}{1-|z|} \le \liminf_{\nu\to+\infty} \frac{1-|f(z_\nu)|}{1-|z_\nu|} \le 1.$$

Hence we can apply Julia lemma (Theorem 2.1.10) with $\sigma = \tau_0$ and $\alpha \le 1$, obtaining a $\tau \in \partial\mathbb{D}$ such that $f(E(\tau_0,R)) \subseteq E(\tau,R)$ for all $R > 0$. Looking at the proof of Theorem 2.1.10, we see that τ can be obtained as limit of (a subsequence of) $f(z_\nu)$; but

$$\lim_{\nu\to\infty} z_\nu = \tau_0 = \lim_{\nu\to\infty} f^{k_\nu}(f(z_0)) = \lim_{\nu\to\infty} f(z_\nu),$$

and thus $\tau = \tau_0$. In particular, by the argument used at the beginning of the proof of Theorem 2.5.1, τ_0 is the Wolff point of f.

Finally, let h be another limit point of $\{f^k\}$. We know that $h \equiv \tilde\tau \in \partial\mathbb{D}$; to conclude the proof, we have to show that $\tilde\tau = \tau_0$. But from $f(E(\tau_0,R)) \subseteq E(\tau_0,R)$ we get

$$\{\tilde\tau\} = h(E(\tau_0,R)) \subseteq \overline{E(\tau_0,R)} \cap \partial\mathbb{D} = \{\tau_0\},$$

and we are done. □

In the next chapter, we shall study in detail how the orbits of a holomorphic self-map of $\mathbb{D}$ converge to the Wolff point and we shall obtain deep results on the holomorphic dynamics in $\mathbb{D}$.

Notes to Section 3.2

The history of the Wolff–Denjoy Theorem 3.2.1 is quite interesting. On December 21, 1925, Wolff [414] presented a first proof, assuming continuity at the boundary. Few weeks later, on January 18, 1926, Wolff [415] succeeded in removing the extra hypothesis, with a brute force approach. Just a few days later, on January 25, 1926, Denjoy [135] published a completely new proof, based on the Fatou Theorem 1.8.11. Finally, after a couple of months, on April 7, 1926, inspired by Denjoy's proof, Wolff [416] discovered the Wolff lemma and the elegant proof we presented. Unfortunately, this proof went somehow unnoticed, and it was rediscovered only much later; see [89] for more on the history of the Wolff–Denjoy theorem.

The second proof we described is due to Vesentini [402]. Other proofs can be found in [398, 298, 131].

[387] shows how to build an (infinite) Blaschke product having a preassigned Wolff point.

Using Theorem 1.6.32, the Wolff–Denjoy theorem can be transferred to simply connected domains in $\mathbb{C}$ bounded by a Jordan curve. More generally, if $D \subset \mathbb{C}$ is a simply connected domain then Carathéodory [94, 96] has constructed a compactification $\hat{D} = D \cup \partial_C D$, where $\partial_C D$ is the set of prime ends of D with a suitable topology, such that every biholomorphism $\phi\colon \mathbb{D} \to D$ extends to a homeomorphism of $\overline{\mathbb{D}}$ with $\hat{D}$ sending $\partial\mathbb{D}$ onto $\partial_C D$; see, e. g., [125, Theorem 14.3.4] or [80, Theorem 4.2.3] for a modern proof. As a consequence, the Wolff–Denjoy theorem says that a sequence of iterates of a holomorphic self-map of D without fixed points converges to a prime end $P \in \partial_C$. See [152] for a couple of conditions ensuring that the sequence of iterates actually converges to a point in the Euclidean boundary of D.

In his famous series of papers [146–148] devoted to holomorphic dynamics, Fatou studied (among many other things) the dynamics of rational functions sending both $\mathbb{D}$ and $\partial\mathbb{D}$ into themselves and, more generally, of *infinite Blaschke products*, that is of functions of the form

$$f(z) = e^{i\theta} \prod_{\nu=0}^{\infty} \frac{z - a_\nu}{1 - \overline{a_\nu} z},$$

where $a_\nu \in \mathbb{D}$ for all $\nu \in \mathbb{N}$ and $\sum_{\nu=0}^{\infty}(1-|a_\nu|) < +\infty$. These functions are also known as *Fatou functions*; see also [399].

More generally, an *inner function* is a holomorphic self-map $f \in \mathrm{Hol}(\mathbb{D}, \mathbb{D})$ such that the radial limit of $f(re^{i\theta})$ as $r \to 1^-$ exists and has modulus 1 for almost all $e^{i\theta} \in \mathbb{S}^1$. Thus an inner function determines a measurable self-map of $\mathbb{S}^1$, whose dynamics is clearly related to the dynamics of f in $\mathbb{D}$. For results on the boundary dynamics of inner functions see, e. g., [1, 2, 137, 232, 72, 151, 329, 122].

3.3 The Heins theorem

In this section, we shall study the dynamics of a function f defined on a multiply connected hyperbolic domain or, more generally, on a hyperbolic Riemann surface. We begin with a lemma about automorphisms without fixed points.

Lemma 3.3.1. *Let X be a hyperbolic Riemann surface, and let $f \in \mathrm{Aut}(X)$ be without fixed points. Assume that $\{f^k\}$ is not compactly divergent. Then X is multiply connected* *and either f is periodic or f is pseudoperiodic and the closure of* $\{f^k\}$ *is the connected component at the identity of* Aut(X), *which is isomorphic to* $\mathbb{S}^1$. *Furthermore, the latter possibility can occur only if X is doubly connected.*

Proof. If X is simply connected, the sequence of iterates of f is compactly divergent, by Proposition 1.4.10; hence X is multiply connected.

If X is multiply but not doubly connected, let $\{f^{k_\nu}\}$ be a converging subsequence, necessarily to an element of Aut(X), by Corollary 1.7.21. Now, Aut(X) is discrete, by Theorem 2.6.2; hence the sequence $\{f^{k_\nu}\}$ must contain only a finite number of distinct elements, and so f is periodic.

Finally, if X is doubly connected we can realize it as $\mathbb{D}^*$ or an annulus $A(r,1)$ with $0 < r < 1$. By Proposition 1.6.38, f should be a rotation around the origin, and the assertion follows from Proposition 1.4.13. □

Then we have a Wolff–Denjoy theorem for hyperbolic Riemann surfaces.

Theorem 3.3.2. *Let X be a hyperbolic Riemann surface, and let* $f \in \mathrm{Hol}(X,X)$. *Then either:*
(i) *f has an attracting fixed point in X, or*
(ii) *f is a periodic automorphism, or*
(iii) *f is a pseudoperiodic automorphism, or*
(iv) *the sequence* $\{f^k\}$ *is compactly divergent.*

Furthermore, the case (iii) *can occur only if X is either simply connected (and f has a fixed point) or doubly connected (and f has no fixed points).*

Proof. If the sequence $\{f^k\}$ is compactly divergent, there is nothing to prove. So, assume $\{f^k\}$ is not compactly divergent; in particular, there is a subsequence $\{f^{k_\nu}\}$ converging to a holomorphic function $h \in \mathrm{Hol}(X,X)$. If $f \in \mathrm{Aut}(X)$, the assertion follows from Theorem 3.1.13 and Lemma 3.3.1. If f is not an automorphism, by Theorem 3.1.5 h is constant, $h \equiv z_0 \in X$, say. But then arguing as in (3.2) we see that z_0 should be a fixed point of f and the assertion follows from Theorem 3.1.13. □

If X is compact, Theorem 3.3.2 drastically simplifies.

Corollary 3.3.3. *Let X be a compact hyperbolic Riemann surface. Then every function* $f \in \mathrm{Hol}(X,X)$ *is either constant or an automorphism, necessarily periodic.*

Proof. Assume f is nonconstant. Then $f(X)$ is open and closed in X, and thus f is surjective. But then the sequence $\{f^k\}$ can neither be compactly divergent (for X is compact) nor converge uniformly to a point in X. Hence, by Theorem 3.3.2 and Corollary 2.6.5, f is a periodic automorphism. □

A further consequence is the *Ritt theorem*, a generalization of Corollary 1.1.15.

Proposition 3.3.4 (Ritt theorem, 1920). *Let X be a noncompact Riemann surface. If* $f \in \mathrm{Hol}(X, X)$ *is such that* $f(X)$ *is relatively compact in X, then f has an attracting fixed point* $z_0 \in X$.

Proof. If X is hyperbolic, the assertion follows from Theorem 3.3.2 because f is not an automorphism and $\{f^k\}$ is not compactly divergent. If X is not hyperbolic, it is biholomorphic to $\mathbb{C}$ or $\mathbb{C}^*$ and then the assertion follows from Liouville theorem or, more generally, from Proposition 1.6.27. □

For a generic hyperbolic Riemann surface, Theorem 3.3.2 is the best statement we can hope for; on the other hand, for a hyperbolic domain we can do something better.

Definition 3.3.5. Let $D \subset \widehat{X}$ be a hyperbolic domain, and take $f \in \mathrm{Hol}(D, D)$. If every limit point of $\{f^k\}$ is a constant belonging to ∂D we say that $\{f^k\}$ *converges to the boundary*. By Corollary 3.1.6, this happens if and only if $\{f^k\}$ is compactly divergent.

Then Theorem 3.3.2 becomes the following.

Theorem 3.3.6. *Let* $D \subset \widehat{X}$ *be a hyperbolic domain, and take* $f \in \mathrm{Hol}(D, D)$. *Then either:*
(i) *f has an attracting fixed point in D, or*
(ii) *the sequence* $\{f^k\}$ *converges to the boundary of D, and the set of limit points is closed and connected, or*
(iii) *f is a periodic automorphism, or*
(iv) *f is a pseudoperiodic automorphism. This latter possibility can occur only if D is either simply or doubly connected.*

Proof. The only thing left to prove is that the set L of limit points when $\{f^k\}$ converges to the boundary is closed and connected. Choose a connected compact subset K of D such that $f(K) \cap K \neq \emptyset$. Then clearly $f^{k+1}(K) \cap f^k(K) \neq \emptyset$ for all $k \in \mathbb{N}$. It follows that both the set

$$L_\nu = \bigcup_{k=\nu}^{\infty} f^k(K)$$

and its closure are connected for all $\nu \in \mathbb{N}$. But now $L = \bigcap_{\nu=1}^{\infty} \overline{L_\nu}$ and so L is closed and connected. □

In some situations, the set L of limit points in case (ii) actually reduces to a point, and thus the sequence of iterates converges. This happens for instance if the self-map extends continuously to the boundary without too many fixed points.

Corollary 3.3.7. *Let* $D \subset \widehat{X}$ *be a hyperbolic domain, and take* $f \in \mathrm{Hol}(D, D)$, *not a periodic or pseudoperiodic automorphism of D. Assume that f extends continuously to the boundary of D, that it has no attracting fixed points in D and that the set of fixed points of f in* ∂D *is totally disconnected. Then the sequence of iterates of f converges to a point of* ∂D.

Proof. The hypotheses ensure that we are in case (ii) of Theorem 3.3.6. Take a subsequence $\{f^{\nu_k}\}$ converging to a point $\tau \in \partial D$. Since f extends continuously to the boundary, we have

$$f(\tau) = f\Big(\lim_{k\to\infty} f^{\nu_k}(z_0)\Big) = \lim_{k\to\infty} f(f^{\nu_k}(z_0)) = \lim_{k\to\infty} f^{\nu_k}(f(z_0)) = \tau,$$

where z_0 is any point of D. Then the set L of limit points is contained in the set of fixed points of f in ∂D, which is totally disconnected; since L is connected it follows that it reduces to a single point and we are done by Lemma 1.7.18. □

However, there are examples of bounded simply connected domains in the plane admitting a holomorphic self-map with a compactly divergent sequence of iterates not convergent to a point in the boundary.

Example 3.3.8. We shall construct a bounded simply connected domain $D \subset \mathbb{C}$ and a $f \in \mathrm{Aut}(D)$ such that the set of limit points of the sequence of iterates $\{f^k\}$ is a nontrivial segment in the boundary of D.

Given a sequence $\{a_k\} \subset (0,1)$ decreasing to 0, let $D \subset \mathbb{C}$ be obtained by removing from the open square $(0,1)\times(0,1)$ the segments $[a_{2k+1},1]\times\{a_{2k+1}\}$ and $[0,1-a_{2k}]\times\{a_{2k}\}$; see Figure 3.1.

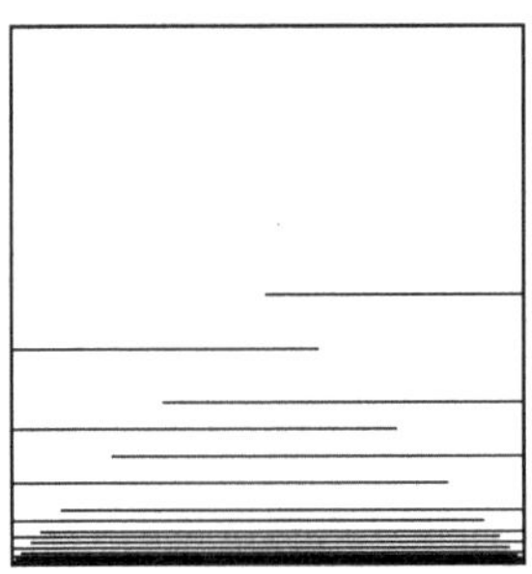

Figure 3.1: A simply connected domain with a bad boundary.

Fix $z_0 \in D$, and let $\pi_D\colon \mathbb{D} \to D$ be a universal covering map with $\pi(0) = z_0$. If $\tau \in \partial D \setminus ([0,1]\times\{0\})$, then Proposition 1.8.12 implies that there exists $\sigma \in \partial\mathbb{D}$ such that π_D has radial limit τ at σ. It follows by (1.12) that $\eta_\tau\colon \mathbb{R}^+ \to D$ given by $\eta_\tau(t) = \pi_D(\tanh(t)\sigma)$ is a geodesic for the Poincaré distance of D with $\eta_\tau(t) \to \tau$ as $t \to +\infty$.

Now choose a sequence $\{\tau_k\} \subset \partial D\setminus([0,1]\times\{0\})$ converging to $\tau_\infty \in (0,1)\times\{0\}$. The sequence $\{\eta_{\tau_k}\} \subset C^0(\mathbb{R}^+, D)$ satisfies the hypotheses of the Ascoli–Arzelà Theorem 1.7.11 when we consider $\mathbb{R}^+$ endowed with the Euclidean distance and D endowed with the Poincaré distance. Indeed, equicontinuity is obvious, and for each $t \in \mathbb{R}^+$ the set $\{\eta_{\tau_k}(t)\}$ is contained in the closure of the Poincaré ball $B_D(z_0, t)$, which is compact in D.

It follows that, up to a subsequence, we can assume that $\{\eta_{\tau_k}\}$ converges in $C^0(\mathbb{R}^+, D)$ to a map $\eta_\infty: \mathbb{R}^+ \to D$. Clearly, η_∞ is still a Poincaré geodesic with $\eta_\infty(0) = z_0$.

Now, since Poincaré geodesics in D are the image via π_D of Poincaré geodesics in $\mathbb{D}$, we can find $\sigma_\infty \in \partial\mathbb{D}$ such that $\eta_\infty(t) = \pi(\tanh(t)\sigma_\infty)$. Let $\gamma \in \mathrm{Aut}(\mathbb{D})$ be the hyperbolic automorphism given by Lemma 1.4.16(iii) such that

$$\gamma(\tanh(t)\sigma_\infty) = \tanh(t+1)\sigma_\infty$$

and put $f = \pi_D \circ \gamma \circ \pi_D^{-1} \in \mathrm{Aut}(D)$; in particular, $f(\eta_\infty(t)) = \eta_\infty(t+1)$ for all $t \in \mathbb{R}^+$.

Put $K = \eta_\infty([0,1])$. By construction, $f^\nu(K) = \eta_\infty([\nu, \nu+1])$; therefore,

$$\bigcup_{k=\nu}^{\infty} f^k(K) = \eta_\infty([\nu, +\infty))$$

for every $\nu \in \mathbb{N}$. Recalling the proof of Theorem 3.3.6, we see that the set L_∞ of limit points of $\{f^\nu\}$ is given by

$$L_\infty = \bigcap_{\nu\in\mathbb{N}} \overline{\eta_\infty([\nu, +\infty))} = \bigcap_{t=0}^{+\infty} \overline{\eta_\infty([t, +\infty))};$$

to end the proof it suffices to show that $L_\infty \supseteq [0,1] \times \{0\}$.

Indeed, since $\eta_\infty([\nu, +\infty)) \cap B_D(z_0, \nu) = \emptyset$ we clearly have $L_\infty \subseteq \partial D$. Moreover, since, by construction, $\tau_\infty \in L$ there must exists a sequence $\{t_\nu\}$ converging to $+\infty$ such that $\mathrm{Im}\, \eta_\infty(t_\nu) \to 0$. The shape of D then forces $\eta_\infty(t)$ to get closer and closer to the left and right boundaries of ∂D as $t \to +\infty$; thus $(0,0)$ and $(1,0)$ must belong to L and, by connectedness, $L \supseteq [0,1] \times \{0\}$ as claimed.

In the previous example, the boundary of the domain had a very complicated topological and geometrical structure. The main goal of this section is to show that if, roughly speaking, the boundary of D is not too wild then a compactly divergent sequence of iterates converges to a point in the boundary as in the Wolff–Denjoy theorem. A first case is when the boundary is totally disconnected.

Corollary 3.3.9. *Let $D \subset \widehat{X}$ be a hyperbolic domain such that ∂D is totally disconnected. Let $f \in \mathrm{Hol}(D, D)$. Then either f is a periodic automorphism or its sequence of iterates converges to a point of $\overline{D}$.*

Proof. Under these assumptions on the boundary, D cannot be either simply or doubly connected. Hence the assertion follows from Theorem 3.3.6. □

Actually, if the boundary consists only of a finite number of points we can be even more precise.

Proposition 3.3.10. *Let $D \subset \widehat{X}$ be a hyperbolic domain such that ∂D is a finite set. Then every $f \in \mathrm{Hol}(D, D)$ is either constant or a periodic automorphisms.*

Proof. Since the boundary consists of a finite number of isolated points the big Picard Theorem 1.7.25 implies that every $f \in \mathrm{Hol}(D,D)$ is the restriction of a $\tilde{f} \in \mathrm{Hol}(\widehat{X},\widehat{X})$ such that $\tilde{f}(D) \subseteq D$.

If $\widehat{X}$ is hyperbolic, the assertion follows from Corollary 3.3.3.

Assume that $\widehat{X}$ is a torus X_τ, and let $\pi_\tau\colon \mathbb{C} \to X_\tau$ be a universal covering map. If $\tilde{f}$ is not constant, its image is open and closed (because X_τ is compact) in X_τ; hence $\tilde{f}$ is surjective. Since $\tilde{f}(D) \subseteq D$, we must have $\partial D \subseteq \tilde{f}(\partial D)$. If there were $z \in \partial D$ such that $f(z) \notin \partial D$, then the cardinality of $f(\partial D) \cap \partial D$ would be strictly less than the cardinality of ∂D, impossible; hence $\tilde{f}|_{\partial D}$ must be a bijection of the finite set ∂D. In particular, there exists $k \geq 1$ such that $\tilde{f}^k|_{\partial D} \equiv \mathrm{id}_{\partial D}$.

Since $\mathrm{Aut}(X_\tau)$ acts transitively on X_τ (Proposition 1.6.38), without loss of generality we can assume that $\pi_\tau(0) \in \partial D$. By Proposition 1.6.26, $\tilde{f}^k$ must have a lifting of the form $\gamma_c^\lambda(z) = \lambda z + c$, with $|\lambda|^2 \in \mathbb{N}^*$ and $c \in \mathbb{C}$; since $\tilde{f}^k(\pi_\tau(0)) = \pi_\tau(0)$, we can assume $c = 0$.

Assume, by contradiction, that $|\lambda|^2 > 1$. Since $\tilde{f}^k|_{\partial D} \equiv \mathrm{id}_{\partial D}$ and $\tilde{f}^k(D) \subseteq D$, the only $p \in X_\tau$ such that $\tilde{f}^k(p) = \pi_\tau(0)$ is $p = \pi_\tau(0)$. However, take $m + n\tau \in \mathbb{Z} \oplus \mathbb{Z}\tau$ not zero and of minimum modulus. Then $z_0 = \frac{1}{\lambda}(m + n\tau) \notin \mathbb{Z} \oplus \mathbb{Z}\tau$ so that $\pi_\tau(z_0) \neq \pi_\tau(0)$, but $\tilde{f}^k(\pi_\tau(z_0)) = \pi_\tau(\gamma_0^\lambda(z_0)) = \pi_\tau(m + n\tau) = \pi_\tau(0)$, contradiction.

So, we must have $|\lambda| = 1$, and then Proposition 1.6.38 ensures that $\tilde{f}^k$ is a periodic automorphism. But then $\tilde{f}$ must necessarily be bijective, and hence an automorphism, necessarily periodic because $\tilde{f}^k$ is, as claimed.

Finally, assume that $\widehat{X} = \widehat{\mathbb{C}}$, so that $\tilde{f}$ is a rational map. If $\deg \tilde{f} = 0$, then $\tilde{f}$ is constant, and we are done. Assume that $\deg f \geq 1$, so that $\tilde{f}$, being not constant, is surjective. Arguing exactly as in the case of the torus, we find that $\tilde{f}$ is a bijection of ∂D with itself, and that there is $k \geq 1$ such that $\tilde{f}^k|_{\partial D} \equiv \mathrm{id}_{\partial D}$.

If $\deg \tilde{f} = 1$, then $\tilde{f}$ and $\tilde{f}^k$ are automorphisms of $\widehat{\mathbb{C}}$. But ∂D contains at least three points, because D is hyperbolic; hence Corollary 1.6.20 implies that $\tilde{f}^k \equiv \mathrm{id}_{\widehat{\mathbb{C}}}$, and so f is a periodic automorphism of D, as claimed.

To conclude the proof, assume by contradiction that $\deg f \geq 2$. We know that every $z_0 \in \partial D$ is a fixed point for $\tilde{f}^k$, and moreover, that $\tilde{f}^k(D) \subseteq D$; so $(\tilde{f}^k)^{-1}(z_0) = \{z_0\}$ for all $z_0 \in \partial D$. Since $\deg \tilde{f}^k = (\deg \tilde{f})^k \geq 2$, it follows that every point of ∂D is a critical point of $\tilde{f}^k$.

Take $z_0 \in \partial D$; without loss of generality, we can assume that $z_0 = 0$, and we can write $\tilde{f}^k(z) = z^2 g(z)$ for a rational function g holomorphic in 0. Then we can find $0 < r, \delta < 1$ such that $|z| < r$ implies $|zg(z)| < \delta$, and thus $|\tilde{f}^k(z)| < \delta|z|$. As a consequence, the sequence $\{\tilde{f}^{\nu k}(z)\}$ converges to 0 as soon as $|z| < r$, and then Corollary 3.3.9 implies that $\tilde{f}^{\nu k} \to 0$ on the whole of D. But we can repeat the same argument for another point of ∂D, and thus the same sequence should converge to two distinct points, impossible. □

A second, quite more important case is when D is of regular type (see Section 1.8). To properly deal with Jordan components in the boundary, we look for a replacement for the horocycles.

We begin with a lemma on the Poincaré distance, whose applicability depends on Theorem 1.8.9.

Lemma 3.3.11. *Let $D \subset \widehat{X}$ be a hyperbolic domain of regular type, $\pi_D\colon \mathbb{D} \to D$ its universal covering map and $\Gamma_D \subset \mathrm{Aut}(\mathbb{D})$ the automorphism group of the covering. Let τ_0 be contained in a Jordan component of ∂D and choose $\tilde{\tau}_0 \in \partial\mathbb{D}$ and a neighborhood U in $\overline{\mathbb{D}}$ of $\tilde{\tau}_0$ such that π_D extends to a homeomorphism between a neighborhood (in $\overline{\mathbb{D}}$) of $\overline{U}$ and its image such that $\pi_D(\tilde{\tau}_0) = \tau_0$. Then for every $z_0 \in \mathbb{D}$ there is a finite subset Γ_0 of Γ_D, depending only on U and z_0, such that for every $w \in U$ we have*

$$\omega_D(\pi_D(z_0), \pi_D(w)) = \min_{\gamma \in \Gamma_0} \omega(\gamma(z_0), w).$$

Proof. Since Γ_D is properly discontinuous on $\mathbb{D}$, by Proposition 1.7.3 for every $w \in U \cap \mathbb{D}$ we can find $\gamma_w \in \Gamma_D$ so that $\omega_D(\pi_D(z_0), \pi_D(w)) = \omega(\gamma_w(z_0), w)$. Let $Z = \{\gamma_w(z_0) \mid w \in U \cap \mathbb{D}\}$; to get the assertion it suffices to show that $\overline{Z} \cap \partial\mathbb{D} = \emptyset$, again because Γ_D is properly discontinuous on $\mathbb{D}$.

Suppose, by contradiction, there is a sequence $\{w_\nu\} \subset U \cap \mathbb{D}$ such that $z_\nu = \gamma_{w_\nu}(z_0) \to \sigma_0 \in \partial\mathbb{D}$; up to a subsequence we can also assume $w_\nu \to \sigma_1 \in \overline{U}$. Since π_D is injective in a neighborhood of $\overline{U}$ and $\pi_D(z_\nu) = \pi_D(z_0)$ for all $\nu \in \mathbb{N}$, it is clear that σ_0 cannot belong to $\overline{U}$; in particular, $\sigma_0 \neq \sigma_1$.

By construction, $\omega(z_\nu, w_\nu) = \omega_D(\pi_D(z_0), \pi_D(w_\nu))$ for all $\nu \in \mathbb{N}$. Hence

$$\begin{aligned}
\omega_D(\pi_D(z_0), \pi_D(0)) &\geq \omega_D(\pi_D(z_0), \pi_D(w_\nu)) - \omega_D(\pi_D(w_\nu), \pi_D(0)) \\
&\geq \omega(z_\nu, w_\nu) - \omega(0, w_\nu) \\
&= \frac{1}{2}\log\left[\frac{|1 - \bar{z}_\nu w_\nu|^2}{1 - |z_\nu|^2} \cdot \left(\frac{1 + |\gamma_{z_\nu}(w_\nu)|}{1 + |w_\nu|}\right)^2\right],
\end{aligned} \tag{3.4}$$

where the last equality is obtained as in the proof of Proposition 2.1.4. Letting $\nu \to +\infty$, the right-hand side of (3.4) diverges, against the fact that it is bounded by $\omega_D(\pi_D(z_0), \pi_D(0))$, contradiction and we are done. □

The idea is to define horocycles in hyperbolic domains of regular type following the model given by Proposition 2.1.4. The main step is contained in the following.

Proposition 3.3.12. *Let $D \subset \widehat{X}$ be a hyperbolic domain of regular type and fix $z_0 \in D$. Then for every $\tau_0 \in \partial D$ contained in a Jordan component of ∂D and for every $z \in D$ the limit*

$$\lim_{w \to \tau_0} [\omega_D(z, w) - \omega_D(z_0, w)]$$

exists and is finite.

Proof. Let $\pi_D\colon \mathbb{D} \to D$ be the universal covering map of D, and let $\Gamma_D \subset \mathrm{Aut}(\mathbb{D})$ be the automorphism group of the covering. Since τ_0 belongs to a Jordan component of ∂D, we can find (Theorem 1.8.9) a point $\tilde{\tau}_0 \in \partial\mathbb{D}$ and a neighborhood U of $\tilde{\tau}_0$ in $\overline{\mathbb{D}}$

such that π_D extends to a homeomorphism of a neighborhood of $\overline{U}$ with its image such that $\pi_D(\tilde{\tau}_0) = \tau_0$.

Choose $z \in D$, and fix $\tilde{z} \in \pi_D^{-1}(z)$ and $\tilde{z}_0 \in \pi_D^{-1}(z_0)$. By Lemma 3.3.11, there are two finite subsets Γ_z and Γ_0 of Γ_D such that for every $\tilde{w} \in U$ we have

$$\omega_D(z, \pi_D(\tilde{w})) - \omega_D(z_0, \pi_D(\tilde{w})) = \min_{\gamma\in\Gamma_z} \max_{\gamma_0\in\Gamma_0} \{\omega(\tilde{z}, \gamma(\tilde{w})) - \omega(\tilde{z}_0, \gamma_0(\tilde{w}))\}. \tag{3.5}$$

Now take $\gamma \in \Gamma_z$ and $\gamma_0 \in \Gamma_0$, and set $a = \gamma^{-1}(0)$ and $a_0 = \gamma_0^{-1}(0)$. Then

$$\begin{aligned}&\lim_{\tilde{w}\to\tilde{\tau}_0} [\omega(\tilde{z}, \gamma(\tilde{w})) - \omega(\tilde{z}_0, \gamma_0(\tilde{w}))] \\ &= \frac{1}{2}\log\left[\frac{|a - \tilde{\tau}_0|^2}{1 - |a|^2} \cdot \frac{1 - |a_0|^2}{|a_0 - \tilde{\tau}_0|^2} \cdot \frac{|\tilde{z} - \gamma(\tilde{\tau}_0)|^2}{1 - |\tilde{z}|^2} \cdot \frac{1 - |\tilde{z}_0|^2}{|\tilde{z}_0 - \gamma_0(\tilde{\tau}_0)|^2}\right];\end{aligned} \tag{3.6}$$

so (3.5) and (3.6) yield the assertion, because $\tilde{w} \to \tilde{\tau}_0$ if and only if $\pi_D(\tilde{w}) \to \tau_0$. □

We can then define horocycles centered at points in Jordan components of the boundary of a hyperbolic domains of regular type.

Definition 3.3.13. Let $D \subset \widehat{X}$ be a hyperbolic domain of regular type; fix $z_0 \in D$ and a point $\tau_0 \in \partial D$ contained in a Jordan component of ∂D. Then the *horocycle* $E_{z_0}(\tau_0, R)$ of *center* τ_0, *pole* z_0 and *radius* $R > 0$ is given by

$$E_{z_0}(\tau_0, R) = \left\{z \in D \,\middle|\, \lim_{w\to\tau_0} [\omega_D(z, w) - \omega_D(z_0, w)] < \frac{1}{2}\log R\right\}.$$

Note that, by Proposition 2.1.4, if we take $D = \mathbb{D}$ and $z_0 = 0$ we recover the classical horocycles in the unit disk.

The proof of our main theorem requires only one feature of the horocycles.

Lemma 3.3.14. *Let $D \subset \widehat{X}$ be a hyperbolic domain of regular type, fix $z_0 \in D$ and choose $\tau_0 \in \partial D$ contained in a Jordan component of ∂D. Then*

$$\overline{E_{z_0}(\tau_0, R)} \cap \partial D = \{\tau_0\}$$

for every $R > 0$.

Proof. First of all, we would like to prove that $\tau_0 \in \overline{E_{z_0}(\tau_0, R)}$.

Let $\pi_D\colon \mathbb{D} \to D$ be a universal covering map of D. Since τ_0 is in a Jordan component, by Theorem 1.8.9 we can find a point $\tilde{\tau}_0 \in \partial\mathbb{D}$ such that π_D extends continuously and injectively to a neighborhood U (in $\overline{\mathbb{D}}$) of $\tilde{\tau}_0$ with $\pi_D(\tilde{\tau}_0) = \tau_0$. Let Γ_0 be the finite subset of the automorphism group of π_D given by Lemma 3.3.11 applied with $z_0 = 0$. For any $\tilde{z} \in \mathbb{D}$, $\tilde{w} \in U \cap D$ and $\gamma \in \Gamma_0$, we have

$$\omega_D(\pi_D(\tilde{z}), \pi_D(\tilde{w})) \le \omega(\tilde{z}, \tilde{w}) = \omega(\gamma^{-1}(\tilde{z}), \gamma^{-1}(\tilde{w})).$$

Recalling Lemma 3.3.11 and the fact that $\tilde{w} \to \tilde{\tau}_0$ if and only if $\pi_D(\tilde{w}) \to \tau_0$, setting $w_0 = \pi_D(0)$ we get

$$\begin{aligned}
&\lim_{w\to\tau_0} [\omega_D(\pi_D(\tilde{z}), w) - \omega_D(w_0, w)] \\
&\quad = \lim_{\tilde{w}\to\tilde{\tau}_0} [\omega_D(\pi_D(\tilde{z}), \pi_D(\tilde{w})) - \omega_D(\pi_D(0), \pi_D(\tilde{w}))] \\
&\quad \le \lim_{\tilde{w}\to\tilde{\tau}_0} [\omega(\tilde{z}, \tilde{w}) - \omega_D(\pi_D(0), \pi_D(\tilde{w}))] \\
&\quad = \max_{\gamma\in\Gamma_0} \lim_{\tilde{w}\to\tilde{\tau}_0} [\omega(\gamma^{-1}(\tilde{z}), \gamma^{-1}(\tilde{w})) - \omega(0, \gamma^{-1}(\tilde{w}))] \\
&\quad = \max_{\gamma\in\Gamma_0} \frac{1}{2} \log \frac{|\gamma^{-1}(\tilde{\tau}_0) - \gamma^{-1}(\tilde{z})|^2}{1 - |\gamma^{-1}(\tilde{z})|^2},
\end{aligned}$$

where we used (2.6). Now setting $a_\gamma = \gamma(0) \in \mathbb{D}$ for $\gamma \in \Gamma_0$, we get

$$\frac{|\gamma^{-1}(\tilde{\tau}_0) - \gamma^{-1}(\tilde{z})|^2}{1 - |\gamma^{-1}(\tilde{z})|^2} = \frac{|\tilde{\tau}_0 - \tilde{z}|^2}{1 - |\tilde{z}|^2} \cdot \frac{1 - |a_\gamma|^2}{|1 - \overline{a_\gamma}\tilde{\tau}_0|^2},$$

and hence

$$\lim_{w\to\tau_0} [\omega_D(\pi_D(\tilde{z}), w) - \omega_D(w_0, w)] \le \frac{1}{2} \log \frac{|\tilde{\tau}_0 - \tilde{z}|^2}{1 - |\tilde{z}|^2} + \max_{\gamma\in\Gamma_0} \frac{1}{2} \log \frac{1 - |a_\gamma|^2}{|1 - \overline{a_\gamma}\tilde{\tau}_0|^2}. \tag{3.7}$$

For $\nu \in \mathbb{N}^*$, set $\tilde{z}_\nu = (1 - 1/\nu)\tilde{\tau}_0$ and $z_\nu = \pi_D(\tilde{z}_\nu)$. Clearly, $\tilde{z}_\nu \to \tilde{\tau}_0$ and $z_\nu \to \tau_0$; moreover,

$$\frac{|\tilde{\tau}_0 - \tilde{z}_\nu|^2}{1 - |\tilde{z}_\nu|^2} = \frac{1}{2\nu - 1}. \tag{3.8}$$

Given $R > 0$, putting together (3.7) and (3.8) we get that $z_\nu \in E_{w_0}(\tau_0, R)$ eventually, and thus $\tau_0 \in \overline{E_{w_0}(\tau_0, R)}$ for all $R > 0$.

Coming back to our $z_0 \in D$, we have

$$\omega_D(z, w) - \omega_D(z_0, w) \le \omega_D(z, w) - \omega_D(w_0, w) + \omega_D(z_0, w_0);$$

it follows that $E_{z_0}(\tau_0, R) \supseteq E_{w_0}(\tau_0, cR)$, where $c = \exp(-2\omega_D(z_0, w_0))$, and thus $\tau_0 \in \overline{E_{z_0}(\tau_0, R)}$ for all $R > 0$ as desired.

Finally, let $\tau_1 \in \overline{E_{z_0}(\tau_0, R)} \cap \partial D$; we have to prove that $\tau_1 = \tau_0$. Choose a sequence $\{z_\nu\} \subset E_{z_0}(\tau_0, R)$ converging to τ_1. Now, by Lemma 3.3.11 for every $\nu \in \mathbb{N}$ we can find a finite subset Z_ν of $\pi_D^{-1}(z_\nu)$ so that for every $\tilde{w} \in U \cap \mathbb{D}$ we have

$$\omega_D(z_\nu, \pi_D(\tilde{w})) = \min_{\tilde{z}\in Z_\nu} \omega(\tilde{z}, \tilde{w}).$$

Choose $\tilde{z}_0 \in \pi_D^{-1}(z_0)$. Since $\omega_D(z_0, \pi_D(\tilde{w})) \le \omega(\tilde{z}_0, \tilde{w})$, we can use (3.6) with $\gamma = \gamma_0 = \mathrm{id}_{\mathbb{D}}$ to get

$$\begin{aligned}\frac{1}{2}\log R &> \lim_{w\to\tau_0}\left[\omega_D(z_\nu,w)-\omega_D(z_0,w)\right]\\ &= \lim_{\tilde w\to\tilde\tau_0}\left[\omega_D(z_\nu,\pi_D(\tilde w))-\omega_D(z_0,\pi_D(\tilde w))\right]\\ &\geq \min_{\tilde z\in Z_\nu}\lim_{\tilde w\to\tilde\tau_0}\left[\omega(\tilde z,\tilde w)-\omega(\tilde z_0,\tilde w)\right]\\ &= \frac{1}{2}\log\frac{1-|\tilde z_0|^2}{|\tilde\tau_0-\tilde z_0|^2}+\min_{\tilde z\in Z_\nu}\frac{1}{2}\log\frac{|\tilde z-\tilde\tau_0|^2}{1-|\tilde z|^2}.\end{aligned}\tag{3.9}$$

Now for every $\nu\in\mathbb{N}$ choose $\tilde z_\nu\in Z_\nu$ realizing the minimum in the right-hand side of (3.9). Since $z_\nu\to\tau_1\in\partial D$, up to a subsequence we can assume $\tilde z_\nu\to\tilde\tau_1\in\partial\mathbb{D}$. But then (3.9) forces $\tilde\tau_1=\tilde\tau_0$ and so $\tau_1=\tau_0$. □

We can now prove a *Wolff lemma* for hyperbolic domains of regular type.

Theorem 3.3.15. *Let $D\subset\widehat X$ be a hyperbolic domain of regular type without point components, fix $z_0\in D$ and take $f\in\mathrm{Hol}(D,D)$ such that $\{f^k\}$ is compactly divergent. Then there exists $\tau\in\partial D$ such that*

$$f(E_{z_0}(\tau,R))\subseteq E_{z_0}(\tau,R)\tag{3.10}$$

for every $R>0$.

Proof. Since $\{f^k\}$ is compactly divergent and ω_D is complete, we necessarily have $\omega_D(f^k(z_0),z_0)\to+\infty$ as $k\to+\infty$. We claim that can find a subsequence $\{f^{k_\nu}\}$ such that for every $\nu\in\mathbb{N}$ we have

$$\omega_D(f^{k_\nu}(z_0),z_0)<\omega_D(f^{k_\nu+1}(z_0),z_0).\tag{3.11}$$

Indeed, let k_ν denote the largest integer k satisfying $\omega_D(f^k(z_0),z_0)\leq\nu$. Then

$$\omega_D(f^{k_\nu}(z_0),z_0)\leq\nu<\omega_D(f^{k_\nu+1}(z_0),z_0)$$

for every $\nu\in\mathbb{N}$, as claimed.

Up to a subsequence, we can assume that $f^{k_\nu}(z_0)\to\tau_0\in\partial D$. Now for every $z\in D$ we have

$$\omega_D(f^{k_\nu+1}(z_0),f(z))\leq\omega_D(f^{k_\nu}(z_0),z).$$

Therefore, by (3.11),

$$\omega_D(f(z),f^{k_\nu+1}(z_0))-\omega_D(z_0,f^{k_\nu+1}(z_0))\leq\omega_D(z,f^{k_\nu}(z_0))-\omega_D(z_0,f^{k_\nu}(z_0)).\tag{3.12}$$

Clearly, we have $\omega_D(f^{k+1}(z_0),f^k(z_0))\leq\omega_D(f(z_0),z_0)$ for all $k\in\mathbb{N}$; so Proposition 1.7.27 implies that $f^{k_\nu+1}(z_0)\to\tau_0$ as $\nu\to+\infty$. Then we can take the limit as $\nu\to+\infty$ in (3.12) obtaining, by Proposition 3.3.12,

$$f(E_{z_0}(\tau_0, R)) \subseteq E_{z_0}(\tau_0, R)$$

for every $R > 0$, and we are done. □

We are finally ready to prove the main dynamical theorem on hyperbolic Riemann surfaces, the *Heins theorem*.

Theorem 3.3.16 (Heins, 1941). *Let $D \subset \widehat{X}$ be a hyperbolic domain of regular type and take a function $f \in \mathrm{Hol}(D, D)$. Then either:*
(i) *the sequence $\{f^k\}$ converges to a point $\tau_0 \in \overline{D}$, and if D is not simply connected, f is not an automorphism, or*
(ii) *f is a periodic automorphism, or*
(iii) *f is a pseudoperiodic automorphism. This latter possibility can occur only if D is either simply or doubly connected.*

Furthermore, in case (i) *if $\tau_0 \in D$ then it is a fixed point of f; if instead $\tau_0 \in \partial D$, then f has no fixed points in D.*

Proof. If D is simply connected, it is bounded by a Jordan curve and the universal covering map of $\widehat{X}$ gives a biholomorphism, continuous up to the boundary, of D with a simply connected plane domain bounded by a Jordan curve; therefore, quoting Theorem 1.6.32, we can apply the Wolff–Denjoy theorem. Hence we are left with the case D multiply connected.

If D is not doubly connected, by Theorem 2.6.9 $\mathrm{Aut}(D)$ is finite, and thus every $f \in \mathrm{Aut}(D)$ is periodic. If D is doubly connected, we know by Proposition 1.6.38 that every $f \in \mathrm{Aut}(D)$ is either periodic or pseudoperiodic. In conclusion, by Theorem 3.3.6 it suffices to show that if $\{f^k\}$ converges to the boundary then the limit point set L contains just one point.

If ∂D has no Jordan components, we can directly apply Corollary 3.3.10. Assume then ∂D has at least one Jordan component. Then, by the big Picard Theorem 1.7.25, every $f \in \mathrm{Hol}(D, D)$ extends holomorphically across the point components of ∂D. The image of a point component must belong to $\overline{D}$, but it cannot belong to a Jordan component of ∂D, by the open mapping theorem; therefore, the extension is a holomorphic self-map of the domain (still of regular type) obtained adding the point components to D. This extension cannot be an automorphism, because f is not; hence, by Theorem 3.3.6, either it has an attracting fixed point (which is necessarily a point component of ∂D), or its sequence of iterates is still compactly divergent. In the former case, $\{f^k\}$ converges to a point component of ∂D; therefore, to complete the proof we can also assume that D is bounded by a finite number of disjoint Jordan curves.

Fix $z_0 \in D$. Since $\{f^k\}$ is compactly divergent, we can apply Theorem 3.3.15 to get a point $\tau_0 \in \partial D$ so that (3.10) holds. It follows that the set L of limit points of $\{f^k\}$ is contained in $\overline{E_{z_0}(\tau_0, R)} \cap \partial D$; but this intersection, by Lemma 3.3.14, is equal to $\{\tau_0\}$, and thus $L = \{\tau_0\}$, as claimed.

Finally, assume $f^k \to \tau_0 \in \overline{D}$. If $\tau_0 \in \partial D$ clearly f cannot have fixed points in D. If $\tau_0 \in D$, we have

$$f(\tau_0) = \lim_{k\to+\infty} f(f^k(\tau_0)) = \lim_{k\to\infty} f^{k+1}(\tau_0) = \tau_0,$$

and thus τ_0 is a fixed point of f. □

We can then define the Wolff point of a holomorphic self-map of a hyperbolic domain of regular type.

Definition 3.3.17. Let $f \in \mathrm{Hol}(D,D)$ be a holomorphic self-map of a hyperbolic domain $D \subset \widehat{X}$ of regular type. If D is simply connected assume that $f \not\equiv \mathrm{id}_D$; if D is multiply connected, assume that $f \notin \mathrm{Aut}(D)$. Then the *Wolff point* $\tau_f \in \overline{D}$ of f is the (necessarily unique, by Corollary 3.1.16) fixed point of f in D if there is one, or the limit in ∂D of its sequence of iterates if f has no fixed points in D (cf. with Definition 2.5.3).

We end this section with a corollary generalizing Proposition 2.6.6.

Corollary 3.3.18. *Let $D \subset \widehat{X}$ be a multiply connected hyperbolic domain of regular type without point components and realize its fundamental group as a subgroup Γ_D of* $\mathrm{Aut}(\mathbb{D})$. *Take $f \in \mathrm{Hol}(D,D)$. Then the following statements are equivalent:*
(i) $f \notin \mathrm{Aut}(D)$;
(ii) $f_*: \pi_1(D) \to \pi_1(D)$ *is nilpotent;*
(iii) *the iterates $\tilde{f}^k$ of any lifting $\tilde{f}$ of f are automorphic under Γ_D for all sufficiently large k.*

Proof. We already know (Proposition 1.6.17) that (ii) $\Longleftrightarrow$ (iii). If $f \in \mathrm{Aut}(D)$, f_* cannot be nilpotent; hence (ii) $\Longrightarrow$ (i). Finally, assume $f \notin \mathrm{Aut}(D)$. Then, by Theorem 3.3.16, the sequence of iterates of f converges to a point of $\overline{D}$. In particular, if K is a compact subset of D, there is a large enough $k \in \mathbb{N}$ such that $f^k(K)$ is contained in a contractible subset of D, because ∂D has no point components.

Now take an element $[\sigma]$ of $\pi_1(D)$. By the previous observation, there is a sufficiently large $k \in \mathbb{N}$ such that $(f_*)^k[\sigma] = [f^k \circ \sigma]$ is trivial. Since $\pi_1(D)$ is finitely generated (by Lemma 1.8.3), this implies that $(f_*)^k$ is trivial for a sufficiently large $k \in \mathbb{N}$. □

Example 3.3.19. This corollary is not true for hyperbolic domains of regular type having a point component in the boundary. For instance, take $D = \mathbb{D}^*$ and $f(z) = z/2$. Then $f \notin \mathrm{Aut}(\mathbb{D}^*)$ but $f_* = \mathrm{id}$ is not nilpotent.

Corollary 3.3.18 reveals an interesting fact: roughly speaking, holomorphic functions do not like topological complications. Besides the automorphisms (which are finite in number), every other $f \in \mathrm{Hol}(D,D)$ gets rid of topological obstructions in a finite number of steps. This may be another reason for the importance of simply connected domains in function theory of one complex variable.

Notes to Section 3.3

This section is mostly inspired by the work of Heins [184, 191]. The paper [184], published in 1941, contains Theorem 3.3.6 and a version of Theorem 3.3.16 for hyperbolic plane domains bounded by a finite number of Jordan curves, obtained by using a clever (and complicated) argument based on the Julia–Wolff–Carathéodory theorem. The complete statement, at least for domains without point components, has been obtained by Heins in 1988 in [191], where it is proved using a different (though essentially equivalent) approach involving Green functions. We preferred to stress the role played by the horocycles because such a metric approach can also be applied in several complex variables. In particular, the proof of Theorem 3.3.15 is adapted from [5].

Versions of Theorem 3.3.6 and Proposition 3.3.7 can also be found in the work of Fatou [146–148]. A proof of a result close to Theorem 3.3.16 can be found in [211].

Theorem 3.3.2 can be expressed using the Alexandroff compactification of the hyperbolic Riemann surface X. Let X be a locally compact noncompact Hausdorff topological space; its *Alexandroff compactification* (or *one-point compactification*) is the topological space $X^\infty = X \cup \{\infty\}$ obtained by adding to X a single point ∞ and saying that a set W^∞ is an open neighborhood of ∞ if and only if $W^\infty = (X \setminus K) \cup \{\infty\}$, where $K \subset X$ is compact. It is clear that a sequence $\{f_k\} \subset C^0(X,X)$ of continuous self-maps of X is compactly divergent if and only if it converges to the constant map ∞ in $C^0(X,X^\infty)$. Therefore, Theorem 3.3.2 can be expressed saying that if X is a hyperbolic Riemann surface and $f \in \mathrm{Hol}(X,X)$ then either f is a periodic automorphism, or f is a pseudoperiodic automorphism, or $\{f^k\}$ converges uniformly on compact sets to a constant in X^∞.

Beardon [41] in 1990 noticed that part of this statement actually depends only on the fact that holomorphic maps contract the Poincaré distance. Indeed, he showed that if (X,d) is a locally compact, noncompact metric space such that every closed bounded subset is compact and $f: X \to X$ is a continuous self-map of X such that $d(f(x),f(y)) < d(x,y)$ for all $x \neq y \in X$ then the sequence of iterates $\{f^k\}$ converges to a constant in X^∞. Beardon also noticed that finer results can possibly be obtained by using a finer compactification, more strictly related to horocycles; this approach has recently turned out to be useful in the study of holomorphic dynamics in several complex variables.

Theorem 3.3.6 too can be expressed in terms of a suitable compactification of X. A *defining sequence* in a Riemann surface X is a sequence $\{\Omega_k\}$ of open connected subsets of X such that $\partial\Omega_k$ is a Jordan curve, $\overline{\Omega_{k+1}} \subset \Omega_k$ for all $k \geq 0$ and $\bigcap_k \Omega_k = \emptyset$. Two defining sequences $\{\Omega_k\}$ and $\{\Omega'_k\}$ are *equivalent* if for all $k \in \mathbb{N}$ there is $h \in \mathbb{N}$ such that $\Omega_k \subset \Omega'_h$ and conversely. The *Stoïlow boundary* $\partial_S X$ is the set of equivalence classes of defining sequences; the *Stoïlow compactification* is $X^S = X \cup \partial_S X$, where a set W^S is an open neighborhood of $\alpha \in \partial_S X$ if and only if $W^S = \Omega \cup \{\alpha\}$, where Ω is a domain in a defining sequence representing α. It is not difficult to see that if X is simply connected then $X^S = X^\infty$ and that if $X \subset \widehat{X}$ is a hyperbolic domain of regular type then $\partial_S X$ is in bijective correspondence with the connected components of the (usual) boundary of X in $\widehat{X}$. In general, the topology on $\partial_S X$ is totally disconnected; therefore, Theorem 3.3.6 can be expressed saying that if X is a hyperbolic Riemann surface and $f \in \mathrm{Hol}(X,X)$ then either f is a periodic automorphism, or f is a pseudoperiodic automorphism, or $\{f^k\}$ converges uniformly on compact sets to a constant in X^S; see [246, 186, 191] and [211].

Proposition 3.3.4 was stated by Farkas [144] in 1884, but the first complete proof is due to Ritt [352] in 1920.

Proposition 3.3.10 is contained in [186] where it is proved, with a different argument, for Riemann surfaces with non-Abelian fundamental group and without nonconstant negative subharmonic functions.

Lárusson [258] has proved a Wolff–Denjoy theorem for some kind of infinitely connected Riemann surfaces, using yet another compactification, the so-called Martin compactification built by using Green functions. Martin compactification is also in the background of [191], even though in that setting it coincides with the Euclidean closure. On the other hand, Poggi–Corradini [326] has constructed examples of hyperbolic domains $D \subset \widehat{\mathbb{C}}$ having an automorphism $\gamma \in \mathrm{Aut}(D)$ such that $\{\gamma^k\}$ converges to a point in the Euclidean closure of D but does not converge to a point in the Martin compactification of D.

For more on different compactifications of Riemann surfaces, see [299] and references therein.

Corollary 3.3.18 is taken from [105]. For more results on the action of holomorphic self-maps of hyperbolic Riemann surfaces on homotopy and homology classes see, e. g., [201, 256, 278, 384, 211, 212], and references therein.

3.4 Stability of the Wolff point

In this section, we discuss how the Wolff point of a holomorphic self-map of a hyperbolic domain of regular type depends on the map. In particular, we shall show that it is stable in the sense that it depends continuously on the self-map.

Definition 3.4.1. Let $D \subset \widehat{X}$ be a hyperbolic domain of regular type. If D is the simply connected set $\mathcal{F} = \overline{\mathrm{Hol}(D,D)} \setminus \{\mathrm{id}_D\}$; if D is a multiply connected set $\mathcal{F} = \overline{\mathrm{Hol}(D,D)} \setminus \mathrm{Aut}(D)$, where in both cases the closure is taken in $C^0(D,\widehat{X})$. Notice that, by a Hurwitz theorem (Corollary A.1.5), $\overline{\mathrm{Hol}(D,D)} = \mathrm{Hol}(D,D) \cup \partial D$. Then the *Heins map* $\tau\colon \mathcal{F} \to \overline{D}$ is given by τ_f if $f \in \mathrm{Hol}(D,D)$ and by $\tau_0 \in \partial D$ if $f \equiv \tau_0$.

Then we have the following natural complement to the Heins theorem.

Theorem 3.4.2. *The Heins map of a hyperbolic domain $D \subset \widehat{X}$ of regular type is continuous.*

Proof. Let $\tau\colon \mathcal{F} \to \overline{D}$ be the Heins map. We have to prove that if $\{f_\nu\} \subset \mathcal{F}$ converges to $f \in \mathcal{F}$ then $\tau(f_\nu) \to \tau_f$.

If ∂D has no Jordan components, by Proposition 3.3.10 $\mathcal{F}$ consists only of constant functions and so there is nothing to prove. So, assume that ∂D has at least a Jordan component. If D has a point component, then it is not simply connected and, as usual by the big Picard Theorem 1.7.25, all elements of $\mathcal{F}$ extend holomorphically across the point components of ∂D. The extension cannot be an automorphism; moreover, if not constant its image cannot contain a point in a Jordan component, by the open mapping theorem. Therefore, without loss of generality we can assume that ∂D has only Jordan components. Again without loss of generality, we can also assume that no f_ν is constant.

Suppose first that f has a fixed point $\tau_f = z_0 \in D$; we claim that, for ν large enough, $\tau(f_\nu) \in D$ and $\tau(f_\nu) \to z_0$.

Since Poincaré balls centered in z_0 are sent into themselves by f and they are simply connected if their radius is small enough (Proposition 1.7.3), we can find two fundamental systems of neighborhoods $\{U_n\}$, $\{V_n\}$ of z_0 in D satisfying the following properties:

- for every $n \in \mathbb{N}$, there exists a biholomorphism $\psi_n\colon U_n \to \mathbb{D}$ with $\psi_n(z_0) = 0$;
- $\overline{V_n} \subset U_n$;

- $f(U_n) \subseteq U_n$ and $f(V_n) \subseteq V_n$;
- for every $n \in \mathbb{N}$, there exists $\nu_0(n) \in \mathbb{N}$ such that if $\nu \geq \nu_0(n)$ then $f_\nu(V_n) \subset U_n$.

Fix $n \in \mathbb{N}$ and for $\nu \geq \nu_0(n)$ define $\phi_\nu, \phi \in \mathrm{Hol}(V_n, \mathbb{C})$ by

$$\phi_\nu(z) = \psi_n(f_\nu(z)) - \psi_n(z)$$

and $\phi(z) = \psi_n(f(z)) - \psi_n(z)$. Clearly, $\phi_\nu \to \phi$ and $\phi(z_0) = 0$; by a Hurwitz theorem (Corollary A.1.3) for ν large enough there must exist $z_\nu \in V_n$ such that $\phi_\nu(z_\nu) = 0$. This means that $z_\nu \in V_n$ is a fixed point of f_ν, and thus $z_\nu = \tau(f_\nu)$. As a consequence, for every $n \in \mathbb{N}$ we have $\tau(f_\nu) \in V_n$ eventually, and thus $\tau(f_\nu) \to \tau_f$ as claimed.

Let now suppose that f has no fixed points in D, so that $\tau_f = \tau_0 \in \partial D$. Fix $z_0 \in D$. For every $\tau \in \overline{D}$, define $\rho_\tau : D \to \mathbb{R}$ by setting

$$\rho_\tau(z) = \lim_{w \to \tau} [\omega_D(z, w) - \omega_D(z_0, w)];$$

by Proposition 3.3.12 the limit exists even when $\tau \in \partial D$. Furthermore, setting $\rho_\nu = \rho_{\tau(f_\nu)}$, we have $\rho_\nu \circ f_\nu \leq \rho_\nu$; this follows from Theorem 1.7.6 if $\tau(f_\nu) \in D$ and by Theorem 3.3.15 if $\tau(f_\nu) \in \partial D$.

Let $\tau \in \overline{D}$ be a limit point of the sequence $\{\tau(f_\nu)\}$, and let $\{\tau(f_{\nu_k})\}$ be a subsequence converging to τ; for simplicity, we put $\tau_k = \tau(f_{\nu_k})$. If $\tau \in D$, then necessarily $\tau_k \in D$ eventually; but then $f_{\nu_k}(\tau_k) = \tau_k$ and passing to the limit we get $f(\tau) = \tau$, against the assumption that f has no fixed points in D.

So, $\tau \in \partial D$; we claim that $\tau = \tau_f$. First of all, we claim that $\rho_{\nu_k} \to \rho_\tau$ as $k \to +\infty$. First of all, notice that if $\tau_k \in \mathbb{D}$ then $\rho_{\tau_k}(z) = \omega_D(z, \tau_k) - \omega_D(z_0, \tau_k)$. Therefore, if $\tau_k \in \mathbb{D}$ for infinitely many $k \in \mathbb{N}$ up to extract a subsequence we can assume $\tau_k \in D$ for all k and then it is obvious that $\rho_{\tau_k} \to \rho_\tau$.

If instead $\tau_k \in \partial D$ eventually, choose (Theorem 1.8.9) a point $\tilde{\tau} \in \partial\mathbb{D}$ and a neighborhood U of $\tilde{\tau}$ in $\overline{\mathbb{D}}$ such that π_D extends to a homeomorphism of a neighborhood of $\overline{U}$ with its image (which is a neighborhood of τ in $\overline{D} \subset \widehat{X}$) such that $\pi_D(\tilde{\tau}) = \tau$. Since $\tau_k \to \tau$, we have $\tau_k \in \pi_D(\overline{U})$ eventually, and thus we can find a unique $\tilde{\tau}_k \in \partial\mathbb{D} \cap \overline{U}$ such that $\pi_D(\tilde{\tau}_k) = \tau_k$. Fix now $z \in D$ and choose $\tilde{z} \in \pi_D^{-1}(z)$ and $\tilde{z}_0 \in \pi_D^{-1}(z_0)$. Then (3.5) and (3.6) imply that

$$\rho_{\nu_k}(z) = \min_{\gamma \in \Gamma_z} \max_{\gamma_0 \in \Gamma_0} \frac{1}{2} \log\left[\frac{|a_\gamma - \tilde{\tau}_k|^2}{1 - |a_\gamma|^2} \cdot \frac{1 - |a_{\gamma_0}|^2}{|a_{\gamma_0} - \tilde{\tau}_k|^2} \cdot \frac{|\tilde{z} - \gamma(\tilde{\tau}_k)|^2}{1 - |\tilde{z}|^2} \cdot \frac{1 - |\tilde{z}_0|^2}{|\tilde{z}_0 - \gamma_0(\tilde{\tau}_k)|^2}\right]$$

where $a_\gamma = \gamma^{-1}(0)$, $a_{\gamma_0} = \gamma_0^{-1}(0)$ and Γ_z, Γ_0 are finite subsets of Γ_D independent of k. A similar formula holds for ρ_τ, with $\tilde{\tau}$ replacing $\tilde{\tau}_k$; since $\tilde{\tau}_k \to \tilde{\tau}$ it follows that $\rho_{\nu_k} \to \rho_\tau$, as claimed.

Assume now that f is not constant. Passing to the limit as $k \to +\infty$ in

$$\rho_{\nu_k} \circ f_{\nu_k} \leq \rho_{\nu_k} \tag{3.13}$$

we get $\rho_\tau \circ f \leq \rho_\tau$. By definition, this yields $f(E_{z_0}(\tau,R)) \subseteq E_{z_0}(\tau,R)$; since $f^k \to \tau_f$, Lemma 3.3.14 forces $\tau = \tau_f$, as claimed.

Finally, assume that $f \equiv \tau_f$. Fix $z \in D$; since $\rho_{\nu_k}(z) \to \rho_\tau(z) < +\infty$, there is a $R > 0$ such that $\rho_{\nu_k}(z) < \frac{1}{2}\log R$ for k large enough. By (3.13), it follows that $f_{\nu_k}(z) \in E_{z_0}(\tau_k, R)$ for all k large enough. Since $\tau_k \to \tau$, $\rho_{\nu_k} \to \rho_\tau$ and $f_{\nu_k}(z) \to \tau_f$ as $k \to +\infty$, we get $\tau_f \in \overline{E_{z_0}(\tau,R)} \cap \partial D$; Lemma 3.3.14 then gives $\tau = \tau_f$ in this case, too.

So, τ_f is the only limit point of the sequence $\{\tau(f_\nu)\}$; it follows that $\tau(f_\nu) \to \tau_f$ and we are done. □

Notes to Section 3.4

Theorem 3.4.2 is proved, in a different way, in [184].

A related topic is the study of the dependence on the parameter of the dynamics of a holomorphic family of holomorphic self-maps. If X is a hyperbolic Riemann surface, a *holomorphic family of holomorphic self-maps* of X is a holomorphic map $F: \mathbb{D} \times X \to X$. If $\lambda \in \mathbb{D}$, we shall set $f_\lambda = F(\lambda, \cdot)$, so that each f_λ is a holomorphic self-map of X depending holomorphically on the parameter λ. Then it is known that:

(i) there exists $\lambda_0 \in \mathbb{D}$ such that $f_{\lambda_0} \in \mathrm{Aut}(X)$ if and only if $f_\lambda \equiv f_{\lambda_0}$ for all $\lambda \in \mathbb{D}$ [156, Proposition V.1.10] and [21, Theorem 2];

(ii) there exists $\lambda_0 \in \mathbb{D}$ such that f_{λ_0} has an attractive fixed point in X if and only if f_λ has an attractive fixed point $z_\lambda \in X$ for all $\lambda \in \mathbb{D}$ [7]; moreover, the function $\lambda \mapsto z_\lambda$ is holomorphic [75];

(iii) there exists $\lambda_0 \in \mathbb{D}$ such that the sequence of iterates $\{(f_{\lambda_0})^k\}$ is compactly divergent if and only if $\{(f_\lambda)^k\}$ is compactly divergent for all $\lambda \in \mathbb{D}$ [7];

(iv) if $D \subset \widehat{X}$ is a hyperbolic domain of regular type and $F: \mathbb{D} \times D \to D$ is a holomorphic family of holomorphic self-maps of D such that $\{(f_{\lambda_0})^k\}$ is compactly divergent for some $\lambda_0 \in \mathbb{D}$ then there exists $\tau_0 \in \partial D$ such that $f_\lambda^k \to \tau_0$ for all $\lambda \in \mathbb{D}$ [7].

Curiously enough, the known proofs all require several complex variables techniques, because they depend on the study of the dynamics of the two-dimensional system $(\lambda, z) \mapsto (\lambda, f_\lambda(z))$. See also [200, 403] for partial results on these topics.

3.5 Models on Riemann surfaces

In the study of dynamical systems, it can be useful to replace a given dynamical system by another system, simpler to study but with essentially the same dynamics, which is usually called a *model* (or a *normal form*) of the original dynamical system. Here, the meaning of "essentially" depends on the context. Typically, the new system is conjugated to the original one via an invertible map whose regularity depends on the kind of dynamical properties one would like to study: the conjugating map is a homeomorphism in topological dynamics, a diffeomorphism in smooth dynamics, a biholomorphism in holomorphic dynamics, and so on. Indeed, we have already effectively used many times this technique when we went from dynamical systems in $\mathbb{D}$ to dynamical systems in $\mathbb{H}^+$ by using the Cayley transform as conjugating biholomorphism.

Sometimes it is not possible to construct a conjugation in the whole space. However, to study dynamical properties it might be enough to focus the attention on the region of space where the dynamics is concentrated. For instance, we have seen that in many cases the sequence of iterates converges to a point τ; this means that, at least asymptotically, we are mostly interested in the dynamics nearby τ, because eventually the orbit of any point will enter a neighborhood of τ. So, to study dynamical properties it might suffices to have a conjugation with a simpler map (still called a model) in a neighborhood of τ.

In this section, we shall start the study of models in our context. The models will be provided by automorphisms of (usually) simply connected (usually noncompact) Riemann surfaces, that we know very well and whose dynamics is easy to study. Here, we shall give general necessary and sufficient conditions for the existence of models; in the next chapter, we shall show that these conditions are satisfied by holomorphic self-maps of $\mathbb{D}$ and we shall also tackle the difficult problem of determining the model associated to a given map.

Let us start by giving the official definition of model we shall use.

Definition 3.5.1. Let $f \in \mathrm{Hol}(X,X)$, where X is a Riemann surface. A domain $A \subseteq X$ is *invariant* for f (or f*-invariant*) if $f(A) \subseteq A$; it is *absorbing* for f (or f*-absorbing* or a *fundamental domain*) if it is f-invariant and

$$X = \bigcup_{\nu \in \mathbb{N}} f^{-\nu}(A).$$

In other words, a f-invariant domain is f-absorbing if it eventually contains the orbit of any point of X.

Definition 3.5.2. Let $f \in \mathrm{Hol}(X,X)$, where X is a Riemann surface. A *semimodel* for f is a triple (X^o, ψ, Φ), where X^o is another Riemann surface, $\psi\colon X \to X^o$ is a holomorphic function and $\Phi \in \mathrm{Aut}(X^o)$, that satisfy

$$\psi \circ f = \Phi \circ \psi \tag{3.14}$$

and

$$\bigcup_{\nu \in \mathbb{N}} \Phi^{-\nu}(\psi(X)) = X^o. \tag{3.15}$$

Notice that ψ cannot be constant, because otherwise the left-hand side of (3.15) would be a countable set. In particular, ψ is an open map and $\psi(X)$ is Φ-absorbing domain. The Riemann surface X^o is the *base*, ψ the *intertwining map* (or *Kœnigs map*) and Φ the *normal form* of the semimodel.

A *model* for f is a semimodel (X^o, ψ, Φ) for which there exists a f-absorbing domain $A \subseteq X$ where ψ is injective.

Remark 3.5.3. If (X^o, ψ, Φ) is a semimodel for $f \in \mathrm{Hol}(X,X)$, then we clearly have $\psi \circ f^\nu = \Phi^\nu \circ \psi$ for all $\nu \in \mathbb{N}$.

The following lemma gives some information regarding absorbing domains in the base of a semimodel.

Lemma 3.5.4. *Let X be a Riemann surface and $f \in \mathrm{Hol}(X,X)$. Assume that (X^o, ψ, Φ) is a semimodel for f.*

(i) *If $A \subseteq X$ is a set where ψ is injective, then $f|_A$ is injective.*
(ii) *If $A \subseteq X$ is a f-absorbing domain, then $\psi(A)$ is a Φ-absorbing domain.*
(iii) *Let $A \subseteq X$ be a f-absorbing domain, and for any $\nu \geq 0$ put $X^o_\nu = \Phi^{-\nu}(\psi(A))$. Then $\nu \leq \mu$ implies $X^o_\nu \subseteq X^o_\mu$, and moreover, $X^o = \bigcup_{\nu \geq 0} X^o_\nu$.*
(iv) *Assume that (X^o, ψ, Φ) is a model for f. If $A \subseteq X$ is a f-invariant domain such that $f|_A$ is injective, then $\psi|_A$ is injective.*
(v) *If f has a fixed point in X, then Φ has a fixed point in X^o. Conversely, if (X^o, ψ, Φ) is a model for f and Φ has a fixed point in X^o, then f has a fixed point in X.*

Proof. (i) Let $z \neq w$ be distinct points of A. Then we have $\psi(z) \neq \psi(w)$, which implies $\Phi(\psi(z)) \neq \Phi(\psi(w))$. Hence $\psi(f(z)) \neq \psi(f(w))$, and thus $f(z) \neq f(w)$.

(ii) Fix $z^o \in X^o$. By definition of semimodel there exists $\nu \geq 0$ and $z \in X$ such that $\Phi^\nu(z^o) = \psi(z) \in \psi(X)$. Since A is f-absorbing, there is $\mu \geq 0$ such that $f^\mu(z) \in A$. Hence

$$\Phi^{\mu+\nu}(z^o) = \Phi^\mu(\Phi^\nu(z^o)) = \Phi^\mu(\psi(z)) = \psi(f^\mu(z)) \in \psi(A),$$

and thus $\psi(A)$ is Φ-absorbing.

(iii) Part (ii) implies that $X^o = \bigcup_{\nu \geq 0} X^o_\nu$. Fix now $\nu \leq \mu$ and take $z^o \in X^o_\nu$. Then there must exist $z \in A$ such that $\Phi^\nu(z^o) = \psi(z)$. But then

$$\Phi^\mu(z^o) = \Phi^{\mu-\nu}(\Phi^\nu(z^o)) = \Phi^{\mu-\nu}(\psi(z)) = \psi(f^{\mu-\nu}(z)) \in \psi(A),$$

and hence $z^o \in X_\mu$, as claimed.

(iv) Let $A_0 \subseteq X$ be a f-absorbing domain where ψ is injective. Let $z \neq w$ be distinct points of A; since A is f-invariant, we have $f^\nu(z) \neq f^\nu(w)$ for all $\nu \in \mathbb{N}$. Choose $\nu \geq 0$ so that $f^\nu(z), f^\nu(w) \in A_0$. Then

$$\Phi^\nu(\psi(z)) = \psi(f^\nu(z)) \neq \psi(f^\nu(w)) = \Phi^\nu(\psi(w)),$$

which implies $\psi(z) \neq \psi(w)$ and we are done.

(v) If $z \in X$ is such that $f(z) = z$, we have $\Phi(\psi(z)) = \psi(f(z)) = \psi(z)$, and hence $\psi(z)$ is a fixed point of Φ.

Conversely, assume that (X^o, ψ, Φ) is a model for f. If $z^o \in X^o$ is a fixed point of Φ, by (ii) it must belong to $\psi(A)$, where $A \subseteq X$ is a f-absorbing domain where ψ is injective. This means that $z^o = \psi(z)$ for a suitable $z \in A$; then

$$\psi(z) = z^o = \Phi(z^o) = \Phi(\psi(z)) = \psi(f(z)),$$

and being $f(z) \in A$ and ψ injective on A we get $f(z) = z$, as claimed. □

We shall need a notion of isomorphism between (semi)models.

Definition 3.5.5. A *morphism* between two semimodels (X^o, ψ, Φ) and $(\widetilde{X}^o, \tilde{\psi}, \widetilde{\Phi})$ for a holomorphic self-map $f \in \mathrm{Hol}(X, X)$ of a Riemann surface X is a holomorphic map $\eta: X^o \to \widetilde{X}^o$ such that $\tilde{\psi} = \eta \circ \psi$ and $\widetilde{\Phi} \circ \eta = \eta \circ \Phi$. A morphism η of semi-models is an *isomorphism* if it is a biholomorphism. It is easy to check that if η is a isomorphism then (X^o, ψ, Φ) is a model if and only if $(\widetilde{X}^o, \tilde{\psi}, \widetilde{\Phi})$ is a model.

Remark 3.5.6. If $\eta: X^o \to \widetilde{X}^o$ is a morphism between the semimodels (X^o, ψ, Φ) and $(\widetilde{X}^o, \tilde{\psi}, \widetilde{\Phi})$, using the fact that Φ and $\widetilde{\Phi}$ are automorphisms it is easy to check that $\widetilde{\Phi}^\nu \circ \eta = \eta \circ \Phi^\nu$ for all $\nu \in \mathbb{Z}$. As a consequence, morphisms of semimodels are always surjective. Indeed,

$$\begin{aligned}\eta(X^o) &= \eta\Big(\bigcup_{\nu\in\mathbb{N}} \Phi^{-\nu}(\psi(X))\Big) = \bigcup_{\nu\in\mathbb{N}} \eta(\Phi^{-\nu}(\psi(X)))\\ &= \bigcup_{\nu\in\mathbb{N}} \widetilde{\Phi}^{-\nu}(\eta(\psi(X))) = \bigcup_{\nu\in\mathbb{N}} \widetilde{\Phi}^{-\nu}(\tilde{\psi}(X)) = \widetilde{X}^o.\end{aligned}$$

Morphisms of semimodels, if they exist, are unique.

Lemma 3.5.7. *Let (X^o, ψ, Φ) and $(\widetilde{X}^o, \tilde{\psi}, \widetilde{\Phi})$ be two semimodels for $f \in \mathrm{Hol}(X, X)$, where X is a Riemann surface. Then there exists at most one morphism from (X^o, ψ, Φ) to $(\widetilde{X}^o, \tilde{\psi}, \widetilde{\Phi})$.*

Proof. Let $\eta_0, \eta_1: X^o \to \widetilde{X}^o$ be two morphisms of semimodels from (X^o, ψ, Φ) to $(\widetilde{X}^o, \tilde{\psi}, \widetilde{\Phi})$; we claim that $\eta_0 \equiv \eta_1$.

By assumption, $\eta_0 \circ \psi = \tilde{\psi} = \eta_1 \circ \psi$. Then, recalling Remark 3.5.6 we have

$$\eta_0 \circ \Phi^{-\nu} \circ \psi = \widetilde{\Phi}^{-\nu} \circ \eta_0 \circ \psi = \widetilde{\Phi}^{-\nu} \circ \eta_1 \circ \psi = \eta_1 \circ \Phi^{-\nu} \circ \psi$$

for all $\nu \in \mathbb{N}$. Given $z^o \in X^o$, the definition of semimodel implies that there exist $z \in X$ and $\nu \in \mathbb{N}$ such that $z^o = \Phi^{-\nu}(\psi(z))$; but then the previous equality yields $\eta_0(z^o) = \eta_1(z^o)$ and we are done. □

Models are, in a sense, the largest possible semimodels.

Lemma 3.5.8. *Assume that $f \in \mathrm{Hol}(X, X)$ admits a model $(\widehat{X}^o, \hat{\psi}, \widehat{\Phi})$. If (X^o, ψ, Φ) is another semimodel for f then there exists a unique morphism of semimodels $\eta: \widehat{X}^o \to X^o$.*

Proof. Since uniqueness follows from Lemma 3.5.7, we have just to prove the existence. For $\nu \in \mathbb{N}$, set $\widehat{X}^o_\nu = \widehat{\Phi}^{-\nu}(\hat{\psi}(A))$, where $A \subseteq X$ is a f-absorbing domain such that $\hat{\psi}|_A$ is injective. By Lemma 3.5.4(iii), $\widehat{X}^o$ is the increasing union of all $\widehat{X}^o_\nu$'s.

Given $\nu \geq 0$, let $\eta_\nu: \widehat{X}^o_\nu \to X^o$ be given by

$$\eta_\nu = \Phi^{-\nu} \circ \psi \circ \hat{\psi}^{-1}|_{\hat{\psi}(A)} \circ \widehat{\Phi}^\nu;$$

since $\widehat{\Phi}^\nu(\widehat{X}_\nu^o) = \hat{\psi}(A)$ and $\hat{\psi}$ is a biholomorphism between A and $\hat{\psi}(A)$, the map η_ν is well-defined and holomorphic.

Now, if $\mu \geq \nu \geq 0$ we have

$$\begin{aligned}\eta_\mu|_{\widehat{X}_\nu^o} &= \Phi^{-\mu} \circ \psi \circ \hat{\psi}^{-1}|_{\hat{\psi}(A)} \circ \widehat{\Phi}^\mu|_{\widehat{X}_\nu^o} = \Phi^{-\mu} \circ \psi \circ \hat{\psi}^{-1}|_{\hat{\psi}(A)} \circ \widehat{\Phi}^{\mu-\nu} \circ \widehat{\Phi}^\nu|_{\widehat{X}_\nu^o} \\ &= \Phi^{-\mu} \circ \psi \circ f^{\mu-\nu} \circ \hat{\psi}^{-1}|_{\hat{\psi}(A)} \circ \widehat{\Phi}^\nu|_{\widehat{X}_\nu^o} \\ &= \Phi^{-\mu} \circ \Phi^{\mu-\nu} \circ \psi \circ \hat{\psi}^{-1}|_{\hat{\psi}(A)} \circ \widehat{\Phi}^\nu|_{\widehat{X}_\nu^o} = \eta_\nu .\end{aligned}$$

Therefore, η_μ restricted to $\widehat{X}_\nu^o$ coincides with η_ν. This means that we can define a holomorphic map $\eta: \widehat{X}^o \to X^o$ by setting $\eta(\hat{z}^o) = \eta_\nu(\hat{z}^o)$ as soon as $\hat{z}^o \in \widehat{X}_\nu^o$. To end the proof, it suffices to show that η is a morphism of semimodels.

First of all, for $z \in X$ choose $\nu \in \mathbb{N}$ such that $\hat{\psi}(z) \in \widehat{X}_\nu^o$. Then we have

$$\begin{aligned}\eta \circ \hat{\psi}(z) &= \Phi^{-\nu} \circ \psi \circ \hat{\psi}^{-1}|_{\hat{\psi}(A)} \circ \widehat{\Phi}^\nu \circ \hat{\psi}(z) = \Phi^{-\nu} \circ \psi \circ \hat{\psi}^{-1}|_{\hat{\psi}(A)} \circ \hat{\psi} \circ f^\nu(z) \\ &= \Phi^{-\nu} \circ \psi \circ f^\nu(z) = \Phi^{-\nu} \circ \Phi^\nu \circ \psi(z) = \psi(z)\end{aligned}$$

and the first condition in the definition of morphism of semimodels holds. Finally, fix $\nu \geq 1$. Then on $\widehat{X}_\nu^o$ we have

$$\begin{aligned}\Phi \circ \eta &= \Phi \circ \Phi^{-\nu} \circ \psi \circ \hat{\psi}^{-1}|_{\hat{\psi}(A)} \circ \widehat{\Phi}^\nu \\ &= \Phi^{-(\nu-1)} \circ \psi \circ \hat{\psi}^{-1}|_{\hat{\psi}(A)} \circ \widehat{\Phi}^{\nu-1} \circ \widehat{\Phi} = \eta_{\nu-1} \circ \widehat{\Phi} = \eta \circ \widehat{\Phi}\end{aligned}$$

and, being ν arbitrary, we are done. □

Corollary 3.5.9. *Assume that $f \in \mathrm{Hol}(X, X)$ admits a model $(\widehat{X}^o, \hat{\psi}, \widehat{\Phi})$. Then:*

(i) *if (X^o, ψ, Φ) is a semimodel for f such that there exists a morphism of semimodels $\eta: X^o \to \widehat{X}^o$, then (X^o, ψ, Φ) is a model, too;*

(ii) *any other model for f is isomorphic to $(\widehat{X}^o, \hat{\psi}, \widehat{\Phi})$. In particular, the base of a model is uniquely determined up to biholomorphisms.*

Proof. (i) Let $A \subseteq X$ be a f-absorbing domain where $\hat{\psi}$ is injective. Then from $\hat{\psi} = \eta \circ \psi$ we get that ψ is injective on A, and hence (X^o, ψ, Φ) is a model, too.

(ii) Assume that (X^o, ψ, Φ) is another model for f. Lemma 3.5.8 provides two morphisms of semimodels, $\eta: X^o \to \widehat{X}^o$ and $\hat{\eta}: \widehat{X}^o \to X^o$. The composition of morphisms clearly still is a morphism; therefore, $\hat{\eta} \circ \eta$ is a morphism of (X^o, ψ, Φ) with itself. But the identity also is a morphism of (X^o, ψ, Φ) with itself; then Lemma 3.5.7 implies $\hat{\eta} \circ \eta = \mathrm{id}_{X^o}$. Analogously, we find that $\eta \circ \hat{\eta} = \mathrm{id}_{\widehat{X}^o}$; hence η is a biholomorphism and the two models are isomorphic. □

The main result of this section is a necessary and sufficient condition for the existence of holomorphic models.

Theorem 3.5.10. *Let X be a Riemann surface and $f \in \mathrm{Hol}(X,X)$. Then there exists a holomorphic model $(\widehat{X}^o, \hat{\psi}, \widehat{\Phi})$ for f if and only if there exists a f-absorbing domain $A \subseteq X$ where f is injective. The model, if it exists, it is unique up to isomorphisms of models. Furthermore, if A is simply connected then the base of the model is biholomorphic either to $\mathbb{D}$ or to $\mathbb{C}$, with the only trivial exception of $X = \widehat{\mathbb{C}} = \widehat{X}^o$ and $f \in \mathrm{Aut}(\widehat{\mathbb{C}})$.*

Proof. The necessity of the condition follows immediately from Lemma 3.5.4(i).

Conversely, let $A \subseteq X$ be a f-absorbing domain where f is injective, and endow $A \times \mathbb{N}$ with the product topology, where $\mathbb{N}$ obviously has the discrete topology. We define an equivalence relation $\sim$ on $A \times \mathbb{N}$ as follows: $(z,\mu) \sim (w,\nu)$ if and only if there exists $\lambda \geq \max\{\mu,\nu\}$ such that $f^{\lambda-\mu}(z) = f^{\lambda-\nu}(w)$. Put $\widehat{X}^o = (A \times \mathbb{N})/\sim$, endowed with the quotient topology and let $\pi: A \times \mathbb{N} \to \widehat{X}^o$ be the canonical projection.

For $\nu \in \mathbb{N}$, define $\psi_\nu: A \to \widehat{X}^o$ by setting $\psi_\nu(z) = \pi(z,\nu)$ and put $\widehat{X}^o_\nu = \psi_\nu(A)$. Clearly, each ψ_ν is continuous and open; moreover, they are injective, because f and all its iterates are injective on A. Thus ψ_ν is a homeomorphism between A and $\widehat{X}^o_\nu$. Since $\widehat{X}^o = \bigcup_{\nu\in\mathbb{N}} \widehat{X}^o_\nu$, it follows that $\widehat{X}^o$ is a Hausdorff and second countable topological space; furthermore, since $\pi(z,\nu) = \pi(f(z),\nu+1)$ for all $z \in A$ it follows that $\widehat{X}^o_\nu \subseteq \widehat{X}^o_{\nu+1}$, and hence $\widehat{X}^o$ is arcwise connected.

We claim that $\mathcal{A} = \{(\widehat{X}^o_\nu, \psi_\nu|^{-1}_{\widehat{X}^o_\nu})\}$ induces a complex atlas on $\widehat{X}^o$. Indeed, it is easy to check that

$$\psi_\nu = \psi_\mu \circ f^{\mu-\nu}|_A \tag{3.16}$$

for all $0 \leq \nu \leq \mu$; hence the change of coordinates $\psi_\mu^{-1} \circ \psi_\nu$ is holomorphic and injective, and thus a biholomorphism between $\psi_\nu^{-1}(\widehat{X}^o_\nu \cap \widehat{X}^o_\mu)$ and $\psi_\mu^{-1}(\widehat{X}^o_\nu \cap \widehat{X}^o_\mu)$. We cannot say that $\mathcal{A}$ is an atlas because A is a domain in the Riemann surface X and not a domain in $\mathbb{C}$; but it is clear that using the fact that X is a Riemann surface we can build an actual complex atlas on $\widehat{X}^o$, thus endowing $\widehat{X}^o$ with a structure of Riemann surface.

Let us now build the automorphism $\widehat{\Phi}$. We start by putting $\widehat{\Phi}|_{\widehat{X}^o_1}(z) = \psi_0 \circ \psi_1^{-1}(z^o)$ for all $z^o \in \widehat{X}^o_1$; notice that $\widehat{\Phi}(\widehat{X}^o_1) = \widehat{X}^o_0$. Now we extend $\widehat{\Phi}$ to $\widehat{X}^o_2$ by putting $\widehat{\Phi}|_{\widehat{X}^o_2} = \psi_1 \circ \psi_2^{-1}$. If $z^o \in \widehat{X}^o_1$, we have

$$\begin{aligned}\widehat{\Phi}|_{\widehat{X}^o_2}(z^o) &= \psi_1 \circ \psi_2^{-1}(z^o) = \psi_1 \circ \psi_2^{-1} \circ \psi_1 \circ \psi_1^{-1}(z^o) = \psi_1 \circ f \circ \psi_1^{-1}(z^o)\\ &= \psi_1 \circ \psi_1^{-1} \circ \psi_0 \circ \psi_1^{-1}(z^o) = \widehat{\Phi}|_{\widehat{X}^o_1}(z^o)\end{aligned}$$

thanks to (3.16). Thus $\widehat{\Phi}|_{\widehat{X}^o_2}$ coincides with $\widehat{\Phi}|_{\widehat{X}^o_1}$ on $\widehat{X}^o_1$.

Arguing in a similar way, we see that putting $\widehat{\Phi}|_{\widehat{X}^o_\nu} = \psi_{\nu-1} \circ \psi_\nu^{-1}$ we define a $\widehat{\Phi}: \widehat{X}^o \to \widehat{X}^o$ holomorphic. Since $\widehat{\Phi}(\widehat{X}^o_\nu) = \widehat{X}^o_{\nu-1}$, we get that $\widehat{\Phi}$ is surjective; since all the ψ_ν's are injective, it follows that $\widehat{\Phi}$ is injective, too, and hence $\widehat{\Phi} \in \mathrm{Aut}(\widehat{X}^o)$.

By construction, and recalling (3.16), we have

$$\widehat{\Phi} \circ \psi_0 = \psi_0 \circ \psi_1^{-1} \circ \psi_0 = \psi_0 \circ f|_A. \tag{3.17}$$

We now define $\hat{\psi}: X \to \widehat{X}^o$ as follows. Given $z \in X$, choose $\nu \in \mathbb{N}$ such that $f^\nu(z) \in A$ and put

$$\hat{\psi}(z) = \widehat{\Phi}^{-\nu}(\psi_0(f^\nu(z))).$$

If $\mu \in \mathbb{N}$ is also such that $f^\mu(z) \in A$, assuming without loss of generality that $\mu \geq \nu$ we have

$$\begin{aligned}\widehat{\Phi}^{-\mu}(\psi_0(f^\mu(z))) &= \widehat{\Phi}^{-\mu}(\psi_0(f^{\mu-\nu}(f^\nu(z)))) \\ &= \widehat{\Phi}^{-\mu}(\widehat{\Phi}^{\mu-\nu}(\psi_0(f^\nu(z)))) = \widehat{\Phi}^{-\nu}(\psi_0(f^\nu(z))),\end{aligned}$$

where we used (3.17). Thus $\hat{\psi}$ is well-defined and clearly holomorphic. Moreover, if $f^\nu(z) \in A$ we have $f^{\nu-1}(f(z)) \in A$, and hence

$$\hat{\psi} \circ f(z) = \widehat{\Phi}^{-(\nu-1)} \circ \psi_0 \circ f^{\nu-1} \circ f(z) = \widehat{\Phi} \circ \widehat{\Phi}^{-\nu} \circ \psi_0 \circ f^\nu(z) = \Phi \circ \hat{\psi}(z),$$

that is $\hat{\psi}$ intertwines f and Φ. Finally, $\hat{\psi}|_A = \psi_0$ is injective, and thus $(\widehat{X}^o, \hat{\psi}, \widehat{\Phi})$ is a model for f, as claimed.

So, we have proved the existence of a model; the uniqueness, up to biholomorphisms, follows from Corollary 3.5.9.

Finally, if A is simply connected every $\widehat{X}^o_\nu$ is simply connected; hence $\widehat{X}^o$, being an increasing union of open simply connected sets, it is simply connected. Moreover, $\widehat{X}^o$ cannot be compact. If it were, we would have $\widehat{X}^o = \widehat{X}^o_\nu$ for some $\nu \in \mathbb{N}$, and hence $\widehat{X}^o$ would be homeomorphic to A, which is impossible because A cannot be compact—unless $A = X$ and X is compact. But this means $X = \widehat{\mathbb{C}}$, and thus $f \in \mathrm{Aut}(\widehat{\mathbb{C}})$ because f is injective on A by Lemma 3.5.4. Excluding this exception, the Riemann uniformization Theorem 1.6.5 implies that $\widehat{X}^o$ is biholomorphic either to $\mathbb{C}$ or to $\mathbb{D}$ and we are done. □

In the next chapter, we shall use this characterization to prove that holomorphic self-maps of $\mathbb{D}$ not having a superattracting fixed point admit models (Corollary 4.5.5).

Notes to Section 3.5

Possibly the first to look for solutions of functional equations of the form (3.14) has been Schröder in 1871; in [369], he asked for solutions of the so-called *Schröder equation* $\psi \circ f = \lambda\psi$, where f was a holomorphic function defined in a neighborhood of the origin satisfying $f(0) = 0$ and $f'(0) = \lambda$ with $|\lambda| \neq 0, 1$. This problem was solved in 1883 by Kœnigs in the paper [244], which in some sense marks the birth of local holomorphic dynamics. The superattracting case $f'(0) = 0$ requires a different functional equation and a different kind of models, as we shall see in Section 4.2, and was solved by Böttcher [71] in 1904.

Apparently, the first one looking for solutions of functional equations of the form $\psi \circ f = \lambda\psi$ with $f \in \mathrm{Hol}(\mathbb{D}, \mathbb{D})$ without fixed points has been Wolff [418] in 1929. A complete solution when f is hyperbolic has been provided by Valiron [398] in 1931; see [83] (and Section 4.3) for a modern exposition of Valiron's work.

When studying the dynamics of rational maps of the Riemann sphere, Fatou [146–148] realized that the correct functional equation to study in the parabolic case is the *Abel equation* $\psi \circ f = \psi + 1$, but he did not explicitly consider the case of parabolic self-maps of the unit disk. This was done in 1979 by Pommerenke with the collaboration of Baker [338, 34]; we shall describe their results in Section 4.4.

In all of these cases, the intertwining map ψ was obtained as limit of a sequence of functions obtained by normalizing the sequence of iterates of f by post-composing with suitable maps, depending on the specific case to be treated. The first one to give a unified treatment of all cases, except for maps with a superattracting fixed point, has been Cowen in 1981. In [126], he realized the importance of the notion of f-absorbing domain and building on the works of Valiron and Pommerenke devised an abstract dynamical way for building the base space of a model (using the so-called *tail space* or *abstract basin of attraction* of a noninvertible dynamical system). The simpler, though still abstract, approach we described here is more recent and is due to Arosio and Bracci [22].

Another source dealing with similar problems with more assumptions on the regularity of f around its Wolff point is [73]; see also Section 4.7.

Whereas the models are unique up to biholomorphisms, the semimodels are not. Discussions of the nonuniqueness of solutions of the Schröder and Abel equations can be found, e. g., in [247, 126, 328, 120].

When $X = \mathbb{D}$, it has been studied the boundary behavior of the intertwining map: see, e. g., [338, 121, 122], and references therein.

Let $D \subset \widehat{\mathbb{C}}$ be a hyperbolic domain, and $f \in \mathrm{Hol}(D, D)$. We say that an isolated point $\sigma \in \partial D$ is an *isolated boundary fixed point* of f if f extends continuously to σ with $f(\sigma) = \sigma$. If $f \in \mathrm{Hol}(D, D)$ is without fixed points in D and without isolated boundary fixed point then Marden and Pommerenke [277] and Bonfert [70] have constructed a domain $H \subseteq \mathbb{C}$, a holomorphic map $\psi: D \to H$ and an automorphism $\Phi \in \mathrm{Aut}(H)$ such that $\psi \circ f = \Phi \circ \psi$. Furthermore, ψ is injective if and only if f is injective; if $f \in \mathrm{Aut}(D)$, then ψ is a biholomorphism between D and H. Finally, H is either hyperbolic or $\mathbb{C}$, and they were able to characterize when H is hyperbolic (if and only if f has positive hyperbolic step; see Definition 4.6.1). However, it is not clear whether the triple (H, ψ, Φ) is a semimodel in our sense, i. e., if (3.15) holds, but it might be so when $H \neq \mathbb{C}$ (cf. the last lines of [70, p. 66]).

In 1999, König studied the existence of a model (that he called *conformal conjugacy*) having a simply connected absorbing domain in [247]. He proved the existence of a model with a simply connected absorbing domain assuming that X is a hyperbolic plane domain $D \subset \widehat{\mathbb{C}}$, that the sequence of iterates of f converges to a point in the boundary of D and that for every $[\sigma] \in \pi_1(D)$ there exists $m \in \mathbb{N}$ such that $f_*^m[\sigma] = 0$ (actually this last condition turns out to be equivalent to the existence of a simply connected absorbing domain; see [37]). In particular, Corollary 3.3.18 implies that holomorphic self-maps without fixed points (not automorphisms) of a multiply connected hyperbolic domain of regular type without point components admit a model with a simply connected base; see also [407] for a similar result. [247] also contains an example of a pair (D, f) not satisfying the hypothesis on $\pi_1(D)$ and without a model with a simply connected base.

Barański, Fagella, Jarque, and Karpińska in [38] proved the existence of "nice" absorbing domains, not necessarily simply connected, for an arbitrary holomorphic self-map f of a hyperbolic domain $D \subset \widehat{\mathbb{C}}$ assuming that the sequence of iterates of f converges to a point τ in the boundary; they also show that the absorbing domain can be chosen so that f is locally injective there. In [37], the same authors showed that simply connected absorbing domains do not always exist. More precisely, assume that $f \in \mathrm{Hol}(D, D)$ is without fixed points and without isolated boundary fixed points. Then they first show that if $\omega_D(f^{k+1}(z), f^k(z)) \to 0$ for some (and hence all) $z \in D$ then there exists a simply connected f-absorbing domain; then they construct examples without any simply connected f-absorbing domain if $\inf_{z\in D} \lim_{k\to+\infty} \omega_D(f^{k+1}(z), f^k(z)) > 0$ or if $\inf_{z\in D} \lim_{k\to+\infty} \omega_D(f^{k+1}(z), f^k(z)) = 0$ but $\omega_D(f^{k+1}(z), f^k(z)) \not\to 0$ for some (and hence all) $z \in D$. They also show that if $\{f^k\}$ converges to an isolated boundary fixed point then there cannot exist simply connected absorbing domains.

More information on models can be found in the notes to Section 4.6.

3.6 Random iteration on Bloch domains

Suggested, among other things, by the development of computer simulations of dynamical systems, a question often studied is the stability of a dynamical system: how does the behavior of an orbit depend on small variations of the initial point or of the map we are iterating? In our situation, with the only exception of the pseudoperiodic automorphisms, all orbits of a given map have the same behavior, and thus there is complete stability with respect to the starting point of the orbit. Furthermore, in Section 3.4 we have seen that the behavior is substantially stable with respect to the map too; for instance, the Wolff point depends continuously on the map.

However, there is another way to tackle the stability issue: it might happen that we only approximately know the map we would like to iterate, and thus instead of composing always the same map we might end up composing different maps close to each other in a suitable sense. The study of sequences of maps so obtained is the subject of study of a theory called *random iteration* or *random dynamics* or *nonautonomous dynamics*. In this and the next section, we shall collect a few results on this topic.

Definition 3.6.1. Let $\{f_\nu\}$ be a sequence of self-maps of a space X. The *left* (or *direct* or *forward*) *iterated function system* (or *composition system*) generated by $\{f_\nu\}$ is the sequence of self-maps $\{L_\nu\}$ given by

$$L_\nu = f_\nu \circ f_{\nu-1} \circ \cdots \circ f_0.$$

The *right* (or *reverse* or *backward*) *iterated function system* generated by $\{f_\nu\}$ is instead the sequence of self-maps $\{R_\nu\}$ given by

$$R_\nu = f_0 \circ f_1 \circ \cdots \circ f_\nu.$$

The sequence $\{f_\nu\}$ is the *generating sequence* of the (left or right) iterated function system. When the generating sequence is contained in a given family $\mathcal{F}$ of self-maps of X, we shall say that the corresponding left or right iterated function system is *in* $\mathcal{F}$.

Clearly, when $f_\nu = f$ for all $\nu \in \mathbb{N}$ both the left and right iterated function systems reduce to the sequence of iterates of the function f; in this sense, the study of the asymptotic behavior of iterated function systems is an extension of the usual iteration theory.

Clearly, to be able to say something meaningful in this generality we need to somewhat restrict the class of functions we use to build an iterated function system. The first case we are going to study is when all functions in the generating sequence $\{f_\nu\}$ have image in the same subdomain. As we shall see, the basic result in this case is of topological nature.

Theorem 3.6.2. *Let $\{f_\nu\} \subset C^0(X,X)$ be a sequence of continuous self-maps of a metric space (X,d). Assume that there exists $0 < \ell < 1$ such that*

$$d(f_\nu(x),f_\nu(y)) \le \ell d(x,y) \tag{3.18}$$

for all $x, y \in X$ and all $\nu \in \mathbb{N}$. Let $\{L_\nu\}$ (resp., $\{R_\nu\}$) be the left (resp., right) iterated function system generated by $\{f_\nu\}$. Then:

(i) *every (pointwise) limit point of $\{L_\nu\}$ (resp., $\{R_\nu\}$) is constant;*
(ii) *if a subsequence $\{L_{\nu_k}\}$ (resp., $\{R_{\nu_k}\}$) converges pointwise to a constant $x_0 \in X$, then it converges to x_0 uniformly on compact subsets;*
(iii) *assume that every f_ν has a (necessarily unique) fixed point $x_\nu \in X$. Then $\{L_\nu\}$ converges uniformly on compact subsets to a constant function $x_\infty \in X$ if and only if the sequence $\{x_\nu\}$ converges to x_∞;*
(iv) *assume that (X,d) is complete. Then if there exists $x_0 \in X$ such that the set $\{f_\nu(x_0)\}$ is bounded then $\{R_\nu\}$ converges uniformly on compact subsets to a constant function.*

Proof. (i) Assume that a subsequence $\{L_{\nu_k}\}$ converges pointwise to a map $g \in C^0(X,X)$. Then for every $x, y \in X$ we have

$$d(g(x),g(y)) = \lim_{k\to+\infty} d(f_{\nu_k}\circ\cdots\circ f_0(x), f_{\nu_k}\circ\cdots\circ f_0(y)) \le \lim_{k\to+\infty} \ell^{\nu_k+1} d(x,y) = 0,$$

and thus g is constant. The same argument applies to $\{R_\nu\}$, too.

(ii) Assume that $L_{\nu_k}(x) \to x_0$ for each $x \in X$. Let $K \subseteq X$ be compact and put $d_0 = \max_{x\in K}\{d(x_0,x)\} < +\infty$. Then

$$\begin{aligned} d(L_{\nu_k}(x),x_0) &\le d(L_{\nu_k}(x),L_{\nu_k}(x_0)) + d(L_{\nu_k}(x_0),x_0) \\ &\le \ell^{\nu_k+1} d(x,x_0) + d(L_{\nu_k}(x_0),x_0) \le \ell^{\nu_k+1} d_0 + d(L_{\nu_k}(x_0),x_0), \end{aligned}$$

for all $x \in K$; thus $L_{\nu_k} \to x_0$ uniformly on K. The same argument works for $\{R_{\nu_k}\}$.

(iii) Assume first that $x_\nu \to x_\infty \in X$. Then for all $x \in X$ we have

$$\begin{aligned} d(L_\nu(x), x_\infty) &\le d(f_\nu\circ\cdots\circ f_0(x), f_\nu(x_\nu)) + d(x_\nu, x_\infty) \\ &\le \ell d(f_{\nu-1}\circ\cdots\circ f_0(x), x_\nu) + d(x_\nu,x_\infty) \\ &\le \cdots \\ &\le \ell^{\nu+1} d(x,x_0) + \sum_{j=0}^{\nu-1} \ell^{\nu-j} d(x_j,x_{j+1}) + d(x_\nu,x_\infty). \end{aligned}$$

The first and third addends in the last line clearly goes to zero as $\nu \to +\infty$. To prove that the second addend goes to zero, too, let $M = \max_j d(x_j,x_{j+1}) < +\infty$, and fix $\varepsilon > 0$. Choose $\nu_0 \in \mathbb{N}$ so that $d(x_j,x_{j+1}) < \varepsilon(1-\ell)/2$ as soon as $j \ge \nu_0$ and $\nu_1 \in \mathbb{N}$ such that $\ell^j < \varepsilon(1-\ell)/(2M)$ as soon as $j \ge \nu_1$. Then if $\nu \ge \nu_0 + \nu_1$ we have

$$\begin{aligned} \sum_{j=0}^{\nu-1} \ell^{\nu-j} d(x_j,x_{j+1}) &= \sum_{j=0}^{\nu-\nu_1} \ell^{\nu-j} d(x_j,x_{j+1}) + \sum_{j=\nu-\nu_1+1}^{\nu-1} \ell^{\nu-j} d(x_j,x_{j+1}) \\ &\le M\sum_{j=0}^{\nu-\nu_1} \ell^{\nu-j} + \frac{\varepsilon(1-\ell)}{2} \sum_{j=\nu-\nu_1+1}^{\nu-1} \ell^{\nu-j} \end{aligned}$$

$$= M \sum_{j=\nu_1}^{\nu} \ell^j + \frac{\varepsilon(1-\ell)}{2} \sum_{j=1}^{\nu_1 - 1} \ell^j$$
$$\leq \frac{M}{1-\ell} \ell^{\nu_1} + \frac{\varepsilon(1-\ell)}{2} \frac{1-\ell^{\nu_1}}{1-\ell} \leq \varepsilon.$$

In this way, we have proved that $L_\nu \to x_\infty$ pointwise—and hence uniformly on compact subsets by (ii)—as claimed.

Conversely, assume that $L_\nu \to x_\infty$ and, by contradiction, that there exist $r > 0$ and a subsequence $\{x_{\nu_k}\}$ such that $d(x_{\nu_k}, x_\infty) > r$ for all k. Fix $x \in X$. Since $L_\nu(x) \to \infty$, we know that

$$d(L_\nu(x), x_\infty) < \frac{1-\ell}{1+\ell} r$$

for ν large enough. Then for k large enough, we have

$$\begin{aligned} d(x_{\nu_k}, x_\infty) - \frac{1-\ell}{1+\ell} r &\leq d(x_{\nu_k}, x_\infty) - d(L_{\nu_k}(x), x_\infty) \\ &\leq d(L_{\nu_k}(x), x_{\nu_k}) \\ &\leq \ell d(L_{\nu_k - 1}(x), x_{\nu_k}) \leq \ell d(L_{\nu_k - 1}(x), x_\infty) + \ell d(x_{\nu_k}, x_\infty) \\ &\leq \frac{1-\ell}{1+\ell} \ell r + \ell d(x_{\nu_k}, x_\infty), \end{aligned}$$

and then $d(x_{\nu_k}, x_\infty) \leq r$, contradiction.

(iv) Since the sequence $\{f_\nu(x_0)\}$ is bounded, there exists $A > 0$ such that $d(x_0, f_\nu(x_0)) \leq A$ for all $\nu \in \mathbb{N}$. Then for every $x \in X$ we have

$$d(x, f_\nu(x)) \leq d(x, x_0) + d(x_0, f_\nu(x_0)) + d(f_\nu(x_0), f_\nu(x)) < 2d(x, x_0) + A;$$

therefore, the sequence $\{d(x, f_\nu(x))\}$ is bounded for every $x \in X$.

Given $x \in X$, for simplicity write $y_\nu = R_\nu(x)$. Then for all $\nu, \mu \in \mathbb{N}$, we have

$$d(y_\nu, y_{\nu+\mu}) \leq \sum_{j=\nu}^{\nu+\mu-1} d(y_j, y_{j+1}) \leq \sum_{j=\nu}^{\nu+\mu-1} \ell^j d(x, f_{j+1}(x)) \leq \frac{\ell^\nu}{1-\ell} (2d(x_0, x) + A).$$

Since X is complete, this implies that for every $x \in X$ the sequence $\{R_\nu(x)\}$ converges; so $\{R_\nu\}$ converges pointwise, and hence uniformly on compact subsets to a constant function by (i) and (ii). □

Remark 3.6.3. If (X, d) is complete, the Banach fixed point theorem (see, e. g., [235, Section 3.1] or [347, Theorem 4.1]) implies that every $f \in C^0(X, X)$ satisfying (3.18) has a fixed point in X, and thus the hypothesis in (iii) is automatically satisfied.

We would like to apply this result to holomorphic self-maps of a hyperbolic Riemann surface. We know that they are not expanding with respect to the Poincaré distance, but this is not enough; we need strict contractions, as in (3.18). This is provided by the notion of Bloch domain introduced in Section 1.11.

Proposition 3.6.4. *Let $\Omega \subset X$ be a Bloch domain in a hyperbolic Riemann surface X. Then there exists $0 < \ell < 1$ such that*

$$\omega_X(f(z), f(w)) \le \ell\, \omega_X(z, w) \tag{3.19}$$

for all $z, w \in X$ and all $f \in \mathrm{Hol}(X, X)$ such that $f(X) \subseteq \Omega$. In particular, every $f \in \mathrm{Hol}(X, \Omega)$ has a (necessarily unique) fixed point in Ω.

Proof. Theorem 1.11.13 says that Ω is a Bloch domain if and only if $\mu(\Omega, X) < 1$, where

$$\mu(\Omega, X) = \sup_{z \in \Omega} \frac{\kappa_X}{\kappa_\Omega}(z).$$

Given $z, w \in X$, let $\sigma: [a, b] \to \Omega$ be a minimal geodesic with respect to κ_Ω connecting $f(z)$ and $f(w)$ (see Proposition 1.9.29). Then

$$\begin{aligned} \omega_X(f(z), f(w)) &\le \int_a^b \kappa_X(\sigma(t); \sigma'(t))\, dt \\ &\le \mu(\Omega, X) \int_a^b \kappa_\Omega(\sigma(t); \sigma'(t))\, dt \\ &= \mu(\Omega, X)\, \omega_\Omega(f(z), f(w)) \le \mu(\Omega, X)\, \omega_X(z, w), \end{aligned}$$

and hence (3.19) holds with $\ell = \mu(\Omega, X)$. The last assertion then follows from the Banach fixed point theorem (see Remark 3.6.3). □

As a consequence we then get precise informations on the dynamics of iterated function systems with value in a Bloch domain.

Corollary 3.6.5. *Let $\Omega \subset X$ be a Bloch domain in a hyperbolic Riemann surface X and let $\{f_\nu\} \subset \mathrm{Hol}(X, X)$ be a sequence of holomorphic self-maps of X such that $f_\nu(X) \subseteq \Omega$ for all $\nu \in \mathbb{N}$. Then:*

(i) *all limit points of the left and right iterated function systems, $\{L_\nu\}$ and $\{R_\nu\}$, generated by $\{f_\nu\}$ are constant;*
(ii) *let $z_\nu \in \Omega$ be the unique fixed point of f_ν. Then $\{L_\nu\}$ converges to a constant $z_\infty \in X$ if and only if $z_\nu \to z_\infty$;*
(iii) *if there is $z_0 \in X$ such that the set $\{f_\nu(z_0)\}$ is relatively compact in X, then $\{R_\nu\}$ converges to a constant function.*

Proof. Since the Poincaré distance of X is complete, by Proposition 3.6.4 and Remark 3.6.3 we can apply Theorem 3.6.2. □

Remark 3.6.6. It might happen that a right iterated function system generated by $\{f_\nu\}$ converges to a constant even though $\{f_\nu(z)\}$ is never relatively compact in X. For example, take $X = \mathbb{D}$, choose a Bloch domain $\Omega \subset \mathbb{D}$ not relatively compact in $\mathbb{D}$ and a sequence $\{z_\mu\} \subset \Omega$ converging to a point of $\partial\mathbb{D}$. Let $\pi_\mu \colon \mathbb{D} \to \Omega$ be a universal covering map with $\pi_\mu(0) = z_\mu$ and put $f_0 \equiv z_0$ and $f_\nu = \pi_{\nu-1}$ for $\nu \geq 1$. Then $\{f_\nu(0)\}$ is not relatively compact in $\mathbb{D}$ but $R_\nu \equiv z_0$ for all $\nu \in \mathbb{N}$, and thus $\{R_\nu\}$ trivially converges.

In particular, if $\Omega \subset\subset X$ is relatively compact then it is a Bloch domain in X (Example 1.11.2) and the assumption in Corollary 3.6.5(iii) is automatically satisfied for every $\{f_\nu\} \subset \mathrm{Hol}(X, \Omega)$; therefore, in this case a right iterated function system always converges to a constant. It turns out that this property actually characterizes relatively compact domains in the disk.

Proposition 3.6.7. *Let $\Omega \subset \mathbb{D}$ be a domain in the unit disk. Then Ω is relatively compact in $\mathbb{D}$ if and only if for every sequence $\{f_\nu\} \subset \mathrm{Hol}(\mathbb{D}, \Omega)$ the corresponding right iterated function system $\{R_\nu\}$ converges, necessarily to a constant function.*

Proof. If Ω is relatively compact in $\mathbb{D}$, the assertion is Corollary 3.6.5(iii).

Conversely, assume that Ω is not relatively compact in $\mathbb{D}$; we would like to find a sequence $\{f_\nu\} \subset \mathrm{Hol}(X, \Omega)$ such that the corresponding right iterated function system $\{R_\nu\}$ does not converge.

Choose two distinct points $p, p_0 \in \Omega$. We claim that we can choose a universal covering map $f_0 \colon \mathbb{D} \to \Omega$ and a point $p_1 \in \Omega$ such that $f_0(p) = p_0$ and $f_0(p_1) = p$. Indeed, let $f \colon \mathbb{D} \to \Omega$ be a universal covering map such that $f(p) = p_0$, and choose $\tilde{p}_1 \in \mathbb{D}$ such that $f(\tilde{p}_1) = p$. Since Ω is connected and not relatively compact in $\mathbb{D}$, it intersects all Poincaré circles centered at p; in particular, we can find $p_1 \in \Omega$ such that $\omega(p_1, p) = \omega(\tilde{p}_1, p)$. By precomposing f with a (non-Euclidean) rotation centered at p, we can then find a universal covering map f_0 such that $f_0(p) = p_0$ and $f_0(p_1) = p$.

Repeating this process, we find a sequence $\{f_\nu\} \subset \mathrm{Hol}(\mathbb{D}, \Omega)$ of universal covering maps and a sequence $\{p_\nu\} \subset \Omega$ of points such that $f_\nu(p) = p_\nu$ and $f_\nu(p_{\nu+1}) = p$ for every $\nu \in \mathbb{N}$. From this, it follows that $R_{2\nu}(p) = p_0$ and $R_{2\nu+1}(p) = p$ for all $\nu \in \mathbb{N}$; since $p \neq p_0$, it follows that the sequence $\{R_\nu\}$ cannot converge. □

With a slightly more complicated construction, we can show that Bloch domains in $\mathbb{D}$ are characterized by the fact that the limit points of right iterated function systems are constant. We need two lemmas; for the first one we need the quantity $R(z; \Omega, \mathbb{D})$ introduced in Definition 1.11.1.

Lemma 3.6.8. *There exists a function $\varepsilon \colon (1, +\infty) \to \mathbb{R}^+$ with $\lim_{t\to+\infty} \varepsilon(t) = 0$ such that for every domain $\Omega \subset \mathbb{D}$, every $z_0 \in \Omega$ with $R(z_0; \Omega, \mathbb{D}) > 1$ and every $z \in \Omega$ with $\omega(z_0, z) < 1$ we have*

$$\omega_\Omega(z_0, z) \leq \big(1 + \varepsilon\big(R(z_0; \Omega, \mathbb{D})\big)\big)\omega(z_0, z).$$

Proof. Without loss of generality, we can assume $z_0 = 0$; for simplicity, let us write R for $R(z_0; \Omega, \mathbb{D})$ and set $r = \tanh R$. Let $B = B_{\mathbb{D}}(0, R)$; by definition, $B \subseteq \Omega$. Moreover, if $\omega(0, z) < 1$ we have $z \in B$, because $R > 1$, and hence $\omega_\Omega(0, z) \le \omega_B(0, z) = \omega(0, |z|/r)$.

For $t \in (0, \tanh 1]$, put

$$h_r(t) = \frac{\omega(0, t/r)}{\omega(0, t)} - 1 \ge 0.$$

Since $h_r(t) \to r^{-1} - 1$ as $t \to 0^+$, we can extend h_r to a continuous function $\tilde{h}_r \colon [0, \tanh 1] \to \mathbb{R}^+$ and define

$$\varepsilon(R) = \max_{t \in [0, \tanh 1]} \tilde{h}_r(t).$$

Since $\tilde{h}_r$ tends uniformly to 0 as $r \to 1$, it follows that $\varepsilon(R) \to 0$ as $R \to +\infty$; furthermore,

$$\omega_\Omega(0, z) \le \omega(0, |z|/r) = (1 + \tilde{h}_r(|z|))\omega(0, |z|) \le (1 + \varepsilon(R))\omega(0, z)$$

and we are done. □

Lemma 3.6.9. *Given $p \in \mathbb{D}$ let $B_p \colon \mathbb{D} \to \mathbb{D}$ be the Blaschke product*

$$B_p(z) = z\frac{z - p}{1 - \bar{p}z}. \tag{3.20}$$

Let $w_0 \in \mathbb{D}$. Then we can order the solutions $w_1, w_2 \in \mathbb{D}$ of the equation $B_p(z) = w_0$ so that $\omega(0, w_1) = \omega(p, w_2)$ and $\omega(0, w_1) \to \omega(0, w_0)$ as $|p| \to 1$.

Proof. The points w_1, w_2 are solutions of the equation

$$z^2 + (\bar{p}w_0 - p)z - w_0 = 0; \tag{3.21}$$

in particular, we have $w_1 w_2 = -w_0$. But then

$$\frac{w_2 - p}{1 - \bar{p}w_2} = \frac{w_0}{w_2} = -w_1,$$

and thus

$$\omega(p, w_2) = \omega(0, -w_1) = \omega(0, w_1).$$

Solving explicitly (3.21), we find

$$2w_{1,2} = p - \bar{p}w_0 \pm \sqrt{p^2 + \bar{p}^2 w_0^2 + 2w_0(2 - |p|^2)}.$$

Then

$$\lim_{|p|\to 1} 2|w_{1,2}| = \lim_{|p|\to 1} 2|pw_{1,2}| = \lim_{|p|\to 1} \left|p^2 - w_0 \pm \sqrt{p^4 + w_0^2 + 2p^2 w_0}\right| = \begin{cases} 2, \\ 2|w_0|, \end{cases}$$

and we are done. □

We can now prove the following.

Proposition 3.6.10. *Let $\Omega \subset \mathbb{D}$ be a domain in the unit disk. Then Ω is a Bloch domain in $\mathbb{D}$ if and only if for every sequence $\{f_\nu\} \subset \mathrm{Hol}(\mathbb{D}, \Omega)$ the corresponding right iterated function system $\{R_\nu\}$ has only constant limit points.*

Proof. If Ω is a Bloch domain, the assertion follows from Corollary 3.6.5.

Conversely, assume that Ω is not a Bloch domain; we would like to find a sequence $\{f_\nu\} \subset \mathrm{Hol}(\mathbb{D}, \Omega)$ such that $\{R_\nu\}$ has a nonconstant limit point.

To do so, fix a sequence $\{\varepsilon_\nu\} \subset \mathbb{R}^+$ such that $\prod_0^\infty (1 + \varepsilon_\nu)^2 \le 2$ and choose two distinct points $p_{-1}, z_{-1} \in \Omega$ with $0 < \omega_\Omega(p_{-1}, z_{-1}) < 1/2$. By induction, we shall construct sequences $\{f_\nu\} \subset \mathrm{Hol}(\mathbb{D}, \Omega)$, $\{p_\nu\}, \{z_\nu\} \subset \Omega$, and $\{w_\nu\} \subset \mathbb{D}$ satisfying the following conditions for all $\nu \in \mathbb{N}$:

$$f_\nu(0) = f_\nu(p_\nu) = p_{\nu-1} \quad \text{and} \quad f_\nu(z_\nu) = f_\nu(w_\nu) = z_{\nu-1}, \tag{3.22}$$

$$\omega(p_\nu, z_\nu) = \omega(0, w_\nu) \le (1 + \varepsilon_\nu)\omega_\Omega(p_{\nu-1}, z_{\nu-1}), \tag{3.23}$$

$$\omega_\Omega(p_\nu, z_\nu) \le (1 + \varepsilon_\nu)\omega(p_\nu, z_\nu). \tag{3.24}$$

Let us start with the case $\nu = 0$. Let $\pi_0 \colon \mathbb{D} \to \Omega$ be a universal covering map such that $\pi_0(0) = p_{-1}$; by Proposition 1.7.3(iv), we can find $w \in \mathbb{D}$ such that $\pi_0(w) = z_{-1}$ and $\omega(0, w) = \omega_\Omega(p_{-1}, z_{-1})$. Given $p \in \mathbb{D}$, let $w_p^1, w_p^2 \in \mathbb{D}$ be the solutions of $B_p(z) = w$, where B_p is the Blaschke product given by (3.20). Set $f_p = \pi_0 \circ B_p$; then

$$f_p(0) = f_p(p) = p_{-1} \quad \text{and} \quad f_p(w_p^1) = f_p(w_p^2) = z_{-1}.$$

Since Ω is not a Bloch domain, we can find $p \in \Omega$ with $|p|$ arbitrarily close to 1 and $R(p; \Omega, \mathbb{D}) > 1$ arbitrarily large. Thus by Lemmas 3.6.9 and 3.6.8 we can find $\tilde{p} \in \Omega$ such that

$$\omega(\tilde{p}, w_{\tilde{p}}^2) = \omega(0, w_{\tilde{p}}^1) \le (1 + \varepsilon_0)\omega(0, w) = (1 + \varepsilon_0)\omega_\Omega(p_{-1}, z_{-1}) < 1$$

and

$$\omega_\Omega(\tilde{p}, w_{\tilde{p}}^2) \le (1 + \varepsilon_0)\omega(\tilde{p}.w_{\tilde{p}}^2) \le (1 + \varepsilon_0)^2\omega_\Omega(p_{-1}, z_{-1}).$$

Notice that $\omega(\tilde{p}, w_{\tilde{p}}^2) < 1$ implies $w_{\tilde{p}}^2 \in \Omega$ because $R(\tilde{p}; \Omega, \mathbb{D}) > 1$. Hence setting $f_0 = f_{\tilde{p}}$, $p_0 = \tilde{p}$, $z_0 = w_{\tilde{p}}^2$, and $w_0 = w_{\tilde{p}}^1$ we have satisfied (3.22)–(3.24) for $\nu = 0$.

Assume now we have constructed our sequences up to $\nu-1$. Starting from $p_{\nu-1}$ and $z_{\nu-1}$ instead of p_{-1} and z_{-1}, we can argue as before to get $f_\nu \in \mathrm{Hol}(\mathbb{D},\Omega)$, $p_\nu \in \Omega$ and $z_\nu, w_\nu \in \mathbb{D}$ such that (3.22) and (3.23) are satisfied. But then

$$\omega(p_\nu, z_\nu) \le (1+\varepsilon_\nu)^2 \omega(p_{\nu-1}, z_{\nu-1}) \le \cdots \le \prod_{\mu=0}^{\nu} (1+\varepsilon_\mu)^2 \omega_\Omega(p_{-1}, z_{-1}) < 1$$

that implies as before that $z_\nu \in \Omega$ and that we can apply Lemma 3.6.8 to get (3.24).

Let $\{R_\nu\}$ be the right iterated function system generated by $\{f_\nu\}$. By (3.22), we have

$$R_\nu(0) = p_{-1} \quad \text{and} \quad R_\nu(w_\nu) = z_{-1} \tag{3.25}$$

for all $\nu \in \mathbb{N}$. Choose a subsequence $\{w_{\nu_k}\}$ converging to a point $w_\infty \in \overline{\mathbb{D}}$; since $\omega(0, w_\nu) = \omega(p_\nu, z_\nu) < 1$ for all $\nu \in \mathbb{N}$, it follows that $w_\infty \in \mathbb{D}$. Now, up to a subsequence we can assume that $\{R_{\nu_k}\}$ converges to a function $R \in \mathrm{Hol}(\mathbb{D}, \mathbb{C})$; but (3.25) then says that $R(0) = p_{-1}$ and $R(w_\infty) = z_{-1}$, and thus R is not constant, as required. □

This latter characterization of Bloch domains does not hold in general hyperbolic Riemann surfaces, as shown in the following example.

Example 3.6.11. Let $X = \widehat{\mathbb{C}} \setminus E$, where $E \subset \widehat{\mathbb{C}}$ is a finite set of cardinality at least three, and $\Omega = X \setminus F$, where $F \subset X$ is another not empty finite set. By Theorem 1.7.25, every $f \in \mathrm{Hol}(X, \Omega)$ is the restriction of a rational map $\tilde{f} \in \mathrm{Hol}(\widehat{\mathbb{C}}, \widehat{\mathbb{C}})$. Since not constant rational maps are surjective, if $\tilde{f}$ were not constant it would satisfy $\tilde{f}(E) = E \cup F$, which is clearly impossible; therefore, $\tilde{f}$, and hence f, is constant.

So, $\mathrm{Hol}(X, \Omega)$ consists only of constant functions and thus every right iterated function system in $\mathrm{Hol}(X, \Omega)$ has only constant limit points, even though Ω is not a Bloch domain in X by Example 1.11.4.

It is interesting to notice that the situation for left iterated function systems is completely different. To prove this, we need the following local version of Proposition 3.6.4.

Lemma 3.6.12. *Let $\Omega \subset X$ be a proper subdomain of a hyperbolic Riemann surface X. Take $z_0 \in \Omega$ and $r > 0$. Then there exists $0 < \ell < 1$ depending only on z_0 and r such that*

$$\omega_X(f(z), f(w)) \le \ell\, \omega_X(z, w)$$

for all $f \in \mathrm{Hol}(X, \Omega)$ and all $z, w \in X$ such that $f(z), f(w) \in B_\Omega(z_0, r)$.

Proof. Take $z, w \in X$ such that $f(z), f(w) \in B_\Omega(z_0, r)$. Let $\sigma\colon [a, b] \to \Omega$ be a minimal geodesic with respect to κ_Ω connecting $f(z)$ with $f(w)$ (see Proposition 1.9.29); clearly, the image of σ is contained in $B = B_\Omega(z_0, 3r)$. We know that $\frac{\kappa_X}{\kappa_\Omega} < 1$ on Ω, by Corollary 1.9.24; since the closure of B is compact in Ω if we put $\ell = \sup_{\zeta \in B} \frac{\kappa_X}{\kappa_\Omega}(\zeta)$ we have $0 < \ell < 1$. Then

$$\omega_X(f(z),f(w)) \le \int_a^b \kappa_X(\sigma(t);\sigma'(t))\,dt \le \ell \int_a^b \kappa_\Omega(\sigma(t);\sigma'(t))\,dt$$
$$= \ell\omega_\Omega(f(z),f(w)) \le \ell\omega_X(z,w)$$

and we are done. □

Proposition 3.6.13. *Let $\Omega \subseteq X$ be a domain in a hyperbolic Riemann surface X. Assume that* $\mathrm{Hol}(X,X)$ *does not consists of constant functions only. Then every left iterated function system in* $\mathrm{Hol}(X,\Omega)$ *has only constant limit points if and only if $\Omega \ne X$. Moreover, the set of limit points can be any closed subset of $\overline{\Omega}$.*

Proof. If $\Omega = X$, taking a left iterated function system generated by a sequence $\{f_\nu\}$ with $f_\nu = \mathrm{id}_X$ eventually, we can get as limit point of $\{L_\nu\}$ any holomorphic self-map of X. So, assume $\Omega \ne X$; we would like to prove that any limit point of a left iterated function system in $\mathrm{Hol}(X,\Omega)$ is constant.

By contradiction, assume there is a sequence $\{f_\nu\} \subseteq \mathrm{Hol}(X,\Omega)$ such that the corresponding left iterated function system $\{L_\nu\}$ has a nonconstant limit point $F \in \mathrm{Hol}(X,X)$; let $\{L_{\nu_j}\}$ denote the subsequence converging to F. Clearly, $F(X) \subset \overline{\Omega}$. Since F is an open map, there exists $z_0 \in X$ such that $F(z_0) \in \Omega$. Put $B = B_X(z_0,1)$ and take $w \in B$. Since

$$\omega_\Omega(L_\nu(z_0),L_\nu(w)) \le \omega_X(z_0,w) < 1$$

for all $\nu \in \mathbb{N}$, it follows that there exists $j_0 \in \mathbb{N}$ so that $L_{\nu_j}(z_0), L_{\nu_j}(w) \in B_\Omega(F(z_0),2)$ for all $j \ge j_0$ and all $w \in B$. Lemma 3.6.12 provides then a $0 < \ell < 1$ such that for all $j > j_0$ and all $w \in B$ we have

$$\omega_X(L_{\nu_j}(z_0),L_{\nu_j}(w)) \le \ell\omega_X(L_{\nu_{j-1}}(z_0),L_{\nu_{j-1}}(w)) \le \cdots$$
$$\le \ell^{j-j_0}\omega_X(L_{\nu_{j_0}}(z_0),L_{\nu_{j_0}}(w)).$$

It follows that $\omega_X(L_{\nu_j}(z_0),L_{\nu_j}(w)) \to 0$ as $j \to +\infty$, and thus $F(w) = F(z_0)$ for all $w \in B$. So, F is constant in a neighborhood of z_0, and hence, being holomorphic, it is constant everywhere, contradiction.

Finally, let $C \subseteq \overline{\Omega}$ be a closed subset, and let $\{c_k\} \subset C$ be a countable dense subset. For each $k \in \mathbb{N}$, choose a sequence $\{p_{k,j}\} \subset \Omega$ such that $\omega_X(p_{k,j},c_k) < 1/(kj)$ for all $k,j \ge 1$. Then the left iteration function system generated by the countable sequence of constant maps $\{p_{k,j}\}$ has C as set of limit points and we are done. □

Notes to Section 3.6

Theorem 3.6.2 and Corollary 3.6.5 for right iterated function systems have been proved by Beardon, Carne, Minda, and Ng [52] in 2004. Preliminary versions of these results, mostly assuming that X is a simply connected domain in the plane and under the stronger condition that Ω is relatively compact in X, were obtained by Gill [161, 163], Baker and Rippon [36], and Lorentzen [271].

On the other hand, curiously enough left iterated function systems have been less studied. In this case preliminary versions of Theorem 3.6.2 and Corollary 3.6.5 were in [164]; the complete statement generalizes the work of Gill [166, 167] and can be found in [10].

Propositions 3.6.7 and 3.6.10 are due to Keen and Lakic [224]. In [223], the same authors show that if $\Omega \subset \mathbb{D}$ is not a Bloch domain and $f \in \mathrm{Hol}(\mathbb{D}, \overline{\Omega})$ then there is a right iterated function system in $\mathrm{Hol}(\mathbb{D}, \Omega)$ having f as accumulation point. Furthermore, if K is either empty or a compact subset of Ω and C is any closed subset of $\overline{\Omega}$ then there is a right iterated function system in $\mathrm{Hol}(\mathbb{D}, \Omega)$ whose full set of accumulation points consists of all the constants in C and all the holomorphic covering maps of $\mathbb{D}$ onto Ω sending 0 in K. In [225] (see also [226, Corollary 13.1.1]), they show that if Ω is a not relatively compact domain in $\mathbb{D}$ then for any finite subset $L \subset \Omega$ there is a right iterated function system in $\mathrm{Hol}(\mathbb{D}, \Omega)$ having as limit points exactly the constants in L. Finally, in [390] Tavakoli proved that if $\Omega \subset \mathbb{D}$ is not relatively compact in $\mathbb{D}$ and $c \in \partial\Omega$ then there is a right iterated function system in $\mathrm{Hol}(\mathbb{D}, \Omega)$ having c as accumulation point (a similar result was proved in [225] under the additional assumption that Ω is a Bloch domain with locally connected boundary).

Example 3.6.11 shows that when $X \neq \mathbb{D}$ there can exist a non-Bloch subdomain $\Omega \subset X$ such that every right iterated function system in $\mathrm{Hol}(X, \Omega)$ has only constant limit points. Keen and Lakic [224] have found a condition on the pair (X, Ω) strictly weaker than the Bloch condition still ensuring that every right iterated function system in $\mathrm{Hol}(X, \Omega)$ has only constant limit points. Given a domain $\Omega \subset X$ of a hyperbolic Riemann surface X, put

$$c_\Omega^X(z; u) = \sup\{\kappa_\Omega\big(f(z); df_z(u)\big) \mid f \in \mathrm{Hol}(X, \Omega)\}$$

and

$$\kappa_\Omega^X(z; u) = \inf\{\kappa_X(w; v) \mid f \in \mathrm{Hol}(X, \Omega),\ w \in X,\ v \in T_w X,\ f(w) = z,\ df_w(v) = u\}$$

for all $z \in \Omega$ and $u \in T_z\Omega$. It is easy to check that the quotient $c_\Omega^X / \kappa_\Omega^X$ depends only on $z \in \Omega$; we shall say that Ω is *cκ-Lipschitz* in X if

$$\sup_{z \in \Omega} \frac{c_\Omega^X}{\kappa_\Omega^X}(z) < 1.$$

Then if Ω is $c\kappa$-Lipschitz in X it turns out (see also [226, Theorem 12.2.1]) that every right iterated function system in $\mathrm{Hol}(X, \Omega)$ has only constant limit points. It is known that every Bloch domain is $c\kappa$-Lipschitz (see, e. g., [226, Theorem 10.4.1, Proposition 12.2.2]). Conversely, Example 3.6.11 gives an example of a $c\kappa$-Lipschitz domain, which is not Bloch; see [226, Example 10.1]. As far as I know, it is still not known whether there exists a non-$c\kappa$-Lipschitz domain $\Omega \subset X$ such that every right iterated function system in $\mathrm{Hol}(X, \Omega)$ has only constant limit points.

Proposition 3.6.13 is again due to Keen and Lakic [226, Theorem 11.2.2].

Several papers deal with random iteration of Möbius maps and automorphisms of the disk; see, e. g., [165, 272, 45, 44, 207, 208], and references therein.

3.7 Random iteration of small perturbations

In this section, we shall discuss the behavior of iterated function systems generated by functions close enough to a given self-map F; in particular, we would like to understand whether the dynamics of the iterated function systems mimics the dynamics of the sequence of iterates of F.

Recalling Theorem 3.3.2, we see that we have three cases to consider: when F has an attracting fixed point, when F is a periodic or pseudoperiodic automorphism, and when the sequence $\{F^k\}$ is compactly divergent.

In the first case, we have a fairly complete result.

Theorem 3.7.1. *Let X be a hyperbolic Riemann surface and let $F \in \mathrm{Hol}(X,X)$ be with an attracting fixed point $z_0 \in X$. Then:*
(i) *there exists a neighborhood $\mathcal{U}$ of F in $\mathrm{Hol}(X,X)$ such that every right iterated function system generated by $\{f_\nu\} \subset \mathcal{U}$ converges to a constant in X;*
(ii) *if $\{f_\nu\} \subset \mathrm{Hol}(X,X)$ is a sequence converging to F, then the left iterated function system generated by $\{f_\nu\}$ converges to z_0.*

Proof. Fix $r > 0$ and let $B = B_X(z_0,r)$. Since $\overline{B}$ is compact and F is not a self-covering of X (see Corollary 3.1.15), by the general Schwarz–Pick lemma (Theorem 1.7.6) there is $0 < k < 1$ such that $\omega_X(F(z),F(w)) \le k\omega_X(z,w)$ for all $z,w \in \overline{B}$. In particular, since $F(z_0) = z_0$, we have $F(\overline{B}) \subseteq \overline{B_X(z_0,kr)} \subset B$. Choose $t \in \mathbb{R}$ such that $kr < t < r$ and put

$$\mathcal{U} = \{h \in \mathrm{Hol}(X,X) \mid h(\overline{B}) \subset B_X(z_0,t)\};$$

clearly, $\mathcal{U}$ is a neighborhood of F. Notice that B is a hyperbolic Riemann surface and $B_X(z_0,t)$ is a Bloch domain in B, because it is relatively compact in B (Example 1.11.2). If $\{f_\nu\} \subset \mathcal{U}$, then, by Corollary 3.6.5(iii), the corresponding right iterated function system converges in B (and hence, by the Vitali Theorem 1.7.19, in X) to a constant contained in $\overline{B_X(z_0,t)}$, and thus (i) is proved.

For (ii), since $f_\nu \to F$ we have $f_\nu \in \mathcal{U}$ for all ν large enough; truncating L_ν by finitely many terms on the right and relabeling we can assume without loss of generality that $f_\nu(\overline{B}) \subset B$ for all $\nu \in \mathbb{N}$. Choose now $z \in \overline{B}$. We have $F^\nu(z), L_\nu(z) \in \overline{B}$ for all $\nu \in \mathbb{N}$; hence

$$\begin{aligned}\omega_X(L_\nu(z),F^\nu(z)) &\le \omega_X(L_\nu(z),F(L_{\nu-1}(z))) + \omega_X(F(L_{\nu-1}(z)),F^\nu(z))\\ &\le \sup_{w\in\overline{B}} \omega_X(f_\nu(w),F(w)) + k\omega_X(L_{\nu-1}(z),F^{\nu-1}(z)).\end{aligned}$$

Repeating this argument, we get by induction

$$\omega_X(L_\nu(z),F^\nu(z)) \le \sum_{j=0}^{\nu} k^j \sup_{w\in\overline{B}} \omega_X(f_{\nu-j}(w),F(w)).$$

Fix $\varepsilon > 0$. Since $0 < k < 1$, and by assumption, $f_\nu \to F$ uniformly on compact subsets, we can find ν_0 large enough so that $k^j < \frac{(1-k)\varepsilon}{4r}$ and $\sup_{w\in\overline{B}} \omega_X(f_j(w),F(w)) < \frac{(1-k)\varepsilon}{2}$ as soon as $j \ge \nu_0/2$. Therefore, if $\nu \ge \nu_0$ we have

$$\omega_X(L_\nu(z),F^\nu(z)) \le \sum_{j=0}^{\nu/2} k^j \sup_{w\in\overline{B}} \omega_X(f_{\nu-j}(w),F(w)) + \sum_{j=\nu/2+1}^{\nu} k^j \sup_{w\in\overline{B}} \omega_X(f_{\nu-j}(w),F(w))$$

$$< \left[\frac{(1-k)\varepsilon}{2} + 2rk^{\nu/2} \right] \sum_{j=0}^{\nu/2} k^j < 2\frac{(1-k)\varepsilon}{2} \frac{1}{1-k} = \varepsilon$$

and so $\omega_X(L_\nu(z), F^\nu(z)) \to 0$ uniformly on $\overline{B}$ as $\nu \to +\infty$. Since Theorem 3.3.2 implies $F^\nu \to z_0$, it follows that $L_\nu \to z_0$ on $\overline{B}$, and hence again by the Vitali theorem, on X, as claimed. □

Remark 3.7.2. In Theorem 3.7.1(i) by changing f_0 one can change the constant limit of the right iterated function system $\{R_\nu\}$.

The next example shows that in Theorem 3.7.1(ii) we cannot replace the hypothesis $f_\nu \to F$ by the hypothesis $\{f_\nu\} \subset \mathcal{U}$ for any neighborhood $\mathcal{U}$ of F.

Example 3.7.3. Let $F \in \mathrm{Hol}(\mathbb{D}, \mathbb{D})$ be given by $F(z) = \frac{1}{2}z$, and let $\mathcal{U} \subset \mathrm{Hol}(\mathbb{D}, \mathbb{D})$ be a neighborhood of F. Given $\delta > 0$ and $\theta \in \mathbb{R}$, put $f_{\delta,\theta}(z) = \frac{1}{2}z + \delta e^{i\theta}$. Clearly, $f_{\delta,\theta} \in \mathrm{Hol}(\mathbb{D}, \mathbb{D})$ as soon as $\delta < 1/2$.

We claim that we can choose δ small enough so that $f_{\delta,\theta} \in \mathcal{U}$ for all $\theta \in \mathbb{R}$. Indeed, fix a compact subset $K \subset \mathbb{D}$ and let $V \subset \mathbb{D}$ be an open neighborhood of $F(K)$ so that $F \in \mathcal{U}(K, V) = \{h \in \mathrm{Hol}(\mathbb{D}, \mathbb{D}) \mid h(K) \subset V\}$. Since $F(K) \cap \partial V = \emptyset$, there is a $\delta_0 > 0$ such that $d(w, F(K)) < \delta_0$ implies $w \in V$. As a consequence, if $\delta \le \delta_0$ we have $f_{\delta,\theta}(K) \subset V$ for all $\theta \in \mathbb{R}$. Since $\mathcal{U}$ contains a finite intersection of sets of the form $\mathcal{U}(K, V)$, the claim follows.

Given $\{\theta_\nu\} \subset \mathbb{R}$, it is easy to check by induction that the left iterated function system $\{L_\nu\}$ generated by $\{f_{\delta,\theta_\nu}\}$ is given by

$$L_\nu(z) = \frac{1}{2^{\nu+1}} z + \delta \sum_{j=0}^{\nu} \frac{1}{2^j} e^{i\theta_{\nu-j}}.$$

For instance, taking $e^{i\theta_j} = (-1)^j$ we get

$$L_\nu(0) = \delta(-1)^\nu \sum_{j=0}^{\nu} \left(-\frac{1}{2}\right)^j$$

that does not converge when $\nu \to +\infty$, and thus $\{L_\nu\}$ cannot be convergent.

Let us now consider the case when $\{F^\nu\}$ is compactly divergent. In this case, if $f_\nu \to F$ fast enough then the dynamics of the left iterated function system generated by $\{f_\nu\}$ is dictated by the dynamics of $\{F^\nu\}$.

Theorem 3.7.4. *Let X be a hyperbolic Riemann surface and let $F \in \mathrm{Hol}(X, X)$ be such that the sequence of iterates $\{F^\nu\}$ is compactly divergent. Then we can find a sequence of neighborhoods $\mathcal{U}_\nu \subset \mathrm{Hol}(X, X)$ of F such that if $f_\nu \in \mathcal{U}_\nu$ for all $\nu \in \mathbb{N}$ then the left iterated function system $\{L_\nu\}$ generated by $\{f_\nu\}$ is compactly divergent. Furthermore, if $X \subset \widehat{X}$ is a hyperbolic domain and $\{F^\nu\}$ converges to a point $\tau \in \partial X$ then $\{L_\nu\}$ converges to τ.*

Proof. Fix a reference point $z_0 \in X$. For $\nu \in \mathbb{N}$, set

$$D_\nu = \overline{B_X(z_0, 1 + \omega_X(F^\nu(z_0), z_0))}.$$

Since $\omega_X(F^\nu(z_0), z_0) \to +\infty$ as $\nu \to +\infty$, we have $X = \bigcup_{\nu\in\mathbb{N}} D_\nu$. Given $\nu \in \mathbb{N}$, choose $z_1, \dots, z_r \in D_\nu$ such that

$$F(D_\nu) \subseteq \bigcup_{j=1}^{r} B_X(F(z_j), (3 \cdot 2^{\nu+1})^{-1}).$$

Put $B_j = B_X(F(z_j), (3 \cdot 2^{\nu+1})^{-1})$, $K_j = D_\nu \cap F^{-1}(\overline{B_j})$ and $\widetilde{B}_j = B_X(F(z_j), 2^{-\nu-2})$; in particular, $D_\nu = \bigcup_{j=1}^{r} K_j$ and if $z \in K_j$ then $\omega_X(F(z), F(z_j)) \leq (3 \cdot 2^{\nu+1})^{-1}$. Finally, put

$$\mathcal{U}_\nu = \bigcap_{j=1}^{r} \{h \in \operatorname{Hol}(X, X) \mid h(K_j) \subset \widetilde{B}_j\};$$

this is a neighborhood of F because $F(K_j) \subseteq \overline{B_j} \subset \widetilde{B}_j$ for all $j = 1, \dots, r$ by construction.

Take $h \in \mathcal{U}_\nu$. If $z \in D_\nu$ we must have $z \in K_j$ for some $j = 1, \dots, r$; then

$$\begin{aligned} \omega_X(h(z), F(z)) &\leq \omega_X(h(z), F(z_j)) + \omega_X(F(z_j), F(z)) \\ &< \frac{1}{2^{\nu+2}} + \frac{1}{3 \cdot 2^{\nu+1}} < \frac{1}{2^{\nu+1}}. \end{aligned} \tag{3.26}$$

Now take $\{f_\nu\} \subset \operatorname{Hol}(X, X)$ with $f_\nu \in \mathcal{U}_\nu$ for all $\nu \in \mathbb{N}$; notice that (3.26) implies that $f_\nu \to F$. We will prove by induction that

$$\omega_X(L_\nu(z_0), F^{\nu+1}(z_0)) < 1 - \frac{1}{2^{\nu+1}}$$

for all $\nu \in \mathbb{N}$. For $\nu = 0$, it follows immediately from (3.26). Assume it holds for $\nu - 1$; then

$$\begin{aligned} \omega_X(L_\nu(z_0), F^{\nu+1}(z_0)) &\leq \omega_X(L_\nu(z_0), F(L_{\nu-1}(z_0))) + \omega_X(F(L_{\nu-1}(z_0)), F^{\nu+1}(z_0)) \\ &\leq \omega_X(L_\nu(z_0), F(L_{\nu-1}(z_0))) + \omega_X(L_{\nu-1}(z_0), F^\nu(z_0)) \\ &< \omega_X(L_\nu(z_0), F(L_{\nu-1}(z_0))) + 1 - \frac{1}{2^\nu}. \end{aligned}$$

Now, since

$$\begin{aligned} \omega_X(L_{\nu-1}(z_0), z_0) &\leq \omega_X(L_{\nu-1}(z_0), F^\nu(z_0)) + \omega_X(F^\nu(z_0), z_0) \\ &< 1 + \omega_X(F^\nu(z_0), z_0), \end{aligned}$$

we have $L_{\nu-1}(z_0) \in D_\nu$. Therefore, since $f_\nu \in \mathcal{U}_\nu$ we can use (3.26) to get

$$\omega_X(L_\nu(z_0), F(L_{\nu-1}(z_0))) = \omega_X(f_\nu(L_{\nu-1}(z_0)), F(L_{\nu-1}(z_0))) < \frac{1}{2^{\nu+1}},$$

and thus

$$\omega_X(L_\nu(z_0),\, F^{\nu+1}(z_0)) < \frac{1}{2^{\nu+1}} + 1 - \frac{1}{2^\nu} = 1 - \frac{1}{2^{\nu+1}}$$

as claimed.

As a consequence for any $z \in X$, we have

$$\begin{aligned}\omega_X(L_\nu(z),\, z_0) &\geq \omega_X(F^{\nu+1}(z_0), z_0) - \omega_X(F^{\nu+1}(z_0), L_\nu(z_0)) - \omega_X(L_\nu(z_0), L_\nu(z))\\ &\geq \omega_X(F^{\nu+1}(z_0), z_0) - 1 - \omega_X(z_0, z) \to +\infty\end{aligned}$$

as $\nu \to +\infty$, and thus $\{L_\nu\}$ is compactly divergent.

Finally, assume that $X \subset \widehat{X}$ is a hyperbolic domain and that $F^\nu \to \tau \in \partial X$. Then since $\omega_X(L_\nu(z_0),\, F^{\nu+1}(z_0)) < 1$ for all $\nu \in \mathbb{N}$ we can apply Proposition 1.7.27 to get $L_\nu(z_0) \to \tau$. Moreover, for any $z \in D$ we also have $\omega_X(L_\nu(z), L_\nu(z_0)) \leq \omega_X(z, z_0)$; so another application of Proposition 1.7.27 yields $L_\nu(z) \to \tau$ for all $z \in X$ and we are done. □

Surprisingly enough, if the convergence of f_ν to F is too slow then the left iterated function system might not be convergent, as the next example shows.

Example 3.7.5. Let $F \in \mathrm{Hol}(\mathbb{H}^+, \mathbb{H}^+)$ be given by $F(w) = w - 1$. For $\nu \in \mathbb{N}$, put

$$\varphi_\nu(w) = \frac{\nu w - 1}{w + \nu} = \frac{w\cos\theta_\nu - \sin\theta_\nu}{w\sin\theta_\nu + \cos\theta_\nu},$$

where $\theta_0 = \pi/2$ and $\theta_\nu = \arctan\frac{1}{\nu}$ for $\nu \geq 1$. Recalling (1.22) we see that each φ_ν is an elliptic automorphism of $\mathbb{H}^+$ fixing i; furthermore, $\varphi_\nu \to \mathrm{id}_{\mathbb{H}^+}$ as $\nu \to +\infty$.

Now, let $g_\nu = \varphi_\nu \circ F \circ \varphi_\nu^{-1}$. By construction, each g_ν is a parabolic automorphism of $\mathbb{H}^+$; furthermore, we have

$$g_\nu(w) = \frac{(\nu^2 + \nu + 1)w - \nu^2}{w + \nu^2 - \nu + 1}$$

and a quick computation shows that $g_\nu(\nu) = \nu$. In particular, for each $w \in \mathbb{H}^+$ as soon as k is sufficiently large we have $g_\nu^k(w)$ arbitrarily close to ν. Moreover, $g_\nu \to F$ as $\nu \to +\infty$.

We shall now build by induction a sequence $\{f_\nu\} \subset \mathrm{Aut}(\mathbb{H}^+)$ generating a left iterated function system $\{L_\nu\}$ and an increasing sequence $\{\nu_j\} \subset \mathbb{N}$ with the following properties:

(a) $\nu_0 = 0$ and $\nu_1 = 1$;
(b) $f_0 = g_0$ and $f_1 = F$;
(c) $\nu_{2j} \geq \nu_{2j-1} + j$ for all $j \geq 1$;

(d) $f_{\nu_{2j-1}+1} = \cdots = f_{\nu_{2j}-j} = g_j$ and $f_{\nu_{2j}-j+1} = \cdots = f_{\nu_{2j+1}} = F$ for all $j \geq 1$;
(e) $|L_{\nu_{2j}}(i)| < 1/2^j$ and $|L_{\nu_{2j+1}}(i)| > j$ for all $j \in \mathbb{N}$.

Since $|g_0(i)| < 1$ and $|F \circ g_0(i)| > 0$, condition (e) is satisfied for $j = 0$. Now choose $n_1 \in \mathbb{N}$ such that $|g_1^{n_1}(L_{\nu_1}(i)) - 1| < 1/2$; in particular, $|(F \circ g_1^{n_1})(L_{\nu_1}(i))| < 1/2$. Choose now $m_1 \in \mathbb{N}$ so that $|(F^{m_1+1} \circ g_1^{n_1})(L_{\nu_1}(i))| > 1$; putting $\nu_2 = \nu_1 + n_1 + 1$, $\nu_3 = \nu_2 + m_1$, $f_{\nu_1+1} = \cdots = f_{\nu_2-1} = g_1$, and $f_{\nu_2} = \cdots = f_{\nu_3} = F$ we get $|L_{\nu_2}(i)| < 1/2$ and $|L_{\nu_3}(i)| > 1$, i. e., conditions (c)–(e) are satisfied for $j = 1$.

Now, given $j \geq 1$ assume by induction that we have found $\nu_0 < \cdots < \nu_{2j-1}$ and $f_0, \ldots, f_{\nu_{2j-1}} \in \mathrm{Aut}(\mathbb{H}^+)$ satisfying (a)–(e). Choose $n_j \in \mathbb{N}$ such that $|g_j^{n_j}(L_{\nu_{2j-1}}(i)) - j| < 1/2^j$; in particular, $|(F^j \circ g_1^{n_j})(L_{\nu_{2j-1}}(i))| < 1/2^j$. Choose now $m_j \in \mathbb{N}$ so that $|(F^{m_j+j} \circ g_j^{n_j})(L_{\nu_{2j-1}}(i))| > j$; putting $\nu_{2j} = \nu_{2j-1} + n_j + j$ and $\nu_{2j+1} = \nu_{2j} + m_j$ and choosing $f_{\nu_{2j-1}+1}, \ldots, f_{\nu_{2j+1}}$ as in (d) we get $|L_{\nu_{2j}}(i)| < 1/2^j$ and $|L_{\nu_{2j+1}}(i)| > j$, as required.

In this way, we have constructed a sequence $\{f_\nu\} \subset \mathrm{Aut}(\mathbb{H}^+)$ converging (very slowly) to F generating a left iterated function system with $L_{\nu_{2j}}(i) \to 0$ and $L_{\nu_{2j+1}}(i) \to \infty$ as $j \to +\infty$; in particular $\{L_\nu\}$ does not converge.

Finally, there is no hope to get a version of Theorem 3.7.4 for right iterated function systems.

Example 3.7.6. Let $F \in \mathrm{Hol}(\mathbb{H}^+, \mathbb{H}^+)$ be given by $F(w) = w + 1$, and define $\{f_\nu\} \subset \mathrm{Hol}(\mathbb{H}^+, \mathbb{H}^+)$ by setting $f_0(w) = i + e^{2\pi i z}$ and $f_\nu = F$ for $\nu \geq 1$. Then $f_\nu \to F$ in the fastest possible way but $R_\nu = f_0$ for all $\nu \in \mathbb{N}$, and thus $\{R_\nu\}$ is not compactly divergent.

We are left with the case when F is a periodic or pseudoperiodic automorphism of X. If the fundamental group of X is not Abelian, Theorem 2.6.2 implies that F is isolated in $\mathrm{Hol}(X, X)$; therefore, the study of random iteration of functions sufficiently close to F reduces to the study of the dynamics of F, which is trivial.

If the fundamental group of X is Abelian, we know that X is biholomorphic either to $\mathbb{D}$ or to $\mathbb{D}^*$ or to an annulus $A(r, 1)$ with $0 < r < 1$ (Theorem 1.6.33). If X is biholomorphic to $\mathbb{D}^*$, then every holomorphic self-map of $\mathbb{D}^*$ extends to a holomorphic self-map of $\mathbb{D}$, and thus random iteration on $\mathbb{D}^*$ reduces to random iteration on $\mathbb{D}$.

If X is biholomorphic to an annulus $A(r, 1)$, then Corollary 2.6.7 says that $\mathrm{Aut}(X)$ is open in $\mathrm{Hol}(X, X)$; in particular, maps sufficiently close to F are automorphisms of X. By Proposition 1.6.38(iv), we know that $\mathrm{Aut}(A(r, 1))$ has two connected components, $A_1 = \{\varphi_{\theta,1} \mid \theta \in \mathbb{R}\}$ and $A_{-1} = \{\varphi_{\theta,-1} \mid \theta \in \mathbb{R}\}$, where $\varphi_{\theta,1}(z) = e^{i\theta} z$ and $\varphi_{\theta,-1}(z) = e^{i\theta} r z^{-1}$. If $F \in A_1$, then every holomorphic self-map of $A(r, 1)$ sufficiently close to F belongs to A_1; since $A_1 \subset \mathrm{Aut}(\mathbb{D})$, in this case too random iteration of holomorphic self-maps close to F is reduced to random iteration on $\mathbb{D}$. If instead $F \in A_{-1}$, then every holomorphic self-map of $A(r, 1)$ sufficiently close to F belongs to A_{-1}. Since it is easy to check that

$$\varphi_{\theta,-1} \circ \varphi_{\eta,-1} = \varphi_{\theta-\eta,1}, \quad \varphi_{\theta,1} \circ \varphi_{\eta,-1} = \varphi_{\theta+\eta,-1} \quad \text{and} \quad \varphi_{\theta,-1} \circ \varphi_{\eta,1} = \varphi_{\theta-\eta,-1},$$

we see that every iterated function system generated by self-maps close enough to F splits in the union of an iterated function system contained in A_1, obtained considering an even number of maps, and of the composition of $\varphi_{0,-1}$ with an iterated function system again contained in A_1, obtained considering an odd number of maps. Thus in this case we are again led to the study of random iteration in $\mathbb{D}$.

Summing up, when F is a periodic or pseudoperiodic automorphism of X for our aims we can safely assume that $X = \mathbb{D}$.

So, let F be an elliptic automorphism of $\mathbb{D}$, and $\{f_\nu\} \subset \mathrm{Hol}(\mathbb{D}, \mathbb{D})$ a sequence of holomorphic self-maps converging to F. It turns out that, without more hypotheses, we can obtain iterated function systems with very different behaviors.

Example 3.7.7. Take $F = \mathrm{id}_{\mathbb{D}}$ and $f_\nu = (1 - \frac{1}{\nu+1})\,\mathrm{id}_{\mathbb{D}}$ for $\nu \in \mathbb{N}$. Then we clearly have $f_\nu \to \mathrm{id}_{\mathbb{D}}$ and $L_\nu, R_\nu \to 0$ as $\nu \to +\infty$.

Example 3.7.8. Take $F = \mathrm{id}_{\mathbb{D}}$ and $f_\nu = e^{2\pi i/(\nu+1)}\,\mathrm{id}_{\mathbb{D}}$ for $\nu \in \mathbb{N}$. Then we again have $f_\nu \to \mathrm{id}_{\mathbb{D}}$ as $\nu \to +\infty$, but this time

$$L_\nu(z) = R_\nu(z) = e^{2\pi i h_\nu} z,$$

where $h_\nu = 1 + \frac{1}{2} + \cdots + \frac{1}{\nu+1}$; we claim that every rotation can be obtained as limit point of $\{L_\nu\} = \{R_\nu\}$. Indeed, take $\theta \in (0, 1]$ and fix $\varepsilon > 0$. Choose $\nu_0 \in \mathbb{N}$ so that $\frac{1}{\nu_0+1} < \varepsilon$, and let $k \in \mathbb{N}$ be such that $\theta + k > h_{\nu_0}$. Since $h_\nu \to +\infty$, we can find $\nu \geq \nu_0$ such that $h_\nu < \theta + k \leq h_{\nu+1}$. It follows that $0 < \theta + k - h_\nu \leq h_{\nu+1} - h_\nu = \frac{1}{\nu+2} < \varepsilon$, and hence

$$|e^{2\pi i\theta} z - L_\nu(z)| = |e^{2\pi i\theta} z - R_\nu(z)| = |e^{2\pi i(\theta+k)} - e^{2\pi i h_\nu}||z| \leq c|\theta + k - h_\nu| < c\varepsilon$$

for a suitable constant $c > 0$. Since we can obtain this with ν arbitrarily large (it suffices to take ν_0 large), we have proved that the rotation $e^{2\pi i\theta} z$ is an accumulation point of $\{L_\nu\} = \{R_\nu\}$, as claimed.

Both of these examples used in an essential way the fact that $f_\nu \to F$ slowly, as measured by the fact that the harmonic series diverges. If instead the convergence is fast enough, we can control the behavior of the iterated function systems.

Theorem 3.7.9. *Let $F \in \mathrm{Aut}(\mathbb{D})$ be $\mathrm{id}_{\mathbb{D}}$ or, more generally, an elliptic automorphism of $\mathbb{D}$. Let $\{f_\nu\} \subset \mathrm{Hol}(\mathbb{D}, \mathbb{D})$ be a sequence of nonconstant holomorphic self-maps of $\mathbb{D}$ for which*

$$\sum_{\nu=0}^{\infty} \omega(f_\nu(a), F(a)) < +\infty \quad \text{and} \quad \sum_{\nu=0}^{\infty} \omega(f_\nu(b), F(b)) < +\infty \tag{3.27}$$

for two distinct points $a, b \in \mathbb{D}$. Then $f_\nu \to F$ as $\nu \to +\infty$ and the sequences $\{F^{-\nu} \circ L_\nu\}$ and $\{R_\nu \circ F^{-\nu}\}$ converge to nonconstant holomorphic self-maps of $\mathbb{D}$.

Proof. An immediate consequence of (3.27) is $f_\nu(a) \to F(a)$ and $f_\nu(b) \to F(b)$; Corollary 1.5.13 then implies that $f_\nu \to F$ uniformly on compact sets, and the first assertion is proved.

Let $K \subset \mathbb{D}$ be a closed Poincaré ball centred in the fixed point of F (or in any point of $\mathbb{D}$ if $F = \mathrm{id}_{\mathbb{D}}$) and containing both a and b; denote by $R > 0$ the Poincaré radius of K. Then Theorem 1.5.12 and (3.27) imply that

$$\sum_{\nu=0}^{+\infty} \sup_{z\in K} \omega(f_\nu(z), F(z)) < +\infty.$$

Since it is easy to see that it suffices to prove the theorem for the iterated function systems generated by the truncated sequence $\{f_\nu\}_{\nu\geq N}$ for a fixed $N \geq 0$, without loss of generality we can assume that

$$\sum_{\nu=0}^{+\infty} \sup_{z\in K} \omega(f_\nu(z), F(z)) < \frac{1}{3}\omega(a, b). \tag{3.28}$$

Let us first study the sequence $\{F^{-\nu} \circ L_\nu\}$. Fix $z \in K$, and put $z_\nu = F^\nu(z)$; clearly, $z_\nu \in K$ for all $\nu \in \mathbb{N}$. We have

$$\begin{aligned}\omega(L_\nu(z), F^{\nu+1}(z)) &\leq \omega(f_\nu \circ \cdots \circ f_0(z), f_\nu \circ \cdots \circ f_1 \circ F(z)) \\ &\quad + \omega(f_\nu \circ \cdots \circ f_1 \circ F(z), F^\nu \circ F(z)) \\ &\leq \omega(f_0(z), F(z)) + \omega(f_\nu \circ \cdots \circ f_1(z_1), F^\nu(z_1)).\end{aligned}$$

Repeating this argument, we get

$$\omega(L_\nu(z), F^{\nu+1}(z)) \leq \sum_{j=0}^{\nu} \omega(f_j(z_j), F(z_j)) < \frac{1}{3}\omega(a, b),$$

where the last inequality follows from (3.28) and holds for all $z \in K$, including $z = a$ or b. We also have

$$\begin{aligned}\omega(L_\nu(z), a) &\leq \omega(L_\nu(z), L_\nu(a)) + \omega(L_\nu(a), F^{\nu+1}(a)) + \omega(F^{\nu+1}(a), a) \\ &< \omega(z, a) + \frac{1}{3}\omega(a, b) + \omega(F^{\nu+1}(a), a) < 5R;\end{aligned}$$

analogously we can prove $\omega(L_\nu(z), b) < 5R$ for all $z \in K$ and $\nu \in \mathbb{N}$. Applying Theorem 1.5.12 to the point $L_{\nu-1}(z)$, where $L_{-1} = \mathrm{id}_{\mathbb{D}}$, we get

$$\omega(L_\nu(z), F \circ L_{\nu-1}(z)) \leq \Lambda[\omega(f_\nu(a), F(a)) + \omega(f_\nu(b), F(b))],$$

where

$$\Lambda = \frac{\exp[\omega(L_{\nu-1}(z), a) + \omega(a, b) + \omega(b, L_{\nu-1}(z))]}{\omega(a, b)} \leq \frac{\exp(12R)}{\omega(a, b)}.$$

As a consequence, we get

$$\sum_{\nu=0}^{+\infty}\omega(F^{-\nu-1}\circ L_\nu(z), F^{-\nu}\circ L_{\nu-1}(z)) = \sum_{\nu=0}^{+\infty}\omega(L_\nu(z), F\circ L_{\nu-1}(z)) \le \frac{2}{3}\exp(12R)$$

for all $z\in K$. So $\{F^{-\nu-1}\circ L_\nu(z)\}$ is a Cauchy sequence uniformly in K; since the radius of K can be chosen arbitrarily large, it follows that the sequence $\{F^{-\nu-1}\circ L_\nu\}$ converges uniformly on compact subsets to a holomorphic map $\Phi\colon \mathbb{D}\to\overline{\mathbb{D}}$. Since we have

$$\omega(F^{-\nu-1}\circ L_\nu(a), a) = \omega(L_\nu(a), F^{\nu+1}(a)) < \frac{1}{3}\omega(a,b),$$

we immediately see that $\Phi(\mathbb{D})\subseteq\mathbb{D}$. Furthermore,

$$\begin{aligned}\omega(F^{-\nu-1}\circ L_\nu(a),\, F^{-\nu-1}\circ L_\nu(b)) &\ge \omega(a,b)-\omega(F^{-\nu-1}\circ L_\nu(a),a)-\omega(F^{-\nu-1}\circ L_\nu(b),b)\\ &\ge \omega(a,b)-\frac{2}{3}\omega(a,b)=\frac{1}{3}\omega(a,b);\end{aligned}$$

thus Φ is not constant and we finally get that $F^{-\nu}\circ L_\nu$ converges to $F\circ\Phi$.

Let us now consider the sequence $\{R_\nu\circ F^{-\nu}\}$. Fix $z\in K$ and put $z_\nu = F^{-\nu-1}(z)$. Then $z_\nu\in K$ for all $\nu\in\mathbb{N}$; moreover, putting $R_{-1}=\mathrm{id}_{\mathbb{D}}$, we have

$$\omega(R_\nu\circ F^{-\nu-1}(z), R_{\nu-1}\circ F^{-\nu}(z)) \le \omega(f_\nu(z_\nu), F(z_\nu)),$$

and hence

$$\sum_{\nu=0}^{+\infty}\omega(R_\nu\circ F^{-\nu-1}(z), R_{\nu-1}\circ F^{-\nu}(z)) \le \frac{1}{3}\omega(a,b)$$

for all $z\in K$. As before, it follows that $\{R_\nu\circ F^{-\nu-1}(z)\}$ is a Cauchy sequence uniformly in K and again, since we can choose K arbitrarily large, we get that $\{R_\nu\circ F^{-\nu-1}\}$ converges uniformly on compact subsets to a holomorphic map $\Psi\colon\mathbb{D}\to\overline{\mathbb{D}}$. We must show that Ψ is not constant. First of all, putting $a_j = F^{-j-1}(a)$ for $j\in\mathbb{Z}$, we have

$$\begin{aligned}\omega(R_\nu\circ F^{-\nu-1}(a), a) &\le \sum_{j=0}^{\nu}\omega(R_j(a_j), R_{j-1}(a_{j-1}))\\ &\le \sum_{j=0}^{\nu}\omega(f_j(a_j), F(a_j)) < \frac{1}{3}\omega(a,b),\end{aligned}$$

because $a_j\in K$ for all $j\in\mathbb{Z}$. In particular, $\Psi(a)\in\mathbb{D}$, and so $\Psi(\mathbb{D})\subseteq\mathbb{D}$. Clearly, we have a similar estimate with b instead of a; hence as before

$$\begin{aligned}\omega(R_\nu\circ F^{-\nu-1}(a),\, R_\nu\circ F^{-\nu-1}(b)) &\ge \omega(a,b)-\omega(R_\nu\circ F^{-\nu-1}(a),a)-\omega(R_\nu\circ F^{-\nu-1}(b),b)\\ &\ge \omega(a,b)-\frac{2}{3}\omega(a,b)=\frac{1}{3}\omega(a,b),\end{aligned}$$

and thus Ψ is not constant. Finally, $\{R_\nu \circ F^{-\nu}\}$ converges to $\Psi \circ F$ and we are done. □

Notes to Section 3.7

Theorems 3.7.1, 3.7.4, and 3.7.9 were originally proved for the case $X = \mathbb{D}$ in [116]; the version we presented comes from [10], where also Example 3.7.5 can be found. Preliminary versions of Theorem 3.7.1(i) are in [162] and in [44, Theorem 4.7]; in these latter results, the authors assume that $F(X)$ is relatively compact in X, but Proposition 3.3.4 then implies that F (is not a self-covering of X and) has a fixed point in X, as required in Theorem 3.7.1. See also [339] for other related results on R_ν when $f_\nu \to f$.

A different approach to random iteration involves semigroups of holomorphic self-maps under composition. Let $D \subset \mathbb{C}$ be a hyperbolic domain. Given a family $\mathcal{F} \subseteq \mathrm{Hol}(D, D)$, we shall denote by $\mathcal{F}_0$ the semigroup generated by all finite compositions of elements of $\mathcal{F}$. Assume that the closure of $\mathcal{F}_0$ in $\mathrm{Hol}(D, \mathbb{C})$ does not contain either the identity of D or a constant in ∂D. Then every right iterated function system in $\mathcal{F}$ converges to a constant in D. This has been proved in [36] when $D = \mathbb{D}$; the general statement is due to Beardon [46]. More generally, in the same paper Beardon has shown that if $\mathcal{F}_0 \cap \partial D = \emptyset$ and $\{R_\nu\}$ is a right iterated function system in $\mathcal{F}$ then either $R_\nu \to c \in D$ or $\{R_\nu\}$ has a nonconstant limit function $\varphi \in \mathrm{Hol}(D, D)$ and all accumulation points of $\{R_\nu\}$ are of the form $\varphi \circ h$ with $h \in \mathrm{Aut}(D)$. Furthermore, in [254] it is shown that $R_\nu \to c \in D$ if and only if either at least one of the functions f_ν is constant or the series

$$\sum_{\nu=0}^{\infty} \left(1 - f_\nu^*(z)\right)$$

diverges at every point of D, where

$$f^*(z) = \lim_{w \to z} \frac{\omega_D(f(z), f(w))}{\omega_D(z, w)}.$$

In [35, 42], it is proved that if $D \subset \mathbb{C}$ is a bounded convex domain and $\{f_\nu\} \subset \mathrm{Hol}(D, D)$ is such that $|f_\nu'| \le 1$ for all $\nu \in \mathbb{N}$ but there exists $z_0 \in D$ such that $\sup_\nu |f_\nu'(z_0)| < 1$ then the right iterated function system $\{R_\nu\}$ generated by $\{f_\nu\}$ converges uniformly to a constant in $\overline{D}$. Furthermore, if D is a disk (or, more generally, a convex domain with the curvature of ∂D uniformly bounded away from both 0 and ∞) then $\mathrm{diam}(R_\nu(D)) = O(\nu^{-1/2})$; see also [351].

Another result by Beardon (see [45]) is the following: if $\{f_\nu\} \subset \mathrm{Hol}(\mathbb{D}, \mathbb{D})$, all f_ν have the same Wolff point $\sigma \in \partial D$ and $\prod_{\nu=0}^{\infty} f_\nu'(\sigma) = 0$ then $R_\nu \to \sigma$. Conversely, if $\{d_\nu\} \subset (0, 1]$ is such that $\prod_{\nu=0}^{\infty} d_\nu > 0$ then there exists $\{f_\nu\} \subset \mathrm{Hol}(\mathbb{D}, \mathbb{D})$ such that each f_ν has σ as Wolff point with $f_\nu'(\sigma) = d_\nu$ and $\{R_\nu\}$ does not converge to σ.

The recent papers [58, 59] contain a detailed study of the asymptotic behavior of left iterated function systems generated by $\{f_\nu\} \subset \mathrm{Hol}(\mathbb{D}, \mathbb{D})$. In particular, they show that exactly one of the following assertion holds (where f_ν^h denotes the hyperbolic derivative of f_ν):

(a) for all $z, w \in \mathbb{D}$, we have $\lim_{\nu \to +\infty} \omega(L_\nu(z), L_\nu(w)) = 0$, and this can happen if and only if we have $\sum_\nu (1 - |f_\nu^h(L_{\nu-1}(z))|) = +\infty$ for one (and hence all) $z \in \mathbb{D}$;
(b) for all $z, w \in \mathbb{D}$ the sequence $\{\omega(L_\nu(z), L_\nu(w))\}$ is either eventually zero or strictly decreasing and admits a positive limit;
(c) for all $z, w \in \mathbb{D}$, the sequence $\{\omega(L_\nu(z), L_\nu(w))\}$ is either eventually zero or eventually constant and positive, and this happens if and only if $|f_\nu^h(L_{\nu-1}(z))| = 1$ eventually for one (and hence all) $z \in \mathbb{D}$.

In particular, if $f_\nu(0) = 0$ for all $\nu \in \mathbb{N}$ we also have

(i) if $\sum_\nu (1 - |f_\nu'(0)|) = +\infty$, then $L_\nu \to 0$;

(ii) if $\sum_\nu (1 - |f_\nu'(0)|) < +\infty$, then for any $z \in \mathbb{D}$ either $L_\nu(z)$ does not converge to 0 or $L_\nu(z) = 0$ eventually.

Furthermore, in case (i) it is also possible to control the rate of convergence to the origin. Indeed, if we have

$$0 < \limsup_{\nu\to\infty} \frac{1}{\nu} \sum_{j=0}^{\nu} |f_j'(0)| = a < 1$$

then there exists $c \in (a, 1)$ such that $|L_\nu(z)| = O(c^\nu)$ for all $z \in \mathbb{D}$. If instead

$$\lim_{\nu\to\infty} \frac{1}{\nu} \sum_{j=0}^{\nu} |f_j'(0)| = 0$$

then for all $z \in \mathbb{D}$ we have $|L_\nu(z)|^{1/\nu} \to 0$ as $\nu \to +\infty$.

In [59], the authors also studied the random dynamics induced on $\partial\mathbb{D}$ in the spirit of [1, 137]. Finally, the paper [153] is devoted to the study of random dynamics generated by Blaschke products and inner functions in $\mathbb{D}$.

4 Discrete dynamics on the unit disk

The previous chapter was mostly devoted to the study of dynamics on general hyperbolic Riemann surfaces, even though we did prove at least one important theorem regarding the dynamics on the unit disk $\mathbb{D}$, namely the Wolff–Denjoy theorem. In this chapter, we shall instead concentrate on the dynamics in $\mathbb{D}$, obtaining deep and detailed results.

We shall encounter two main interrelated themes: the study of how the orbits approach the Wolff point and the study of the possible models (in the sense of Definition 3.5.2) a holomorphic self-map can have. When the Wolff point of $f \in \mathrm{Hol}(\mathbb{D}, \mathbb{D})$ is in $\mathbb{D}$, then the theory is relatively simple. If the derivative $f'(\tau)$ at the Wolff point satisfies $0 < |f'(\tau)| < 1$ (*attracting elliptic* case), then the orbits essentially behave like the orbits of the linear map $z \mapsto f'(\tau)z$ in $\mathbb{D}$, which is the model for our f. When instead $f'(\tau) = 0$ (*superattracting elliptic* case) the model is given by a power map, and even though the intertwining map in general is not defined on the whole of $\mathbb{D}$, we shall anyway be able to understand how the orbits approach the Wolff point.

When the Wolff point τ belongs to the boundary, the situation becomes more complicated—and more interesting. In the *hyperbolic* case, i. e., when the angular derivative $f'(\tau)$ at the Wolff point satisfies $0 < f'(\tau) < 1$, it turns out that the orbits converge to the Wolff point nontangentially with a precise slope; furthermore, the model will be again given by the multiplication by $f'(\tau)$ but on the upper half-plane, not in $\mathbb{D}$. When f is *parabolic*, i. e., $f'(\tau) = 1$, it turns out that we have two different cases to consider. When the *hyperbolic step*, that is the limit of the Poincaré distance between two consecutive points in an orbit, is positive, then we shall see that the orbits approach the Wolff point tangentially; moreover, there are two possible models, both on $\mathbb{H}^+$ but one given by $w \mapsto w+1$ and the other by $w \mapsto w-1$. When instead the hyperbolic step is zero, then there is only one model, given again by $w \mapsto w+1$ but this time on $\mathbb{C}$; furthermore, there are examples where the orbits converge tangentially to the Wolff point and examples where the orbits converge nontangentially to the Wolff point. To understand when a given parabolic map has positive or zero hyperbolic step and, in the latter case, decide how the orbits converge to the Wolff point is a problem not yet completely solved.

The second part of this chapter will be devoted to three applications of the theory seen so far. First of all, we shall study more in detail the boundary repelling fixed points, obtaining some nontrivial relations among the angular derivatives there. We shall then study in detail the backward dynamics, i. e., the asymptotic behavior of orbits in the past, not in the future. We shall be able to give quite a complete description of backward orbits when the Poincaré distance between two consecutive points in the orbit is bounded above: these backward orbits either converge nontangentially along a precise slope to a boundary repelling fixed point or they converge tangentially to the Wolff point (and this can happen only for parabolic map). We shall use backward orbits to build a model for f at a boundary repelling fixed point and, in

https://doi.org/10.1515/9783110601978-004

turn, we shall use the model to prove the existence of backward orbits converging to a given boundary repelling fixed point along any given nontangential slope. Finally, we shall come back to another recurring theme of this book, the relation between commuting functions and common fixed points, showing that (with just one exception) two commuting functions without fixed points in the disk must share the same Wolff point.

More precisely, Section 4.1 will be devoted to the study of the dynamics of attracting elliptic self-maps of $\mathbb{D}$, proving in particular the classical Kœnigs theorem. In Section 4.2, we shall instead study the dynamics in the presence of a superattracting point, obtaining the Böttcher theorem. Section 4.3 will deal with the dynamics of hyperbolic maps, proving that the orbits converge to the Wolff point along a precise nontangential slope. Section 4.4 will be devoted to the dynamics of parabolic maps and we shall encounter for the first time the two classes of parabolic maps mentioned above. In Section 4.5, building on the work done so far, we shall prove that (with the exception of the superattracting case) every holomorphic self-map has a model in the sense of Definition 3.5.2; in Section 4.6, we shall be able to identify those models by using the hyperbolic step and the related notion of divergence rate. In Section 4.7, we shall discuss how to tell apart the positive hyperbolic step and the zero hyperbolic step cases for parabolic self-maps assuming a bit of regularity at the Wolff point. Section 4.8 will be devoted to prove a few formulas due to Cowen and Pommerenke on the angular derivative at boundary repelling fixed points. These formulas will be applied in Section 4.9 where we shall study in detail the dynamics of backward orbits and their close relations with boundary repelling fixed points. Finally, in Section 4.10 we shall prove the Behan theorem about common fixed points in the boundary for commuting holomorphic maps, as well as a couple of other results about commuting holomorphic self-maps of the disk.

4.1 Elliptic dynamics

Let $f \in \mathrm{Hol}(\mathbb{D},\mathbb{D}) \setminus \{\mathrm{id}_{\mathbb{D}}\}$ with Wolff point $\tau_f \in \overline{\mathbb{D}}$. If $\tau_f \in \mathbb{D}$, the Schwarz–Pick lemma implies that $|f'(\tau_f)| \le 1$, with equality if and only if f is an elliptic automorphism. On the other hand, if $\tau_f \in \partial\mathbb{D}$ then Corollary 2.5.5 implies that $0 < f'(\tau_f) \le 1$. This suggests the following definition (cf. also Definition 2.5.6).

Definition 4.1.1. Let $f \in \mathrm{Hol}(\mathbb{D},\mathbb{D}) \setminus \{\mathrm{id}_{\mathbb{D}}\}$ with Wolff point $\tau_f \in \overline{\mathbb{D}}$. We say that f is:
- *elliptic* if $\tau_f \in \mathbb{D}$;
- *hyperbolic* if $\tau_f \in \partial\mathbb{D}$ and $0 < f'(\tau_f) < 1$;
- *parabolic* if $\tau_f \in \partial\mathbb{D}$ and $f'(\tau_f) = 1$.

Moreover, if f is elliptic we shall say that it is *attracting* if $0 < |f'(\tau_f)| < 1$ and that it is *superattracting* if $f'(\tau_f) = 0$.

We begin studying attracting elliptic functions, which is the easiest case. We shall see that the dynamics is modeled on the dynamics of the linear map $F(z) = f'(\tau_f)z$; in particular, we shall obtain a model (in the sense of Definition 3.5.2) of the form $(\mathbb{C}, \psi, F)$ and we shall show that the orbits approach the Wolff point in a way comparable to the way the orbits of F approach the origin. This is the content of the *Kœnigs theorem.*

Theorem 4.1.2 (Kœnigs, 1883). *Let $f \in \mathrm{Hol}(\mathbb{D}, \mathbb{D})$ be attracting elliptic. Let $\tau_f \in \mathbb{D}$ be its Wolff point, and $\lambda = f'(\tau_f) \in \mathbb{D}^*$ its derivative at the Wolff point. Then:*

(i) *there exists a simply connected f-absorbing domain $A \subset \mathbb{D}$ containing τ_f and such that $f|_A$ is injective;*

(ii) *there exists a unique $\psi \in \mathrm{Hol}(\mathbb{D}, \mathbb{C})$, with $\psi(\tau_f) = 0$ and $\psi'(\tau_f) = 1$, solving the following functional equation:*

$$\psi \circ f = \lambda \psi. \tag{4.1}$$

Furthermore, it is possible to choose the simply connected absorbing domain A of part (i) *so that $\psi|_A$ is injective and $\psi(A)$ is a F-absorbing domain, where $F \in \mathrm{Aut}(\mathbb{C})$ is given by $F(w) = \lambda w$.*

(iii) *for every $z \in \mathbb{D} \setminus \{\tau_f\}$, we have*

$$\lim_{\nu \to +\infty} \frac{f^{\nu+1}(z) - \tau_f}{f^\nu(z) - \tau_f} = \lambda.$$

Proof. Let us first assume $\tau_f = 0$; at the end of the proof, we shall show how to deal with a general $\tau_f \in \mathbb{D}$.

By assumption, we can write $f(z) = \lambda z(1 + zp_1(z))$ for a suitable $p_1 \in \mathrm{Hol}(\mathbb{D}, \mathbb{C})$. Since $0 < |\lambda| < 1$, we can choose $\varepsilon > 0$ and $0 < \lambda_1 < 1$ with $\lambda_1^2 < |\lambda| < \lambda_1$ so that, setting $A = D(0, \varepsilon)$, we have that $f|_A$ is injective and $|\lambda||1 + zp_1(z)| < \lambda_1$ for all $z \in A$. In particular,

$$|f^\nu(z)| \le \lambda_1^\nu |z| \tag{4.2}$$

for all $z \in A$ and $\nu \in \mathbb{N}$; thus A is invariant (and obviously simply connected). Moreover, since we already know (Theorem 3.1.13) that all orbits converge to 0, it automatically follows that A is f-absorbing, and thus (i) is proved.

For $\nu \in \mathbb{N}$, put $\psi_\nu = \lambda^{-\nu} f^\nu(z)$. We clearly have $\psi_\nu(0) = 0$ and $\psi_\nu'(0) = 1$ for all $\nu \in \mathbb{N}$. Furthermore, if $z \in A$ we have

$$\begin{aligned} |\psi_{\nu+1}(z) - \psi_\nu(z)| &= \frac{1}{|\lambda|^{\nu+1}} |f(f^\nu(z)) - \lambda f^\nu(z)| = \frac{1}{|\lambda|^\nu} |f^\nu(z)|^2 |p_1(f^\nu(z))| \\ &\le M\varepsilon^2 \left(\frac{\lambda_1^2}{|\lambda|}\right)^\nu, \end{aligned}$$

where $M = \sup_{z\in A}|p_1(z)| < +\infty$. Since $\lambda_1^2/|\lambda| < 1$, the telescopic series with general term $\psi_{\nu+1} - \psi_\nu$ is converging on A. As a consequence, the sequence $\{\psi_\nu\}$ is converging, too; let us denote by $\psi \in \mathrm{Hol}(A, \mathbb{C})$ its limit.

Clearly, we have $\psi(0) = 0$ and $\psi'(0) = 1$. In particular, reducing ε if necessary we can assume that ψ is injective on A. Furthermore, since $\psi_\nu(f(z)) = \lambda\psi_{\nu+1}(z)$ for all $z \in A$ passing to the limit we get $\psi \circ f = \lambda\psi$ in A. We shall now use this relation to extend ψ to the whole of $\mathbb{D}$.

Indeed, take $z \in \mathbb{D}$. Since A is f-absorbing, we can find $\nu \in \mathbb{N}$ such that $f^\nu(z) \in A$; then we set

$$\psi(z) = \lambda^{-\nu}\psi(f^\nu(z)).$$

If $\mu \in \mathbb{N}$ is also such that $f^\mu(z) \in A$ then, assuming without loss of generality that $\mu \geq \nu$, using (4.1) in A we have

$$\lambda^{-\mu}\psi(f^\mu(z)) = \lambda^{-\mu}\psi(f^{\mu-\nu}(f^\nu(z))) = \lambda^{-\mu}\lambda^{\mu-\nu}\psi(f^\nu(z)) = \lambda^{-\nu}\psi(f^\nu(z)).$$

So, the value of $\psi(z)$ does not depend on ν, and thus we have defined a function $\psi\colon \mathbb{D} \to \mathbb{C}$, which is holomorphic because if $f^\nu(z_0) \in A$ then $f^\nu(z) \in A$ for all z close enough to z_0.

Clearly, $\psi(A)$ is F-invariant, because $F(\psi(A)) = \psi(f(A)) \subseteq \psi(A)$; furthermore, being a neighborhood of the origin it is also F-absorbing, because for any $w \in \mathbb{C}$ we can find ν so large that $\lambda^\nu w \in \psi(A)$.

In this way, we have proved the existence of ψ as required by (ii). For the uniqueness, assume that $\psi_1 \in \mathrm{Hol}(\mathbb{D}, \mathbb{C})$ is another solution of (4.1) satisfying $\psi_1(0) = 0$ and $\psi_1'(0) = 1$. Then in a neighborhood of the origin we have

$$F \circ \psi \circ \psi_1^{-1} = \psi \circ f \circ \psi_1^{-1} = \psi \circ \psi_1^{-1} \circ F;$$

therefore, setting $\eta = \psi \circ \psi_1^{-1}$ we have

$$\eta(\lambda z) = \lambda\eta(z) \tag{4.3}$$

for all z small enough. But we know that $\eta(0) = 0$ and $\eta'(0) = 1$; therefore, expanding in power series both members of (4.3) and comparing the coefficients of each term we get $\eta(z) = z$. So, $\psi \equiv \psi_1$ in a neighborhood of the origin, and hence everywhere.

We are left with proving (iii). Clearly, we can assume that $z \in A \setminus \{0\}$. Then

$$\frac{f^{\nu+1}(z)}{f^\nu(z)} = \lambda\frac{f^{\nu+1}(z)}{\lambda^{\nu+1}}\frac{\lambda^\nu}{f^\nu(z)} \to \lambda\frac{\psi(z)}{\psi(z)} = \lambda,$$

as claimed.

Finally, let us consider the case $\tau_f \neq 0$. Let $\gamma \in \mathrm{Aut}(\mathbb{D})$ be given by

$$\gamma(z) = \frac{z - \tau_f}{1 - \overline{\tau_f} z}.$$

Then $\tilde{f} = \gamma \circ f \circ \gamma^{-1}$ is attracting elliptic with Wolff point at the origin and $\tilde{f}'(0) = f'(\tau_f) = \lambda$. If $\tilde{A}$ and $\tilde{\psi}$ satisfy (i) and (ii) for $\tilde{f}$, it is then clear that $A = \gamma^{-1}(\tilde{A})$ and $\psi = \tilde{\psi} \circ \gamma$ satisfy (i) and (ii) for f. Moreover, writing $w = \gamma(z)$ we have

$$\begin{aligned} \frac{f^{\nu+1}(z) - \tau_f}{f^\nu(z) - \tau_f} &= \frac{\gamma \circ f^{\nu+1}(z)}{\gamma \circ f^\nu(z)} \frac{1 - \overline{\tau_f} f^{\nu+1}(z)}{1 - \overline{\tau_f} f^\nu(z)} \\ &= \frac{\tilde{f}^{\nu+1}(w)}{\tilde{f}^\nu(w)} \frac{1 - \overline{\tau_f} f^{\nu+1}(z)}{1 - \overline{\tau_f} f^\nu(z)} \to \lambda \frac{1 - |\tau_f|^2}{1 - |\tau_f|^2} = \lambda \end{aligned} \tag{4.4}$$

and we are done. □

Corollary 4.1.3. *Let $F \in \mathrm{Hol}(\mathbb{D}, \mathbb{D})$ be attracting elliptic with Wolff point $\tau_f \in \mathbb{D}$. Then f admits the model $(\mathbb{C}, \psi, F)$ where $F \in \mathrm{Aut}(\mathbb{C})$ is given by $F(w) = f'(\tau_f)w$.*

Proof. It follows immediately from Theorem 4.1.2. □

Remark 4.1.4. Even though we presented a simple ad hoc proof, Theorem 4.1.2(ii) actually is an immediate consequence of part (i) and Theorem 3.5.10. Indeed, the latter theorem, that can be applied thanks to Theorem 4.1.2(i), yields the existence of a model (X^o, ψ, F) for f, where $X^o = \mathbb{C}$ or $\mathbb{D}$. Since f has a fixed point $\tau_f \in \mathbb{D}$, Lemma 3.5.4 implies that F has a fixed-point $\tau^o = \psi(\tau_f) \in X^o$. Differentiating at τ_f the identity $\psi \circ f = F \circ \psi$, we get $\psi'(\tau_f) f'(\tau_f) = F'(\tau^o)\psi'(\tau_f)$. Since ψ is, by definition of model, injective in a neighborhood of τ_f, we have $\psi'(\tau_f) \neq 0$ and so $F'(\tau^o) = f'(\tau_f)$. But $|f'(\tau_f)| < 1$; therefore, necessarily $X^o = \mathbb{C}$ and, up to a biholomorphism, $F(w) = f'(\tau_f)w$ for all $w \in \mathbb{C}$. The rest of the assertions in Theorem 4.1.2(ii) follows from Lemma 3.5.4.

Remark 4.1.5. The map $\psi \colon \mathbb{D} \to \mathbb{C}$ is not necessarily surjective: for instance, if $f(z) = \frac{1}{2}z$, then $\psi = \mathrm{id}_{\mathbb{D}}$. A sufficient condition for having $\psi(\mathbb{D}) = \mathbb{C}$ is $A \subseteq f^\nu(\mathbb{D})$ for all $\nu \in \mathbb{N}$, where $A \subset \mathbb{D}$ is the f-absorbing domain given by Theorem 4.1.2(i). Indeed, take $w \in \mathbb{C}$. We can find $\nu \in \mathbb{N}$ such that $\lambda^\nu w = \psi(\zeta)$ for some $\zeta \in A$. Choose $z \in \mathbb{D}$ such that $f^\nu(z) = \zeta$; then it is clear that $\psi(z) = w$. This condition is satisfied for instance when f is surjective.

The upshot of this theorem is that we can change variables in a neighborhood of the attracting Wolff point τ_f so that in the new coordinates the function f is exactly given by the multiplication by $f'(\tau_f)$. Furthermore, we can extend the coordinate map to a function defined on the whole of $\mathbb{D}$ (though not injective anymore) still transforming f into the multiplication by $f'(\tau_f)$. In this sense, the model is able to describe the asymptotic behavior of all orbits of f.

***Notes to Section* 4.1**

As already remarked in the notes to Section 3.5, (4.1) is called *Schröder equation*, because Schröder has possibly been the first one to consider it, in 1871 [369]. Theorem 4.1.2 is a modern exposition of the solution provided in 1883 by Kœnigs [244]. See also [199] for a different look at the Schröder equation with applications to one-parameter semigroups.

4.2 Superattracting dynamics

The superattracting elliptic case has slightly different features, mainly because the function f is never injective in a neighborhood of its Wolff point, and thus it cannot have a model in the sense of Theorem 3.5.10. However, we shall still be able to change variables so that in the new coordinates f will be expressed in a simple form; but in general it will not be possible to extend the coordinate map to the whole of $\mathbb{D}$. To express our results, we need a couple of definitions.

Definition 4.2.1. Let $f \in \mathrm{Hol}(\mathbb{D}, \mathbb{D})$ and let

$$f(z) = a_0 + a_1(z - z_0) + a_2(z - z_0)^2 + \cdots$$

be the power series expansion of f at a point $z_0 \in \mathbb{D}$. The *multiplicity* $m_f^1(z_0)$ of f at z_0 is given by $m_f^1(z_0) = \min\{k \mid a_k \neq 0\}$. More generally, given $\nu \geq 1$ the *ν-multiplicity* $m_f^\nu(z_0)$ of f at z_0 is the multiplicity of f^ν at z_0, i. e., $m_f^\nu(z_0) = m_{f^\nu}^1(z_0)$.

Clearly, we have $f(0) = 0$ if and only if $m_f^1(0) \geq 1$ and 0 is superattracting if and only if $m_f^1(0) \geq 2$.

We shall now prove the superattracting version of Theorem 4.1.2, the *Böttcher theorem*.

Theorem 4.2.2 (Böttcher, 1904). *Let $f \in \mathrm{Hol}(\mathbb{D}, \mathbb{D})$ be superattracting elliptic. Let $\tau_f \in \mathbb{D}$ be its Wolff point, and $m \geq 2$ the multiplicity of $f - \tau_f$ at τ_f. Then:*

(i) *there exists a simply connected f-absorbing domain $A \subset \mathbb{D}$ containing τ_f and a never vanishing holomorphic function $\psi \in \mathrm{Hol}(A, \mathbb{C})$ with $\psi(\tau_f) = 1$ such that the function $\varphi(z) = z\psi(z)$ is the unique solution of the functional equation*

$$\varphi \circ f(z) = \varphi(z)^m$$

satisfying $\varphi(\tau_f) = 0$ and $\varphi'(\tau_f) = 1$;

(ii) *for every $z \in A \setminus \{\tau_f\}$, we have*

$$\lim_{\nu \to +\infty} \left[\frac{f^{\nu+1}(z) - \tau_f}{f^\nu(z) - \tau_f} \right]^{1/m^\nu} = \varphi(z)^{m-1}.$$

Proof. As we have seen in the proof of Theorem 4.1.2, recalling in particular (4.4), without loss of generality we can assume that $\tau_f = 0$.

Write $f(z) = a_m z^m h_1(z)$ for suitable $a_m \neq 0$ and $h_1 \in \mathrm{Hol}(\mathbb{D}, \mathbb{C})$ with $h_1(0) = 1$. Up to a linear conjugation $z \mapsto \mu z$ with $\mu^{m-1} = a_m$, we can also assume $a_m = 1$.

By induction, it is easy to see that we can write $f^\nu(z) = z^{m^\nu} h_\nu(z)$ for a suitable $h_\nu \in \mathrm{Hol}(\mathbb{D}, \mathbb{C})$ with $h_\nu(0) = 1$. Furthermore, the equalities $f \circ f^\nu = f^{\nu+1} = f^\nu \circ f$ yield

$$h_\nu(z)^m h_1(f^\nu(z)) = h_{\nu+1}(z) = h_1(z)^{m^\nu} h_\nu(f(z)). \tag{4.5}$$

Fix $0 < \delta < 1$ and choose $1 > \varepsilon > 0$ such that $M\varepsilon < \delta$, where $M = \max_{z \in \overline{D(0,\varepsilon)}} |h_1(z)|$; we can also assume that $h_1(z) \neq 0$ for all $z \in \overline{D(0, \varepsilon)}$. Put $A = D(0, \varepsilon)$. Since

$$|f(z)| \leq M|z|^m < \delta|z|^{m-1}$$

for all $z \in \overline{A}$, we have $f(A) \subset A$. Since we already know (Theorem 3.1.13) that all orbits of f converge to the origin, it follows that A is f-absorbing (and obviously simply connected).

We also remark that (4.5) implies that each h_ν is never vanishing on $\overline{A}$. So, for every $\nu \geq 1$ we can choose a unique ψ_ν holomorphic and never vanishing in A such that $\psi_\nu(0) = 1$ and $\psi_\nu(z)^{m^\nu} = h_\nu(z)$ on A.

We claim that the sequence $\{\psi_\nu\}$ converges to a holomorphic function ψ on A. Indeed, we have

$$\begin{aligned}\left|\frac{\psi_{\nu+1}(z)}{\psi_\nu(z)}\right| &= \left|\frac{\psi_{\nu+1}(z)^{m^{\nu+1}}}{\psi_\nu(z)^{m^{\nu+1}}}\right|^{1/m^{\nu+1}} = \left|\frac{h_{\nu+1}(z)}{h_\nu(z)^m}\right|^{1/m^{\nu+1}} = |h_1(f^\nu(z))|^{1/m^{\nu+1}} \\ &= |1 + O(|f^\nu(z)|)|^{1/m^{\nu+1}} = 1 + \frac{1}{m^{\nu+1}} O(|f^\nu(z)|) = 1 + O\left(\frac{1}{m^{\nu+1}}\right);\end{aligned}$$

so the telescopic product $\prod_\nu (\psi_{\nu+1}/\psi_\nu)$ converges to a holomorphic function ψ/ψ_1 uniformly in A, which implies that ψ_ν converges to ψ. Clearly, $\psi(0) = 1$ and hence the Hurwitz theorem (see Corollary A.1.4) implies that ψ is never vanishing because the ψ_ν are never vanishing.

Set $\varphi_\nu(z) = z\psi_\nu(z)$, so that $\varphi_\nu(0) = 0$ and $\varphi_\nu'(0) = 1$. Clearly, φ_ν converges to $\varphi(z) = z\psi(z)$. Since $\varphi'(0) = 1$, up to possibly reducing ε, we can assume that φ is a biholomorphism with its image. Furthermore, $\varphi_\nu(z)^{m^\nu} = f^\nu(z)$ on A; hence we have

$$\begin{aligned}\varphi_\nu(f(z))^{m^\nu} &= f(z)^{m^\nu} \psi_\nu(f(z))^{m^\nu} = z^{m^\nu+1} h_1(z)^{m^\nu} h_\nu(f(z)) \\ &= z^{m^{\nu+1}} h_{\nu+1}(z) = [\varphi_{\nu+1}(z)^m]^{m^\nu},\end{aligned}$$

and thus $\varphi_\nu \circ f = [\varphi_{\nu+1}]^m$, because $\varphi_{\nu+1}(0) = \varphi_\nu(0) = 1$. Passing to the limit, we get $\varphi \circ f = \varphi^m$, as claimed.

If $\tilde{\varphi} \in \mathrm{Hol}(A, \mathbb{C})$ satisfies the same conditions as φ, we must have

$$\tilde{\varphi} \circ \varphi^{-1}(w^m) = [\tilde{\varphi} \circ \varphi^{-1}(w)]^m$$

for all w in a neighborhood of the origin; comparing the series expansions at the origin we get $\tilde{\varphi} \circ \varphi^{-1}(z) = az$ with $a^{m-1} = 1$; hence $a = 1$ and $\tilde{\varphi} \equiv \varphi$ because $\tilde{\varphi}'(0) = \varphi'(0) = 1$.

Finally, in A we have

$$\frac{f^{\nu+1}(z)}{f^{\nu}(z)} = z^{m^{\nu}(m-1)} \frac{\psi_{\nu+1}(z)^{m^{\nu+1}}}{\psi_{\nu}(z)^{m^{\nu}}}.$$

Therefore, we have a consistent way of computing the m^{ν}-th root, obtaining

$$\left[\frac{f^{\nu+1}(z)}{f^{\nu}(z)}\right]^{1/m^{\nu}} = z^{m-1} \frac{\psi_{\nu+1}(z)^{m}}{\psi_{\nu}(z)}.$$

Taking the limit as $\nu \to +\infty$, we then obtain (ii). □

To obtain a necessary and sufficient condition for the existence of an extension of φ to the whole disk, we need another definition.

Definition 4.2.3. Let $f \in \mathrm{Hol}(\mathbb{D}, \mathbb{D})$, $f \not\equiv 0$, with $m_1(0) \geq 2$. The *multiplicity set* of f is given by

$$Q_f = \left\{ \frac{m_f^{\nu}(z)}{m_f^1(0)^{\nu}} \;\middle|\; z \in \mathbb{D},\ \nu \geq 1 \right\}.$$

The next lemma contains the basic properties of the multiplicities and of the multiplicity set.

Lemma 4.2.4. *Let $f \in \mathrm{Hol}(\mathbb{D}, \mathbb{D})$, $f \not\equiv 0$, be with $m_f^1(0) \geq 2$. Then:*
(i) *$m_f^{\nu}(z) = 0$ for all $\nu \geq 1$ except for a countable number of exceptional $z \in \mathbb{D}$;*
(ii) *$m_f^{\nu}(0) = m_f^1(0)^{\nu}$ for all $\nu \geq 1$;*
(iii) *if $m_f^1(z_0) \geq 1$ then $m_f^{\nu}(z_0) = m_f^1(z_0) m_f^1(0)^{\nu-1}$ for all $\nu \geq 1$;*
(iv) *given $z_0 \in \mathbb{D}$, a neighborhood $U \subset \mathbb{D}$ of z_0 and $\varphi \in \mathrm{Hol}(U, \mathbb{D})$ injective in U, if $f(\varphi(z_0)) \in \varphi(U)$ then we have $m^1_{\varphi^{-1}\circ f\circ\varphi}(z_0) = m_f^1(\varphi(z_0))$;*
(v) *$\{0, 1\} \subseteq Q_f \subseteq \mathbb{Q}^+$;*
(vi) *if the set $\{z \in \mathbb{D} \mid f^{\nu}(z) = 0$ for some $\nu \geq 1\}$ is finite then Q_f is finite.*

Proof. (i) The zero set of a holomorphic function not identically zero is countable.

(ii) If $f(z) = az^m + O(z^{m+1})$ with $a \neq 0$ it is easy to prove by induction that $f^{\nu}(z) = a_{\nu} z^{m^{\nu}} + O(z^{m^{\nu}+1})$ for a suitable $a_{\nu} \neq 0$.

(iii) It follows by induction noting that if $g(z) = b_j(z - z_0)^j + O((z - z_0)^{j+1})$ and $f(z) = a_m z^m + O(z^{m+1})$ with $b_j, a_m \neq 0$, so that $m = m_f^1(0)$ and $j = m_g^1(z_0)$, we have

$$f(g(z)) = a_m b_j^m (z - z_0)^{mj} [1 + O(z - z_0)]^m + O((z - z_0)^{mj+1}),$$

i. e., $m^1_{f\circ g}(z_0) = m_g^1(z_0) m_f^1(0)$.

(iv) Writing

$$\varphi(z)=\varphi(z_0)+a_1(z-z_0)+O((z-z_0)^2)\quad\text{and}\quad f(z)=b_m(z-\varphi(z_0))^m+O((z-\varphi(z_0))^{m+1})$$

with $a_1, b_m \neq 0$ we have

$$(\varphi^{-1}\circ f\circ\varphi)(z)=(\varphi^{-1}\circ f\circ\varphi)(z_0)+a_1^{m-1}b_m(z-z_0)^m+O(|z-z_0|^{m+1})$$

as claimed.

(v) It follows from (i), (ii), and the definition of Q_f.

(vi) It follows from (ii) and (iii). □

Then we have the following.

Proposition 4.2.5 (Cowen, 1982). *Let $f \in \mathrm{Hol}(\mathbb{D},\mathbb{D})$ be superattracting elliptic with Wolff point $\tau_f \in \mathbb{D}$ and let $m \geq 2$ be the multiplicity of $f - \tau_f$ at τ_f. Then there is a holomorphic function $\varphi \in \mathrm{Hol}(\mathbb{D},\mathbb{C})$, with $\varphi(\tau_f) = 0$ and $\varphi'(\tau_f) = 1$, solving the functional equation $\varphi \circ f = \varphi^m$ on the whole of $\mathbb{D}$ if and only if $Q_f \subseteq \mathbb{N}$.*

Proof. As usual, we can assume $\tau_f = 0$ and $f(z) = z^m + O(z^{m+1})$, because the multiplicities are invariant under conjugation (Lemma 4.2.4). Let $A = D(0,\varepsilon)$ and $\varphi \in \mathrm{Hol}(A,\mathbb{C})$ be given by Theorem 4.2.2; we shall write $\varphi(z) = z\psi(z)$ where $\psi \in \mathrm{Hol}(A,\mathbb{C})$ is never vanishing. By the uniqueness of φ, it suffices to show that φ can be extended holomorphically to $\mathbb{D}$ if and only if $Q_f \subseteq \mathbb{N}$.

The main point is the following. Since A is f-absorbing, for every $z_0 \in \mathbb{D}$ we can find $\nu \in \mathbb{N}$ and a simply connected $U \subset \mathbb{D}$ of z_0 such that $f^\nu(U) \subset A$; we can also assume that f^ν does not vanish on $U \setminus \{z_0\}$. So, $\varphi \circ f^\nu$ is defined on U; moreover, in U we can write $f^\nu(z) = (z-z_0)^j h_\nu(z)$, with $j = m_f^\nu(z_0)$ and h_ν never vanishing on U. Then

$$\varphi(f^\nu(z)) = f^\nu(z)\psi(f^\nu(z)) = (z-z_0)^j k_\nu(z),$$

where $k_\nu = (\psi\circ f^\nu)h_\nu$ is never vanishing on U. Since we can always find a m^ν-th root of a never vanishing holomorphic function on a simply connected domain, we see that we can find a well-defined m^ν-th root of $\varphi\circ f^\nu$ on U if and only if m^ν divides $j = m_f^\nu(z_0)$.

So, assume that $Q_f \subseteq \mathbb{N}$. Given $0 < r < 1$, since A is f-absorbing we can find $\nu \geq 0$ such that $f^\nu(\overline{D(0,r)}) \subset A$. The previous argument shows that we can find a m^ν-th root η_r of $\varphi \circ f^\nu$ on $D(0,r)$, unique up to the multiplication by a m^ν-th root of unity. Since around the origin we have $\varphi(f^\nu(z)) = z^{m^\nu}h(z)$ with $h(0) = 1$, we can choose η_r in a unique way so that $\eta_r'(0) = 1$.

Notice that if $\mu \geq \nu$ we have

$$\varphi(f^\mu(z)) = \varphi(f^{\mu-\nu}(f^\nu(z))) = [\varphi(f^\nu(z))]^{m^{\mu-\nu}} = [\eta_r(z)^{m^\nu}]^{m^{\mu-\nu}} = \eta_r(z)^{m^\mu}; \tag{4.6}$$

this means that η_r does not depend on the choice of ν.

We then extend φ to the whole of $\mathbb{D}$ by setting $\varphi(z) = \eta_r(z)$ as soon as $|z| < r$. The uniqueness of η_r and (4.6) imply that we have a well-defined extension of φ to the whole of $\mathbb{D}$ still satisfying $\varphi \circ f = \varphi^m$.

Conversely, assume we have a $\varphi \in \mathrm{Hol}(\mathbb{D}, \mathbb{C})$, with $\varphi(0) = 0$ and $\varphi'(0) = 1$, such that $\varphi \circ f = \varphi^m$. Let $z_0 \in \mathbb{D}$ be a zero of f^ν and write $f^\nu(z) = (z - z_0)^j \alpha(z)$, where $j = m_f^\nu(z_0) \geq 1$ and $\alpha(z_0) \neq 0$; moreover, write $\varphi(z) = z\psi(z) = (z - z_0)^\ell \beta(z)$ with $\ell \in \mathbb{N}$ and $\psi(0), \beta(z_0) \neq 0$. Then

$$(z - z_0)^j \alpha(z) \psi(f^\nu(z)) = \varphi(f^\nu(z)) = \varphi(z)^{m^\nu} = (z - z_0)^{\ell m^\nu} \beta(z)^{m^\nu}.$$

Comparing multiplicities in both sides we get $m_f^\nu(z_0) = \ell m^\nu$ and so $m_f^\nu(z_0)/m^\nu \in \mathbb{N}$. Since z_0 was a generic zero of f^ν, we have $Q_f \subseteq \mathbb{N}$ and we are done. □

Notes to Section 4.2

Theorem 4.2.2 has been proved by Böttcher [71] in 1904. Proposition 4.2.5 has been proved by Cowen [127] in a slightly more general form in 1982.

4.3 Hyperbolic dynamics

The study of the dynamics nearby the Wolff point τ when it belongs to the boundary of ∂D is much more complicated, because we cannot work in a full neighborhood of τ. However, in this and the next two sections we shall be able to get models and versions of Theorem 4.1.2 for hyperbolic and parabolic maps.

The main goal of this section is to prove that in the hyperbolic case the orbits converge to the Wolff point along a precise slope. We start with a lemma valid in the parabolic case, too.

Lemma 4.3.1. *Let $f \in \mathrm{Hol}(\mathbb{D}, \mathbb{D})$ be with Wolff point $\tau_f \in \partial\mathbb{D}$. Assume that there exists $z_0 \in \mathbb{D}$ such that its orbit $\{f^\nu(z_0)\}$ converges to τ_f nontangentially. Then for every compact $K \subset \mathbb{D}$ the sequence $\{f^\nu(K)\}$ converges uniformly to τ_f nontangentially, i. e., there exists $M > 1$ such that $f^\nu(K) \subset K(\tau_f, M)$ for all $\nu \in \mathbb{N}$.*

Proof. Let $M_0 > 1$ be such that $f^\nu(z_0) \in K(\tau_f, M_0)$ for all $\nu \in \mathbb{N}$, and choose $R > 0$ such that $\omega(z, z_0) < R$ for all $z \in K$. Then we have

$$\begin{aligned}\omega(f^\nu(z), w) &\leq \omega(f^\nu(z_0), w) + \omega(f^\nu(z), f^\nu(z_0)) \leq \omega(f^\nu(z_0), w) + \omega(z, z_0)\\ &< \omega(f^\nu(z_0), w) + R\end{aligned}$$

for all $\nu \in \mathbb{N}$, $z \in K$ and $w \in \mathbb{D}$. Hence

$$\begin{aligned}&\lim_{w \to \tau_f} [\omega(f^\nu(z), w) - \omega(0, w)] + \omega(0, f^\nu(z))\\ &\quad \leq \lim_{w \to \tau_f} [\omega(f^\nu(z_0), w) - \omega(0, w)] + \omega(0, f^\nu(z_0)) + 2R < M_0 + 2R\end{aligned}$$

for all $\nu \in \mathbb{N}$ and $z \in K$ and the assertion follows from Lemma 2.2.2. □

We can now prove that in the hyperbolic case the orbits converge nontangentially to the Wolff point.

Proposition 4.3.2. *Let $f \in \mathrm{Hol}(\mathbb{D}, \mathbb{D})$ be hyperbolic with Wolff point $\tau_f \in \partial\mathbb{D}$ and angular derivative $0 < f'(\tau_f) < 1$. Then for any $z_0 \in \mathbb{D}$ the orbit $\{f^\nu(z_0)\}$ converges to τ_f nontangentially. Moreover, the convergence is uniform on compact subsets.*

Proof. Without loss of generality, we can assume $\tau_f = 1$. Let us then transfer the setting to $\mathbb{H}^+$. Putting $F = \Psi \circ f \circ \Psi^{-1}$, where Ψ is the Cayley transform, we can assume that $F \in \mathrm{Hol}(\mathbb{H}^+, \mathbb{H}^+)$ has Wolff point at ∞ with angular derivative $F'(\infty) = \frac{1}{f'(\tau_f)} > 1$ (see Remark 2.1.17). Fix $w_0 \in \mathbb{H}^+$ and put $w_\nu = F^\nu(w_0)$. Let $\gamma_\nu \in \mathrm{Aut}(\mathbb{H}^+)$ be given by

$$\gamma_\nu(w) = \frac{w - \operatorname{Re} w_\nu}{\operatorname{Im} w_\nu} \tag{4.7}$$

so that $\gamma_\nu(w_\nu) = i$; put $q_\nu = \gamma_\nu(w_{\nu+1})$. Then we have

$$\omega_{\mathbb{H}^+}(q_\nu, i) = \omega_{\mathbb{H}^+}(\gamma_\nu(w_{\nu+1}), \gamma_\nu(w_\nu)) = \omega_{\mathbb{H}^+}(w_{\nu+1}, w_\nu) \le \omega_{\mathbb{H}^+}(w_1, w_0). \tag{4.8}$$

On the other hand, the Julia lemma for $\mathbb{H}^+$ (Theorem 2.1.19) implies that $\operatorname{Im} F(w) \ge F'(\infty) \operatorname{Im} w$. In particular, $\operatorname{Im} w_\nu \ge F'(\infty)^\nu \operatorname{Im} w_0 \to +\infty$; therefore, we can assume without loss of generality that $\operatorname{Im} w_\nu > 1$ for all ν. Moreover,

$$\operatorname{Im} q_\nu = \frac{\operatorname{Im} w_{\nu+1}}{\operatorname{Im} w_\nu} \ge F'(\infty) > 1. \tag{4.9}$$

The inequalities (4.8) and (4.9) implies that the set $\{q_\nu - i \mid \nu \in \mathbb{N}\}$ is contained in a compact set of $\mathbb{H}^+$; therefore, we can find $\varepsilon \in (0, 1)$ such that

$$q_\nu \in i + K(\infty, \varepsilon) \tag{4.10}$$

for all $\nu \in \mathbb{N}$, where $K(\infty, \varepsilon)$ is the Stolz region with vertex ∞ and amplitude ε. Choosing ε small enough we can also assume that $w_0 \in K(\infty, \varepsilon)$. Now, since $\gamma_\nu^{-1}(w) = (\operatorname{Im} w_\nu)w + \operatorname{Re} w_\nu$, applying γ_ν^{-1} to (4.10) we get

$$w_{\nu+1} \in w_\nu + (\operatorname{Im} w_\nu)K(\infty, \varepsilon) \subseteq w_\nu + K(\infty, \varepsilon)$$

for all $\nu \in \mathbb{N}$, because $tK(\infty, \varepsilon) \subseteq K(\infty, \varepsilon)$ for all $t > 1$, by Proposition 2.2.7(i). Furthermore, it is not too difficult to check that $K(\infty, \varepsilon) + K(\infty, \varepsilon) \subseteq K(\infty, \varepsilon)$. Indeed, take $w_1, w_2 \in K(\infty, \varepsilon)$; then

$$\sqrt{|w_1 + w_2|^2 + 1 - \varepsilon^2} \le \sqrt{|w_1|^2 + 1 - \varepsilon^2} + \sqrt{|w_2|^2 + 1 - \varepsilon^2} \le \frac{1}{\varepsilon} \operatorname{Im}(w_1 + w_2)$$

and thus $w_1 + w_2 \in K(\infty, \varepsilon)$, again by Proposition 2.2.7.(i). Arguing by induction it then follows that $w_\nu \in w_0 + K(\infty, \varepsilon) \subseteq K(\infty, \varepsilon)$ for all $\nu \in \mathbb{N}$, which exactly means that w_ν tends to ∞ nontangentially.

The final assertion follows from Lemma 4.3.1. □

We can then prove the following analogue of Theorem 4.1.2(iii).

Corollary 4.3.3. *Let $f \in \mathrm{Hol}(\mathbb{D}, \mathbb{D})$ be hyperbolic with Wolff point $\tau_f \in \partial\mathbb{D}$ and angular derivative $0 < f'(\tau_f) < 1$. Then for any $z_0 \in \mathbb{D}$ we have*

$$\lim_{\nu\to+\infty} \frac{f^{\nu+1}(z_0) - \tau_f}{f^\nu(z_0) - \tau_f} = \lim_{\nu\to\infty} f'(f^\nu(z_0)) = f'(\tau_f).$$

Proof. It follows from the Julia–Wolff-Carathéodory Theorem 2.3.2 that can be applied because the previous proposition ensures that $f^\nu(z_0) \to \tau_f$ nontangentially. □

Actually, we have a much more precise result: the orbits converge to the Wolff point with a precise slope.

Theorem 4.3.4. *Let $f \in \mathrm{Hol}(\mathbb{D}, \mathbb{D})$ be hyperbolic type with Wolff point $\tau_f \in \partial\mathbb{D}$ and angular derivative $0 < f'(\tau_f) < 1$. Then for every $z \in \mathbb{D}$ there exists $\theta_z \in (-\pi/2, \pi/2)$ such that*

$$\lim_{\nu\to+\infty} \frac{\tau_f - f^\nu(z)}{|\tau_f - f^\nu(z)|} = \tau_f e^{i\theta_z}. \tag{4.11}$$

Moreover, we have

$$\lim_{\nu\to+\infty} \omega(f^\nu(z), f^{\nu+1}(z)) = \frac{1}{2}\log\frac{|f'(\tau_f) + e^{-2i\theta_z}| + 1 - f'(\tau_f)}{|f'(\tau_f) + e^{-2i\theta_z}| - 1 + f'(\tau_f)} > 0. \tag{4.12}$$

Proof. We can clearly assume $\tau_f = 1$. First of all, notice that if $w = re^{i\phi} \in \mathbb{H}^+$ with $\phi \in (0, \pi)$, setting $z = \Psi^{-1}(w) = (w - i)/(w + i) \in \mathbb{D}$ we have

$$1 - z = \frac{2i}{w + i} = \frac{2(1 + ire^{-i\phi})}{|re^{i\phi} + i|^2} = \frac{2\sqrt{1 + r^2}}{|re^{i\phi} + i|^2}\frac{r^{-1} + ie^{-i\phi}}{\sqrt{r^{-2} + 1}} \tag{4.13}$$

and

$$\frac{1 - z}{|1 - z|} = \frac{r^{-1} + ie^{-i\phi}}{\sqrt{r^{-2} + 1}}. \tag{4.14}$$

Therefore, w tends to ∞ in such a way that $\arg(w) \to \phi_0 \in (0, \pi)$ if and only if $z \to 1$ in such a way that $\frac{1-z}{|1-z|} \to e^{i\theta_0}$ with $\theta_0 = \frac{\pi}{2} - \phi_0 \in (-\pi/2, \pi/2)$.

Summing up, putting $F = \Psi \circ f \circ \Psi^{-1} \in \mathrm{Hol}(\mathbb{H}^+, \mathbb{H}^+)$ as usual, to prove (4.11) it suffices to prove that $\arg(F^\nu(w))$ converges to $\phi \in (0, \pi)$ when $\nu \to \infty$.

Fix $w_0 \in \mathbb{H}^+$ and put $w_\nu = F^\nu(w_0)$. Our goal is then to prove that the sequence $\left\{\frac{\operatorname{Re} w_\nu}{\operatorname{Im} w_\nu} = \cot\arg(w_\nu)\right\}$ converges to a finite limit.

Since we know that $w_\nu \to \infty$ nontangentially (Proposition 4.3.2), the Julia–Wolff–Carathéodory theorem for $\mathbb{H}^+$ (Corollary 2.3.4) yields $w_{\nu+1} = F'(\infty)w_\nu + o(1)w_\nu$, and hence

$$\frac{w_{\nu+1}}{\operatorname{Im} w_\nu} = F'(\infty)\frac{w_\nu}{\operatorname{Im} w_\nu} + o(1)\frac{w_\nu}{\operatorname{Im} w_\nu}.$$

Since $w_\nu \to +\infty$ nontangentially, the quotient $w_\nu/\operatorname{Im} w_\nu = i + (\operatorname{Re} w_\nu)/(\operatorname{Im} w_\nu)$ is bounded away from 0 and ∞; therefore, taking the imaginary part of the previous equality we get

$$\frac{\operatorname{Im} w_{\nu+1}}{\operatorname{Im} w_\nu} = F'(\infty) + o(1).$$

We remark that

$$\frac{w_{\nu+1}}{w_\nu} = \frac{\operatorname{Im} w_{\nu+1}}{\operatorname{Im} w_\nu}\frac{i + \frac{\operatorname{Re} w_{\nu+1}}{\operatorname{Im} w_{\nu+1}}}{i + \frac{\operatorname{Re} w_\nu}{\operatorname{Im} w_\nu}};$$

so if $(\operatorname{Re} w_{\nu_j})/(\operatorname{Im} w_{\nu_j})$ has a limit ℓ for a subsequence $\{w_{\nu_j}\}$ then $(\operatorname{Re} w_{\nu_j+1})/(\operatorname{Im} w_{\nu_j+1})$ converges to the same limit ℓ.

Let $\gamma_\nu \in \operatorname{Aut}(\mathbb{H}^+)$ be again given by (4.7) and set again

$$q_\nu = \gamma_\nu(w_{\nu+1}) = \frac{\operatorname{Re} w_{\nu+1} - \operatorname{Re} w_\nu}{\operatorname{Im} w_\nu} + i\frac{\operatorname{Im} w_{\nu+1}}{\operatorname{Im} w_\nu};$$

in particular, $\operatorname{Im} q_\nu \to F'(\infty) > 1$. Moreover,

$$\operatorname{Re} q_\nu = \frac{\operatorname{Re} w_{\nu+1}}{\operatorname{Im} w_{\nu+1}}\frac{\operatorname{Im} w_{\nu+1}}{\operatorname{Im} w_\nu} - \frac{\operatorname{Re} w_\nu}{\operatorname{Im} w_\nu};$$

therefore, $\{q_\nu\}$ converges if and only if $\{\frac{\operatorname{Re} w_\nu}{\operatorname{Im} w_\nu}\}$ converges. Indeed, if $\{\frac{\operatorname{Re} w_\nu}{\operatorname{Im} w_\nu}\}$ converges to ℓ then clearly $\{q_\nu\}$ converges to $\ell(F'(\infty) - 1) + iF'(\infty)$. Conversely, if $\{\frac{\operatorname{Re} w_\nu}{\operatorname{Im} w_\nu}\}$ does not converge then (since this sequence is bounded) we can find a subsequence $\{\frac{\operatorname{Re} w_{\nu_j}}{\operatorname{Im} w_{\nu_j}}\}$ converging to ℓ and a subsequence $\{\frac{\operatorname{Re} w_{\mu_j}}{\operatorname{Im} w_{\mu_j}}\}$ converging to $\ell' \neq \ell$. But then the previous remark shows that $\{q_{\nu_j}\}$ converges to $\ell(F'(\infty) - 1) + iF'(\infty)$ and $\{q_{\mu_j}\}$ converges to $\ell'(F'(\infty) - 1) + iF'(\infty)$, and so $\{q_\nu\}$ does not converge.

So, we have to prove that the sequence $\{q_\nu\}$ actually converges. The first observation is that the sequence $\omega_{\mathbb{H}^+}(q_\nu, i) = \omega_{\mathbb{H}^+}(w_{\nu+1}, w_\nu)$ is a decreasing sequence and thus it admits a limit d_∞; moreover, we have $d_\infty > 0$ because, as we noticed, $\operatorname{Im} q_\nu$ is eventually strictly larger than 1 and so q_ν cannot converge to i.

It follows that all limit points of the sequence $\{q_\nu\}$ must belong to the intersection between the line $\{\operatorname{Im} w = F'(\infty)\}$ and the boundary of the Poincaré ball of center i and radius d_∞. If this intersection is just one point q_∞, then $q_\nu \to q_\infty$, as claimed. Otherwise, the intersection consists of two points, q_+ and q_-. Assume, by contradiction, that $\{q_\nu\}$ does not converge. This means that there exist disjoint neighborhoods $U_\pm$ of $q_\pm$ so that $\{q_\nu\}$ intersects infinitely many times both U_+ and U_-. But then we can construct a subsequence $\{q_{\nu_j}\}$ such that $q_{\nu_j} \in U_+$ and $q_{\nu_j+1} \in U_-$ for all $j \in \mathbb{N}$ and this is impossible because, by the previous remark, q_{ν_j} and q_{ν_j+1} must have the same limit.

So, we have proved that $\{q_\nu\}$ converges; this implies that $\{\frac{\operatorname{Re} w_\nu}{\operatorname{Im} w_\nu}\}$ converges to a finite limit, that implies that $\{\cot \arg w_\nu\}$ converges to a finite limit, that it turns imply that $\arg(w_\nu) \to \phi \in (0,\pi)$ and (4.11) holds.

To prove (4.12), we come back to $\mathbb{D}$. Fix $z \in \mathbb{D}$ and put $z_\nu = f^\nu(z)$. Formula (2.44) applied with $w = z_{\nu+1}$ and $z = z_\nu$ yields

$$\frac{z_{\nu+1} - z_\nu}{1 - \overline{z_\nu} z_{\nu+1}} = -\frac{1}{\overline{z_\nu}}\left[\frac{1}{\overline{\tau_f} z_\nu}\frac{\overline{\tau_f} - \overline{z_\nu}}{|\overline{\tau_f} - \overline{z_\nu}|}\frac{|\tau_f - z_\nu|}{\tau_f - z_\nu} + \frac{\tau_f - z_{\nu+1}}{\tau_f - z_\nu}\right]^{-1}\left[\frac{\tau_f - z_{\nu+1}}{\tau_f - z_\nu} - 1\right].$$

Recalling Corollary 4.3.3 and (4.11), if we pass to the limit as $\nu \to \infty$ we get

$$\lim_{\nu\to+\infty} \frac{z_{\nu+1} - z_\nu}{1 - \overline{z_\nu} z_{\nu+1}} = \tau_f \frac{1 - f'(\tau_f)}{e^{-2i\theta_z} + f'(\tau_f)}$$

and then (4.12) follows because

$$\lim_{\nu\to+\infty} \omega(f^{\nu+1}(z), f^\nu(z)) = \lim_{\nu\to+\infty} \tanh^{-1}\left|\frac{z_{\nu+1} - z_\nu}{1 - \overline{z_\nu} z_{\nu+1}}\right|$$

and $0 < f'(\tau_f) < 1$. □

Actually, one can push further the computations made in these proofs to show that the sequence

$$\psi_\nu(w) = \gamma_\nu \circ F^\nu(w) = \frac{F^\nu(w) - \operatorname{Re} w_\nu}{\operatorname{Im} w_\nu}$$

converges to a solution ψ of the functional equation $\psi \circ F = F'(\infty)\psi$, so that $\tilde{\psi} = \psi \circ \Psi$ solves the functional equation $\tilde{\psi} \circ f = f'(\tau_f)^{-1}\tilde{\psi}$; we can also build in this way a model for f. However, this approach becomes much more complicated for parabolic self-maps; thus in Section 4.6 we shall describe an unified approach to the construction of models that will work both in the hyperbolic and in the parabolic case.

Notes to Section 4.3

The study of the way orbits of a hyperbolic self-map approach the Wolff point has been started by Wolff [418] in 1929; in particular, he proved Proposition 4.3.2. The more precise Theorem 4.3.4 has been proved by Valiron [398] in 1931; see also [399, Chapitre VI]. Actually, Wolff and Valiron worked with the normalized sequence $\frac{1}{|w_\nu|}F^\nu$ instead of $\gamma_\nu \circ F^\nu$; our approach follows [83] and it is inspired by [338]. See also [81, 330].

4.4 Parabolic dynamics

In the previous section, we saw that the orbits of a hyperbolic self-map converge nontangentially to the Wolff point. For parabolic self-maps, the situation is more complicated: there are parabolic self-maps with orbits converging nontangentially to the Wolff point and parabolic self-maps with orbits converging tangentially to the Wolff point.

Example 4.4.1. Let $F_1 \in \mathrm{Hol}(\mathbb{H}^+, \mathbb{H}^+)$ given by $F_1(z) = z + i$. Then it is clear that F_1 is parabolic with Wolff point at infinity and its orbits converge nontangentially to ∞.

On the other hand, take $F_2 \in \mathrm{Hol}(\mathbb{H}^+, \mathbb{H}^+)$ given by $F_2(z) = z + 1$. Then every orbit of F_2 converges to ∞ eventually leaving every Stolz region with vertex at ∞, which means that F_2 is parabolic with Wolff point at infinity and its orbits converge tangentially to ∞.

One of the goals of this and the next section is to give conditions for understanding when the orbits of a parabolic map converge tangentially to the Wolff point; as we shall see, this is a delicate question and in this section we shall be able only to find a sufficient condition. This will be a byproduct of the main goal of this section, which is to show that Corollary 4.3.3 holds for parabolic self-maps, too.

We shall need a lemma.

Lemma 4.4.2. *Let $F \in \mathrm{Hol}(\mathbb{H}^+, \mathbb{H}^+)$. Then*

$$\left|\frac{F(w_2) - F(w_1)}{w_2 - w_1}\right|^2 \le \frac{(\operatorname{Im} F(w_2))(\operatorname{Im} F(w_1))}{(\operatorname{Im} w_2)(\operatorname{Im} w_1)} \tag{4.15}$$

for all $w_1, w_2 \in \mathbb{H}^+$, with equality for a pair of distinct points (and hence everywhere) if and only if $F \in \mathrm{Aut}(\mathbb{H}^+)$.

Proof. The Schwarz–Pick lemma in $\mathbb{H}^+$ (Proposition 1.3.7) says that

$$\left|\frac{F(w_2) - F(w_1)}{F(w_2) - \overline{F(w_1)}}\right| \le \left|\frac{w_2 - w_1}{w_2 - \overline{w_1}}\right| \tag{4.16}$$

for all $w_1, w_2 \in \mathbb{H}^+$, with equality if and only if $F \in \mathrm{Aut}(\mathbb{H}^+)$. Now, a quick computation yields

$$\left|\frac{w_2 - \overline{w_1}}{w_2 - w_1}\right|^2 - 1 = \frac{4(\operatorname{Im} w_2)(\operatorname{Im} w_1)}{|w_2 - w_1|^2}; \tag{4.17}$$

from this it follows that (4.16) is equivalent to (4.15) and we are done. □

To state and prove the main result of this section, notice that if $F \in \mathrm{Hol}(\mathbb{H}^+, \mathbb{H}^+)$ then by the Schwarz–Pick lemma the sequence $\{\omega_{\mathbb{H}^+}(F^{\nu+1}(w_0), F^\nu(w_0))\}$ is decreasing, and hence admits a finite nonnegative limit for any $w_0 \in \mathbb{H}^+$. Then we have the following.

Theorem 4.4.3 (Pommerenke, 1979). *Let $F \in \mathrm{Hol}(\mathbb{H}^+, \mathbb{H}^+)$ be parabolic with Wolff point ∞. Then*

$$\lim_{\nu\to+\infty} \frac{F^{\nu+1}(w_0)}{F^\nu(w_0)} = \lim_{\nu\to+\infty} \frac{\operatorname{Im} F^{\nu+1}(w_0)}{\operatorname{Im} F^\nu(w_0)} = \lim_{\nu\to+\infty} F'(F^\nu(w_0)) = 1 \tag{4.18}$$

for all $w_0 \in \mathbb{H}^+$. Furthermore, if

$$\lim_{\nu\to+\infty} \omega_{\mathbb{H}^+}(F^{\nu+1}(w_0), F^\nu(w_0)) > 0 \tag{4.19}$$

then either $\arg F^\nu(w_0) \to 0$ *for all $w_0 \in \mathbb{H}^+$ such that* (4.19) *holds or* $\arg F^\nu(w_0) \to \pi$ *for all $w_0 \in \mathbb{H}^+$ such that* (4.19) *holds.*

Proof. Given $w_0 \in \mathbb{H}^+$, put $w_\nu = F^\nu(w_0)$ and

$$q_\nu = \frac{w_{\nu+1} - w_\nu}{w_{\nu+1} - \overline{w_\nu}} = \frac{w_{\nu+1} - w_\nu}{w_{\nu+1} - w_\nu + 2i \operatorname{Im} w_\nu}. \tag{4.20}$$

Since

$$\omega_{\mathbb{H}^+}(w_{\nu+1}, w_\nu) = \omega(0, |q_\nu|), \tag{4.21}$$

the Schwarz–Pick lemma implies that

$$|q_\nu| \le |q_{\nu-1}| \le \cdots \le |q_0|;$$

hence

$$\begin{aligned} \left|\frac{w_{\nu+1}}{w_\nu} - 1\right| &= \frac{|w_{\nu+1} - w_\nu|}{\operatorname{Im} w_\nu} \frac{\operatorname{Im} w_\nu}{|w_\nu|} = \left|\frac{2q_\nu}{1 - q_\nu}\right| \frac{\operatorname{Im} w_\nu}{|w_\nu|} \\ &\le \frac{2|q_\nu|}{1 - |q_\nu|} \frac{\operatorname{Im} w_\nu}{|w_\nu|} \le \frac{2|q_0|}{1 - |q_0|} \frac{\operatorname{Im} w_\nu}{|w_\nu|}. \end{aligned} \tag{4.22}$$

Now, assume that the subsequence $\{w_{\nu_j}\}$ converges nontangentially to ∞. Then $w_{\nu_j+1}/w_{\nu_j} \to 1$ by the Julia–Wolff–Carathéodory theorem (see Corollary 2.3.4). If instead $\{w_{\nu_j}\}$ converges tangentially to ∞, i. e., it tends to ∞ eventually leaving any Stolz region with vertex ∞, we have $(\operatorname{Im} w_{\nu_j})/|w_{\nu_j}| \to 0$ by Proposition 2.2.7(i); so (4.22) implies $w_{\nu_j+1}/w_{\nu_j} \to 1$ in this case, too. Assume, by contradiction, that the whole sequence $\{w_{\nu+1}/w_\nu\}$ does not converge to 1. Then we can find a subsequence $\{w_{\nu_j+1}/w_{\nu_j}\}$ having a positive distance from 1; up to a subsequence, we can then assume that $\{w_{\nu_j+1}/w_{\nu_j}\}$ admits a limit $\ell \in \widehat{\mathbb{C}} \setminus \{1\}$. But either $\{w_{\nu_j}\}$ converges nontangentially to ∞ or up to a subsequence we can assume that it converges tangentially to ∞. In both cases, we get $w_{\nu_j+1}/w_{\nu_j} \to 1 \neq \ell$, contradiction. So, $w_{\nu+1}/w_\nu \to 1$ and we have computed the first limit in (4.18).

For the other two limits, let us write $F(w) = w+p(w)$. Proposition 2.5.8 implies that $\operatorname{Im} p(w) \geq 0$ for all $w \in \mathbb{H}^+$. If we have equality at one point, then F is an automorphism and in that case the assertion is obvious; so we can assume that $p \in \operatorname{Hol}(\mathbb{H}^+, \mathbb{H}^+)$. Notice moreover that Corollary 2.3.4 also yields $p'(\infty) = 0$. Then the Schwarz–Pick lemma for $\mathbb{H}^+$ (see Proposition 1.3.7) yields

$$|p'(w_\nu)| \leq \frac{\operatorname{Im} p(w_\nu)}{\operatorname{Im} w_\nu} \leq \frac{|p(w_\nu)|}{\operatorname{Im} w_\nu} = \frac{|w_{\nu+1} - w_\nu|}{\operatorname{Im} w_\nu} = \left|\frac{2q_\nu}{1-q_\nu}\right|. \tag{4.23}$$

Furthermore,

$$F'(w_\nu) - \frac{\operatorname{Im} w_{\nu+1}}{\operatorname{Im} w_\nu} = p'(w_\nu) - \frac{\operatorname{Im} w_{\nu+1} - \operatorname{Im} w_\nu}{\operatorname{Im} w_\nu},$$

and thus

$$\left|F'(w_\nu) - \frac{\operatorname{Im} w_{\nu+1}}{\operatorname{Im} w_\nu}\right| \leq \left|\frac{4q_\nu}{1-q_\nu}\right|.$$

Therefore, if $|q_\nu| \to 0$ we immediately have $p'(w_\nu) \to 0$; hence $F'(w_\nu) \to 1$ and the previous formula yields the middle limit in (4.18), too. Notice that $|q_\nu| \to 0$ if and only if $\omega_{\mathbb{H}^+}(w_{\nu+1}, w_\nu) \to 0$, by (4.21).

Now assume that (4.19) holds, so that $|q_\nu|$ has a positive limit. Define $\phi_\nu \colon \mathbb{H}^+ \to \mathbb{C}$ by setting

$$\phi_\nu(w) = \frac{F^{\nu+1}(w) - w_{\nu+1}}{F^{\nu+1}(w) - \overline{w_{\nu+1}}} \frac{F^\nu(w) - \overline{w_\nu}}{F^\nu(w) - w_\nu};$$

notice that $|\phi_\nu| \leq 1$ by the Schwarz–Pick lemma for $\mathbb{H}^+$ and that

$$\phi_\nu(w_0) = F'(w_\nu)\frac{\operatorname{Im} w_\nu}{\operatorname{Im} w_{\nu+1}}, \quad \phi_\nu(F(w_0)) = \frac{q_{\nu+1}}{q_\nu}.$$

Since $|q_\nu|$ has a positive limit, we have $|\phi_\nu(F(w_0))| \to 1$; hence the Montel theorem and the maximum principle imply that $|\phi_\nu(w)| \to 1$ for all $w \in \mathbb{H}^+$. In particular,

$$|\phi_\nu(w_0)| = |F'(w_\nu)|\frac{\operatorname{Im} w_\nu}{\operatorname{Im} w_{\nu+1}} \to 1,$$

that we can rewrite as

$$|1 + p'(w_\nu)|^2 = (1 + \varepsilon_\nu)\left(1 + \frac{\operatorname{Im} p(w_\nu)}{\operatorname{Im} w_\nu}\right)^2$$

with $\varepsilon_\nu \to 0$. Recalling that $|p'(w_\nu)| \le (\operatorname{Im} p(w_\nu))/(\operatorname{Im} w_\nu)$ we get

$$\frac{1}{2}\varepsilon_\nu\left(1+\frac{\operatorname{Im} p(w_\nu)}{\operatorname{Im} w_\nu}\right)^2 \le \operatorname{Re} p'(w_\nu) - \frac{\operatorname{Im} p(w_\nu)}{\operatorname{Im} w_\nu} \le 0.$$

Since

$$\left|\frac{\operatorname{Im} p(w_\nu)}{\operatorname{Im} w_\nu}\right| = \left|\frac{2q_\nu}{1-q_\nu}\right| \le \frac{2|q_\nu|}{1-|q_\nu|} \le \frac{2|q_0|}{1-|q_0|},$$

we obtain

$$\lim_{\nu\to+\infty}\left[\operatorname{Re} p'(w_\nu) - \frac{\operatorname{Im} p(w_\nu)}{\operatorname{Im} w_\nu}\right] = 0$$

and, using again $|p'(w_\nu)| \le (\operatorname{Im} p(w_\nu))/(\operatorname{Im} w_\nu)$, that

$$\lim_{\nu\to\infty} \operatorname{Im} p'(w_\nu) = 0.$$

Putting everything together, we have proved that

$$\lim_{\nu\to+\infty}\left[F'(w_\nu) - \frac{\operatorname{Im} w_{\nu+1}}{\operatorname{Im} w_\nu}\right] = \lim_{\nu\to+\infty}\left[p'(w_\nu) - \frac{\operatorname{Im} p(w_\nu)}{\operatorname{Im} w_\nu}\right] = 0, \tag{4.24}$$

and thus

$$\lim_{\nu\to+\infty} F'(w_\nu)\frac{\operatorname{Im} w_\nu}{\operatorname{Im} w_{\nu+1}} = 1,$$

because $\frac{\operatorname{Im} w_{\nu+1}}{\operatorname{Im} w_\nu}$ is bounded below by 1. In particular, $\phi_\nu(w_0) \to 1$, and hence, again by the Montel theorem and the maximum principle, $\phi_\nu(w) \to 1$ for all $w \in \mathbb{H}^+$.

To conclude the proof of (4.18), it suffices to show that $\frac{\operatorname{Im} w_{\nu+1}}{\operatorname{Im} w_\nu}$ tends to 1. Since $\frac{\operatorname{Im} w_{\nu+1}}{\operatorname{Im} w_\nu}$ is bounded below by 1, we can assume that there is $a \in [1, +\infty]$ and a subsequence $\{\nu_j\}$ such that

$$\frac{\operatorname{Im} w_{\nu_j+1}}{\operatorname{Im} w_{\nu_j}} \to a \ge 1.$$

If we prove that $a = 1$ for any such subsequence we are done. Put

$$\psi_\nu(w) = \frac{F^\nu(w) - \operatorname{Re} w_\nu}{\operatorname{Im} w_\nu}.$$

Clearly, $\psi_\nu \in \operatorname{Hol}(\mathbb{H}^+, \mathbb{H}^+)$; since $\psi_\nu(w_0) = i$ for all $\nu \in \mathbb{N}$ by the Montel theorem we can also assume that $\psi_{\nu_j} \to \psi \in \operatorname{Hol}(\mathbb{H}^+, \mathbb{H}^+)$. Now, we have

$$\phi_\nu(w) - 1 = \frac{2i(\psi_{\nu+1}(w)/\psi_\nu(w) - 1)}{(\psi_\nu(w) - i)(\psi_{\nu+1}(w)/\psi_\nu(w) + i)};$$

therefore, the fact that $\phi_\nu \to 1$ implies that $\psi_{\nu_j+1} \to \psi$, too.

A quick computation shows that

$$\operatorname{Im}\psi_{\nu_j}(F(w)) = \operatorname{Im}\left[\frac{F^{\nu_j+1}(w) - \operatorname{Re} w_{\nu_j}}{\operatorname{Im} w_{\nu_j}}\right] = \frac{\operatorname{Im} w_{\nu_j+1}}{\operatorname{Im} w_{\nu_j}} \operatorname{Im}\psi_{\nu_j+1}(w).$$

Passing to the limit as $j \to +\infty$, we get

$$\operatorname{Im}\psi(F(w)) = a \operatorname{Im}\psi(w); \tag{4.25}$$

in particular a must be finite. From this, it follows that

$$\operatorname{Im}\psi(w_\nu) = a^\nu \operatorname{Im}\psi(w_0) = a^\nu.$$

Applying (4.15) to ψ with $w_2 = w_\nu$ and $w_1 = w_0$, we get

$$\begin{aligned}(a^\nu - 1)^2 &= (\operatorname{Im}\psi(w_\nu) - \operatorname{Im}\psi(w_0))^2 \le |\psi(w_\nu) - \psi(w_0)|^2 \le \frac{\operatorname{Im}\psi(w_\nu)}{\operatorname{Im} w_\nu}|w_\nu - w_0|^2 \\ &\le \frac{a^\nu}{\operatorname{Im} w_0}|w_\nu - w_0|^2 \le \frac{a^\nu}{\operatorname{Im} w_0}(|w_0| + |w_\nu|)^2,\end{aligned}$$

where we have used $\operatorname{Im} w_\nu \ge \operatorname{Im} w_0$. Dividing by a^ν, applying the 2ν-th root and taking the limsup as $\nu \to +\infty$, we get

$$\begin{aligned}1 \le a^{1/2} &= a^{1/2} \limsup_{\nu\to+\infty}\left(1 - \frac{1}{a^\nu}\right)^{1/\nu} \le \limsup_{\nu\to+\infty}(|w_0| + |w_\nu|)^{1/\nu} \\ &\le \limsup_{\nu\to+\infty} \frac{|w_0| + |w_{\nu+1}|}{|w_0| + |w_\nu|} = 1\end{aligned}$$

where we used a standard property of lim sup (see Lemma A.4.1) and the first limit in (4.18). So, $a = 1$ and (4.18) is completely proved.

We are left with proving the last assertion. To do so, we first prove that the sequence $\{\psi_\nu\}$ actually converges. Assume, by contradiction, that this is not the case; then we can find another subsequence $\{\psi_{\tilde{\nu}_k}\}$ converging to $\tilde{\psi} \in \operatorname{Hol}(\mathbb{H}^+, \mathbb{H}^+)$. The fact that $|\phi_\nu| \le 1$ implies that the sequence

$$\left|\frac{F^\nu(w) - w_\nu}{F^\nu(w) - \overline{w_\nu}}\right| = \left|\frac{\psi_\nu(w) - i}{\psi_\nu(w) + i}\right|$$

is decreasing, and hence it converges to a limit. This means that there must exist $\alpha \in \mathbb{R}$ such that

$$\frac{\tilde{\psi}(w)-i}{\tilde{\psi}(w)+i}=e^{i\alpha}\frac{\psi(w)-i}{\psi(w)+i}. \tag{4.26}$$

Now, since $a = 1$, (4.25) implies that $\operatorname{Im}(\psi\circ F-\psi)\equiv 0$; therefore, there must exist $b\in\mathbb{R}$ such that

$$\psi(F(w))=\psi(w)+b, \tag{4.27}$$

where

$$b=\lim_{j\to+\infty}\operatorname{Re}[\psi_{\nu_j}(F(w_0))-\psi_{\nu_j}(w_0)]=\lim_{j\to+\infty}\frac{\operatorname{Re}w_{\nu_j+1}-\operatorname{Re}w_{\nu_j}}{\operatorname{Im}w_{\nu_j}}.$$

Then (4.27) implies that $\psi(w_\nu)=i+\nu b$ for all $\nu\in\mathbb{N}$, and hence

$$\frac{\psi(w_\nu)-i}{\psi(w_\nu)+i}=\frac{\nu b}{\nu b+2i}\longrightarrow 1$$

as $\nu\to\infty$. But the same argument can be applied to $\tilde{\psi}$; recalling (4.26) we find $e^{i\alpha}=1$, and hence $\tilde{\psi}\equiv\psi$.

So, the whole sequence $\{\psi_\nu\}$ converges to ψ and, in particular,

$$b=\lim_{\nu\to+\infty}\frac{\operatorname{Re}w_{\nu+1}-\operatorname{Re}w_\nu}{\operatorname{Im}w_\nu}. \tag{4.28}$$

Now, we already noticed that the assumption (4.19) is equivalent to saying that $\{q_\nu\}$ is bounded away from 0. Then

$$\frac{|\operatorname{Re}w_{\nu+1}-\operatorname{Re}w_\nu|}{\operatorname{Im}w_\nu}\geq\frac{|w_{\nu+1}-w_\nu|}{\operatorname{Im}w_\nu}-\left(\frac{\operatorname{Im}w_{\nu+1}}{\operatorname{Im}w_\nu}-1\right)=\left|\frac{2q_\nu}{1-q_\nu}\right|-\left(\frac{\operatorname{Im}w_{\nu+1}}{\operatorname{Im}w_\nu}-1\right)$$

is eventually bounded away from zero, too. Hence $b\neq 0$ and

$$\frac{\operatorname{Re}w_{\nu+1}-\operatorname{Re}w_\nu}{\operatorname{Im}w_{\nu+1}-\operatorname{Im}w_\nu}=\frac{\operatorname{Re}w_{\nu+1}-\operatorname{Re}w_\nu}{\operatorname{Im}w_\nu}\,\frac{1}{\frac{\operatorname{Im}w_{\nu+1}}{\operatorname{Im}w_\nu}-1}\longrightarrow\pm\infty,$$

where the sign is the sign of b. Since $\{\operatorname{Im}w_{\nu+1}\}$ is a positive increasing sequence diverging to $+\infty$, the Stolz–Cesàro theorem (see Proposition A.4.4) implies that $\frac{\operatorname{Re}w_\nu}{\operatorname{Im}w_\nu}\to\pm\infty$, where the sign depends only on the sign of b and not on w_0. This exactly means that $\arg(w_\nu)\to 0$ when $b>0$ or $\arg(w_\nu)\to\pi$ when $b<0$, as claimed.

To conclude the proof, let $\tilde{w}_0\in\mathbb{H}^+$ be another point such that (4.19) holds. Put $\tilde{w}_\nu=F^\nu(\tilde{w}_0)$ and then build $\tilde{\psi}\in\operatorname{Hol}(\mathbb{H}^+,\mathbb{H}^+)$ satisfying $\tilde{\psi}(F(w))=\tilde{\psi}(w)+\tilde{b}$, with $\tilde{b}\in\mathbb{R}^*$, as limit of $\tilde{\psi}_\nu(w)=(F^\nu(w)-\operatorname{Re}\tilde{w}_\nu)/(\operatorname{Im}\tilde{w}_\nu)$ as before; we need to prove that b and $\tilde{b}$ have the same sign. But indeed

$$\frac{\tilde{b}}{b} = \frac{\tilde{\psi}(F(i)) - \tilde{\psi}(i)}{\tilde{\psi}(F(i)) - \tilde{\psi}(i)} = \lim_{\nu\to+\infty} \frac{\operatorname{Im} w_\nu}{\operatorname{Im} \tilde{w}_\nu} > 0$$

and we are done. □

Corollary 4.4.4. *Let $f \in \operatorname{Hol}(\mathbb{D}, \mathbb{D})$ be parabolic with Wolff point $\tau_f \in \partial\mathbb{D}$. Then*

$$\lim_{\nu\to+\infty} \frac{f^{\nu+1}(z_0) - \tau_f}{f^\nu(z_0) - \tau_f} = \lim_{\nu\to+\infty} f'(f^\nu(z_0)) = 1$$

for all $z_0 \in \mathbb{D}$. Furthermore, let

$$E = \left\{z_0 \in \mathbb{D} \mid \lim_{\nu\to+\infty} \omega(f^{\nu+1}(z_0), f^\nu(z_0)) > 0\right\};$$

then the orbit of each $z_0 \in E$ converges to τ_f tangentially. More precisely, we have

$$\lim_{\nu\to+\infty} \frac{\tau_f - f^\nu(z_0)}{|\tau_f - f^\nu(z_0)|} = e^{\pm i\pi/2} = \pm i,$$

where the sign is the same for all $z_0 \in E$.

Proof. Without loss of generality, we can assume $\tau_f = 1$. Let $\Psi\colon \mathbb{D} \to \mathbb{H}^+$ be the Cayley transform, and put $F = \Psi \circ f \circ \Psi^{-1}$. Then $F \in \operatorname{Hol}(\mathbb{H}^+, \mathbb{H}^+)$ has Wolff point at ∞ with $F'(\infty) = 1$. Moreover, if we set $w_0 = \Psi(z_0)$ then $f^\nu(z_0) = \Psi^{-1}(F^\nu(w_0))$ and

$$\frac{f^{\nu+1}(z_0) - 1}{f^\nu(z_0) - 1} = \frac{F^\nu(w_0) + i}{F^{\nu+1}(w_0) + i} = \frac{F^\nu(w_0)}{F^{\nu+1}(w_0)} \frac{1 + i/F^\nu(w_0)}{1 + i/F^{\nu+1}(w_0)}.$$

Moreover,

$$f'(f^\nu(z_0)) = \left(\frac{F^\nu(w_0) + i}{F^{\nu+1}(w_0) + i}\right)^2 F'(F^\nu(w_0));$$

hence the assertions follow from Theorem 4.4.3 and (4.14). □

Remark 4.4.5. In Corollary 4.6.9(i), we shall show that $\lim_\nu \omega(f^{\nu+1}(z_0), f^\nu(z_0)) > 0$ for some $z_0 \in \mathbb{D}$ if and only if $\lim_\nu \omega(f^{\nu+1}(z), f^\nu(z)) > 0$ for all $z \in \mathbb{D}$; in other words, either $E = \emptyset$ or $E = \mathbb{D}$. Therefore, in the latter case either all orbits converge to τ with angle $\pi/2$ or all orbits converge to τ with angle $-\pi/2$.

When $\lim_\nu \omega(f^{\nu+1}(z_0), f^\nu(z_0)) = 0$ then the orbits might tend to the boundary tangentially or nontangentially.

Example 4.4.6. Let $F \in \operatorname{Hol}(\mathbb{H}^+, \mathbb{H}^+)$ be given by $F(w) = w + i$. Then it is easy to check that F is parabolic with Wolff point ∞, that $\lim_\nu \omega_{\mathbb{H}^+}(F^{\nu+1}(w_0), F^\nu(w_0)) = 0$ and that $F^\nu(w_0) \to \infty$ radially for all $w_0 \in \mathbb{H}^+$.

Example 4.4.7. Let $F \in \mathrm{Hol}(\mathbb{H}^+, \mathbb{H}^+)$ be given by

$$F(w) = \left(\sqrt{(\sqrt{z}-1)^2 - 1} + 1\right)^2.$$

In other words, F is conjugated to the translation by -1 on the image of $\mathbb{H}^+$ under the injective map $\Phi\colon \mathbb{H}^+ \to \mathbb{C}$ given by $\Phi(w) = (\sqrt{w} - 1)^2$; in particular, F has no fixed points in $\mathbb{H}^+$. Taking $w_0 = (1+i)^2$ it is easy to check that $w_\nu = F^\nu(w_0) = (1 + i\sqrt{\nu+1})^2 = -\nu + 2i\sqrt{\nu+1}$. Clearly, $F^\nu(w_0) \to +\infty$ and so ∞ is the Wolff point of F. Furthermore, $\Phi'(w) \to 1$ as $w \to +\infty$, and hence $F'(\infty) = 1$, i. e., F is parabolic. Next,

$$\frac{\operatorname{Re} w_\nu}{\operatorname{Im} w_\nu} = -\frac{\nu}{2\sqrt{\nu+1}} \to -\infty$$

and so $w_\nu \to \infty$ tangentially. Finally,

$$\frac{w_\nu - w_{\nu-1}}{w_\nu - \overline{w_{\nu-1}}} = \frac{-\frac{1}{\sqrt{\nu}} + 2i\left(\sqrt{1+\frac{1}{\nu}} - 1\right)}{-\frac{1}{\sqrt{\nu}} + 2i\left(\sqrt{1+\frac{1}{\nu}} + 1\right)} \to 0,$$

and hence $\lim_\nu \omega_{\mathbb{H}^+}(F^{\nu+1}(w_0), F^\nu(w_0)) = 0$.

In the proof of Theorem 4.4.3, we have seen that if $\lim_\nu \omega_{\mathbb{H}^+}(F^{\nu+1}(w_0), F^\nu(w_0)) > 0$ then the sequence

$$\psi_\nu(w) = \frac{F^\nu(w) - \operatorname{Re} w_\nu}{\operatorname{Im} w_\nu}$$

converges to a solution of the Abel equation $\psi \circ F = \psi + b$ with $b \in \mathbb{R}^*$. Unfortunately, if $\lim_\nu \omega_{\mathbb{H}^+}(F^{\nu+1}(w_0), F^\nu(w_0)) = 0$ this is not anymore true: it is possible to prove (see [338]) that in that case the sequence $\{\psi_\nu\}$ converges to the constant 1. Instead of dealing with this case in yet another way, in the next section we shall use a unified approach relying on Theorem 3.5.10.

Remark 4.4.8. Assume that $F \in \mathrm{Hol}(\mathbb{H}^+, \mathbb{H}^+)$ is parabolic with Wolff point at ∞ and assume that $w_0 \in \mathbb{H}^+$ satisfies (4.19). Keeping the notation introduced in the proof of Theorem 4.4.3, let us define

$$\widehat{\psi}_\nu(w) = \frac{F^\nu(w) - w_\nu}{w_{\nu+1} - w_\nu}.$$

We clearly have

$$\widehat{\psi}_\nu = (\psi_\nu(w) - i)\frac{\operatorname{Im} w_\nu}{w_{\nu+1} - w_\nu}.$$

Now,

$$\frac{w_{\nu+1} - w_\nu}{\operatorname{Im} w_\nu} = \frac{\operatorname{Re} w_{\nu+1} - \operatorname{Re} w_\nu}{\operatorname{Im} w_\nu} + i\left(\frac{\operatorname{Im} w_{\nu+1}}{\operatorname{Im} w_\nu} - 1\right);$$

therefore, recalling (4.18) and (4.28) we see that $\widehat{\psi}_\nu \to \widehat{\psi} \in \operatorname{Hol}(\mathbb{H}^+, \mathbb{C})$ given by

$$\widehat{\psi}(w) = \frac{1}{b}(\psi(w) - i).$$

In particular, $\hat{\psi}$ is a solution of the Abel equation $\widehat{\psi} \circ F = \widehat{\psi} + 1$.

Notes to Section 4.4

This section is adapted from the work of Pommerenke [338]. In a subsequent paper, Baker and Pommerenke [34] proved that even when $\lim_\nu \omega_{\mathbb{H}^+}(F^{\nu+1}(w_0), F^\nu(w_0)) = 0$ the sequence $\{\frac{F^\nu - w_\nu}{w_{\nu+1} - w_\nu}\}$ converges to a solution of the Abel equation; in particular, the Abel equation is solvable for all parabolic self-maps. We shall obtain this result in a different way in the next two sections.

Building on [338, 34], Contreras, Díaz-Madrigal, and Pommerenke [120] in 2007 showed that the sequence $\{\frac{f^\nu - z_\nu}{z_{\nu+1} - z_\nu}\}$, where $z_\nu = f^\nu(z_0)$ as usual, converges to a solution ψ of the Abel equation $\psi \circ f = \psi + 1$ for all parabolic self-maps $f \in \operatorname{Hol}(\mathbb{D}, \mathbb{D})$ with Wolff point 1.

[121] study other sequences potentially converging to a solution of the Abel equation under some condition on the smoothness of f at the Wolff point.

4.5 Models on the unit disk

The goal of this section is to prove that every $f \in \operatorname{Hol}(\mathbb{D}, \mathbb{D})$ which is not superattracting elliptic admits a model in the sense of Definition 3.5.2. When f is attracting elliptic, we know this already (Corollary 4.1.3); so we shall focus on hyperbolic and parabolic maps.

The idea is to apply Theorem 3.5.10; therefore, we must build a simply connected f-absorbing domain where f is injective. To do so, we need a couple of lemmas that might be interesting on their own.

Lemma 4.5.1. *Let $\Omega \subset \mathbb{C}$ be a convex domain and take $f \in \operatorname{Hol}(\Omega, \mathbb{C})$ such that $\operatorname{Re} f' \geq 0$ on Ω. Then f is either constant or injective.*

Proof. If there is $z_0 \in \Omega$ so that $\operatorname{Re} f'(z_0) = 0$, then by the minimum principle for harmonic functions f' is constant, and hence f is either constant or injective. We henceforth can suppose that $\operatorname{Re} f' > 0$ everywhere.

Assume, by contradiction, that $f(z_1) = f(z_2)$ for two distinct points $z_1, z_2 \in \Omega$. Integrating along the segment σ from z_1 to z_2, we obtain

$$0 = f(z_2) - f(z_1) = \int_\sigma f'(\zeta)\,d\zeta = (z_2 - z_1)\int_0^1 f'(z_1 + t(z_2 - z_1))\,dt.$$

Hence $z_1 \neq z_2$ implies

$$0 = \operatorname{Re}\int_0^1 f'(z_1 + t(z_2 - z_1))\,dt = \int_0^1 \operatorname{Re} f'(z_1 + t(z_2 - z_1))\,dt > 0,$$

contradiction. □

Lemma 4.5.2. *Let $\Omega \subset \mathbb{D}$ be a subdomain of $\mathbb{D}$ and let $\widehat{\Omega}$ be given by the union of Ω with the compact connected components of $\mathbb{C} \setminus \Omega$. Then:*
(i) *$\widehat{\Omega} \subseteq \mathbb{D}$;*
(ii) *$\widehat{\Omega}$ is open and simply connected;*
(iii) *if $f \in \operatorname{Hol}(\mathbb{D}, \mathbb{D})$ is such that $f(\Omega) \subseteq \Omega$, then $f(\widehat{\Omega}) \subseteq \widehat{\Omega}$;*
(iv) *if $f \in \operatorname{Hol}(\mathbb{D}, \mathbb{D})$ is injective on Ω, then it is injective on $\widehat{\Omega}$, too.*

Proof. (i) Since $\Omega \subset \mathbb{D}$ and $\mathbb{C}\setminus\mathbb{D}$ is connected and unbounded, the compact connected components of $\mathbb{C} \setminus \Omega$ are contained in $\mathbb{D}$.

(ii) By construction, $\mathbb{C}\setminus\widehat{\Omega}$ is the unique unbounded connected component of $\mathbb{C}\setminus\Omega$. Therefore, $\mathbb{C} \setminus \widehat{\Omega}$ is closed and connected, and hence $\widehat{\Omega}$ is open and simply connected (see, e. g., [301, p. 145]).

(iii) It is well known that $\widehat{\Omega}$ is the holomorphic hull of Ω in $\mathbb{D}$, where the holomorphic hull of $B \subset \mathbb{D}$ is given by

$$\hat{B} = \left\{z \in \mathbb{D} \;\middle|\; |f(z)| \le \sup_{\zeta\in B}|f(\zeta)| \text{ for all } f \in \operatorname{Hol}(\mathbb{D}, \mathbb{C})\right\};$$

see, e. g., [124, pp. 201–202]. Let $U \subset \mathbb{D}$ be a compact connected component of $\mathbb{C} \setminus \Omega$; by construction, $\partial U \subseteq \overline{\Omega} \cap \mathbb{D}$. By the maximum principle, we have

$$f(U) \subseteq \widehat{\partial f(U)} = \widehat{f(\partial U)} \subseteq \widehat{f(\overline{\Omega} \cap \mathbb{D})} \subseteq \widehat{\overline{\Omega} \cap \mathbb{D}} = \overline{\widehat{\Omega}} \cap \mathbb{D};$$

since U was a generic compact connected component of $\mathbb{C}\setminus\Omega$ and f is open or constant, we get $f(\widehat{\Omega}) \subseteq \widehat{\Omega}$, as claimed.

(iv) Assume, by contradiction, that there are $z_1, z_2 \in \widehat{\Omega}$ with $z_1 \neq z_2$ such that $f(z_1) = f(z_2) = w_0$. Since Ω is connected and $\widehat{\Omega}$ is simply connected, we can find a Jordan curve $\gamma\colon [0, 1] \to \Omega$ containing z_1 and z_2 in its interior and disjoint from $f^{-1}(w_0)$. Since f is injective on Ω, the curve $f\circ\gamma$ is still a Jordan curve; therefore, the Jordan curve theorem (see, e. g., [11, pp. 79–85]) implies that the winding number $\operatorname{Ind}(f \circ \gamma, w_0)$ of $f \circ \gamma$ with respect to w_0 has absolute value at most 1. But the argument principle (see, e. g., [301, pp. 76, 308]) implies that $\operatorname{Ind}(f \circ \gamma, w_0)$ is equal to the number of zeroes of $f - w_0$ in the interior of γ, that by assumption is at least 2, contradiction. □

The next lemma is the main technical point of the proof.

Lemma 4.5.3. *Let $f \in \operatorname{Hol}(\mathbb{D}, \mathbb{D})$ be hyperbolic or parabolic with Wolff point $\tau_f \in \partial\mathbb{D}$. Given $M > 1$ and $R > 0$, put*

$$G_{M,R} = K(\tau_f, M) \cup \bigcup_{\nu \in \mathbb{N}} f^\nu(B_{\mathbb{D}}(0,R)).$$

Then there exists $0 < \rho < 1$ *such that* f *is injective on* $G_{M,R} \cap \{z \in \mathbb{D} \mid |\tau_f - z| < \rho\}$.

Proof. Let us first consider the hyperbolic case. By Proposition 4.3.2, all orbits of f converge to τ_f nontangentially; therefore, by Lemma 4.3.1, we can find $M_1 > M$ such that $f^\nu(B_{\mathbb{D}}(0,R)) \subset K(\tau_f, M_1)$ for all $\nu \in \mathbb{N}$. Furthermore, we know that $f'(z) \to f'(\tau_f) > 0$ nontangentially; hence we can find $0 < \rho < 1$ such that $\operatorname{Re} f' > 0$ on $\Omega = K(\tau_f, M_1) \cap \{|\tau_f - z| < \rho\}$. Since Ω is convex, f is injective on Ω, by Lemma 4.5.1; thus we are done because $G_{M,R} \cap \{|\tau_f - z| < \rho\} \subseteq \Omega$ by construction.

To study the parabolic case, let us transfer the stage to $\mathbb{H}^+$, as usual. Let $F \in \operatorname{Hol}(\mathbb{H}^+, \mathbb{H}^+)$ be with Wolff point at ∞ and assume that $F'(\infty) = 1$. Write $F(w) = w + p(w)$; Corollary 2.3.4 and Proposition 2.5.8 imply that $p'(\infty) = 0$ and that $\operatorname{Im} p(w) \geq 0$ for all $w \in \mathbb{H}^+$, with equality in one point (and hence everywhere) if and only if p is a real constant. In this case, $F \in \operatorname{Aut}(\mathbb{H}^+)$ and the statement is obvious; so we can assume that $p \in \operatorname{Hol}(\mathbb{H}^+, \mathbb{H}^+)$.

Lemma 4.4.2 applied to p yields

$$\left|\frac{F(w_2) - F(w_1)}{w_2 - w_1}\right| = \left|1 + \frac{p(w_2) - p(w_1)}{w_2 - w_1}\right|$$
$$\geq 1 - \left|\frac{p(w_2) - p(w_1)}{w_2 - w_1}\right| \geq 1 - \left(\frac{\operatorname{Im} p(w_2) \operatorname{Im} p(w_1)}{(\operatorname{Im} w_2)(\operatorname{Im} w_1)}\right)^{1/2};$$

so to prove the assertion it suffices to show that $\operatorname{Im} p(w)/\operatorname{Im} w \to 0$ as $w \to +\infty$ inside

$$\Psi(G_{M,R}) = K(\infty, \varepsilon) \cup \bigcup_{\nu \in \mathbb{N}} F^\nu(B_{\mathbb{H}^+}(i, R)),$$

where $\varepsilon = 1/M$.

First of all, notice that if $w \in K(\infty, \varepsilon)$ then $\varepsilon |w| < \operatorname{Im} w$ by Proposition 2.2.7(i); therefore, if $w \to \infty$ inside $K(\infty, \varepsilon)$ we have

$$\frac{p(w)}{\operatorname{Im} w} = \frac{p(w)}{w} \frac{w}{\operatorname{Im} w} \to 0$$

because $p(w)/w \to p'(\infty) = 0$ as $w \to \infty$ nontangentially and $|w|/\operatorname{Im} w < 1/\varepsilon$. Therefore, $\operatorname{Im} p(w)/\operatorname{Im} w \to 0$ as $w \to \infty$ inside $K(\infty, \varepsilon)$ and this case is done.

For $\nu \in \mathbb{N}$, put $w_\nu = F^\nu(i)$. If $w_\nu \to \infty$ nontangentially then the balls $F^\nu(B_{\mathbb{H}^+}(i, R))$ are contained in a Stolz region at ∞ (Lemma 4.3.1) and we can argue as above. Unfortunately, we know that in the parabolic case the orbits might converge to ∞ tangentially; so we need a different approach.

Notice that, by the Schwarz–Pick lemma, $F^\nu(B_{\mathbb{H}^+}(i, R)) \subseteq B_{\mathbb{H}^+}(w_\nu, R)$; we need then some information on the Euclidean shape of the Poincaré balls in $\mathbb{H}^+$. Take $w \in B_{\mathbb{H}^+}(w_0, R)$; then

$$\left|\frac{w - w_0}{w - \overline{w_0}}\right| < r,$$

where $r = \tanh R \in (0,1)$. Recalling (4.17), we see that this is equivalent to

$$\frac{|w - w_0|^2}{(\operatorname{Im} w)^2} < \frac{4r^2}{1 - r^2}\frac{\operatorname{Im} w_0}{\operatorname{Im} w}. \tag{4.29}$$

In particular, this implies that

$$\left(1 - \frac{\operatorname{Im} w_0}{\operatorname{Im} w}\right)^2 = \frac{|\operatorname{Im} w - \operatorname{Im} w_0|^2}{(\operatorname{Im} w)^2} \le \frac{|w - w_0|^2}{(\operatorname{Im} w)^2} < \frac{4r^2}{1 - r^2}\frac{\operatorname{Im} w_0}{\operatorname{Im} w},$$

and this can happen if and only if

$$\frac{1 - r}{1 + r} < \frac{\operatorname{Im} w_0}{\operatorname{Im} w} < \frac{1 + r}{1 - r}.$$

Putting this estimate in (4.29), we also obtain

$$\frac{|w - w_0|}{\operatorname{Im} w} < \frac{2r}{1 - r}.$$

Let us apply these estimates to the ball $B_{\mathbb{H}^+}(w_\nu, R)$. Recalling (4.15), we have

$$\begin{aligned}
\operatorname{Im} p(w) - 2\operatorname{Im} p(w_\nu) &\le \frac{|\operatorname{Im} p(w) - \operatorname{Im} p(w_\nu)|^2}{\operatorname{Im} p(w)} \le \frac{|p(w) - p(w_\nu)|^2}{\operatorname{Im} p(w)} \\
&\le \frac{\operatorname{Im} p(w_\nu)}{(\operatorname{Im} w)(\operatorname{Im} w_\nu)}|w - w_\nu|^2.
\end{aligned}$$

Therefore, if $w \in B_{\mathbb{H}^+}(w_\nu, R)$ we get

$$\begin{aligned}
\frac{\operatorname{Im} p(w)}{\operatorname{Im} w} &\le 2\frac{\operatorname{Im} p(w_\nu)}{\operatorname{Im} w} + \frac{|w - w_\nu|^2}{(\operatorname{Im} w)^2}\frac{\operatorname{Im} p(w_\nu)}{\operatorname{Im} w_\nu} = \left[2\frac{\operatorname{Im} w_\nu}{\operatorname{Im} w} + \frac{|w - w_\nu|^2}{(\operatorname{Im} w)^2}\right]\frac{\operatorname{Im} p(w_\nu)}{\operatorname{Im} w_\nu} \\
&\le 2\frac{1 + r^2}{(1 - r)^2}\frac{\operatorname{Im} p(w_\nu)}{\operatorname{Im} w_\nu} = 2\frac{1 + r^2}{(1 - r)^2}\left(\frac{\operatorname{Im} w_{\nu+1}}{\operatorname{Im} w_\nu} - 1\right)
\end{aligned}$$

and we are done thanks to Theorem 4.4.3. □

Figures 4.1 and 4.2 contain examples of possible shapes for $G_{M,R}$ in the hyperbolic and parabolic cases.

We have all we need to prove the existence of simply connected f-absorbing domains where f is injective.

Theorem 4.5.4 (Cowen, 1982). *Let $f \in \operatorname{Hol}(\mathbb{D}, \mathbb{D})$ be hyperbolic or parabolic with Wolff point $\tau_f \in \partial\mathbb{D}$. Then there exists a simply connected f-absorbing domain $A \subset \mathbb{D}$ with $\tau_f \in \partial A$ such that $f|_A$ is injective.*

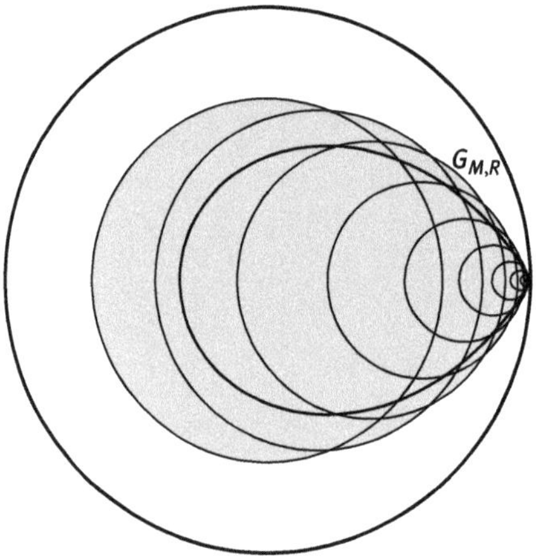

Figure 4.1: An example of domain $G_{M,R}$ in the hyperbolic case.

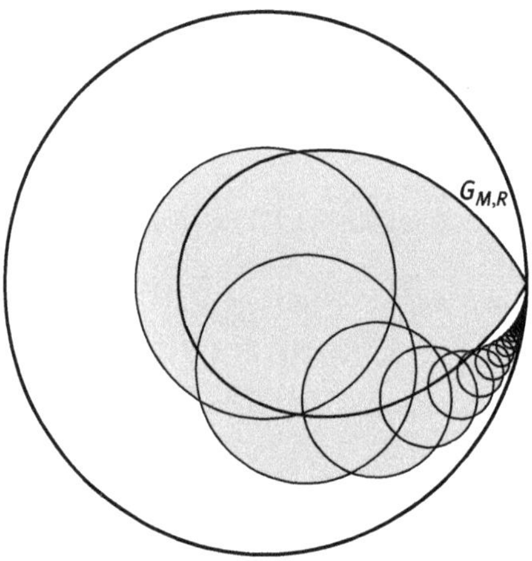

Figure 4.2: An example of domain $G_{M,R}$ in the parabolic case.

Proof. For $k \geq 2$, put $K_k = \overline{B_{\mathbb{D}}(0,k)}$ and $K_k^\circ = B_{\mathbb{D}}(0,k)$; let k_0 be the smallest integer so that $f(0) \in K_{k_0}^\circ$. Since $f^\nu \to \tau_f$ uniformly on compact sets, for any $0 < \rho < 1$ we can find $\nu_0 \in \mathbb{N}$ such that $\bigcup_{\nu \geq \nu_0} f^\nu(K_{k_0})$ is contained in $\{|\tau_f - z| < \rho\}$; therefore, by Lemma 4.5.3 we can assume that f is injective on $U_0 = \bigcup_{\nu \geq \nu_0} f^\nu(K_{k_0}^\circ)$. We clearly have $f(U_0) \subseteq U_0$; furthermore, since by construction $f(0) \in K_{k_0}^\circ \cap f(K_{k_0}^\circ)$ we have $f^{\nu+1}(0) \in f^\nu(K_{k_0}^\circ) \cap f^{\nu+1}(K_{k_0}^\circ)$ for all $\nu \in \mathbb{N}$, and hence U_0 is connected. Moreover, notice that for any $\mu \geq \nu_0$ we have $U_0 \cap f^\mu(K_{k_0+1}^\circ) \neq \emptyset$ and that the set

$$U_0 \setminus \bigcup_{\nu \geq \mu} f^\nu(K_{k_0+1}^\circ) \subseteq \bigcup_{\nu=\nu_0}^{\mu-1} f^\nu(K_{k_0}^\circ)$$

is relatively compact in $\mathbb{D}$.

Assume now that we have constructed integers $\nu_0 \le \nu_1 < \cdots \le \nu_{m-1}$ and open sets $U_0 \subset U_1 \subset \cdots \subset U_{m-1} \subset \mathbb{D}$ satisfying the following properties for each $0 \le j \le m-1$:
(a) U_j is connected;
(b) $f(U_j) \subseteq U_j$;
(c) f is injective on U_j;
(d) $\bigcup_{\nu \ge \nu_j} f^\nu(K^\circ_{k_0+j}) \subseteq U_j$;
(e) for any $\mu \ge \nu_j$, we have $U_j \cap f^\mu(K^\circ_{k_0+j+1}) \ne \emptyset$; and
(f) for any $\mu \ge \nu_j$, the set $U_j \setminus \bigcup_{\nu \ge \mu} f^\nu(K^\circ_{k_0+j+1})$ is relatively compact in $\mathbb{D}$.
Choose now $\tilde{\nu}_m \ge \nu_{m-1}$ so that f is injective on $\bigcup_{\nu \ge \tilde{\nu}_m} f^\nu(K_{k_0+m})$. By property (f), the set

$$L = \overline{U_{m-1} \setminus \bigcup_{\nu \ge \tilde{\nu}_m} f^\nu(K^\circ_{k_0+m})} \subset \mathbb{D}$$

is compact; so $f(L)$, too, is compact. Since $\{f^\nu\}$ converges uniformly on K_{k_0+m} to $\tau_f \notin f(L)$, we can find $\nu_m \ge \tilde{\nu}_m$ such that $f(L) \cap \bigcup_{\nu \ge \nu_m} f^\nu(K^\circ_{k_0+m}) = \emptyset$; put

$$U_m = U_{m-1} \cup \bigcup_{\nu \ge \nu_m} f^\nu(K^\circ_{k_0+m}).$$

We claim that ν_m and U_m satisfy properties (a)–(e) for $j = m$. Property (a) follows immediately from properties (a) and (e) for $j = m-1$. Property (b) follows from property (b) for $m-1$. Properties (d) and (e) are clear by construction. Since for any $\mu \ge \nu_m$ we have

$$U_m \setminus \bigcup_{\nu \ge \mu} f^\nu(K^\circ_{k_0+m+1}) \subseteq \left(U_{m-1} \setminus \bigcup_{\nu \ge \mu} f^\nu(K^\circ_{k_0+m+1})\right) \cup \bigcup_{\nu=\nu_m}^{\mu-1} f^\nu(K^\circ_{k_0+m}),$$

property (f) follows from property (f) for $m-1$. We are left with property (c). Take $z, w \in U_m$ with $z \ne w$. If $z, w \in \bigcup_{\nu \ge \tilde{\nu}_m} f^\nu(K^\circ_{k_0+m})$, then $f(z) \ne f(w)$ by construction; if $z, w \in U_{m-1}$, then $f(z) \ne f(w)$ by property (c) for $m-1$. So, we can assume that $z \in L$ and $w \in \bigcup_{\nu \ge \nu_m} f^\nu(K^\circ_{k_0+m})$; notice that the latter union starts from ν_m and not from $\tilde{\nu}_m$. But by construction $f(L)$ is disjoint from $\bigcup_{\nu \ge \nu_m} f^\nu(K^\circ_{k_0+m})$, which is f-invariant; therefore, $f(z) \ne f(w)$ and property (c) is proved.

Put $A' = \bigcup_{j \in \mathbb{N}} U_j$. Since A' is the increasing union of the U_j, it follows immediately that A' is connected, f-invariant, f-absorbing, and that f is injective on A'. The fact that A' is f-absorbing also implies that $\tau_f \in \partial A'$.

Finally, let A be the union of A' with the compact connected components of $\mathbb{C} \setminus A'$. By Lemma 4.5.2, $A \subseteq \mathbb{D}$ is simply connected and f-invariant; moreover, f is injective on A. Since clearly $\tau_f \in \partial A$ and A is f-absorbing (because A' is), we are done. □

Corollary 4.5.5. *Let $f \in \mathrm{Hol}(\mathbb{D}, \mathbb{D})$ be attracting elliptic, hyperbolic, or parabolic. Then there exists a model (X^o, ψ, Φ) for f, unique up to isomorphisms of models, with base $X^o = \mathbb{D}$ or $X^o = \mathbb{C}$.*

Proof. It follows from Corollary 4.1.3, Theorem 3.5.10, and Theorem 4.5.4. □

A natural question now is: how is it possible to identify the base of the model and the automorphism Φ starting from f? We shall address this question in the next section.

Notes to Section 4.5

Lemma 4.5.1 has been proved by Wolff [419] and Noshiro [308] in 1934; see also [410]. It is the first step in the characterization, due to Grunsky [176], of the domains *convex in the vertical direction*, i. e., such that the intersection with any vertical line is connected (possibly empty).

Theorem 4.5.4 in the hyperbolic case was implicit in the work of Valiron [398, 399]; however, the explicit statement and the unified proof for both the hyperbolic and the parabolic case has been given by Cowen [126] in 1981. Cowen has also shown that, when the orbit of f converges to the Wolff point nontangentially, the f-absorbing domain A can be chosen to have an inner tangent at τ (see Definition 2.4.8). The paper [120] contains a similar construction.

4.6 The hyperbolic step

In the previous section, we saw that if $f \in \mathrm{Hol}(\mathbb{D},\mathbb{D})$ is not superattracting elliptic then it admits a model. A natural question is whether we can understand the model just by looking at the function f. The answer is essentially affirmative; to show how, we need a couple of definitions and preliminary results.

Definition 4.6.1. Let X be a hyperbolic Riemann surface and take $f \in \mathrm{Hol}(X,X)$ and $z \in X$. For any $\mu \geq 1$ the Schwarz–Pick lemma implies that the sequence $\{\omega_X(f^\nu(z), f^{\nu+\mu}(z))\}_{\nu\in\mathbb{N}} \subset \mathbb{R}^+$ is decreasing, and hence it has a limit $s_\mu^f(z) \in \mathbb{R}^+$, that we shall call the *hyperbolic μ-step* of f at z. When the function f is clear by the context we shall write s_μ instead of s_μ^f. If $\mu = 1$, we shall write $s^f(z)$ instead of $s_1^f(z)$ and we shall call $s^f(z)$ the *hyperbolic step* of f at z. We shall say that f has *positive hyperbolic step* if there exists $z_0 \in \mathbb{D}$ such that $s^f(z_0) > 0$; otherwise, we say that f has a *zero hyperbolic step*. If X is an elliptic or parabolic Riemann surface, we shall put $s_\mu^f \equiv 0$ for all $\mu \geq 1$ and $f \in \mathrm{Hol}(X,X)$.

Remark 4.6.2. In the literature, parabolic maps with a positive hyperbolic step are sometimes called of *simply parabolic type* or of *parabolic II type* or of *automorphic type*. Analogously, parabolic maps with a zero hyperbolic step are sometimes called of *doubly parabolic type* or of *parabolic I type* or of *nonautomorphic type*.

The next lemma gives some basic properties of the hyperbolic step.

Lemma 4.6.3. *Let $f \in \mathrm{Hol}(\mathbb{D},\mathbb{D}) \setminus \{\mathrm{id}_\mathbb{D}\}$. Then*
(i) *if $f \in \mathrm{Aut}(\mathbb{D})$, then f has positive hyperbolic step;*
(ii) *if $f \notin \mathrm{Aut}(\mathbb{D})$ is elliptic, then it has zero hyperbolic step;*
(iii) *if the model of f has $\mathbb{D}$ as base, then f has positive hyperbolic step.*

Proof. (i) If $f \in \mathrm{Aut}(\mathbb{D})$, we have $\omega(f^\nu(z), f^{\nu+1}(z)) = \omega(z, f(z))$ for all $z \in \mathbb{D}$. Hence $s^f(z) = \omega(z, f(z))$ is strictly positive except when z is a fixed point of f.

(ii) In this case, all orbits of f converge to the Wolff point $\tau_f \in \mathbb{D}$; therefore, $\omega(f^\nu(z), f^{\nu+1}(z)) \to \omega(\tau_f, \tau_f) = 0$ for all $z \in \mathbb{D}$.

(iii) Let $(\mathbb{D}, \psi, \Phi)$ be the model for f. Then for all $z \in \mathbb{D}$ we have

$$\begin{aligned}\omega(f^\nu(z), f^{\nu+1}(z)) &\geq \omega(\psi \circ f^\nu(z), \psi \circ f^{\nu+1}(z)) \\ &= \omega(\Phi^\nu \circ \psi(z), \Phi^{\nu+1} \circ \psi(z)) = \omega(\psi(z), \Phi \circ \psi(z));\end{aligned}$$

thus $s^f(z) \geq \omega(\psi(z), F \circ \psi(z)) > 0$ as soon as $\psi(z)$ is not a fixed point of Φ, and we know that Φ has at most one fixed point by Lemma 3.5.4(v). □

We shall see later that all hyperbolic self-maps of $\mathbb{D}$ have positive hyperbolic step (see Corollary 4.6.9). On the other hand, parabolic self-maps of $\mathbb{D}$ can have either positive or zero hyperbolic step. Indeed, a parabolic automorphism has a positive hyperbolic step by Lemma 4.6.3(i); examples of parabolic self-maps with a zero hyperbolic step can be found in Examples 4.4.6 and 4.4.7 (actually in Example 4.4.7 we only proved that the hyperbolic step is zero at one point, but we shall see in Corollary 4.6.9 that this implies that it is zero everywhere).

The next lemma will be fundamental in the sequel.

Lemma 4.6.4. *Let X be a hyperbolic Riemann surface and $f \in \mathrm{Hol}(X, X)$. Then for every $z \in X$ the sequences $\{\omega_X(z, f^\nu(z))\}$ and $\{s_\nu^f(z)\}$ are subadditive. Furthermore, the limit of the sequence $\{\frac{1}{\nu}\omega_X(z, f^\nu(z))\}$ does not depend on $z \in X$ and we have*

$$\lim_{\nu\to+\infty} \frac{\omega_X(z, f^\nu(z))}{\nu} = \lim_{\nu\to+\infty} \frac{\omega_X(w, f^\nu(z))}{\nu} = \lim_{\nu\to+\infty} \frac{s_\nu^f(z)}{\nu} \tag{4.30}$$

for all $z, w \in X$.

Proof. For all $\mu, \nu \in \mathbb{N}$, we have

$$\begin{aligned}\omega_X(z, f^{\mu+\nu}(z)) &\leq \omega_X(z, f^\mu(z)) + \omega_X(f^\mu(z), f^{\mu+\nu}(z)) \\ &\leq \omega_X(z, f^\mu(z)) + \omega_X(z, f^\nu(z)),\end{aligned}$$

and hence $\{\omega_X(z, f^\nu(z))\}$ is subadditive.

Analogously, for all $\mu, \nu, \lambda \in \mathbb{N}$ we have

$$\begin{aligned}\omega_X(f^\lambda(z), f^{\lambda+\mu+\nu}(z)) &\leq \omega_X(f^\lambda(z), f^{\lambda+\mu}(z)) + \omega_X(f^{\lambda+\mu}(z), f^{\lambda+\mu+\nu}(z)) \\ &\leq \omega_X(f^\lambda(z), f^{\lambda+\mu}(z)) + \omega_X(f^\lambda(z), f^{\lambda+\nu}(z));\end{aligned}$$

taking the limit as $\lambda \to +\infty$ we get $s_{\mu+\nu}^f(z) \leq s_\mu^f(z) + s_\nu^f(z)$, as claimed.

Now, the Fekete lemma (see Lemma A.4.3) implies that the sequences $\frac{1}{\nu}\omega_X(z, f^\nu(z))$ and $\frac{1}{\nu}s_\nu^f(z)$ have a nonnegative limit for all $z \in X$. Since

$$\begin{aligned}\omega_X(z,f^\nu(z)) &\le \omega_X(z,w) + \omega_X(w,f^\nu(w)) + \omega_X(f^\nu(w),f^\nu(z))\\ &\le \omega_X(w,f^\nu(w)) + 2\omega_X(z,w)\end{aligned}$$

for all $z, w \in X$, it is clear that the limit of the sequence $\{\frac{1}{\nu}\omega_X(z,f^\nu(z))\}$ does not depend on $z \in X$.

The first equality in (4.30) now follows from the estimates

$$\omega_X(z,f^\nu(z)) - \omega_X(z,w) \le \omega_X(w,f^\nu(z)) \le \omega_X(z,f^\nu(z)) + \omega_X(z,w).$$

Finally, since we clearly have $\omega_X(z,f^\nu(z)) \ge s_\nu^f(z)$, it follows that

$$\lim_{\nu\to+\infty} \frac{s_\nu^f(z)}{\nu} \le \lim_{\nu\to+\infty} \frac{\omega_X(z,f^\nu(z))}{\nu}.$$

Now fix $\mu \ge 1$. Then for all $\nu \ge 1$, we have

$$\frac{\omega_X(z,f^{\mu\nu}(z))}{\mu\nu} \le \frac{1}{\mu\nu}\sum_{\lambda=0}^{\nu-1} \omega_X(f^{\lambda\mu}(z),f^{(\lambda+1)\mu}(z));$$

thus taking the limit as $\nu \to +\infty$ the Cesàro means theorem (Corollary A.4.5) yields

$$\lim_{\nu\to+\infty} \frac{\omega_X(z,f^\nu(z))}{\nu} = \lim_{\nu\to+\infty} \frac{\omega_X(z,f^{\mu\nu}(z))}{\mu\nu} \le \frac{s_\mu^f(z)}{\mu}.$$

Letting $\mu \to +\infty$ we then get the equality between the first and the third member of (4.30) and we are done. □

Definition 4.6.5. Let X be a hyperbolic Riemann surface and $f \in \mathrm{Hol}(X,X)$. The *divergence rate* $c(f) \in \mathbb{R}^+$ is defined by

$$c(f) = \lim_{\nu\to+\infty} \frac{\omega_X(z,f^\nu(z))}{\nu}$$

for any $z \in X$; the previous lemma implies that the limit exists and does not depend on z. If X is an elliptic or parabolic Riemann surface, we shall put $c(f) = 0$ for all $f \in \mathrm{Hol}(X,X)$.

If $f \in \mathrm{Hol}(X,X)$ has a fixed point in X, we clearly have $c(f) = 0$. When $X = \mathbb{D}$ and $f \in \mathrm{Hol}(\mathbb{D},\mathbb{D})$ has no fixed point, we can explicitly compute the divergence rate.

Proposition 4.6.6. *Let $f \in \mathrm{Hol}(\mathbb{D},\mathbb{D})$ be with Wolff point $\tau_f \in \partial\mathbb{D}$. Then*

$$c(f) = \frac{1}{2}\log\frac{1}{f'(\tau_f)}.$$

In particular, $c(f) > 0$ iff f is hyperbolic and $c(f) = 0$ iff f is parabolic.

Proof. Put $z_\nu = f^\nu(0)$; we know that $z_\nu \to \tau_f$. Then the Julia–Wolff–Carathéodory Theorem 2.3.2 and Proposition 2.1.15 yield

$$\liminf_{\nu\to+\infty} \frac{1-|z_{\nu+1}|}{1-|z_\nu|} = \liminf_{\nu\to+\infty} \frac{1-|f(z_\nu)|}{1-|z_\nu|} \geq f'(\tau_f);$$

so a standard property of lim inf (see Lemma A.4.1) implies

$$\liminf_{\nu\to+\infty}(1-|z_\nu|)^{1/\nu} \geq f'(\tau_f).$$

Thus

$$c(f) = \lim_{\nu\to+\infty} \frac{\omega(0,z_\nu)}{\nu} = \frac{1}{2}\lim_{\nu\to+\infty} \log\left(\frac{1+|z_\nu|}{1-|z_\nu|}\right)^{1/\nu} \leq \frac{1}{2}\log\frac{1}{f'(\tau_f)}.$$

In particular, if f is parabolic we get $c(f) = 0$ and we are done.

Assume then that f is hyperbolic and take $0 < R < 1$. Since the point of the closed horocycle $\overline{E(\tau_f, R)}$ closest to 0 is $\frac{1-R}{1+R}\tau_f$, we have

$$\inf_{z\in E(\tau_f,R)} \omega(0,z) = \omega\left(0, \frac{1-R}{1+R}\tau_f\right) = \frac{1}{2}\log\frac{1}{R}.$$

Since $0 \in \partial E(\tau_f, 1)$, the Julia lemma (Theorem 2.1.10) implies that $z_\nu \in \overline{E(\tau_f, f'(\tau_f)^\nu)}$ for all $\nu \geq 1$; then

$$\omega(0,z_\nu) \geq \frac{1}{2}\log\frac{1}{f'(\tau_f)^\nu} = \frac{\nu}{2}\log\frac{1}{f'(\tau_f)}.$$

Dividing by ν and letting $\nu \to +\infty$, we get $c(f) \geq \frac{1}{2}\log\frac{1}{f'(\tau_f)}$ and we are done. □

The reason we introduced hyperbolic step and the divergence rate in this context is because they behave well when passing to a model. More precisely, we have the following statement, where when X^o is not hyperbolic we put $\omega_{X^o} \equiv 0$.

Lemma 4.6.7. *Let X be a hyperbolic Riemann surface. Take $f \in \mathrm{Hol}(X,X)$ and let (X^o, ψ, Φ) be a semimodel for f. Then:*

(i) *we have $c(\Phi) \leq c(f)$ and $s_\mu^\Phi(\psi(z)) \leq s_\mu^f(z)$ for all $\mu \in \mathbb{N}$ and $z \in X$;*

(ii) *if (X^o, ψ, Φ) is a model with $X^o = \mathbb{C}$ or $X^o = \mathbb{D}$, we have*

$$\omega_{X^o}(\psi(z), \psi(w)) \geq \lim_{\nu\to+\infty} \omega_X(f^\nu(z), f^\nu(w))$$

for all $z, w \in X$;

(iii) *if (X^o, ψ, Φ) is a model with $X^o = \mathbb{C}$ or $X^o = \mathbb{D}$, then $c(\Phi) = c(f)$ and $s_\mu^\Phi(\psi(z)) = s_\mu^f(z)$ for all $\mu \in \mathbb{N}$ and $z \in X$.*

Proof. (i) If X^o is not hyperbolic, the assertions are obvious. If instead X^o is hyperbolic, from $\psi \circ f = \Phi \circ \psi$ we get

$$c(\Phi) = \lim_{\nu\to+\infty} \frac{\omega_{X^o}(\psi(z), \Phi^\nu(\psi(z)))}{\nu} = \lim_{\nu\to+\infty} \frac{\omega_{X^o}(\psi(z), \psi(f^\nu(z)))}{\nu}$$
$$\leq \lim_{\nu\to+\infty} \frac{\omega_X(z, f^\nu(z))}{\nu} = c(f).$$

Analogously,

$$s_\mu^\Phi(\psi(z)) = \lim_{\nu\to+\infty} \omega_{X^o}(\Phi^\nu(\psi(z)), \Phi^{\mu+\nu}(\psi(z)))$$
$$= \lim_{\nu\to+\infty} \omega_{X^o}(\psi(f^\nu(z)), \psi(f^{\mu+\nu}(z)))$$
$$\leq \lim_{\nu\to+\infty} \omega_X(f^\nu(z), f^{\mu+\nu}(z)) = s_\mu^f(z).$$

(ii) Let $A \subseteq X$ be a f-absorbing domain where ψ is injective; in particular, ψ is an isometry between ω_A and $\omega_{\psi(A)}$. Setting $X_\nu^o = \Phi^{-\nu}(\psi(A))$, Lemma 3.5.4(iii) implies that X^o is the increasing union of the X_ν^o. In particular, given $z, w \in X$, we have $\psi(z), \psi(w) \in X_\nu^o$ for all ν large enough. By Proposition 1.9.31, we then have

$$\omega_{X^o}(\psi(z), \psi(w)) = \lim_{\nu\to+\infty} \omega_{X_\nu^o}(\psi(z), \psi(w)) = \lim_{\nu\to+\infty} \omega_{\psi(A)}(\Phi^\nu(\psi(z)), \Phi^\nu(\psi(w)))$$
$$= \lim_{\nu\to+\infty} \omega_{\psi(A)}(\psi(f^\nu(z)), \psi(f^\nu(w))) = \lim_{\nu\to+\infty} \omega_A(f^\nu(z), f^\nu(w))$$
$$\geq \lim_{\nu\to+\infty} \omega_X(f^\nu(z), f^\nu(w)),$$

as claimed.

(iii) Using (ii) and the fact that Φ is an isometry for ω_{X^o}, we have

$$s_\mu^f(z) = \lim_{\nu\to+\infty} \omega_X(f^\nu(z), f^{\mu+\nu}(z)) = \lim_{\nu\to+\infty} \omega_X(f^\nu(z), f^\nu(f^\mu(z)))$$
$$\leq \omega_{X^o}(\psi(z), \psi(f^\mu(z))) = \omega_{X^o}(\psi(z), \Phi^\mu(\psi(z)))$$
$$= \lim_{\nu\to+\infty} \omega_{X^o}(\Phi^\nu(\psi(z)), \Phi^{\mu+\nu}(\psi(z))) = s_\mu^\Phi(\psi(z))$$

and the equality follows from (i). The equality of the divergence rates then is a consequence of (4.30). □

We are finally able to identify the model for any $f \in \mathrm{Hol}(\mathbb{D}, \mathbb{D})$ not superattracting elliptic. In the statement, for simplicity we use $\mathbb{H}^+$ instead of $\mathbb{D}$ as base of the models.

Theorem 4.6.8. *Let $f \in \mathrm{Hol}(\mathbb{D}, \mathbb{D})$ and assume that f is not a superattracting elliptic. Then:*
(i) *f is an elliptic automorphism if and only if its model is $(\mathbb{D}, \psi, \Phi)$ with $\Phi(z) = \lambda z$, where $|\lambda| = 1$ and λ is the derivative of f at its Wolff point;*
(ii) *f is attracting elliptic if and only if its model is $(\mathbb{C}, \psi, \Phi)$ with $\Phi(w) = \lambda w$, where $0 < |\lambda| < 1$ and λ is the derivative of f at its Wolff point;*

(iii) *f is hyperbolic if and only if its model is $(\mathbb{H}^+, \psi, \Phi)$ with $\Phi(w) = \beta^{-1}w$, where $\beta \in (0,1)$ is the angular derivative of f at its Wolff point;*
(iv) *f is parabolic with positive hyperbolic step if and only if its model is $(\mathbb{H}^+, \psi, \Phi)$ with $\Phi(w) = w + 1$ or $\Phi(w) = w - 1$;*
(v) *f is parabolic with zero hyperbolic step if and only if its model is $(\mathbb{C}, \psi, \Phi)$, with $\Phi(w) = w + 1$.*

Proof. First of all, assume that f is an elliptic automorphism with fixed point $\tau \in \mathbb{D}$ and let $\psi \in \mathrm{Aut}(\mathbb{D})$ be an automorphism sending τ to 0. Then $\psi \circ f \circ \psi^{-1}(z) = f'(\tau)z$, i. e., $\psi \circ f = \Phi \circ \psi$. Taking the trivial f-absorbing domain $A = \mathbb{D}$, we have proved that $(\mathbb{D}, \psi, \Phi)$ is a (trivial) model for f.

Assume now that f is of attracting elliptic type with Wolff point $\tau_f \in \mathbb{D}$. In particular, by Lemma 4.6.3 f has zero hyperbolic step, and thus the base of its model (that exists by Corollary 4.5.5) is $\mathbb{C}$. Moreover, by Lemma 3.5.4(v) the normal form $\Phi \in \mathrm{Aut}(\mathbb{C})$ must have a fixed point, that up to post-composing ψ with a biholomorphism of $\mathbb{C}$ we can assume to be the origin; in particular, $\Phi(w) = \lambda w$ for some $\lambda \in \mathbb{C}^*$. Differentiating $\psi \circ f = \Phi \circ \psi$, we get

$$\psi'(f(z))f'(z) = \Phi'(\psi(z))\psi'(z) = \lambda\psi'(z)$$

for all $z \in \mathbb{D}$. Taking $z = \tau_f$, we get $\psi'(\tau_f)f'(\tau_f) = \lambda\psi'(\tau_f)$; since τ_f must belong to any f-absorbing domain—and in particular to the f-absorbing domain where ψ is injective—we have $\psi'(\tau_f) \neq 0$, and hence $\lambda = f'(\tau_f)$, as claimed.

Conversely, if the model for f is as in cases (i) or (ii) then Lemma 3.5.4(v) implies that f has a fixed point $\tau \in \mathbb{D}$. The argument just used shows that $f'(\tau) = \lambda$, and thus f is an elliptic automorphism in case (i) and is attracting elliptic in case (ii).

The next observation is that the models listed in cases (iii)–(v) are all the possible models for functions without fixed points in $\mathbb{D}$ (i. e., not elliptic). Indeed, if $(\mathbb{C}, \psi, \Phi)$ is such a model then Φ cannot have fixed points in $\mathbb{C}$; thus up to a conjugation we can assume $\Phi(w) = w + 1$ and we are in case (v). Analogously, if $(\mathbb{H}^+, \psi, \Phi)$ is such a model, up to a conjugation we can assume that the Wolff point of Φ is ∞. Then $\Phi(w) = \beta^{-1}w$ for some $0 < \beta < 1$ or $\Phi(w) = w + b$ for some $b \in \mathbb{R}^*$. In the former situation, we are in case (iii); in the latter situation, an easy computation shows that Φ is conjugated either to $w \mapsto w + 1$ or to $w \mapsto w - 1$ and these two automorphisms are not conjugated; therefore, we are in case (iv).

Using the hyperbolic step and the divergence rate, we can tell these models apart. Indeed, by definition the model in case (v) has zero hyperbolic step and zero divergence rate. The models in case (iii) have positive hyperbolic step, by Lemma 4.6.3(iii); moreover,

$$c(\Phi) = \lim_{\nu \to +\infty} \frac{\omega_{\mathbb{H}^+}(i, \beta^{-\nu} i)}{\nu} = \frac{1}{2}\log\frac{1}{\beta} > 0,$$

i. e., the divergence rate is positive. Finally, the models in case (iv) have again positive hyperbolic step but the divergence rate is 0 because

$$c(\Phi) = \lim_{\nu\to\infty} \frac{\omega_{\mathbb{H}^+}(i, i \pm \nu)}{\nu} = \lim_{\nu\to+\infty} \frac{1}{\nu} \tanh^{-1}\left|\frac{\nu}{\pm\nu + 2i}\right|$$
$$= \lim_{\nu\to+\infty}\left[\frac{1}{\nu}\log\nu + \frac{1}{2\nu}\log\frac{\sqrt{1+4\nu^{-2}}+1}{\frac{1}{2}+O(\nu^{-2})}\right] = 0.$$

This is enough to complete the proof. If f is hyperbolic with angular derivative β at the Wolff point, Proposition 4.6.6 says that $c(f) = \frac{1}{2}\log\frac{1}{\beta} > 0$. By Lemma 4.6.7(iii), the model should have the same divergence rate, and thus the previous analysis shows that we necessarily are in case (iii).

If instead f is parabolic with positive hyperbolic step, again the combination of Proposition 4.6.6 and Lemma 4.6.7(iii) implies that we must be in case (iv); analogously, if f is parabolic with zero hyperbolic step then we must be in case (v). □

We can finally keep an old promise and show that $s^f(z_0) > 0$ for some $z_0 \in \mathbb{D}$ if and only if $s^f(z) > 0$ for all $z \in \mathbb{D}$.

Corollary 4.6.9. *Let $f \in \mathrm{Hol}(\mathbb{D},\mathbb{D}) \setminus \mathrm{Aut}(\mathbb{D})$. Then:*
(i) *there exists $z_0 \in \mathbb{D}$ such that $s^f(z_0) > 0$ if and only if $s^f(z) > 0$ for all $z \in \mathbb{D}$;*
(ii) *if f is hyperbolic with Wolff point $\tau_f \in \partial\mathbb{D}$, then f has positive hyperbolic step and*

$$\inf_{z\in\mathbb{D}} s^f(z) = \frac{1}{2}\log\frac{1}{f'(\tau_f)} > 0;$$

(iii) *if f is parabolic, then*

$$\inf_{z\in\mathbb{D}} s^f(z) = 0;$$

(iv) *if f is parabolic with zero hyperbolic step, we have*

$$\lim_{\nu\to+\infty} \omega(f^\nu(z), f^\nu(w)) = 0$$

for all $z, w \in \mathbb{D}$.

Proof. (i) We already know (Lemma 4.6.3) that if f is elliptic, then $s^f(z) = 0$ for all $z \in \mathbb{D}$. Assume then that f is hyperbolic or parabolic and there exists $z_0 \in \mathbb{D}$ with $s^f(z_0) > 0$; let (X^o, ψ, Φ) the model for f. By Lemma 4.6.7, we have $s^\Phi(\psi(z_0)) = s^f(z_0) > 0$; in particular, X^o cannot be $\mathbb{C}$. But then we have $s^\Phi(\psi(z)) > 0$ for all $z \in \mathbb{D}$, because Φ is an automorphism without fixed points, and thus $s^f(z) > 0$ for all $z \in \mathbb{D}$, again by Lemma 4.6.7.

(ii) The hyperbolic automorphism $\Phi(w) = \lambda w$ of $\mathbb{H}^+$, where $\lambda = f'(\tau_f)^{-1} > 1$, has positive hyperbolic step, and hence $s^f(z) > 0$ for all $z \in \mathbb{D}$ thanks to Theorem 4.6.8 and Lemma 4.6.7(iii). Now, if $w = re^{i\theta} \in \mathbb{H}^+$ we have

$$s^\Phi(w) = \lim_{\nu\to+\infty} \omega_{\mathbb{H}^+}(\lambda^\nu w, \lambda^{\nu+1} w) = \omega_{\mathbb{H}^+}(w, \lambda w) = \tanh^{-1}\left|\frac{\lambda - 1}{\lambda - e^{-2i\theta}}\right|.$$

Therefore,

$$\inf_{w\in\mathbb{H}^+} s^\Phi(w) = \tanh^{-1}\frac{\lambda-1}{\lambda+1} = \frac{1}{2}\log\lambda;$$

the infimum is realized when $\theta = \pi/2$. By definition of (semi)model the set $\psi(\mathbb{D})$ must be Φ-absorbing; hence for every $\theta \in (0,\pi)$ there must exist $z \in \mathbb{D}$ such that $\psi(z)$ has argument equal to θ. Since $s^\Phi(\psi(z)) = s^f(z)$ and $s^\Phi(w)$ depends only on the argument of w, we have

$$\inf_{z\in\mathbb{D}} s^f(z) = \inf_{w\in\mathbb{H}^+} s^\Phi(w) = \frac{1}{2}\log\frac{1}{f'(\tau_f)},$$

as claimed.

(iii) If f has zero hyperbolic step, the assertion is obvious. Assume then that f has positive hyperbolic step; by Theorem 4.6.8 the model for f is $(\mathbb{H}^+, \psi, \Phi)$ with $\Phi(w) = w \pm 1$. Now, if $w \in \mathbb{H}^+$ we have

$$s^\Phi(w) = \omega_{\mathbb{H}^+}(w, w\pm 1) = \tanh^{-1}\frac{1}{\sqrt{1+4(\operatorname{Im} w)^2}},$$

and hence $\inf_{w\in\mathbb{H}^+} s^\Phi(w) = 0$. Arguing as in the previous case, we then find $\inf_{z\in\mathbb{D}} s^f(z) = 0$ and we are done.

(iv) Since, by Theorem 4.6.8, the base of the model of f is $\mathbb{C}$, the statement follows from Lemma 4.6.7(ii). □

Corollary 4.6.10. *Let $f \in \mathrm{Hol}(\mathbb{D},\mathbb{D})$ be parabolic. Assume that some orbit converges nontangentially to the Wolff point. Then f has zero parabolic step.*

Proof. If f has positive hyperbolic step, then all orbits converge tangentially to the Wolff point, by Corollary 4.6.9(i) and Corollary 4.4.4. □

When $f \in \mathrm{Hol}(\mathbb{D},\mathbb{D})$ is parabolic with a positive hyperbolic step, we saw that the orbits converge tangentially to the Wolff point either all with angle $\pi/2$ or all with angle $-\pi/2$. We have also seen that for parabolic self-maps with a zero hyperbolic step the slope of convergence of the orbits to the Wolff point, assuming that it is well-defined, might depend on the map. However, it does not depend on the orbit.

Corollary 4.6.11. *Let $f \in \mathrm{Hol}(\mathbb{D},\mathbb{D})$ be parabolic with zero hyperbolic step and Wolff point $\tau_f \in \partial\mathbb{D}$. Assume that there exists $z_0 \in \mathbb{D}$ such that $\frac{\tau_f - f^\nu(z_0)}{|\tau_f - f^\nu(z_0)|} \to \sigma \in \partial\mathbb{D}$ as $\nu \to +\infty$. Then $\frac{\tau_f - f^\nu(z)}{|\tau_f - f^\nu(z)|} \to \sigma \in \partial\mathbb{D}$ for all $z \in \mathbb{D}$.*

Proof. It follows from Corollary 4.6.9(iv) and Lemma 2.2.17(ii). □

We have shown that in the model for f the base and the normal form are uniquely determined up to isomorphisms of models. We end this section by proving that also the intertwining maps are substantially unique.

Corollary 4.6.12. *Let $f \in \mathrm{Hol}(\mathbb{D},\mathbb{D})$ be not superattracting elliptic. Let (X^o, ψ, Φ) and $(X^o, \tilde{\psi}, \Phi)$ be two models for f as in Theorem 4.6.8. Then:*
(i) *if f is an elliptic automorphism, then $\tilde{\psi} = \lambda\psi$ for some $\lambda \in \mathbb{S}^1$;*
(ii) *if f is attracting elliptic, then $\tilde{\psi} = a\psi$ for some $a \in \mathbb{C}^*$;*
(iii) *if f is hyperbolic, then $\tilde{\psi} = a\psi$ for some $a > 0$;*
(iv) *if f is parabolic with a positive hyperbolic step, then $\tilde{\psi} = \psi + b$ for some $b \in \mathbb{R}$;*
(v) *if f is parabolic with a zero hyperbolic step, then $\tilde{\psi} = \psi + b$ for some $b \in \mathbb{C}$.*

Proof. By Corollary 3.5.9, there must exist $\gamma \in \mathrm{Aut}(X^o)$ such that $\Phi \circ \gamma = \gamma \circ \Phi$ and $\tilde{\psi} = \gamma \circ \psi$.

In case (i), we have $X^o = \mathbb{D}$ and $\Phi(z) = \lambda_0 z$ with $|\lambda_0| = 1$. Then Proposition 1.4.12 implies $\gamma(z) = \lambda z$ with $|\lambda| = 1$, as claimed.

In case (ii), we have $X^o = \mathbb{C}$ and $\Phi(w) = \lambda w$ with $0 < |\lambda| < 1$. Then Corollary 1.6.21(vi) yields $\gamma(w) = aw$ with $a \in \mathbb{C}^*$, as claimed.

In cases (iii) and (iv), we have $X = \mathbb{H}^+$. Proposition 1.4.12 then implies that γ must have the same fixed points as Φ; therefore, $\gamma(w) = aw$ with $a > 0$ in case (iii) and $\gamma(w) = w + b$ with $b \in \mathbb{R}$ in case (iv).

Finally, in case (v) we have $X = \mathbb{C}$ and $\Phi(w) = w + 1$. Applying again Corollary 1.6.21(vi), we get $\gamma(w) = w + b$ for some $b \in \mathbb{C}$ and we are done. □

Notes to Section 4.6

The use of the hyperbolic step and the rate of convergence for classifying models is adapted from [80, Chapter 9]. It would be interesting to have a direct proof of Corollary 4.6.9 not depending on models.

To find a version of Theorem 4.6.8 for an arbitrary hyperbolic Riemann surface X is delicate. If $\tilde{f} \in \mathrm{Hol}(\mathbb{D},\mathbb{D})$ is the lift of a $f \in \mathrm{Hol}(X,X)$ without fixed points, we shall say that f is *hyperbolic* (resp. *positive parabolic, zero parabolic*) if $\tilde{f}$ is hyperbolic (resp., parabolic with positive hyperbolic step, parabolic with zero hyperbolic step). Bonfert [70] has shown that if $D \subset \mathbb{C}$ is a hyperbolic domain then a holomorphic self-map $f \in \mathrm{Hol}(D,D)$ is zero parabolic if and only if it has a zero hyperbolic step (at one point, and hence at all points).

König [247] proved an analogous of Theorem 4.6.8 for a hyperbolic domain $D \subset \widehat{\mathbb{C}}$ and $f \in \mathrm{Hol}(D,D)$ such that $f^k \to \infty \in \partial D$ and with a simply connected absorbing domain. To do so, he replaces the hyperbolic step by an Euclidean analogue. More precisely, put

$$s_\nu^f(z) = \frac{|f^{\nu+1}(z) - f^\nu(z)|}{d(f^\nu(z), \partial D)}.$$

Then he shows that f is zero parabolic if and only if $S_\nu^f(z) \to 0$ as $\nu \to +\infty$ for all $z \in D$; that f is positive parabolic if and only if $\liminf_\nu S_\nu^f(z) > 0$ for all $z \in D$ but $\inf_z \limsup_\nu S_\nu^f(z) = 0$; that f is hyperbolic if and only if $\inf_z \inf_\nu S_\nu^f(z) > 0$.

[37] studies a similar problem when $D \subset \widehat{\mathbb{C}}$ is hyperbolic and $f \in \mathrm{Hol}(D, D)$ has no fixed points in D nor isolated boundary fixed points (we say that an isolated point $\tau \in \partial D$ is an *isolated boundary fixed point* of f if f extends continuously to τ with $f(\tau) = \tau$; notice that, since D is hyperbolic, the existence of a continuous extension of f to τ is automatic, thanks to Theorem 1.7.25). The authors show that f is zero parabolic if and only if $S_\nu^f(z) \to 0$ for one (and hence all) $z \in D$. They also show that if $\inf_{z\in D} s^f(z) > 0$ then f is hyperbolic and ask whether the converse is true.

More generally, let $f \in \mathrm{Hol}(X,X)$ be a holomorphic self-map of a hyperbolic Riemann surface X. If f has an attracting but not superattracting fixed point $z_0 \in X$, then we can clearly find a simply connected absorbing domain for f where f is injective (it suffices to take a sufficiently small Poincaré ball centered at z_0), and thus by Theorem 3.5.10 we can construct a model, which is of the kind given by Theorem 4.6.8(ii). When f is a periodic or pseudoperiodic automorphism, there is no need to build a model; so the only remaining interesting case is when $\{f^\nu\}$ is compactly divergent.

The more general construction in this case is due to Marden and Pommerenke [277]. Let $\tilde{f} \in \mathrm{Hol}(\mathbb{D}, \mathbb{D})$ be a lift of f with respect to the universal covering map $\pi\colon \mathbb{D} \to X$. Let $(\mathbb{X}^o, \tilde{\psi}, \tilde{\Phi})$ be the model of $\tilde{f}$, with $\mathbb{X}^o = \mathbb{C}$ is f is zero parabolic and $\mathbb{X}^o = \mathbb{H}^+$ otherwise. Let $\Gamma \subset \mathrm{Aut}(\mathbb{D})$ be the covering group of π. Then Marden and Pommerenke have constructed a properly discontinuous group $\Gamma_\infty \subset \mathrm{Aut}(\mathbb{X}^o)$ acting freely (with $\Gamma_\infty = \{\mathrm{id}_{\mathbb{C}}\}$ if $\mathbb{X}^o = \mathbb{C}$) such that $\tilde{\Phi} \circ \Gamma_\infty \circ \tilde{\Phi}^{-1} \subseteq \Gamma_\infty$; and a group homomorphism $\Theta\colon \Gamma \to \Gamma_\infty$ such that $\tilde{\psi} \circ \gamma = \Theta(\gamma) \circ \tilde{\psi}$ for all $\gamma \in \Gamma$. As a consequence, if we set $X^o = \mathbb{X}^o/\Gamma_\infty$ and denote by $\tilde{\pi}\colon \mathbb{X}^o \to X^o$ the associated covering map, we see that $\tilde{\psi}$ induces a holomorphic map $\psi\colon X \to X^o$ and $\tilde{\Phi}$ induces a self-covering map $\Phi\colon X^o \to X^o$ so that the following diagram commutes:

$$
\begin{array}{ccccccc}
\mathbb{D} & & & \xrightarrow{\tilde{\psi}} & & & \mathbb{X}^o \\
\downarrow \pi & \searrow^{\tilde{f}} & & & & \swarrow^{\tilde{\Phi}} & \downarrow \tilde{\pi} \\
 & & \mathbb{D} & \xrightarrow{\tilde{\psi}} & \mathbb{X}^o & & \\
 & & \downarrow \pi & & \downarrow \tilde{\pi} & & \\
 & & X & \xrightarrow{\psi} & X^o & & \\
 & \nearrow^{f} & & & & \nwarrow^{\Phi} & \\
X & & & \xrightarrow{\psi} & & & X^o
\end{array}
$$

Furthermore, if $\tilde{A} \subset \mathbb{D}$ is a simply connected $\tilde{f}$-absorbing domain where $\tilde{f}$ is injective and such that $\tilde{\psi}(\tilde{A})$ is $\tilde{\Phi}$-absorbing, then $A = \pi(\tilde{A})$ is f-absorbing and $\psi(A)$ is Φ-absorbing.

The triple (X^o, ψ, Φ) in general is not a model because Φ might not be an automorphism and A might not be simply connected. However, König [247] has given a sufficient condition for this construction to give an actual model. We say that f satisfies *condition* (K) if for every closed curve σ in X there is $\nu \geq 1$ such that $f^\nu \circ \sigma$ is homotopic to a constant in X. Then König has shown that if f is zero parabolic or satisfies condition (K) then $\Gamma_\infty = \{\mathrm{id}_{\mathbb{X}^o}\}$ and (X^o, ψ, Φ) becomes an actual model, so that A is a simply connected f-absorbing domain (in [247] a model is called *conformal conjugacy*). Notice that Corollary 3.3.18 implies that condition (K) is satisfied in multiply connected hyperbolic domains of regular type without points component; so in this case we do have a true model for f. See also [407] for a related result.

When $X = D \subset \mathbb{C}$ is a hyperbolic domain in the plane something more can be said. Let $f \in \mathrm{Hol}(D, D)$. Bonfert [70], improving the construction of Marden–Pommerenke, has shown that if f is not zero parabolic and it does not have an isolated boundary fixed point, then one can find a hyperbolic domain $D^o \subset \mathbb{C}$, a holomorphic map $\psi: D \to D^o$ and an automorphism $\Phi \in \mathrm{Aut}(D^o)$ so that $\psi \circ f = \Phi \circ \psi$. In [37], it is shown that when f is zero parabolic and $f^\nu \to \tau \in \partial D$ then f automatically satisfies condition (K).

König [247] has also constructed an example of a hyperbolic domain $D \subset \mathbb{C}$ having a $f \in \mathrm{Hol}(D, D)$ such that $f^k \to \tau \in \partial D$ and f has no simply connected absorbing domains. Extending König's arguments, Barański, Fagella, Jarque, and Karpińska in [37] have proved that when $D \subset \mathbb{C}$ and $f^k \to \tau \in \partial D$ then a simply connected f-absorbing domain always exists if f is zero parabolic (essentially because Condition (K) holds), but there are examples of hyperbolic and positive parabolic maps where a simply connected f-absorbing domain cannot exist. In the same paper, they also show that if τ_0 is an isolated boundary fixed point then there cannot exist simply connected f-absorbing domains. On the positive side, the same authors in [38] proved the existence of "nice" absorbing domains, not necessarily simply connected, independently of the type of f; such an absorbing domain can also be chosen so that f is locally (but not necessarily globally) injective there.

Corollary 4.6.9(ii) and (iii) can also be found in [60].

Corollary 4.6.10 has been proved in [73, Theorem 6.1] without recurring to models. See also [126, Theorem 3.5].

Adapting the techniques of [80, Chapter 17] it is possible to prove that all orbits of a parabolic self-map $f \in \mathrm{Hol}(\mathbb{D}, \mathbb{D})$ with positive hyperbolic step converge to the Wolff point τ tangentially with angle $\pm\pi/2$ if and only if the model for f is $(\mathbb{H}^+, \psi, \Phi^\mp)$, where $\Phi^\mp(w) = w \mp 1$. Furthermore, adapting [80, Example 17.5.4] it is possible to construct a parabolic self-map with zero hyperbolic step such that the orbits do not converge to the Wolff point with a precise slope.

We introduced the hyperbolic step as the limit of a decreasing sequence computed starting from an orbit. Another construction can be used to study in more detail the differences between maps with positive hyperbolic step and maps with zero hyperbolic step. Given $\tau \in \partial\mathbb{D}$, let $h_\tau: \mathbb{D} \to \mathbb{R}$ be given by

$$h_\tau(z) = \frac{1}{2} \log \frac{|\tau - z|^2}{1 - |z|^2} = \lim_{\zeta \to \tau} \big[\omega(z, \zeta) - \omega(0, \zeta)\big].$$

Notice that

$$h_\tau(z) - \omega(z, w) \le h_\tau(w) \le h_\tau(z) + \omega(z, w)$$

for all $z, w \in \mathbb{D}$. If $f \in \mathrm{Hol}(\mathbb{D}, \mathbb{D})$ has a Wolff point $\tau_f \in \partial\mathbb{D}$ then the Julia lemma gives $h_{\tau_f}(f(z)) \le h_{\tau_f}(z) - \frac{1}{2} \log \frac{1}{\beta}$, where $\beta = f'(\tau_f) \in (0, 1]$. In the hyperbolic case, this implies that $h_{\tau_f}(f^\nu(z)) \to -\infty$ as $\nu \to +\infty$ for all $z \in \mathbb{D}$. In the parabolic case, we can instead only infer that the sequence $\{h_{\tau_f}(f^\nu(z))\}$ is not increasing, and hence it admits a limit $\ell(z) \in [-\infty, +\infty)$. We shall say that f has *finite height* (or *finite shift*) if $\ell(z)$ is finite for some (and hence all, by the previous estimate; see also [121, Proposition 3.2]) $z \in \mathbb{D}$; otherwise, we shall say that f has *infinite height* (or *infinite shift*). Notice that if $\Psi_{\tau_f}: \mathbb{D} \to \mathbb{H}^+$ is the generalized Cayley transform such that $\Psi_{\tau_f}(\tau_f) = \infty$ then $h_{\tau_f}(\Psi_{\tau_f}^{-1}(w)) = \frac{1}{2} \log \frac{1}{\mathrm{Im}\, w}$; therefore, being of finite height when read in $\mathbb{H}^+$ is equivalent to requiring that the imaginary parts of any orbit does not converge to $+\infty$, i. e., the orbit stays at finite height.

Poggi-Corradini [329, Proposition 4.1] has proved that a parabolic map with zero hyperbolic step has necessarily infinite height; see also [121, Proposition 3.3] and Remark 4.7.2. If instead f is parabolic with positive hyperbolic step then (see [121, Theorem 4.1]) f has finite height if and only if $f \in C_A^2(\tau)$; see Definition 4.7.3 for the meaning of $C_A^2(\tau)$. In particular, if $f \in C_A^2(\tau)$ then f is parabolic with a zero hyperbolic step if and only if it has infinite height; see also Proposition 4.7.1.

4.7 Parabolic type and boundary smoothness

In the last section, we saw that parabolic self-maps of $\mathbb{D}$ fall in two categories having different dynamical behavior: positive hyperbolic step and zero hyperbolic step. So it is interesting to have some procedure to decide to which category a given parabolic self-map belongs.

In this section, we collect a few results of this kind, assuming a bit of regularity at the Wolff point. The main technical step is the following.

Proposition 4.7.1. *Let $F \in \mathrm{Hol}(\mathbb{H}^+, \mathbb{H}^+)$ be of the form $F(w) = w + i\alpha + \eta(w)$ with $\alpha \in \mathbb{C}$ and*

$$\lim_{w\to\infty} \eta(w) = 0.$$

Then:
(i) *F is parabolic with Wolff point at infinity;*
(ii) *$\frac{1}{\nu}F^\nu(w_0) \to i\alpha$ as $\nu \to +\infty$ for every $w_0 \in \mathbb{H}^+$;*
(iii) *$\operatorname{Re}\alpha \geq 0$;*
(iv) *for each $w_0 \in \mathbb{H}^+$, the sequence $\{\operatorname{Im} F^\nu(w_0)\}$ is not decreasing;*
(v) *if $\alpha = 0$, then F has zero hyperbolic step;*
(vi) *F has zero hyperbolic step if and only if $\operatorname{Im} F^\nu(w_0) \to +\infty$ for some (and hence all) $w_0 \in \mathbb{H}^+$;*
(vii) *if $\operatorname{Re}\alpha > 0$, then F has zero hyperbolic step;*
(viii) *if $\alpha \neq 0$, then the orbit $\{F^\nu(w_0)\}$ tends to ∞ nontangentially if and only if $\operatorname{Re}\alpha > 0$.*

Proof. (i) We clearly have

$$\lim_{w\to\infty} \frac{F(w)}{w} = 1.$$

Recalling Remark 2.1.17, Corollaries 2.3.16 and 2.5.5 then imply that ∞ is the Wolff point of F with $F'(\infty) = 1$; in particular, F is parabolic.

(ii) Write $w_\nu = F^\nu(w_0)$. By (i), we know that $w_\nu \to \infty$, and hence $\eta(w_\nu) \to 0$. The Cesàro means theorem (Corollary A.4.5) then yields

$$\lim_{\nu\to+\infty} \frac{1}{\nu} \sum_{j=0}^{\nu-1} \eta(w_j) = 0.$$

Since

$$w_\nu = w_0 + i\nu\alpha + \sum_{j=0}^{\nu-1} \eta(w_j)$$

we then get

$$\lim_{nu\to\infty} \frac{w_\nu}{\nu} = i\alpha,$$

as claimed.

(iii) Assume, by contradiction, that $\operatorname{Re}\alpha < 0$. Take $w = x - \frac{i}{2}\operatorname{Re}\alpha \in \mathbb{H}^+$ with $x \in \mathbb{R}$. Then

$$\operatorname{Im} F(w) = \frac{1}{2}\operatorname{Re}\alpha + \operatorname{Im}\eta(w)$$

and hence $\operatorname{Im} F(w) < 0$ when x is large enough, impossible.

(iv) It follows from the Wolff lemma in $\mathbb{H}^+$ (Proposition 2.5.8).

(v) Fix $w_0 \in \mathbb{H}^+$ and put again $w_\nu = F^\nu(w_0)$. By (i), $w_\nu \to \infty$ as $\nu \to +\infty$; therefore,

$$\lim_{\nu\to+\infty} |w_{\nu+1} - w_\nu| = |\alpha|. \tag{4.31}$$

In particular, the Euclidean distance between w_ν and $w_{\nu+1}$ is bounded by a constant independent of ν.

Now let $\gamma_\nu \in \operatorname{Aut}(\mathbb{H}^+)$ be given by

$$\gamma_\nu(w) = \frac{\operatorname{Im} w_\nu}{\operatorname{Im} w_0}(w - \operatorname{Re} w_0) + \operatorname{Re} w_\nu.$$

Since $\gamma_\nu(w_0) = w_\nu$, it follows that $B_{\mathbb{H}^+}(w_\nu, \varepsilon) = \gamma_\nu(B_{\mathbb{H}^+}(w_0, \varepsilon))$ for all $\varepsilon > 0$; in particular, the Euclidean diameter of $B_{\mathbb{H}^+}(w_\nu, \varepsilon)$ is given by $\frac{\operatorname{Im} w_\nu}{\operatorname{Im} w_0}$ times the Euclidean diameter of $B_{\mathbb{H}^+}(w_0, \varepsilon)$, and thus it does not decrease, by (iii).

Assume now $\alpha = 0$. Then (4.31) implies that the Euclidean distance between $w_{\nu+1}$ and w_ν tends to zero. On the other hand, since for every $\varepsilon > 0$ the Euclidean diameter of $B_{\mathbb{H}^+}(w_\nu, \varepsilon)$ does not decrease as $\nu \to +\infty$, we have $w_{\nu+1} \in B_{\mathbb{H}^+}(w_\nu, \varepsilon)$, i. e., $\omega_{\mathbb{H}^+}(w_{\nu+1}, w_\nu) < \varepsilon$, for all ν large enough. Since ε is arbitrary, it follows that F has zero hyperbolic step, as claimed.

(vi) Assume that $\operatorname{Im} w_\nu \to +\infty$ as $\nu \to +\infty$. This means that for any $\varepsilon > 0$ the Euclidean diameter of $B_{\mathbb{H}^+}(w_\nu, \varepsilon)$ tends to $+\infty$, while we know that the Euclidean distance between w_ν and $w_{\nu+1}$ is bounded. It follows that $w_{\nu+1} \in B_{\mathbb{H}^+}(w_\nu, \varepsilon)$, i. e., $\omega_{\mathbb{H}^+}(w_{\nu+1}, w_\nu) < \varepsilon$ eventually; since ε is arbitrary, it follows again that F has a zero hyperbolic step.

Assume instead that $\{\operatorname{Im} w_\nu\}$ is bounded, and that $\alpha \neq 0$. Write

$$M = \sup_\nu \frac{\operatorname{Im} w_\nu}{\operatorname{Im} w_0} < +\infty.$$

The previous argument shows that the Euclidean diameter of $B_{\mathbb{H}^+}(w_\nu, \varepsilon)$ is at most M times the Euclidean diameter of $B_{\mathbb{H}^+}(w_0, \varepsilon)$, that tends to 0 as $\varepsilon \to 0$. On the other hand, (4.31) implies that $|w_{\nu+1} - w_\nu| > \frac{1}{2}|\alpha| > 0$ for ν large enough; this means that we can choose $\varepsilon_0 > 0$ so that $w_{\nu+1} \notin B_{\mathbb{H}^+}(w_\nu, \varepsilon_0)$, i. e., $\omega_{\mathbb{H}^+}(w_\nu, w_{\nu+1}) \geq \varepsilon_0$, eventually. But then $s^F(w_0) \geq \varepsilon_0$ and F has a positive hyperbolic step, as claimed.

It remains to prove that if $\alpha = 0$ then $\{\operatorname{Im} w_\nu\}$ cannot be bounded. Assume, by contradiction, that $\operatorname{Im} w_\nu \to L < +\infty$; take another $\tilde{w}_0 \in \mathbb{H}^+$ and put $\tilde{w}_\nu = F^\nu(\tilde{w}_0)$. Since F has a zero hyperbolic step, by Corollary 4.6.9 we know that $\omega_{\mathbb{H}^+}(w_\nu, \tilde{w}_\nu) \to 0$ as $\nu \to +\infty$. Recalling the definition of $\omega_{\mathbb{H}^+}$ we find that

$$\left|1 + \frac{\overline{w_\nu} - w_\nu}{\tilde{w}_\nu - \overline{w_\nu}}\right| = \left|\frac{\tilde{w}_\nu - w_\nu}{\tilde{w}_\nu - \overline{w_\nu}}\right| \to 0 \Longleftrightarrow \frac{2i \operatorname{Im} w_\nu}{\tilde{w}_\nu - \overline{w_\nu}} \to 1.$$

Reversing the roles of w_ν and $\tilde{w}_\nu$, we get $\frac{2i \operatorname{Im} \tilde{w}_\nu}{w_\nu - \overline{\tilde{w}_\nu}} \to 1$; taking the ratio of the two limits and then the modulus, we find

$$\frac{\operatorname{Im} w_\nu}{\operatorname{Im} \tilde{w}_\nu} \to 1.$$

But if we take $\tilde{w}_0$ such that $\operatorname{Im} \tilde{w}_0 > L$ we have $\operatorname{Im} \tilde{w}_\nu > L$ for all ν, contradiction.

(vii) We have $\operatorname{Im} w_{\nu+1} = \operatorname{Im} w_\nu + \operatorname{Re} \alpha + \operatorname{Im} \eta(w_\nu)$. Since $w_\nu \to \infty$, it follows that $\operatorname{Im} w_{\nu+1} \geq \operatorname{Im} w_\nu + \frac{1}{2} \operatorname{Re} \alpha$ for ν large enough; the assumption $\operatorname{Re} \alpha > 0$ then implies that $\operatorname{Im} w_\nu \to +\infty$, and the assertion follows from (vi).

(viii) By (ii), we know that $\arg F^\nu(w_0)$ converges to $\arg(i\alpha) \in [0, \pi]$. This means that the convergence is nontangential if and only if $\arg(i\alpha) \notin \{0, \pi\}$, i. e., if and only if $\operatorname{Re} \alpha > 0$. □

Remark 4.7.2. The argument used to prove Proposition 4.7.1(vi) actually shows that if $F \in \operatorname{Hol}(\mathbb{H}^+, \mathbb{H}^+)$ has Wolff point ∞ and it is parabolic with zero hyperbolic step then, without assuming any regularity at the Wolff point, we have $\operatorname{Im} F^\nu(w) \to +\infty$ as $\nu \to +\infty$ for all $w \in \mathbb{H}^+$ and $(\operatorname{Im} F^\nu(w))/(\operatorname{Im} F^\nu(\tilde{w})) \to 1$ as $\nu \to +\infty$ for all $w, \tilde{w} \in \mathbb{H}^+$.

To apply the previous proposition to self-maps of $\mathbb{D}$, we need a definition.

Definition 4.7.3. Given $n \in \mathbb{N}$ and $\varepsilon \in [0, 1)$ we shall say that $f \in \operatorname{Hol}(\mathbb{D}, \mathbb{D})$ *belongs to the class* $C^{n+\varepsilon}(\tau)$ at the point $\tau \in \partial\mathbb{D}$ if there exist $a_0, \ldots, a_n \in \mathbb{C}$ and $\eta \in \operatorname{Hol}(\mathbb{D}, \mathbb{C})$ such that

$$f(z) = \sum_{j=0}^{n} \frac{a_j}{j!}(z - \tau)^j + \eta(z) \tag{4.32}$$

with

$$\lim_{z \to \tau} \frac{\eta(z)}{|z - \tau|^{n+\varepsilon}} = 0.$$

We shall instead write $f \in C_A^{n+\varepsilon}(\tau)$ if (4.32) holds with

$$K\text{-}\lim_{z \to \tau} \frac{\eta(z)}{|z - \tau|^{n+\varepsilon}} = 0.$$

In both cases, we shall write $f(\tau) = a_0, f'(\tau) = a_1, f''(\tau) = a_2, f'''(\tau) = a_3$, and $f^{(j)}(\tau) = a_j$ for $j \geq 4$.

Clearly, $f \in C^0(\tau)$ if and only if it extends continuously to $\mathbb{D} \cup \{\tau\}$. If τ_f is the Wolff point of f, we saw that $f \in C^1_A(\tau_f)$ with $0 < f'(\tau_f) \leq 1$. Our aim will be to see whether we can identify the hyperbolic step by looking at $f''(\tau_f)$ when $f \in C^2(\tau_f)$.

Lemma 4.7.4. *Let $f \in \mathrm{Hol}(\mathbb{D}, \mathbb{D})$ be parabolic with Wolff point $\tau_f \in \partial\mathbb{D}$ and such that $f \in C^2(\tau_f)$. Let $\Psi_{\tau_f} \colon \mathbb{D} \to \mathbb{H}^+$ be the generalized Cayley transform such that $\Psi_{\tau_f}(\tau_f) = \infty$ and put $F = \Psi_{\tau_f} \circ f \circ \Psi_{\tau_f}^{-1} \in \mathrm{Hol}(\mathbb{H}^+, \mathbb{H}^+)$. Then*

$$F(w) = w + i\tau_f f''(\tau_f) + \eta(w) \tag{4.33}$$

with $\eta(w) \to 0$ as $w \to \infty$.

Proof. By definition, we have

$$\tau_f - z = \frac{2i\tau_f}{\Psi_{\tau_f}(z) + i} \quad \text{and} \quad \tau_f - \Psi_{\tau_f}^{-1}(w) = \frac{2i\tau_f}{w + i}. \tag{4.34}$$

Then

$$\begin{aligned} F(w) + i &= \frac{2i\tau_f}{\tau_f - f(\Psi_{\tau_f}^{-1}(w))} \\ &= \frac{2i\tau_f}{\tau_f - \Psi_{\tau_f}^{-1}(w)} \frac{1}{1 - \frac{f''(\tau_f)}{2}(\tau_f - \Psi_{\tau_f}^{-1}(w)) + o(\tau_f - \Psi_{\tau_f}^{-1}(w))} \\ &= \frac{2i\tau_f}{\tau_f - \Psi_{\tau_f}^{-1}(w)} + i\tau_f f''(\tau_f) + o(1) = w + i + i\tau_f f''(\tau_f) + o(1) \end{aligned}$$

and we are done. □

Corollary 4.7.5. *Let $f \in \mathrm{Hol}(\mathbb{D}, \mathbb{D})$ be parabolic with Wolff point $\tau_f \in \partial\mathbb{D}$. Assume that $f \in C^2(\tau_f)$. Then:*
(i) $\mathrm{Re}(\tau_f f''(\tau_f)) \geq 0$;
(ii) *if $f''(\tau_f) = 0$ or $\mathrm{Re}(\tau_f f''(\tau_f)) > 0$, then f has a zero hyperbolic step.*
(iii) *if $\mathrm{Re}(\tau_f f''(\tau_f)) > 0$, then all the orbits of f converge nontangentially to the Wolff point with the same slope.*

Proof. It follows from Lemma 4.7.4, Proposition 4.7.1, and Corollary 4.6.11. □

It is natural to conjecture that $\tau_f f''(\tau_f) = ia$ with $a \in \mathbb{R}^*$ should imply that f has positive hyperbolic step; surprisingly, this is in general not true.

Example 4.7.6. Choose $0 < \alpha < 1$ and let $f_\alpha \in \mathrm{Hol}(\mathbb{D}, \mathbb{D})$ be given by $f_\alpha = \Psi^{-1} \circ F_\alpha \circ \Psi$, where $F_\alpha \in \mathrm{Hol}(\mathbb{H}^+, \mathbb{H}^+)$ is given by

$$F_\alpha(w) = w - 1 + \frac{1}{\alpha}\frac{i^{1+\alpha}}{(w+i)^\alpha}.$$

First of all, since $\arg(w+i)^{-\alpha} \in (-\alpha\pi, 0)$ we have

$$\arg(i^{1+\alpha}(w+i)^{-\alpha}) \in \left(\frac{1-\alpha}{2}\pi, \frac{1+\alpha}{2}\pi\right) \subset (0,\pi); \tag{4.35}$$

so $F_\alpha(\mathbb{H}^+) \subseteq \mathbb{H}^+$ and $f_\alpha \in \mathrm{Hol}(\mathbb{D},\mathbb{D})$.

Arguing as in the proof of Proposition 4.7.1(i) we immediately see that F_α is parabolic with Wolff point ∞ and so f_α is parabolic with Wolff point 1. Furthermore,

$$\begin{aligned} 1 - f_\alpha(z) &= \frac{2i}{F_\alpha(\Psi(z)) + i} = \frac{2i}{\Psi(z) - 1 + \frac{1}{\alpha}\frac{i^{1+\alpha}}{(\Psi(z)+i)^\alpha} + i} \\ &= \frac{2i}{\frac{2i}{1-z} - 1 + \frac{1}{\alpha}\frac{i^{1+\alpha}}{(\frac{2i}{1-z})^\alpha}} = \frac{1-z}{1 - \frac{1-z}{2i} + \frac{1}{\alpha} i^{1+\alpha}(\frac{1-z}{2i})^{1+\alpha}} \\ &= 1 - z + \frac{(1-z)^2}{2i} + \frac{2i^\alpha}{\alpha}\left(\frac{1-z}{2i}\right)^{2+\alpha} + O((1-z)^3) \end{aligned}$$

i. e.,

$$f_\alpha(z) = z + \frac{i}{2}(z-1)^2 + O((z-1)^{2+\alpha}).$$

This means that $f \in C^2(1) \setminus C^3(1)$ and $f''(1) = i$.

Now we would like to prove that F_α (and hence f_α) has a zero hyperbolic step. By Proposition 4.7.1(vi) it suffices to show that $\operatorname{Im} F_\alpha^\nu(w_0) \to +\infty$ for any $w_0 \in \mathbb{H}^+$. Put $w_\nu = F_\alpha^\nu(w_0)$, as usual. Then

$$w_\nu = w_0 - \nu + \frac{i^{1+\alpha}}{\alpha}\sum_{j=0}^{\nu-1}\frac{1}{(w_j+i)^\alpha},$$

and recalling (4.35) we get

$$\operatorname{Im} w_\nu \ge \operatorname{Im} w_0 + \frac{\sin(\frac{1+\alpha}{2}\pi)}{\alpha}\sum_{j=0}^{\nu-1}\frac{1}{|w_j+i|^\alpha}.$$

Now, Proposition 4.7.1(ii) yields $\frac{1}{\nu}w_\nu \to -1$. It follows that there is $C > 0$ such that

$$\operatorname{Im} w_\nu \ge \operatorname{Im} w_0 + \frac{\sin(\frac{1+\alpha}{2}\pi)}{\alpha}\left[\frac{1}{|w_0+i|^\alpha} + \frac{1}{C}\sum_{j=1}^{\nu-1}\frac{1}{j^\alpha}\right],$$

and so $\operatorname{Im} w_\nu \to +\infty$, as claimed.

Notice that all orbits of f converge to 1 tangentially by Lemma 4.7.1(viii).

However, assuming a little bit more of regularity at the Wolff point we can show that $\tau_f f''(\tau_f) = ia$ with $a \in \mathbb{R}^*$ implies positive hyperbolic step. The main computation is contained in the following proposition.

Proposition 4.7.7. *Let $F \in \mathrm{Hol}(\mathbb{H}^+, \mathbb{H}^+)$ be of the form*

$$F(w) = w + a + \frac{b}{w+i} + \eta(w) \tag{4.36}$$

with $a \in \mathbb{R}$ and

$$\lim_{w\to\infty} |w|^{1+\varepsilon}\eta(w) = 0$$

for some $\varepsilon > 0$. Then:

(i) *$b \in \mathbb{R}$;*
(ii) *for every $w_0 \in \mathbb{H}^+$ the sequence $\{\operatorname{Im} F^\nu(w_0)\}$ is bounded.*

Proof. (i) Take $w = x + i$, with $x \in \mathbb{R}$. Then

$$x(\operatorname{Im} F(w) - \operatorname{Im} w) = \frac{x^2 \operatorname{Im} b - 2x \operatorname{Re} b}{x^2 + 4} + x \operatorname{Im} \eta(w).$$

Proposition 4.7.1(iv) implies that $\operatorname{Im} F(w) \geq \operatorname{Im} w$. Therefore, letting $x \to +\infty$ we get $\operatorname{Im} b \geq 0$ whereas letting $x \to -\infty$ we get $\operatorname{Im} b \leq 0$; so $b \in \mathbb{R}$.

(ii) Set $w_\nu = F^\nu(w_0)$. We have

$$w_\nu = w_0 + \nu a + \sum_{j=0}^{\nu-1} \frac{b}{w_j + i} + \sum_{j=0}^{\nu-1} \eta(w_j),$$

and hence

$$\operatorname{Im} w_\nu \leq \operatorname{Im} w_0 + \sum_{j=0}^{\nu-1} \frac{|b|}{|w_j + i|} + C \sum_{j=0}^{\nu-1} \frac{1}{|w_j|^{1+\varepsilon}}$$

for a suitable $C > 0$. Recalling Proposition 4.7.1(ii) we find constants $C_h > 0$ for $h = 1, \dots, 4$ such that

$$\operatorname{Im} w_\nu \leq C_0 + C_1 \sum_{j=1}^{\nu-1} \frac{|b|}{j} + C_2 \sum_{j=1}^{\nu-1} \frac{1}{j^{1+\varepsilon}} \leq C_3 + C_4 \log \nu$$

for all $\nu \geq 1$. Therefore,

$$
\begin{aligned}
\operatorname{Im} w_\nu &= \operatorname{Im} w_0 + \operatorname{Im} \sum_{j=0}^{\nu-1} \frac{b}{w_j + i} + \operatorname{Im} \sum_{j=0}^{\nu-1} \eta(w_j) \\
&\le \operatorname{Im} w_0 + \operatorname{Im} \eta(w_0) + \operatorname{Im} \frac{b}{w_0 + i} + \sum_{j=1}^{\nu-1} \frac{(1 + \operatorname{Im} w_j)|b|}{|w_j + i|^2} + C \sum_{j=1}^{\nu-1} \frac{1}{|w_j|^{1+\varepsilon}} \\
&\le C_5 + |b| \sum_{j=1}^{\nu-1} \frac{C_6 + C_7 \log j}{j^2} + C_8 \sum_{j=1}^{\nu-1} \frac{1}{j^{1+\varepsilon}}
\end{aligned}
$$

for suitable $C_5, C_6, C_7, C_8 \ge 0$. Since both sums on the write converge as $\nu \to +\infty$, we get that $\operatorname{Im} w_\nu$ is bounded, as claimed. □

We then have the following characterization of positive hyperbolic step.

Theorem 4.7.8. *Let $f \in \operatorname{Hol}(\mathbb{D}, \mathbb{D})$ be parabolic with Wolff point $\tau_f \in \partial\mathbb{D}$. Assume that $f \in C^{3+\varepsilon}(\tau_f)$ and that $f''(\tau_f) \neq 0$. Then the following assertions are equivalent:*
(i) *f has positive hyperbolic step;*
(ii) *every orbit of f converges to τ_f tangentially;*
(iii) $\operatorname{Re}(\tau_f f''(\tau_f)) = 0$.

Proof. We already know (Corollary 4.6.10) that (i) implies (ii); we also know that (ii) implies (iii), by Proposition 4.7.1(viii), Lemma 4.7.4, and Corollary 4.7.5; it remains to prove that (iii) implies (i).

Assume then that $\tau_f f''(\tau_f) = ia$ for some $a \in \mathbb{R}^*$; we have to prove that f has positive hyperbolic step. Let $\Psi_{\tau_f} \colon \mathbb{D} \to \mathbb{H}^+$ be again the generalized Cayley transform sending τ_f to ∞ and put $F = \Psi_{\tau_f} \circ f \circ \Psi_{\tau_f}^{-1}$. Since

$$
f(z) = z - \frac{a}{2i\tau_f}(z - \tau_f)^2 + \frac{f'''(\tau_f)}{6}(z - \tau_f)^3 + o(|z - \tau_f|^{3+\varepsilon}),
$$

recalling (4.34) we have

$$
\begin{aligned}
F(w) + i &= \frac{2i\tau_f}{\tau_f - f(\Psi_{\tau_f}^{-1}(w))} \\
&= \frac{2i\tau_f}{\tau_f - \Psi_{\tau_f}^{-1}(w)} \, \frac{1}{1 + \frac{a}{2i\tau_f}(\tau_f - \Psi_{\tau_f}^{-1}(w)) + \frac{f'''(\tau_f)}{6}(\tau_f - \Psi_{\tau_f}^{-1}(w))^2 + o(|\tau_f - \Psi_{\tau_f}^{-1}(w)|^{2+\varepsilon})} \\
&= \frac{2i\tau_f}{\tau_f - \Psi_{\tau_f}^{-1}(w)} - a - i\left[\frac{a^2}{2\tau_f} + \frac{\tau_f f'''(\tau_f)}{3}\right](\tau_f - \Psi_{\tau_f}^{-1}(w)) + o\left(\left|\tau_f - \Psi_{\tau_f}^{-1}(w)\right|^{1+\varepsilon}\right) \\
&= w + i - a + \frac{a^2 + \frac{2}{3}\tau_f^2 f'''(\tau_f)}{w + i} + o(|w|^{-1-\varepsilon}).
\end{aligned}
\tag{4.37}
$$

Therefore, F is of the form (4.36), and the assertion follows from Propositions 4.7.7 and 4.7.1(vi). □

Putting together Theorem 4.7.8 and Corollary 4.7.5, we have a complete description of the way the orbits of a parabolic map $f \in C^{3+\varepsilon}(\tau_f)$ converge to the Wolff point $\tau_f \in \partial\mathbb{D}$ if $f''(\tau_f) \neq 0$.

If $f''(\tau_f) = 0$, then necessarily $f'''(\tau_f) \neq 0$ by Theorem 2.7.4. Therefore, we can write

$$f(z) = z + \frac{f'''(\tau_f)}{6}(z - \tau_f)^3 + o(|z - \tau_f|^{3+\varepsilon}). \tag{4.38}$$

Notice that (4.37) and Proposition 4.7.7 imply that $\tau_f^2 f'''(\tau_f) \in \mathbb{R}^*$.

We can then describe the slope of approach of the orbits even when $f''(\tau_f) = 0$.

Proposition 4.7.9. *Let $f \in \mathrm{Hol}(\mathbb{D}, \mathbb{D})$ be parabolic with Wolff point $\tau_f \in \partial\mathbb{D}$. Assume that $f \in C^{3+\varepsilon}(\tau_f)$ with $f''(\tau_f) = 0$. Then all the orbits of f converge to τ_f tangentially if $\tau_f^2 f'''(\tau_f) > 0$ and radially if $\tau_f^2 f'''(\tau_f) < 0$.*

Proof. Write f as in (4.38); we already noticed that $\tau_f^2 f'''(\tau_f)$ must be real and not zero. Let $\gamma \in \mathrm{Aut}(\mathbb{C})$ be given by $\gamma(z) = \tau_f(z + 1)$ and set $\hat{f} = \gamma^{-1} \circ f \circ \gamma$. Then $\hat{f}$ is a parabolic self-map of the disk $D = D(-1, 1)$ with Wolff point at the origin and such that

$$\hat{f}(z) = z + az^3 + o(|z|^{3+\varepsilon}) = z(1 + az^2 + o(|z|^{2+\varepsilon})), \tag{4.39}$$

where $a = \frac{1}{6}\tau_f^2 f'''(\tau_f) \in \mathbb{R}^*$. Define $\Phi\colon \mathbb{C}^* \to \mathbb{C}$ by setting

$$\Phi(z) = -\frac{1}{2az^2}.$$

It is easy to check that Φ is a biholomorphism between $\Sigma^- = \{z \in \mathbb{C} \mid \operatorname{Re} z < 0\} \supset D$ and $\mathbb{C} \setminus \mathbb{R}^-$, with inverse

$$\Phi^{-1}(w) = \frac{1}{(-2aw)^{1/2}},$$

where we choose the square root with $\operatorname{Re}(-2aw)^{1/2} < 0$ so that $\Phi^{-1}(w) \in \Sigma^-$. Notice that we have $w \in \Phi(D)$ if and only if

$$\left|\frac{1}{(-2aw)^{1/2}} + 1\right|^2 < 1 \iff |\operatorname{Re}(-aw)^{1/2}| > \frac{1}{2\sqrt{2}}.$$

Let $F = \Phi \circ \hat{f} \circ \Phi^{-1}$. By construction, F is a holomorphic self-map of $\Phi(D)$ with Wolff point at ∞, but we can be more precise. Indeed, using (4.39) we get

$$\begin{aligned} F(w) &= \Phi\left(\hat{f}\left(\frac{1}{(-2aw)^{1/2}}\right)\right) = \Phi\left(\frac{1}{(-2aw)^{1/2}}\left(1 - \frac{1}{2w} + o(|w|^{-(2+\varepsilon)/2})\right)\right) \\ &= w\left(1 - \frac{1}{2w} + o(|w|^{-(2+\varepsilon)/2})\right)^{-2} = w + 1 + o(|w|^{-\varepsilon/2}). \end{aligned}$$

In particular, if $|w|$ is large enough (that implies that z is close to the origin) we have

$$\operatorname{Re} F(w) > \operatorname{Re} w + \frac{1}{2},$$

which immediately implies that $F^\nu(w) \to \infty$ as $\nu \to +\infty$, which is not surprising because we already knew that $\hat{f}^\nu(z) \to 0$. But we have something more. Indeed, putting as usual $w_\nu = F^\nu(w)$, we have

$$\frac{F^\nu(w)}{\nu} = \frac{w}{\nu} + 1 + \frac{1}{\nu}\sum_{j=0}^{\nu-1} o(|w_j|^{-\varepsilon/2}).$$

Recalling the Cesàro means theorem (Corollary A.4.5), we then get

$$\lim_{\nu\to+\infty} \frac{F^\nu(w)}{\nu} = 1.$$

This implies that $\arg(F^\nu(w)) \to 0$ as $\nu \to +\infty$. Coming back to D, since

$$\arg(\Phi(z)) = \begin{cases} \pi - 2\arg(z) & \text{if } a > 0, \\ -2\arg(z) & \text{if } a < 0, \end{cases}$$

we get that $\arg(\hat{f}^\nu(z)) \to \pm\pi/2$ if $a > 0$ and that $\arg(\hat{f}^\nu(z)) \to \pi$ if $a < 0$, as claimed. □

Notes to Section 4.7

Most of this section, with the exception of Proposition 4.7.9, comes from [73]. In particular, Corollary 4.7.5 is [73, Theorem 4.4] and Theorem 4.7.8 is [73, Theorem 4.15]. See also [121].

The proof of Proposition 4.7.9 is inspired by the proof of the classical Leau–Fatou flower theorem in local holomorphic dynamics; see, e. g., [287, Chapter 10]. A version of Proposition 4.7.9 can be found in [67].

Contreras, Dìaz-Madrigal, and Pommerenke in [121] (see also [119]) have studied in even more details the relationships between hyperbolic step and smoothness at the Wolff point. For instance, they proved that if f has positive hyperbolic step then either $f''(\tau) = \infty$ or $f \in C^2_A(\tau)$ with $\tau f''(\tau) = ia$ for some $a \in \mathbb{R}^*$. Moreover, $f \in C^2_A(\tau)$ if and only if f has finite height.

The final part of the proof of Proposition 4.7.1(vi) and Remark 4.7.2 come from [329]. Putting this together with results from [121], we see that $f \in C^2_A(\tau)$ parabolic has a zero hyperbolic step if and only if it has infinite height, and thus we get Proposition 4.7.1(vi) just assuming that the nontangential limit of η at ∞ vanishes.

Another characterization of positive hyperbolic step contained in [121] is the following: let $f \in \operatorname{Hol}(\mathbb{D}, \mathbb{D})$ be parabolic with Wolff point $\tau_f \in \partial\mathbb{D}$. Assume that $f \in C^2_A(\tau_f)$. Then f has positive hyperbolic step if and only if the following two conditions hold:

(i) $\tau_f f''(\tau_f) = ia$ with $a \in \mathbb{R}^*$;
(ii) we have

$$\int_i^{\epsilon\infty+i} \operatorname{Re}\big(\phi(t) - t\big)\, dt < +\infty,$$

where $\epsilon = \text{sgn}(a) \in \{+1, -1\}$ and $\phi = \Psi_{\tau_f} \circ f \circ \Psi_{\tau_f}^{-1}$, where $\Psi_{\tau_f} \in \text{Hol}(\mathbb{D}, \mathbb{H}^+)$ is the generalized Cayley transform sending τ_f to ∞.

Assume that $f \in \text{Hol}(\mathbb{D}, \mathbb{D})$ is a parabolic inner function with Wolff point $\tau_f \in \partial\mathbb{D}$. If $f \in C^2(\tau_f)$, using (4.33) it is not difficult to see that $\text{Re}(\tau_f f''(\tau_f)) = 0$. If moreover $f \in C^{3+\varepsilon}(\tau_f)$, then Theorem 4.7.8 implies that f has positive hyperbolic step if and only if $f''(\tau_f) \neq 0$. Under these assumptions, assuming moreover that f can be holomorphically extended to a neighborhood of τ_f and denoting by $\iota \in \mathbb{R}$ the residue of $1/(\text{id} - f)$ at τ_f, Bergweiler [60] has shown that: if $s^f(z) > 0$ then

$$\omega\big(f^{\nu+1}(z), f^{\nu}(z)\big) = s^f(z) + \frac{c}{\nu^3} + O\left(\frac{1}{\nu^4}\right)$$

where $c = \frac{\iota-1}{3\tanh(\frac{s^f(z)}{2})} \geq 0$, and $c = 0$ if and only if f is injective; whereas if $s^f(z) = 0$ then

$$\omega\big(f^{\nu+1}(z), f^{\nu}(z)\big) = \frac{1}{2\nu} + \left(\frac{\iota}{4} - \frac{3}{8}\right)\frac{\log \nu}{\nu^2} + O\left(\frac{1}{\nu^2}\right).$$

The proof strongly depends on the asymptotic expression, due to Kimura [234], of the intertwining map in Theorem 4.6.8; this expression is valid when f extends holomorphically to a neighborhood of τ_f.

Another line of research involves Blaschke and interpolating sequences. A sequence $\{z_\nu\} \subset \mathbb{D}$ is a *Blaschke sequence* if

$$\sum_{\nu} \big(1 - |z_\nu|\big) < +\infty;$$

it is an *interpolating sequence* if for every sequence $\{w_\nu\} \subset \mathbb{D}$ there is $g \in \text{Hol}(\mathbb{D}, \mathbb{D})$ such that $g(z_\nu) = w_\nu$ for every $\nu \in \mathbb{N}$. It is known that every interpolating sequence is a Blaschke sequence but the converse is not true (see, e. g., [102] or [198, Chapter 10]).

Cowen [128] has shown that a parabolic map $f \in \text{Hol}(\mathbb{D}, \mathbb{D})$ has positive hyperbolic step if and only if some (and hence any) orbit of f is interpolating. In particular, if f is parabolic with a positive hyperbolic step then every orbit is a Blaschke sequence converging tangentially to the Wolff point; it is not known whether the converse is also true. It is however known that if $f \in C^2(\tau_f)$ is parabolic with $\text{Re}(\tau_f f''(\tau_f)) > 0$ (and thus it has a zero hyperbolic step) then the orbits are not Blaschke sequences; see [73, Lemma 4.5] and [119].

4.8 Boundary fixed points

Recall that a boundary fixed point of a $f \in \text{Hol}(\mathbb{D}, \mathbb{D})$ is a point $\sigma \in \partial\mathbb{D}$ such that $f(\sigma) = \sigma$, where $f(\sigma)$ is the nontangential limit of f at σ (see Definition 2.3.14). In Remark 2.3.15, we saw that if σ is a boundary fixed point then we can define the derivative $f'(\sigma)$ of f at σ by setting $f'(\sigma) = \beta_f(\sigma) \in (0, +\infty]$; in particular, $f'(\sigma)$ is the nontangential limit of f' at σ when $\beta_f(\sigma) < +\infty$.

Definition 4.8.1. Let $f \in \text{Hol}(\mathbb{D}, \mathbb{D})$ be a holomorphic self-map of the unit disk. We say that $\sigma \in \partial\mathbb{D}$ is a *boundary repelling fixed point* if it is a boundary fixed point with $f'(\sigma) > 1$. Given $A > 1$, we shall set

$$\text{Fix}_A(f) = \{\sigma \in \partial\mathbb{D} \mid f(\sigma) = \sigma \text{ and } f'(\sigma) \le A\}.$$

Corollaries 2.3.16 and 2.5.5 say that if f has a fixed point in $\mathbb{D}$, then all boundary fixed points are repelling, and that if f has no fixed points in $\mathbb{D}$ then exactly one boundary fixed point is not repelling, the Wolff point of f. Furthermore, we have $f'(\sigma_1)f'(\sigma_2) \ge 1$ for all pairs of boundary fixed points (Theorem 2.3.13 contains a more precise estimate for boundary contact points).

In this section, we shall prove a precise quantitative generalization of these facts that we shall use in the next section to study the backward dynamics of a holomorphic self-map of $\mathbb{D}$.

We shall need two lemmas. The first one concerns Blaschke products (see Definition 1.5.5).

Lemma 4.8.2. *For $n \ge 1$, let $\sigma_1, \dots, \sigma_n \in \partial\mathbb{D}$ be distinct points, $a_1, \dots, a_n \in \mathbb{R}^+$ and $B \in \text{Hol}(\mathbb{D}, \mathbb{D})$ a Blaschke product of degree $d \ge 0$. If $B \not\equiv 1$ define $\phi\colon \mathbb{D} \to \mathbb{C}$ by*

$$\frac{1 + \phi(z)}{1 - \phi(z)} = \sum_{j=1}^{n} a_j \frac{\sigma_j + z}{\sigma_j - z} + \frac{1 + B(z)}{1 - B(z)}.$$

Then ϕ is a Blaschke product of degree $n + d$.

Proof. A quick computation yields

$$\phi(z) = \frac{2B(z) + (1 - B(z))S(z)}{2 + (1 - B(z))S(z)},$$

where

$$S(z) = \sum_{j=1}^{n} a_j \frac{\sigma_j + z}{\sigma_j - z};$$

in particular ϕ is a rational function of degree $n + d$ because numerator and denominator have no common factors.

When $z \in \mathbb{D}$, we have

$$\begin{aligned}
\text{Re}\,\frac{1 + \phi(z)}{1 - \phi(z)} &= \sum_{j=1}^{n} a_j \,\text{Re}\,\frac{\sigma_j + z}{\sigma_j - z} + \text{Re}\,\frac{1 + B(z)}{1 - B(z)} \\
&= \sum_{j=1}^{n} a_j \frac{1 - |z|^2}{|\sigma_j - z|^2} + \frac{1 - |B(z)|^2}{|1 - B(z)|^2} > 0;
\end{aligned}$$

this yields $\phi(z) \in \mathbb{D}$, and hence $\phi(\mathbb{D}) \subseteq \mathbb{D}$.

If $\sigma \in \partial\mathbb{D}$ is different from σ_j, we have

$$\frac{\sigma_j + \sigma}{\sigma_j - \sigma} = 2i \frac{\text{Im}(\sigma\overline{\sigma_j})}{|\sigma_j - \sigma|^2} \in i\mathbb{R};$$

thus if $\sigma \neq \sigma_1, \dots \sigma_n$ we have $S(\sigma) = ia \in i\mathbb{R}$, and hence setting $B(\sigma) = e^{i\phi} \in \partial\mathbb{D}$ we have

$$\phi(\sigma) = \frac{2e^{i\phi} + (1 - e^{i\phi})ia}{2 + (1 - e^{i\phi})ia} \in \partial\mathbb{D}.$$

To deal with $\sigma = \sigma_j$, we write

$$\phi(z) = \frac{a_j(\sigma_j + z)(1 - B(z)) + (\sigma_j - z)(S_j(z) + 2B(z))}{a_j(\sigma_j + z)(1 - B(z)) + (\sigma_j - z)(S_j(z) + 2)}$$

where $S_j(\sigma_j) = 0$; therefore, we get $\phi(\sigma_j) = 1$, also when $B(\sigma_j) = 1$.

Summing up, we have proved that ϕ is a Blaschke product of degree $n + d$ and we are done. □

The second lemma will be part of the core of the proof.

Lemma 4.8.3. *Let $g \in \mathrm{Hol}(\mathbb{D}, \mathbb{D})$. Assume there are distinct points $\sigma_1, \dots, \sigma_n \in \partial\mathbb{D}$ such that $g(\sigma_j) = 1$ for $j = 1, \dots, n$. Put*

$$h(z) = \frac{1 + g(z)}{1 - g(z)} - \sum_{j=1}^{n} \frac{1}{\beta_g(\sigma_j)} \frac{\sigma_j + z}{\sigma_j - z},$$

Then $\mathrm{Re}\, h(z) \geq 0$ for all $z \in \mathbb{D}$.

Proof. First of all, notice that we can assume $\beta_g(\sigma_j) < +\infty$ for all $j = 1, \dots, n$, because the points with $\beta_g(\sigma_j) = +\infty$ do not contribute to the sum.

Let us proceed by induction on n. For $n = 1$, the definition of boundary dilation and Lemma 2.2.14 yield

$$\mathrm{Re}\, \frac{1 + g(z)}{1 - g(z)} = \frac{1 - |g(z)|^2}{|1 - g(z)|^2} \geq \frac{1}{\beta_g(\sigma_1)} \frac{1 - |z|^2}{|\sigma_1 - z|^2} = \frac{1}{\beta_g(\sigma_1)} \mathrm{Re}\, \frac{\sigma_1 + z}{\sigma_1 - z},$$

and hence

$$\mathrm{Re}\left[\frac{1 + g(z)}{1 - g(z)} - \frac{1}{\beta_g(\sigma_1)} \frac{\sigma_1 + z}{\sigma_1 - z}\right] \geq 0$$

as required.

Assume the assertion holds for $n - 1$. Put

$$h_1(z) = \frac{1 + g(z)}{1 - g(z)} - \sum_{j=1}^{n-1} \frac{1}{\beta_g(\sigma_j)} \frac{\sigma_j + z}{\sigma_j - z}$$

and

$$g_1(z) = \frac{h_1(z) - 1}{h_1(z) + 1}.$$

If $\operatorname{Re} h_1(z_0) = 0$ for some $z_0 \in \mathbb{D}$, the minimum principle yields $h_1 \equiv i\eta$ for some $\eta \in \mathbb{R}$ and then we cannot have $g(\sigma_n) = 1$ because σ_n is different from $\sigma_1, \ldots, \sigma_{n-1}$. Therefore, we can assume that $\operatorname{Re} h_1(z) > 0$ for all $z \in \mathbb{D}$. Then $g_1 \in \operatorname{Hol}(\mathbb{D}, \mathbb{D})$ and it satisfies $g_1(\sigma_n) = 1$ and $\beta_{g_1}(\sigma_n) = g_1'(\sigma_n) = \beta_g(\sigma_n)$. So, the case $n = 1$ yields

$$\operatorname{Re}\left[\frac{1+g(z)}{1-g(z)} - \sum_{j=1}^{n} \frac{1}{\beta_g(\sigma_j)} \frac{\sigma_j + z}{\sigma_j - z}\right] = \operatorname{Re}\left[\frac{1+g_1(z)}{1-g_1(z)} - \frac{1}{\beta_{g_1}(\sigma_n)} \frac{\sigma_n + z}{\sigma_n - z}\right] \geq 0,$$

as desired. □

Now we can prove the following.

Theorem 4.8.4 (Cowen–Pommerenke, 1982). *Let $f \in \operatorname{Hol}(\mathbb{D}, \mathbb{D}) \setminus \{\mathrm{id}_{\mathbb{D}}\}$ be with Wolff point $\tau_f \in \overline{\mathbb{D}}$. Assume that $\sigma_1, \ldots, \sigma_n \in \partial\mathbb{D}$ are boundary fixed points of f distinct from τ_f. Then:*

(i) *if $\tau_f \in \mathbb{D}$ then*

$$\sum_{j=1}^{n} \frac{1}{f'(\sigma_j) - 1} \leq \frac{1 - |f'(\tau_f)|^2}{|1 - f'(\tau_f)|^2}; \tag{4.40}$$

(ii) *if $\tau_f \in \partial\mathbb{D}$ and $0 < f'(\tau_f) < 1$, then*

$$\sum_{j=1}^{n} \frac{1}{f'(\sigma_j) - 1} \leq \frac{f'(\tau_f)}{1 - f'(\tau_f)}; \tag{4.41}$$

(iii) *if $\tau_f \in \partial\mathbb{D}$ and $f'(\tau_f) = 1$, then*

$$\sum_{j=1}^{n} \frac{|\tau_f - \sigma_j|^2}{f'(\sigma_j) - 1} \leq 2 \operatorname{Re}\left(\frac{\tau_f}{f(0)} - 1\right). \tag{4.42}$$

Moreover, equality holds if and only if f is a Blaschke product of degree $n + 1$.

Proof. Clearly, we can assume $f'(\sigma_j) = \beta_f(\sigma_j) < +\infty$ for all $j = 1, \ldots, n$.

(i) Let $\gamma \in \operatorname{Aut}(\mathbb{D})$ be given by $\gamma(z) = (z + \tau_f)/(1 + \overline{\tau_f} z)$, and put $\tilde{f} = \gamma^{-1} \circ f \circ \gamma$ and $\tilde{\sigma}_j = \gamma^{-1}(\sigma_j) \in \partial\mathbb{D}$ for $j = 1, \ldots, n$. Then $\tilde{f}(0) = 0$, $\tilde{f}'(0) = f'(\tau_f)$, $\tilde{f}(\tilde{\sigma}_j) = \tilde{\sigma}_j$ and $\tilde{f}'(\tilde{\sigma}_j) = f'(\sigma_j)$ for all $j = 1, \ldots, n$, thanks to Proposition 2.4.7. So, without loss of generality we can assume $\tau_f = 0$.

Put $g(z) = z^{-1} f(z)$. Since f is not an elliptic automorphism (because it is not the identity and it has fixed points in $\partial\mathbb{D}$), the Schwarz lemma implies that $g \in \operatorname{Hol}(\mathbb{D}, \mathbb{D})$. Clearly, $g(\sigma_j) = 1$ for $j = 1, \ldots, n$; therefore, Lemma 4.8.3 yields

$$\frac{1 - |g(z)|^2}{|1 - g(z)|^2} - \sum_{j=1}^{n} \frac{1}{\beta_g(\sigma_j)} \frac{1 - |z|^2}{|\sigma_j - z|^2} = \operatorname{Re}\left[\frac{1+g(z)}{1-g(z)} - \sum_{j=1}^{n} \frac{1}{\beta_g(\sigma_j)} \frac{\sigma_j + z}{\sigma_j - z}\right] \geq 0.$$

Since $g(0) = f'(0)$ and $\beta_g(\sigma_j) = \sigma g'(\sigma_j) = f'(\sigma_j) - 1$, putting $z = 0$ we obtain exactly (4.40).

Let us denote by $h(z)$ the expression in the square brackets in the previous formula. If we have equality in (4.40), the minimum principle implies that h is a purely imaginary constant. It follows that g is a rational function of degree n sending $\partial\mathbb{D}$ into itself, because the real part of $(1+g)/(1-g)$ is identically zero on $\partial\mathbb{D}$; hence g is a Blaschke product of degree n and $f(z) = zg(z)$ is a Blaschke product of degree $n+1$.

Conversely, if f is a Blaschke product of degree $n+1$ with $f(0) = 0$ then g is a Blaschke product of degree n. It follows that $\operatorname{Re} h|_{\partial\mathbb{D}} \equiv 0$ and then $\operatorname{Re} h(0) = 0$, by the mean value property of harmonic functions, which gives the equality in (4.40).

(ii) Put

$$\phi(z) = \frac{\tau_f + f(z)}{\tau_f - f(z)} - \frac{\tau_f + z}{\tau_f - z} \quad \text{and} \quad g(z) = \frac{1-\phi(z)}{1+\phi(z)}. \tag{4.43}$$

Since $f'(\tau_f) = \beta_f(\tau_f) < 1$, the definition of boundary dilation implies that $\operatorname{Re}\phi > 0$; therefore, $g \in \operatorname{Hol}(\mathbb{D},\mathbb{D})$. Moreover, $\phi(\sigma_j) = 0$ and so $g(\sigma_j) = 1$ for all $j = 1,\dots,n$. A quick computation also shows that

$$g'(\sigma_j) = \frac{4\tau_f(1 - f'(\sigma_j))}{(\tau_f - \sigma_j)^2} \quad \Longrightarrow \quad \beta_g(\sigma_j) = \frac{4(f'(\sigma_j) - 1)}{|\tau_f - \sigma_j|^2}.$$

We can now apply Lemma 4.8.3 to get

$$\operatorname{Re} h(z) = \operatorname{Re}\left[\frac{1+g(z)}{1-g(z)} - \sum_{j=1}^{n} \frac{|\tau_f - \sigma_j|^2}{4(f'(\sigma_j) - 1)} \frac{\sigma_j + z}{\sigma_j - z}\right] \geq 0.$$

Harnack inequality (Theorem A.3.2) then yields $\operatorname{Re} h(z) \geq \frac{1-|z|}{1+|z|} \operatorname{Re} h(0)$; therefore, taking $z = r\tau_f$ with $0 < r < 1$ and dividing by $1 - r$ we get

$$\frac{1}{1-r} \operatorname{Re} \frac{1}{\phi(r\tau_f)} - \sum_{j=1}^{n} \frac{|\tau_f - \sigma_j|^2}{4(f'(\sigma_j) - 1)} \frac{1+r}{|\sigma_j - r\tau_f|^2} \geq \frac{1}{1+r} \operatorname{Re} h(0) \geq 0. \tag{4.44}$$

Now,

$$\begin{aligned}
(1-r)\phi(r\tau_f) &= (1-r)\left[\frac{2\tau_f(f(r\tau_f) - r\tau_f)}{(\tau_f - f(r\tau_f))\tau_f(1-r)}\right] = 2\frac{f(r\tau_f) - r\tau_f}{\tau_f - f(r\tau_f)} \\
&= 2\left[\frac{\tau_f - r\tau_f}{\tau_f - f(r\tau_f)} - 1\right] \to 2\left[\frac{1}{f'(\tau_f)} - 1\right] = \frac{2(1 - f'(\tau_f))}{f'(\tau_f)}
\end{aligned}$$

as $r \to 1^-$, thanks to Theorem 2.3.2. Therefore, letting $r \to 1^-$ in (4.44) we get

$$\frac{f'(\tau_f)}{2(1-f'(\tau_f))} - \sum_{j=1}^{n} \frac{1}{2(f'(\sigma_j)-1)} \geq \frac{1}{2}\operatorname{Re} h(0) \geq 0, \tag{4.45}$$

i. e., (4.41).

If we have equality in (4.41), then (4.45) implies $\operatorname{Re} h(0) = 0$. Then arguing as in case (i) we see that g is a Blaschke product of degree n. Since we have

$$\frac{1+\overline{\tau_f}f(z)}{1-\overline{\tau_f}f(z)} = \frac{\tau_f+z}{\tau_f-z} + \frac{1-g(z)}{1+g(z)},$$

Lemma 4.8.2 applied with $B = -g$ implies that $\overline{\tau_f}f$ (and hence f) is a Blaschke product of degree $n+1$.

Conversely, assume that f is a Blaschke product of degree $n+1$. Then $\operatorname{Re}\phi|_{\partial\mathbb{D}} \equiv 0$, which implies that g is a Blaschke product, necessarily of degree n because of Lemma 4.8.2. Therefore, $\operatorname{Re} h|_{\partial\mathbb{D}} \equiv 0$; hence $\operatorname{Re} h(0) = 0$ and, by the minimum principle, $\operatorname{Re} h \equiv 0$. Repeating the argument going from (4.44) to (4.45), then we get equality in (4.41).

(iii) Define again ϕ and g as in (4.43). The Julia lemma (Theorem 2.1.10) implies that $\operatorname{Re}\phi(z) \geq 0$ with equality in one point if and only if $f \in \operatorname{Aut}(\mathbb{D})$. But since a parabolic automorphism has only one fixed point, its Wolff point, this cannot happen; therefore, we have $\operatorname{Re}\phi > 0$ everywhere. We can then argue as in case (ii) obtaining (4.44). Putting $r = 0$, we get

$$\frac{1}{2}\operatorname{Re}\left(\frac{\tau_f}{f(0)} - 1\right) - \sum_{j=1}^{n} \frac{|\tau_f - \sigma_j|^2}{4(f'(\sigma_j)-1)} \geq \operatorname{Re} h(0) \geq 0,$$

which is exactly (4.42).

If we have equality in (4.42), then necessarily $\operatorname{Re} h(0) = 0$ and then arguing as in case (ii) we find that f is a Blaschke product of degree $n+1$. Conversely, if f is a Blaschke product of degree $n+1$ then arguing as in (ii) we get again $\operatorname{Re} h \equiv 0$, and putting $r = 0$ in (4.44) we finally obtain the equality in (4.42). □

These estimates implies that the number of boundary repelling fixed points with dilation coefficient bounded by $A > 1$ is (with one potential exception) finite.

Corollary 4.8.5. *Let $f \in \operatorname{Hol}(\mathbb{D},\mathbb{D})$ and $A > 1$.*
(i) *If $\tau_f \in \mathbb{D}$ or $\tau_f \in \partial\mathbb{D}$ and $0 < f'(\tau_f) < 1$, then $\operatorname{Fix}_A(f)$ is finite;*
(ii) *if $\tau_f \in \partial\mathbb{D}$ and $f'(\tau_f) = 1$, then $\operatorname{Fix}_A(f) \cap \{\sigma \in \partial\mathbb{D} \mid |\tau_f - \sigma| \geq \varepsilon\}$ is finite for all $\varepsilon > 0$.*

Proof. If $\sigma \in \operatorname{Fix}_A(f)$, then

$$\frac{1}{f'(\sigma)-1} \geq \frac{1}{A-1},$$

and the assertions follow from Theorem 4.8.4. □

Finally, we also get a statement concerning boundary contact points.

Corollary 4.8.6. *Let $f \in \mathrm{Hol}(\mathbb{D},\mathbb{D})$. Assume there are $\tau \in \partial\mathbb{D}$ and distinct boundary contact points $\sigma_1,\dots,\sigma_n \in \partial\mathbb{D}$ such that $f(\sigma_j) = \tau$ for $j = 1,\dots,n$. Then*

$$\sum_{j=1}^{n} \frac{1}{|f'(\sigma_j)|} \leq \frac{1-|f(0)|^2}{|\tau - f(0)|^2}.$$

Moreover, equality holds if and only if f is a Blaschke product of degree $n+1$.

Proof. Put $g(z) = \overline{\tau} z f(z)$. Then $g(0) = 0$ and $g(\sigma_j) = \sigma_j$ for $j = 1,\dots,n$; moreover, $g'(\sigma_j) = 1+\overline{\tau}\sigma_j f'(\sigma_j) = 1+\beta_f(\sigma_j) = 1+|f'(\sigma_j)|$, by (2.46). The assertion then follows from Theorem 4.8.4(i) applied to g. □

Notes to Section 4.8

The content of this section comes from [130]. Sharper estimates can be obtained by assuming that f is univalent; see, e. g., [130, 20]. See also [69] and references therein for some more precise estimates with applications to the Nevanlinna–Pick interpolation problem.

A different proof of Theorem 4.8.4 has been given by Li [266] using Hilbert spaces methods.

See also [9] for a generalization of (4.40) to the case of multiple fixed points.

An extension of Corollary 4.8.6 can be found in [389]. See also [340, 341, 178].

4.9 Backward dynamics

Up to now, we have studied the forward dynamics, i. e., the asymptotic behavior of the forward orbits obtained by iterating a given self-map. In a sense, studying forward dynamics is like looking into the future and trying to understand what will happen to a given point. But what about the past? Are we able to understand from where a given point is coming? This is the main goal of this section: to understand the asymptotic behavior of orbits ending (instead of starting) at a given point.

To give a formal definition of the object of study of this section, let us remark that a forward orbit for a map f starting from a point x_0 is a sequence $\{x_\nu\}$ such that $x_{\nu+1} = f(x_\nu)$ for all $\nu \in \mathbb{N}$. Reversing this approach, we can define the notion of backward orbit ending at a given point.

Definition 4.9.1. Let $f: X \to X$ be a self-map of a set X. A *backward orbit* (or *backward iteration sequence*) is a sequence $\{x_\nu\} \subset X$ such that $f(x_{\nu+1}) = x_\nu$ for each $n \in \mathbb{N}$. The point x_0 is the *root* of the backward orbit.

If f is invertible, then the only backward orbit with root x_0 is the forward orbit of the inverse: $x_\nu = f^{-\nu}(x_0)$. If instead f is not invertible, then a root may have more than one backward orbits (if f is not injective) or none at all (if f is not surjective).

In this section, we shall study the asymptotic behavior of a particularly important class of backward orbits of holomorphic self-maps of the unit disk $\mathbb{D}$. If $\{z_\nu\}$ is a forward orbit of a self-map $f \in \mathrm{Hol}(\mathbb{D},\mathbb{D})$, in the last few sections we have already used many times the fact that the sequence $\{\omega(z_\nu, z_{\nu+1})\}$ is not increasing, and hence it has a finite limite. On the other hand, if $\{z_\nu\} \subset \mathbb{D}$ is a *backward* orbit of f then the Schwarz–Pick lemma yields

$$\omega(z_\nu, z_{\nu+1}) = \omega\big(f(z_{\nu+1}), f(z_{\nu+2})\big) \le \omega(z_{\nu+1}, z_{\nu+2});$$

therefore, the sequence $\{\omega(z_\nu, z_{\nu+1})\}$ this time is not decreasing and thus it still has a limit, possibly equal to $+\infty$.

Definition 4.9.2. The *hyperbolic step* $A \in [1, +\infty]$ of a backward orbit $O = \{z_\nu\} \subset \mathbb{D}$ of a holomorphic self-map $f \in \mathrm{Hol}(\mathbb{D}, \mathbb{D})$ is given by

$$\frac{1}{2}\log A = \lim_{\nu\to+\infty} \omega(z_\nu, z_{\nu+1}) = \sup_{\nu\in\mathbb{N}} \omega(z_\nu, z_{\nu+1}).$$

We shall say that O has *bounded step* if $A < +\infty$.

We shall see that the existence of backward orbits with bounded step is related to the existence of boundary fixed points with finite angular derivative; so the restriction to consider only backward orbits with bounded step is a natural one.

We begin our investigations considering the trivial cases.

Proposition 4.9.3. *Let $\{z_\nu\} \subset \mathbb{D}$ be a backward orbit of a self-map $f \in \mathrm{Hol}(\mathbb{D}, \mathbb{D})$. Assume that $\{z_\nu\}$ has an accumulation point in $\mathbb{D}$. Then one and only one of the following cases is possible:*
(i) *f is an elliptic automorphism and $z_\nu = f^{-\nu}(z_0)$ for all $\nu \in \mathbb{N}$;*
(ii) *f is elliptic but not an automorphism and $z_\nu = \tau_f$ for all $\nu \in \mathbb{N}$, where $\tau_f \in \mathbb{D}$ is the fixed point of f.*

Proof. If f is an elliptic automorphism, then clearly we are in case (i). So, assume that f is not an elliptic automorphism; we would like to prove that if $O = \{z_\nu\}$ has an accumulation point $p \in \mathbb{D}$ then we are in case (ii).

First of all we claim that the Wolff point τ_f of f belongs to $\overline{O}$. If $p = \tau_f$ this is obvious; so assume $p \ne \tau_f$ and put

$$d = \inf_{\nu\in\mathbb{N}}\{|z_\nu - \tau_f|\}.$$

We need to prove that $d = 0$. Assume, by contradiction, that $d > 0$. Since $p \in \mathbb{D}$ and τ_f is the Wolff point of f, there exists $\mu \in \mathbb{N}$ such that $|f^\mu(p) - \tau_f| < d/2$. Choose a sequence $\{z_{\nu_k}\} \subseteq O$ converging to p. Then $z_{\nu_k-\mu} = f^\mu(z_{\nu_k}) \to f^\mu(p)$ as $k \to +\infty$; choosing k large enough, we then have $\nu_k > \mu$ and $|z_{\nu_k-\mu} - f^\mu(p)| < d/2$, which implies $|z_{\nu_k-\mu} - \tau_f| < d$, against the definition of d.

So, $\tau_f \in \overline{O}$; we claim that $\tau_f \in \mathbb{D}$. By contradiction, assume $\tau_f \in \partial\mathbb{D}$. Let $R_0, R_1, R_2 > 0$ be so that $f^j(p) \in \partial E(\tau_f, R_j)$ for $j = 0, 1, 2$; then the Wolff lemma (Theorem 2.5.1) implies

$$\overline{E(\tau_f, R_2)} \subset E(\tau_f, R_1) \subset \overline{E(\tau_f, R_1)} \subset E(\tau_f, R_0)$$

and in particular $R_2 < R_1 < R_0$ because the equality would imply $f \in \mathrm{Aut}(\mathbb{D})$, which is impossible because $p \in \mathbb{D}$. Therefore, we can find a neighborhood U of p and a neighborhood V of $f^2(p)$ such that $U \cap \overline{E(\tau_f, R_1)} = \emptyset$ and $V \subset E(\tau_f, R_1)$. Since $z_{\nu_k} \to p$, we have $z_{\nu_k - 2} \to f^2(p)$. Therefore, we can find $\hat{\nu}_0 < \hat{\nu}_1$ such that $z_{\hat{\nu}_0} \in U$ and $z_{\hat{\nu}_1} \in V$. But, again by the Wolff lemma, $f(E(\tau_f, R_1)) \subseteq E(\tau_f, R_1)$; therefore, $z_{\hat{\nu}_0} = f^{\hat{\nu}_1 - \hat{\nu}_0}(z_{\hat{\nu}_1}) \in E(\tau_f, R_1)$, contradiction.

So $\tau_f \in \mathbb{D}$. Now, notice that for any $\nu \in \mathbb{N}$ we have

$$\omega(\tau_f, z_\nu) = \omega(f(\tau_f), f(z_{\nu+1})) \le \omega(\tau_f, z_{\nu+1}).$$

Therefore, the sequence $\{\omega(\tau_f, z_\nu)\}$ is not decreasing; on the other hand, we know that a subsequence of $\{z_\nu\}$ converges to τ_f. These two facts together imply $z_\nu = \tau_f$ for all $\nu \in \mathbb{N}$ and we are in case (ii). □

So the only interesting backward orbits are the ones accumulating to the boundary. Actually, in this case backward orbits with bounded step actually converge to a point in the boundary.

Theorem 4.9.4. *Let $\{z_\nu\} \subset \mathbb{D}$ be a backward orbit with bounded step $A \ge 1$ for a holomorphic self-map $f \in \mathrm{Hol}(\mathbb{D}, \mathbb{D})$. Assume that $\{z_\nu\}$ has no accumulation points in $\mathbb{D}$. Then $\{z_\nu\}$ converges to a boundary fixed point $\sigma \in \partial\mathbb{D}$ with $f'(\sigma) \le A$.*

Proof. Choose a subsequence $\{z_{\nu_k}\}$ converging to a point $\sigma \in \partial\mathbb{D}$. Then

$$\begin{aligned} |\omega(0, z_{\nu_k}) - \omega(0, f(z_{\nu_k}))| &\le \omega(z_{\nu_k}, f(z_{\nu_k})) \\ &= \omega(z_{\nu_k}, z_{\nu_k - 1}) \le \sup_{\nu \in \mathbb{N}} \omega(z_\nu, z_{\nu-1}) = \frac{1}{2}\log A < +\infty \end{aligned} \tag{4.46}$$

because $\{z_\nu\}$ has bounded step A. Proposition 2.1.21 then implies that $\beta_f(\sigma) \le A$; in particular, f has nontangential limit $\tau \in \partial\mathbb{D}$, by Lemma 2.2.14.

We claim that $f(z_{\nu_k}) \to \tau$. If not, up to a subsequence we can assume that $f(z_{\nu_k}) \to \tau_1 \ne \tau$; notice that $\tau_1 \in \partial\mathbb{D}$ because $f(z_{\nu_k}) = z_{\nu_k - 1}$ and so τ_1 is an accumulation point of the backward orbit. Therefore, arguing as in the first part of the proof of the Julia lemma (Theorem 2.1.10) we obtain

$$\frac{|\tau_1 - f(z)|^2}{1 - |f(z)|^2} \le \alpha \frac{|\sigma - z|^2}{1 - |z|^2}$$

for all $z \in \mathbb{D}$, where $\alpha \in \mathbb{R}^+$. But this implies that $\beta_f(\sigma, \tau_1) < +\infty$ and then $\tau_1 = \tau$ by Lemma 2.1.12, contradiction.

So, we have $z_{\nu_k} \to \sigma$ and $f(z_{\nu_k}) \to \tau$. Since

$$\omega(z_{\nu_k}, f(z_{\nu_k})) = \omega(z_{\nu_k}, z_{\nu_k - 1}) \le \frac{1}{2}\log A$$

it follows by Proposition 1.7.27 that $\tau = \sigma$.

We have then proved that every accumulation point of $\{z_\nu\}$ is a boundary fixed point of f, either repelling or equal to the Wolff point of f (Corollaries 2.3.16 and 2.5.5). Furthermore, the Julia–Wolff–Carathéodory theorem yields $f'(\sigma) = \beta_f(\sigma) \leq A$. In particular, Corollary 4.8.5 implies that either the set of accumulation points of $\{z_\nu\}$ is finite or f is parabolic and in this case the accumulation points of $\{z_\nu\}$ different from τ_f are all isolated.

If $z_\nu \to \tau_f$, we are done. If $\{z_\nu\}$ does not converge to τ_f, then it must have an accumulation point $\sigma \neq \tau_f$, necessarily isolated. We claim that $\{z_\nu\}$ converges to σ. If, by contradiction, $\{z_\nu\}$ does not converge to σ we can find a neighbourhood U of σ and a neighbourhood U_1 of the other accumulation points of $\{z_\nu\}$ with $\overline{U} \cap \overline{U_1} = \emptyset$ such that $\{z_\nu\}$ is eventually contained in $U \cup U_1$ and intersects infinitely many times both U and U_1. But then we can construct a subsequence $\{z_{\nu_k}\}$ such that $z_{\nu_k} \in U$ and $z_{\nu_k+1} \in U_1$ for all k. Up to a subsequence, we can assume that both $\{z_{\nu_k}\}$ and $\{z_{\nu_k+1}\}$ are converging, necessarily to distinct points; but this would contradict Proposition 1.7.27 because $\omega(z_{\nu_k}, z_{\nu_k+1}) \leq \frac{1}{2}\log A$. Hence $z_\nu \to \sigma$ and we are done in this case, too. □

So, backward orbits with bounded step necessarily converge to a boundary fixed point, either repelling or equal to the Wolff point. It turns out that the convergence to a repelling fixed point is necessarily nontangential whereas the convergence to the Wolff point is necessarily tangential, and this latter case can happen only for parabolic self-maps. To prove these statements, we need a technical lemma giving a condition ensuring nontangential convergence.

Lemma 4.9.5. *Let $\{z_\nu\} \subset \mathbb{D}$ be a sequence converging to a point $\tau \in \partial\mathbb{D}$ such that*

$$\liminf_{\nu\to+\infty} \frac{1 - |z_{\nu+1}|}{|z_{\nu+1} - z_\nu|} \geq a > 0 \tag{4.47}$$

and

$$\liminf_{\nu\to+\infty} \frac{1 - |z_\nu|}{1 - |z_{\nu+1}|} \geq C > 1. \tag{4.48}$$

Then

$$\liminf_{\nu\to+\infty} \frac{1 - |z_\nu|}{|\tau - z_\nu|} \geq \frac{1}{2}(C-1)a > 0,$$

i.e., $z_\nu \in K(\tau, M)$ eventually for any $M > \frac{2}{(C-1)a}$. In particular, $\{z_\nu\}$ converges to τ nontangentially.

Proof. Since

$$|z_{\nu+1}| - |z_\nu| = (1 - |z_{\nu+1}|)\left(\frac{1 - |z_\nu|}{1 - |z_{\nu+1}|} - 1\right),$$

we have

$$\liminf_{\nu\to+\infty}\frac{|z_{\nu+1}|-|z_\nu|}{1-|z_{\nu+1}|}\ge C-1>0$$

and then

$$\liminf_{\nu\to+\infty}\frac{|z_{\nu+1}|-|z_\nu|}{|z_{\nu+1}-z_\nu|}\ge (C-1)a.$$

As a consequence, there exists a ν_0 such that if $\nu\ge\nu_0$ then

$$|z_{\nu+1}|-|z_\nu|\ge\frac{(C-1)a}{2}|z_{\nu+1}-z_\nu|.$$

Using telescopic sums, we see that if $\mu>\nu>\nu_0$ then

$$|z_\mu|-|z_\nu|\ge\frac{(C-1)a}{2}|z_\mu-z_\nu|.$$

Letting first μ go to $+\infty$ and then taking the lim inf as $\nu\to+\infty$ we get the assertion. □

Theorem 4.9.6. *Given $f\in\mathrm{Hol}(\mathbb{D},\mathbb{D})$, let $\{z_\nu\}\subset\mathbb{D}$ be a backward orbit for f with bounded step $A\ge1$ and without accumulation points in $\mathbb{D}$. Let $\sigma\in\partial\mathbb{D}$ be the boundary fixed point such that $z_\nu\to\sigma$. Then:*
(i) *if σ is repelling, then z_ν converges to σ nontangentially;*
(ii) *if σ is the Wolff point τ_f of f, then z_ν converges to τ_f tangentially either with angle $\pi/2$ or with angle $-\pi/2$;*
(iii) *if $\sigma=\tau_f$, then f is parabolic.*

Proof. We start with a preliminary estimate. Put $r=(A-1)/(A+1)\in(0,1)$ so that

$$\tanh^{-1}\left|\frac{z_\nu-z_{\nu+1}}{1-\overline{z_\nu}z_{\nu+1}}\right|=\omega(z_\nu,z_{\nu+1})\le\frac12\log\frac{1+r}{1-r}=\tanh^{-1}r \tag{4.49}$$

for all $\nu\in\mathbb{N}$. It follows that

$$\frac12\log\frac{1+|z_{\nu+1}|}{1-|z_{\nu+1}|}=\omega(0,z_{\nu+1})\le\omega(0,z_\nu)+\omega(z_\nu,z_{\nu+1})\le\frac12\log\left(\frac{1+|z_\nu|}{1-|z_\nu|}\frac{1+r}{1-r}\right),$$

i. e.,

$$|z_{\nu+1}|\le\frac{|z_\nu|+r}{1+r|z_\nu|}\quad\Longrightarrow\quad 1-|z_{\nu+1}|\ge\frac{1-r}{1+r}(1-|z_\nu|).$$

Using this together with (4.49), we get

$$1-r^2 \le 1-\left|\frac{z_{\nu+1}-z_\nu}{1-\overline{z_\nu}z_{\nu+1}}\right|^2 = \frac{(1-|z_\nu|^2)(1-|z_{\nu+1}|^2)}{|1-\overline{z_\nu}z_{\nu+1}|^2}$$
$$\le 4r^2\frac{(1-|z_\nu|)(1-|z_{\nu+1}|)}{|z_\nu-z_{\nu+1}|^2} \le 4r^2\frac{1+r}{1-r}\frac{(1-|z_{\nu+1}|)^2}{|z_\nu-z_{\nu+1}|^2},$$

and hence

$$\frac{1-|z_{\nu+1}|}{|z_\nu-z_{\nu+1}|} \ge \frac{1-r}{2r} > 0. \tag{4.50}$$

In particular, (4.47) holds with $a=(1-r)/2r$.

(i) Assume now that σ is a repelling fixed point. Then

$$\liminf_{\nu\to+\infty}\frac{1-|z_\nu|}{1-|z_{\nu+1}|} = \liminf_{\nu\to+\infty}\frac{1-|f(z_{\nu+1})|}{1-|z_{\nu+1}|} \ge \liminf_{z\to\sigma}\frac{1-|f(z)|}{1-|z|} = \beta_f(\sigma) > 1,$$

where we used Proposition 2.1.15. Therefore, the hypotheses of Lemma 4.9.5 are satisfied and $z_\nu \to \sigma$ nontangentially.

(ii) Assume now that $z_n \to \tau_f$. The Wolff lemma implies that

$$\frac{1-|z_{\nu+1}|^2}{|\tau_f-z_{\nu+1}|^2} \le \frac{1-|z_\nu|^2}{|\tau_f-z_\nu|^2};$$

in particular the sequence $\{(1-|z_\nu|^2)/|\tau_f-z_\nu|^2\}$ is not increasing and converges to some $\ell \ge 0$. Therefore,

$$\frac{1-|z_\nu|}{|\tau_f-z_\nu|} = \frac{1-|z_\nu|^2}{|\tau_f-z_\nu|^2}\frac{|\tau_f-z_\nu|}{1+|z_\nu|} \to \ell\cdot 0 = 0 \tag{4.51}$$

as $\nu\to+\infty$; it follows (Lemma 2.2.15) that $z_n \to \tau_f$ tangentially. To complete the proof, we must show that either $\frac{\tau_f-z_\nu}{|\tau_f-z_\nu|} \to i\tau_f$ or $\frac{\tau_f-z_\nu}{|\tau_f-z_\nu|} \to -i\tau_f$. Suppose, by contradiction, that both $i\tau_f$ and $-i\tau_f$ are accumulation points of the sequence $\left\{\frac{\tau_f-z_\nu}{|\tau_f-z_\nu|}\right\}$. Since they are the only possible accumulation points of this sequence, we can find a subsequence $\{z_{\nu_k}\}$ such that $\frac{\tau_f-z_{\nu_k}}{|\tau_f-z_{\nu_k}|} \to i\tau_f$ and $\frac{\tau_f-z_{\nu_k+1}}{|\tau_f-z_{\nu_k+1}|} \to -i\tau_f$. Now we can rewrite (4.50) as

$$\frac{1-|z_{\nu+1}|}{|\tau_f-z_{\nu+1}|} \ge \frac{1-r}{2r}\left|1-\frac{\tau_f-z_\nu}{\tau_f-z_{\nu+1}}\right|;$$

therefore, (4.51) yields

$$\lim_{k\to+\infty}\frac{\tau_f-z_{\nu_k+1}}{\tau_f-z_{\nu_k}} = 1.$$

But then

$$\lim_{k\to+\infty}\frac{\tau_f - z_{\nu_k+1}}{|\tau_f - z_{\nu_k+1}|}\frac{|\tau_f - z_{\nu_k}|}{\tau_f - z_{\nu_k}} = 1 \neq -1 = \frac{-i\tau_f}{i\tau_f},$$

contradiction and we are done.

(iii) If f is elliptic, then obviously $\sigma \neq \tau_f$; so we have to prove that if f is hyperbolic then $\sigma \neq \tau_f$. Assume, by contradiction, that z_ν converges to the Wolff point $\tau_f \in \partial\mathbb{D}$ and let $\beta = f'(\tau_f)$, with $0 < \beta < 1$ because f is hyperbolic. The Julia lemma ensures that

$$\frac{1-|z_{\nu+1}|^2}{|\tau_f - z_{\nu+1}|^2} \leq \beta\frac{1-|z_\nu|^2}{|\tau_f - z_\nu|^2}. \tag{4.52}$$

Iterating this estimate, we get

$$\lim_{\nu\to+\infty}\frac{1-|z_\nu|^2}{|\tau_f - z_\nu|^2} = 0,$$

and hence

$$\lim_{\nu\to+\infty}\frac{1-|z_\nu|}{|\tau_f - z_\nu|} = \left[\lim_{\nu\to+\infty}\frac{1-|z_\nu|^2}{|\tau_f - z_\nu|^2}\frac{1-|z_\nu|}{1+|z_\nu|}\right]^{1/2} = 0. \tag{4.53}$$

In particular, $z_\nu \to \tau_f$ tangentially, by Lemma 2.2.15.

On the other hand, arguing as in (ii) from (4.53) we get

$$\lim_{\nu\to+\infty}\frac{\tau_f - z_\nu}{\tau_f - z_{\nu+1}} = 1.$$

Rewriting (4.52) as

$$\frac{1-|z_\nu|^2}{1-|z_{\nu+1}|^2} \geq \frac{1}{\beta}\left|\frac{\tau_f - z_\nu}{\tau_f - z_{\nu+1}}\right|^2,$$

we then obtain

$$\liminf_{\nu\to+\infty}\frac{1-|z_\nu|}{1-|z_{\nu+1}|} = \liminf_{\nu\to+\infty}\frac{1-|z_\nu|^2}{1-|z_{\nu+1}|^2} \geq \frac{1}{\beta} > 1.$$

Recalling (4.50) we see that the hypotheses of Lemma 4.9.5 are satisfied; but then z_ν must converge nontangentially to τ_f, against (4.53). □

We now know well the asymptotic behavior of backward orbits with bounded step. Later on (Theorem 4.9.13) we shall also prove that backward orbits with bounded step converging to a repelling fixed point do so with a precise slope; but now we turn to the question of the existence of backward orbits converging to a given repelling fixed point. We need a preliminary result.

Lemma 4.9.7. *Let $f \in \mathrm{Hol}(\mathbb{D},\mathbb{D})$ and let $\sigma \in \partial\mathbb{D}$ be different from the Wolff point of f. Assume there are $A > 1$ and a sequence $\{z_\nu\} \subset \mathbb{D}$ converging to σ such that*

$$\liminf_{\nu\to+\infty} \omega(z_\nu, f(z_\nu)) = \frac{1}{2}\log A < +\infty.$$

Then σ is a boundary repelling fixed point of f with $f'(\sigma) \le A$.

Proof. Put $R_\nu = \omega(z_\nu, f(z_\nu))$ and $R = \frac{1}{2}\log A$. Since $f(z_\nu) \in \partial B_{\mathbb{D}}(z_\nu, R_\nu)$, Proposition 1.2.7(ii) yields

$$1 - |f(z_\nu)| \le \frac{1+\tanh R_\nu}{1-(\tanh R_\nu)|z_\nu|}(1-|z_\nu|).$$

Therefore,

$$\begin{aligned}\liminf_{z\to\sigma}\frac{1-|f(z)|}{1-|z|} &\le \liminf_{\nu\to+\infty}\frac{1-|f(z_\nu)|}{1-|z_\nu|}\\ &\le \liminf_{\nu\to+\infty}\frac{1+\tanh R_\nu}{1-(\tanh R_\nu)|z_\nu|} = \frac{1+\tanh R}{1-\tanh R} = A.\end{aligned}$$

Since, by Proposition 1.7.27, we know that $f(z_\nu) \to \sigma$, we can apply the Julia–Wolff–Carathéodory Theorem 2.3.2 to deduce that σ is a boundary fixed point with $f'(\sigma) \le A$. Finally, since σ is not the Wolff point of f we must have $f'(\sigma) > 1$, and we are done. □

Theorem 4.9.8. *Let $\sigma \in \partial\mathbb{D}$ be a boundary repelling fixed point of $f \in \mathrm{Hol}(\mathbb{D},\mathbb{D})$ with $1 < f'(\sigma) < +\infty$. Then there exists a backward orbit $\{z_\nu\} \subset \mathbb{D}$ converging to σ with hyperbolic step equal to $f'(\sigma)$.*

Proof. For the sake of simplicity, put $A = f'(\sigma)$. Since σ is not the Wolff point of f, by Corollary 4.8.5 we can find a small Euclidean disk $D = D(\sigma, \varepsilon)$ of center σ such that $\overline{D} \cap \mathrm{Fix}_A(f) = \{\sigma\}$; we can also assume that $\overline{D}$ does not contain the Wolff point $\tau_f \in \overline{\mathbb{D}}$ of f.

For all $k \in \mathbb{N}$, put $a_k = \frac{A^k-1}{A^k+1}\sigma$. Clearly $a_k \to \sigma$ as $k \to +\infty$; therefore, we can find $k_0 \ge 1$ such that $a_k \in D$ for all $k \ge k_0$. Put $t_\nu = a_{k_0+\nu}$, and denote by ℓ_ν the geodesic segment connecting t_ν and $f(t_\nu)$. Since $f^k(t_\nu) \to \tau_f \notin \overline{D}$, we eventually have $f^k(t_\nu) \notin \overline{D}$. Since $\bigcup_{j=0}^{k-1} f^j(\ell_\nu)$ is a path connecting $t_\nu \in D$ with $f^k(t_\nu) \notin \overline{D}$, we can find a smallest integer $j_\nu \ge 0$ such that $f^{j_\nu}(\ell_\nu)$ intersects $J = \partial D \cap \mathbb{D}$.

Put $E_\nu = E(\sigma, A^{-(k_0+\nu)})$; a quick computation shows that $t_\nu \in \partial E_\nu$. Moreover, the Julia lemma implies that $f(E_\nu) \subseteq E_{\nu-1}$ and, more generally, $f^k(E_\nu) \subseteq E_{\nu-k}$. Since $\overline{E_j} \subset D$ as soon as $j \ge 0$ and $\ell_\nu \subset \overline{E_{\nu-1}}$, it follows that $j_\nu \ge \nu$; in particular, $j_\nu \to +\infty$ as $\nu \to +\infty$.

We now claim that the sequence $\{w_\nu = f^{j_\nu}(t_\nu)\} \subset \overline{D}$ is relatively compact in $\mathbb{D}$. Assume, by contradiction, that this is not the case. Then we can extract a subsequence

$\{w_{\nu_j}\}$ converging to a point $\tau \in \partial\mathbb{D} \cap \overline{D}$. Since, by construction, there exists $p \in \ell_\nu$ such that $f^{j_\nu}(p) \in J$ we have

$$\omega(w_\nu, J) \le \omega(f^{j_\nu}(t_\nu), f^{j_\nu}(p)) \le \omega(t_\nu, p) \le \omega(t_\nu, f(t_\nu)) \longrightarrow \frac{1}{2}\log A < +\infty,$$

by Corollary 2.3.10. It follows that τ must be one of the end points of J; in particular, $\tau \neq \sigma$. Moreover,

$$\liminf_{j\to+\infty} \omega(w_{\nu_j}, f(w_{\nu_j})) \le \lim_{j\to+\infty} \omega(t_{\nu_j}, f(t_{\nu_j})) = \frac{1}{2}\log A$$

by Corollary 2.3.10, and so Lemma 4.9.7 implies that τ is either the Wolff point of f or a boundary repelling fixed point with $f'(\tau) \le A$, against the choice of D.

So, we can find an infinite set $I_0 \subseteq \mathbb{N}$ such that $\{f^{j_\nu}(t_\nu)\}_{\nu\in I_0}$ converges to a point $z_0 \in \mathbb{D}$; this will be the root of our backward orbit. To obtain other points in the backward orbit, we shall construct by induction a family $\{I_k\}_{k\in\mathbb{N}}$ of infinite subsets of $\mathbb{N}$ with $I_k \subseteq I_{k-1}$ (where $I_{-1} = \mathbb{N}$) such that the sequence $\{f^{j_\nu-k}(t_\nu)\}_{\nu\in I_k}$ converges to a point $z_k \in \mathbb{D}$. For $k = 0$, the set I_0 will do; so assume we have constructed $I_0, \dots, I_k$, and put $\tilde{I}_k = \{\nu \in I_k \mid j_\nu \ge k+1\} \neq \emptyset$. When $\nu \in \tilde{I}_k$, we have

$$\omega(f^{j_\nu-(k+1)}(t_\nu), f^{j_\nu-k}(t_\nu)) \le \omega(t_\nu, f(t_\nu)) \to \frac{1}{2}\log A;$$

since $\{f^{j_\nu-k}(t_\nu)\}_{\nu\in I_k}$ is relatively compact in $\mathbb{D}$, it follows that $\{f^{j_\nu-(k+1)}(t_\nu)\}_{\nu\in\tilde{I}_k}$ is relatively compact in $\mathbb{D}$, too, and hence we can find $I_{k+1} \subseteq \tilde{I}_k$ such that $\{f^{j_\nu-(k+1)}(t_\nu)\}_{\nu\in\tilde{I}_{k+1}}$ converges to $z_{k+1} \in \mathbb{D}$, as desired.

The sequence $\{z_k\}$ is a backward orbit: indeed,

$$f(z_{k+1}) = \lim_{\nu\in I_{k+1}} f(f^{j_\nu-(k+1)}(t_\nu)) = \lim_{\nu\in I_{k+1}} f^{j_\nu-k}(t_\nu) = z_k.$$

Furthermore,

$$\begin{aligned}\omega(z_{k+1}, z_k) &= \lim_{\nu\in I_{k+1}} \omega(f^{j_\nu-(k+1)}(t_\nu), f^{j_\nu-k}(t_\nu)) \\ &\le \lim_{\nu\in I_{k+1}} \omega(t_\nu, f(t_\nu)) = \frac{1}{2}\log A;\end{aligned}$$

in particular, this shows that $\{z_k\}$ has bounded step at most A. So, to conclude the proof it suffices to show that $z_k \to \sigma$, because then Theorem 4.9.4 will imply that the bounded step of $\{z_k\}$ is exactly A.

First of all, notice that by construction $f^{j_\nu-k}(t_\nu) \in D$ as soon as $j_\nu > k \ge 0$; therefore, $z_k \in \overline{D} \cap \mathbb{D}$ for all $k \in \mathbb{N}$. Assume, by contradiction, that $\{z_k\}$ does not converge to σ. Then we can find a subsequence $K = \{z_{k_\nu}\}$ converging to a point $\tau \in \overline{D} \cap \overline{\mathbb{D}} \setminus \{\sigma\}$; in particular, τ cannot be the Wolff point of f. If $\tau \in \mathbb{D}$, since K is relatively compact in $\mathbb{D}$, we have $f^j(K) \cap K = \emptyset$ for all j large enough and this is impossible because K is an infinite subset of a backward orbit. Then $\tau \in \partial\mathbb{D} \cap \overline{D}$. Applying again Lemma 4.9.7, we see that τ must be a boundary repelling fixed point with $f'(\tau) \le A$; by the choice of D it follows $\tau = \sigma$, contradiction and we are done. □

A related question is whether a parabolic map always has a backward orbit converging to the Wolff point. The answer in general is negative, as shown by the following example.

Example 4.9.9. Let $F \in \mathrm{Hol}(\mathbb{H}^+, \mathbb{H}^+)$ be given by $F(w) = w + i$. Then F is parabolic, the only boundary fixed point is the Wolff point ∞ and F has no backward orbits at all. Indeed, if $\{w_\nu\}$ were a backward orbit we would have $w_\nu + \nu i = F^\nu(w_\nu) = w_0$ for all $\nu \in \mathbb{N}$; but then $w_0 - \nu i \in \mathbb{H}^+$ for all $\nu \in \mathbb{N}$, impossible.

Now let $\sigma \in \partial\mathbb{D}$ be a boundary repelling fixed point with $f'(\sigma) < +\infty$. Our next goal is to prove that backward orbits with bounded step converging to σ do so along a precise slope and that for every nontangential slope there is a backward obit converging to σ with that slope. To do so, we shall construct a sort of model of f at σ, obtaining a result similar to what we got in Section 4.5 at the Wolff point. We start with a lemma.

Lemma 4.9.10. *Fix $A > 1$ and $\sigma \in \partial\mathbb{D}$. Let $\gamma \in \mathrm{Aut}(\mathbb{D})$ be given by*

$$\gamma(z) = \frac{z - \sigma\frac{A-1}{A+1}}{1 - \overline{\sigma}\frac{A-1}{A+1}z}. \tag{4.54}$$

Then:

(i) $\gamma(\pm\sigma) = \pm\sigma$;

(ii) *for all $\nu \in \mathbb{N}$, we have*

$$\gamma^\nu(z) = \frac{z - \sigma\frac{A^\nu-1}{A^\nu+1}}{1 - \overline{\sigma}\frac{A^\nu-1}{A^\nu+1}z}; \tag{4.55}$$

(iii) *every backward orbit $\{z_\nu\}$ of γ converges to σ in such a way that*

$$\lim_{\nu\to+\infty} \frac{\sigma - z_\nu}{|\sigma - z_\nu|} = \frac{\sigma - z_0}{|\sigma - z_0|}\frac{|1 + \overline{\sigma}z_0|}{1 + \overline{\sigma}z_0} \in \partial\mathbb{D} \setminus \{\pm i\sigma\};$$

(iv) *conversely, for any $\tau \in \partial\mathbb{D} \setminus \{\pm i\sigma\}$ we can find a backward orbit $\{z_\nu\}$ of γ such that*

$$\lim_{\nu\to+\infty} \frac{\sigma - z_\nu}{|\sigma - z_\nu|} = \tau.$$

Proof. The verification of (i) is trivial. Next, for all $\nu \ge 1$ we have

$$\frac{\gamma(z) - \sigma\frac{A^{\nu-1}-1}{A^{\nu-1}+1}}{1 - \overline{\sigma}\frac{A^{\nu-1}-1}{A^{\nu-1}+1}\gamma(z)} = \frac{z - \sigma\frac{A^\nu-1}{A^\nu+1}}{1 - \overline{\sigma}\frac{A^\nu-1}{A^\nu+1}z},$$

and hence by induction we get (ii). In particular, given a backward orbit $\{z_\nu\}$ we get

$$z_\nu = \gamma^{-\nu}(z_0) = \frac{z_0 + \sigma\frac{A^\nu-1}{A^\nu+1}}{1+\overline{\sigma}\frac{A^\nu-1}{A^\nu+1}z_0}.$$

From this, it follows that $z_\nu \to \sigma$ and that

$$\frac{\sigma - z_\nu}{|\sigma - z_\nu|} = \frac{\sigma - z_0}{|\sigma - z_0|}\frac{\left|1+\frac{A^\nu-1}{A^\nu+1}\overline{\sigma}z_0\right|}{1+\frac{A^\nu-1}{A^\nu+1}\overline{\sigma}z_0} \to \frac{\sigma - z_0}{|\sigma - z_0|}\frac{|1+\overline{\sigma}z_0|}{1+\overline{\sigma}z_0}.$$

Since

$$\frac{\sigma - z_0}{|\sigma - z_0|}\frac{|1+\overline{\sigma}z_0|}{1+\overline{\sigma}z_0} = \sigma\frac{1-|z_0|^2-2i\,\mathrm{Im}(\overline{\sigma}z_0)}{|1-(\overline{\sigma}z_0)^2|} \neq \pm i\sigma$$

for any $z_0 \in \mathbb{D}$, we have proved (iii).

Finally, write $\tau = \sigma e^{i\theta}$ with $\theta \neq \pm\pi/2$, and choose $r \in (-1, 1)$ such that

$$\begin{cases} \dfrac{1-r^2}{1+r^2} = \cos\theta, \\ -\dfrac{2r}{1+r^2} = \sin\theta. \end{cases}$$

Then taking $z_0 = i\sigma r$ we get a backward orbit satisfying (iv) and we are done. □

Next, we can characterize backward orbits with a bounded step exactly equal to the angular derivative.

Proposition 4.9.11. *Given $f \in \mathrm{Hol}(\mathbb{D},\mathbb{D})$, let $\{z_\nu\} \subset \mathbb{D}$ be a backward orbit with bounded step $A > 1$ for f converging to a boundary repelling fixed point $\sigma \in \partial\mathbb{D}$ with $f'(\sigma) < +\infty$. Then $A = f'(\sigma)$ if and only if $z_\nu \to \sigma$ radially, i. e., if and only if*

$$\lim_{\nu\to+\infty}\frac{|\sigma - z_\nu|}{1-|z_\nu|} = 1. \tag{4.56}$$

Proof. Assume first that $A = f'(\sigma)$. Put $R_\nu = \omega(z_\nu, z_{\nu+1})$ so that $R_\nu \to \frac{1}{2}\log f'(\sigma)$. Then Proposition 1.2.7(ii) yields

$$\frac{1-|z_\nu|}{1-|z_{\nu+1}|} \leq \frac{1+\tanh R_\nu}{1-(\tanh R_\nu)|z_{\nu+1}|}$$

and, therefore,

$$\limsup_{\nu\to+\infty}\frac{1-|z_\nu|}{1-|z_{\nu+1}|} \leq f'(\sigma).$$

On the other hand, Proposition 2.1.15 yields

$$\liminf_{\nu\to+\infty}\frac{1-|z_\nu|}{1-|z_{\nu+1}|}\geq f'(\sigma),$$

and hence we have proved that

$$\lim_{\nu\to+\infty}\frac{1-|z_\nu|}{1-|z_{\nu+1}|}=f'(\sigma). \tag{4.57}$$

Now, $R_\nu \leq \frac{1}{2}\log f'(\sigma)$ becomes

$$\left|\frac{z_{\nu+1}-z_\nu}{1-\overline{z_\nu}z_{\nu+1}}\right|\leq\frac{f'(\sigma)-1}{f'(\sigma)+1}.$$

Squaring both sides, subtracting 1 and recalling that $|z_\nu|<1$ we obtain

$$\frac{f'(\sigma)}{(f'(\sigma)+1)^2}\leq\frac{(1-|z_\nu|)(1-|z_{\nu+1}|)}{|1-\overline{z_\nu}z_{\nu+1}|^2}\leq\left(\frac{f'(\sigma)-1}{f'(\sigma)+1}\right)^2\frac{(1-|z_\nu|)(1-|z_{\nu+1}|)}{|z_{\nu+1}-z_\nu|^2}.$$

Multiplying both sides by

$$\frac{(|z_{\nu+1}|-|z_\nu|)^2}{(1-|z_\nu|)(1-|z_{\nu+1}|)}=\left(1-\frac{1-|z_{\nu+1}|}{1-|z_\nu|}\right)\left(\frac{1-|z_\nu|}{1-|z_{\nu+1}|}-1\right)$$

we get

$$\frac{f'(\sigma)}{(f'(\sigma)-1)^2}\left(1-\frac{1-|z_{\nu+1}|}{1-|z_\nu|}\right)\left(\frac{1-|z_\nu|}{1-|z_{\nu+1}|}-1\right)\leq\left(\frac{|z_{\nu+1}|-|z_\nu|}{|z_{\nu+1}-z_\nu|}\right)^2.$$

Recalling (4.57), we then obtain

$$\liminf_{\nu\to+\infty}\frac{|z_{\nu+1}|-|z_\nu|}{|z_{\nu+1}-z_\nu|}\geq 1.$$

This means that for every $\varepsilon>0$ there is ν_0 such that if $\nu\geq\nu_0$ then

$$|z_{\nu+1}|-|z_\nu|\geq(1-\varepsilon)|z_{\nu+1}-z_\nu|.$$

Therefore, for every $\mu>\nu\geq\nu_0$ we have

$$\begin{aligned}|z_\mu|-|z_\nu|&=\sum_{j=\nu}^{\mu-1}(|z_{j+1}|-|z_j|)\geq(1-\varepsilon)\sum_{j=\nu}^{\mu-1}|z_{j+1}-z_j|\\&\geq(1-\varepsilon)\left|\sum_{j=\nu}^{\mu-1}(z_{j+1}-z_j)\right|=(1-\varepsilon)|z_\mu-z_\nu|.\end{aligned}$$

Letting $\mu\to+\infty$, we get

$$1 - |z_\nu| \geq (1-\varepsilon)|\sigma - z_\nu|$$

for all $\nu \geq \nu_0$, and hence

$$1 \geq \limsup_{\nu + \infty} \frac{1-|z_\nu|}{|\sigma - z_\nu|} \geq \liminf_{\nu + \infty} \frac{1-|z_\nu|}{|\sigma - z_\nu|} \geq 1 - \varepsilon.$$

Since $\varepsilon > 0$ is arbitrary, we have proved (4.56), that means that $z_\nu \to \sigma$ radially thanks to Lemma 2.2.15.

Conversely, assume that (4.56) holds; then Lemma 2.2.15 yields

$$\lim_{\nu \to +\infty} \frac{\sigma - z_\nu}{|\sigma - z_\nu|} = \sigma. \tag{4.58}$$

We can then apply Corollary 2.3.10 to infer that

$$\frac{1}{2}\log A = \lim_{\nu \to +\infty} \omega(z_\nu, z_{\nu+1}) = \frac{1}{2}\log f'(\sigma),$$

and hence $A = f'(\sigma)$ as claimed. □

Then we have the following.

Theorem 4.9.12 (Poggi–Corradini, 2000). *Let $f \in \mathrm{Hol}(\mathbb{D}, \mathbb{D})$ and let $\sigma \in \partial\mathbb{D}$ be a boundary repelling fixed point of f with $1 < f'(\sigma) < +\infty$. Put $A = f'(\sigma)$ and $a = \sigma(A-1)/(A+1)$. Then there exists $\psi \in \mathrm{Hol}(\mathbb{D}, \mathbb{D})$ such that:*
(i) *ψ has nontangential limit σ at σ;*
(ii) *$\psi \circ \gamma = f \circ \psi$, where $\gamma \in \mathrm{Aut}(\mathbb{D})$ is given by (4.54);*
(iii) *we have*

$$K\text{-}\lim_{z\to\sigma} \arg \frac{\sigma - \psi(z)}{\sigma - z} = K\text{-}\lim_{z\to\sigma} \arg \psi'(z) = 0.$$

Moreover, ψ is unique up to precomposition with an automorphism of $\mathbb{D}$ commuting with γ and satisfies
(iv) *there is a simply connected domain $\Omega \subset \mathbb{D}$ such that $\psi|_\Omega$ is injective and $\psi(\Omega)$ eventually contains any sequence converging nontangentially to σ.*

Proof. Let $\{z_\nu\}$ be a backward orbit of f converging to σ with bounded step equal to A, as given by Theorem 4.9.8; in particular $z_\nu \to \sigma$ radially, by Proposition 4.9.11. Put $\gamma_\nu(z) = \frac{z+z_\nu}{1+\overline{z_\nu}z}$; then $f^\nu \circ \gamma_{\nu+k}(0) = z_k$ for all $k, \nu \in \mathbb{N}$ and $\gamma_\nu \to \sigma$ as $\nu \to +\infty$.

We start by proving three technical claims.

Claim 1: For every $k \in \mathbb{N}$, we have $\gamma_{\nu+k}^{-1} \circ \gamma_\nu \to \gamma^k$ (and hence $\gamma_\nu^{-1} \circ \gamma_{\nu+k} \to \gamma^{-k}$) as $\nu \to +\infty$.

We know, by Proposition 4.9.11, that $z_\nu \to \sigma$ radially. Then (4.58) yields

$$\frac{1 - \overline{\sigma} z_\nu}{1 - \sigma\overline{z_\nu}} = \frac{\overline{\sigma}}{\sigma}\frac{\sigma - z_\nu}{\overline{\sigma} - \overline{z_\nu}} = \frac{\overline{\sigma}}{\sigma}\frac{\sigma - z_\nu}{|\sigma - z_\nu|}\frac{|\overline{\sigma} - \overline{z_\nu}|}{\overline{\sigma} - \overline{z_\nu}} \to 1 \tag{4.59}$$

as $\nu \to +\infty$. Furthermore, f^k has angular derivative A^k at σ, by Proposition 2.4.7, and hence

$$\frac{\sigma - z_\nu}{\sigma - z_{\nu+k}} = \frac{\sigma - f^k(z_{\nu+k})}{\sigma - z_{\nu+k}} \to A^k \tag{4.60}$$

as $\nu \to +\infty$ by the Julia–Wolff–Carathéodory Theorem 2.3.2, again because $z_\nu \to \sigma$ radially. Now, we have

$$\gamma_{\nu+k}^{-1} \circ \gamma_\nu(z) = \frac{\left(\frac{1-z_{\nu+k}\overline{z_\nu}}{1-\overline{\sigma}z_{\nu+k}}\right)z + \frac{z_\nu - z_{\nu+k}}{1-\overline{\sigma}z_{\nu+k}}}{\left(\frac{\overline{z_\nu}-\overline{z_{\nu+k}}}{1-\overline{\sigma}z_{\nu+k}}\right)z + \frac{1-\overline{z_{\nu+k}}z_\nu}{1-\overline{\sigma}z_{\nu+k}}}.$$

Using (4.59) and (4.60), we get

$$\begin{aligned}
\frac{1-z_{\nu+k}\overline{z_\nu}}{1-\overline{\sigma}z_{\nu+k}} &= 1 + \frac{z_{\nu+k}}{\sigma}\frac{1-\sigma\overline{z_\nu}}{1-\overline{\sigma}z_\nu}\frac{\sigma - z_\nu}{\sigma - z_{\nu+k}} \to 1 + A^k;\\
\frac{z_\nu - z_{\nu+k}}{1-\overline{\sigma}z_{\nu+k}} &= \sigma\left(1 - \frac{\sigma - z_\nu}{\sigma - z_{\nu+k}}\right) \to \sigma(1-A^k);\\
\frac{\overline{z_\nu}-\overline{z_{\nu+k}}}{1-\overline{\sigma}z_{\nu+k}} &= \overline{\sigma}\left(\frac{1-\sigma\overline{z_{\nu+k}}}{1-\overline{\sigma}z_{\nu+k}} - \frac{1-\sigma\overline{z_\nu}}{1-\overline{\sigma}z_\nu}\frac{\sigma - z_\nu}{\sigma - z_{\nu+k}}\right) \to \overline{\sigma}(1-A^k);\\
\frac{1-\overline{z_{\nu+k}}z_\nu}{1-\overline{\sigma}z_{\nu+k}} &= \overline{\sigma}z_\nu\frac{1-\sigma\overline{z_{\nu+k}}}{1-\overline{\sigma}z_{\nu+k}} + \frac{\sigma - z_\nu}{\sigma - z_{\nu+k}} \to 1 + A^k.
\end{aligned}$$

Hence

$$\gamma_{\nu+k}^{-1} \circ \gamma_\nu(z) \to \frac{(1+A^k)z + \sigma(1-A^k)}{\overline{\sigma}(1-A^k)z + (1+A^k)} = \frac{z - \sigma\frac{A^k-1}{A^k+1}}{1-\overline{\sigma}\frac{A^k-1}{A^k+1}z} = \gamma^k(z),$$

by (4.55) and Claim 1 is proved, because the inversion is continuous in $\mathrm{Aut}(\mathbb{D})$.

Claim 2: For every $k \in \mathbb{N}$, we have $\gamma_{\nu+k}^{-1} \circ \gamma^{-k} \circ \gamma_\nu \to \mathrm{id}_{\mathbb{D}}$ as $\nu \to +\infty$.

Recalling (4.55), it is easy to check that

$$\gamma_\nu^{-1} \circ \gamma^{-k} \circ \gamma_\nu(z) = \frac{\left(\frac{1-|z_\nu|^2}{1-\overline{\sigma}z_\nu} + \frac{A^k-1}{A^k+1}\frac{\sigma\overline{z_\nu}-\overline{\sigma}z_\nu}{1-\overline{\sigma}z_\nu}\right)z + \frac{A^k-1}{A^k+1}\sigma(1+\overline{\sigma}z_\nu)}{\frac{1-|z_\nu|^2}{1-\overline{\sigma}z_\nu} - \frac{A^k-1}{A^k+1}\frac{\sigma\overline{z_\nu}-\overline{\sigma}z_\nu}{1-\overline{\sigma}z_\nu} + \frac{A^k-1}{A^k+1}\overline{\sigma}(1+\sigma\overline{z_\nu})z}.$$

Since, by (4.59), we have

$$\begin{aligned}
\frac{1-|z_\nu|^2}{1-\overline{\sigma}z_\nu} &= 1 + \overline{\sigma}z_\nu\frac{1-\sigma\overline{z_\nu}}{1-\overline{\sigma}z_\nu} \to 2,\\
\frac{\sigma\overline{z_\nu}-\overline{\sigma}z_\nu}{1-\overline{\sigma}z_\nu} &= 1 - \frac{1-\sigma\overline{z_\nu}}{1-\overline{\sigma}z_\nu} \to 0
\end{aligned}$$

as $\nu \to +\infty$, it follows that $\gamma_\nu^{-1} \circ \gamma^{-k} \circ \gamma_\nu \to \gamma^{-k}$. Therefore, using Claim 1 we get

$$\gamma_{\nu+k}^{-1} \circ \gamma^{-k} \circ \gamma_\nu = (\gamma_{\nu+k}^{-1} \circ \gamma_\nu) \circ (\gamma_\nu^{-1} \circ \gamma^{-k} \circ \gamma_\nu) \to \gamma^k \circ \gamma^{-k} = \mathrm{id}_{\mathbb{D}}$$

and Claim 2 is proved.

Claim 3: For all $z \in \mathbb{D}$, we have $\omega(f^\nu \circ \gamma_\nu(z), f^{\nu+1} \circ \gamma_{\nu+1}(z)) \to 0$ as $\nu \to +\infty$.

Indeed, for any $z \in \mathbb{D}$ we have

$$\begin{aligned}\omega(f^\nu \circ \gamma_\nu(z), f^{\nu+1} \circ \gamma_{\nu+1}(z)) &\le \omega(\gamma_\nu(z), f \circ \gamma_{\nu+1}(z)) \\ &\le \omega(\gamma_\nu(z), f \circ \gamma^{-1} \circ \gamma_\nu(z)) + \omega(f \circ \gamma^{-1} \circ \gamma_\nu(z), f \circ \gamma_{\nu+1}(z)) \\ &\le \omega(\gamma_\nu(z), f \circ \gamma^{-1} \circ \gamma_\nu(z)) + \omega(\gamma_{\nu+1}^{-1} \circ \gamma^{-1} \circ \gamma_\nu(z), z).\end{aligned} \tag{4.61}$$

The second addend goes to 0 by Claim 2. For the first addend, since $z_\nu \to \sigma$ radially, using (4.58) and (4.59) we get

$$\frac{\sigma - \gamma_\nu(z)}{|\sigma - \gamma_\nu(z)|} = \frac{|1 + \overline{z_\nu} z|}{1 + \overline{z_\nu} z} \sigma \frac{\overline{z_\nu} - \overline{\sigma}}{|\overline{z_\nu} - \overline{\sigma}|} \frac{z - \sigma \frac{1 - \overline{\sigma} z_\nu}{1 - \sigma \overline{z_\nu}}}{\left| z - \sigma \frac{1 - \overline{\sigma} z_\nu}{1 - \sigma \overline{z_\nu}} \right|} \to \sigma \frac{|\sigma + z|}{\sigma + z} \frac{\sigma - z}{|\sigma - z|}.$$

Denoting by $\tau_z \in \partial\mathbb{D}$ the latter limit, notice that $\tau_z \neq \pm i\sigma$. Indeed,

$$\frac{|\sigma + z|}{\sigma + z} \frac{\sigma - z}{|\sigma - z|} = \pm i \iff \frac{|\sigma + z|}{|\sigma - z|} = \pm i \frac{1 - |z|^2 + 2i \operatorname{Im}(\overline{\sigma} z)}{|1 - (\overline{\sigma} z)^2|},$$

and taking the imaginary part of both sides we get $1 - |z|^2 = 0$, impossible. In particular, $\gamma_\nu(z) \to \sigma$ nontangentially and $\sigma^2 \overline{\tau_z}^2 \neq -1$ for all $z \in \mathbb{D}$.

Since $(\gamma^{-1})'(\sigma) = A^{-1}$, by using the chain rule for angular derivatives (Proposition 2.4.7), we see that $f \circ \gamma^{-1}$ has angular derivative 1 at σ, i. e., $\beta_{f \circ \gamma^{-1}}(\sigma, \sigma) = 1$. Recalling Corollary 2.3.10, we then get

$$\lim_{\nu \to +\infty} \omega(\gamma_\nu(z), f \circ \gamma^{-1} \circ \gamma_\nu(z)) = 0,$$

and the proof of Claim 3 is complete.

We can now start with the main argument. Let $\psi \in \mathrm{Hol}(\mathbb{D}, \mathbb{C})$ be an accumulation point of the sequence $\{f^\nu \circ \gamma_\nu\}$. We shall prove that ψ is as desired; actually we shall show that the whole sequence $\{f^\nu \circ \gamma_\nu\}$ converges to ψ. Notice that $\psi(0) = z_0$ and thus $\psi(\mathbb{D}) \subseteq \mathbb{D}$, by a Hurwitz theorem (Corollary A.1.5).

Choose a subsequence $\{f^{\nu_k} \circ \gamma_{\nu_k}\}$ converging to ψ. Claim 3 implies that $\{f^{\nu_k+1} \circ \gamma_{\nu_k+1}\}$ converges to ψ, too. But

$$f^{\nu_k+1} \circ \gamma_{\nu_k+1} = f \circ (f^{\nu_k} \circ \gamma_{\nu_k}) \circ (\gamma_{\nu_k}^{-1} \circ \gamma_{\nu_k+1});$$

taking the limit as $k \to +\infty$ and using Claim 1 we get

$$\psi = f \circ \psi \circ \gamma^{-1},$$

and thus (ii) is proved.

To prove (i), for $k \in \mathbb{N}$ put $a_k = \sigma(A^k - 1)/(A^k + 1)$; notice that $a_1 = a$ and $\gamma^k(a_k) = 0$ by (4.55). We claim that $f^\nu \circ \gamma_\nu(a_k) \to z_k$ as $\nu \to +\infty$. Indeed,

$$\begin{aligned}\omega(f^\nu \circ \gamma_\nu(a_k), z_k) &= \omega(f^\nu \circ \gamma_\nu(a_k), f^\nu(z_{\nu+k})) \\ &\leq \omega(a_k, \gamma_\nu^{-1} \circ \gamma_{\nu+k}(0)) = \omega(\gamma^{-k}(0), \gamma_\nu^{-1} \circ \gamma_{\nu+k}(0)) \to 0\end{aligned}$$

as $\nu \to +\infty$, by Claim 1. Thus $\psi(a_k) = z_k$ for all $k \in \mathbb{N}$. Put $g_\nu = \gamma_\nu^{-1} \circ \psi \circ \gamma^{-\nu}$; we claim that $g_\nu \to \mathrm{id}_\mathbb{D}$ as $\nu \to +\infty$. Indeed, we have

$$g_\nu(0) = \gamma_\nu^{-1}(\psi(a_\nu)) = \gamma_\nu^{-1}(z_\nu) = 0$$

and

$$g_\nu(a) = \gamma_\nu^{-1} \circ \psi(a_{\nu+1}) = \gamma_\nu^{-1}(z_{\nu+1}) = \gamma_\nu^{-1}(\gamma_{\nu+1}(0)) \to \gamma^{-1}(0) = a,$$

again by Claim 1. Therefore, any accumulation point of $\{g_\nu\}$ must have 0 and a as fixed points, and hence it must be the identity. So, $\mathrm{id}_\mathbb{D}$ is the only accumulation point of $\{g_\nu\}$, which is equivalent to saying that $g_\nu \to \mathrm{id}_\mathbb{D}$, as claimed. In particular, we get

$$\begin{aligned}\omega(f^\nu \circ \gamma_\nu(z), \psi(z)) &= \omega(f^\nu \circ \gamma_\nu(z), \psi \circ \gamma^\nu \circ \gamma^{-\nu}(z)) \\ &= \omega(f^\nu \circ \gamma_\nu(z), f^\nu \circ \psi \circ \gamma^{-\nu}(z)) \\ &\leq \omega(\gamma_\nu(z), \psi \circ \gamma^{-\nu}(z)) = \omega(z, g_\nu(z)) \to 0\end{aligned}$$

as $\nu \to +\infty$, and thus $f^\nu \circ \gamma_\nu \to \psi$.

Denoting by ℓ_ν the segment from a_ν to $a_{\nu+1}$, since $\psi(\ell_\nu) = \gamma_\nu \circ g_\nu(\ell_0)$ it follows that $\psi(\ell_\nu) \to \sigma$ as $\nu \to +\infty$. This means that ψ has radial limit σ at σ and (i) follows by the Lindelöf Theorem 2.4.2.

To prove (iii), notice that since $g_\nu \to \mathrm{id}_\mathbb{D}$ we also have

$$\lim_{\nu \to +\infty} \omega(\psi(\gamma^{-\nu}(z)), \gamma_\nu(z)) = 0$$

for all $z \in \mathbb{D}$. Since $\gamma_\nu(z) \to \sigma$ nontangentially Lemma 2.2.17 yields

$$\lim_{\nu \to +\infty} \frac{\sigma - \psi(\gamma^{-\nu}(z))}{\sigma - \gamma_\nu(z)} = 1. \tag{4.62}$$

Now,

$$\arg(\sigma - \gamma_\nu(z)) = \arg\left(\sigma - \frac{z + z_\nu}{1 + \overline{z_\nu} z}\right) = \arg \frac{\frac{\sigma - z_\nu}{|\sigma - z_\nu|} - \sigma \frac{\overline{\sigma} - \overline{z_\nu}}{|\sigma - z_\nu|} z}{1 + \overline{z_\nu} z},$$

and hence

$$\lim_{\nu\to\infty} \arg(\sigma - \gamma_\nu(z)) = \arg \frac{\sigma - z}{1 + \overline{\sigma} z},$$

because $z_\nu \to \sigma$ radially. Analogously,

$$\arg(\sigma - \gamma^{-\nu}(z)) = \arg \frac{\sigma - z}{1 + \overline{a_\nu} z},$$

and hence

$$\lim_{\nu\to+\infty} \arg(\sigma - \gamma^{-\nu}(z)) = \arg \frac{\sigma - z}{1 + \overline{\sigma} z}.$$

Recalling (4.62), we then get

$$\lim_{\nu\to\infty} \arg \frac{\sigma - \psi(\gamma^{-\nu}(z))}{\sigma - \gamma^{-\nu}(z)} = 0, \tag{4.63}$$

uniformly on compact subsets of $\mathbb{D}$.

Now let $\{w_k\} \subset \mathbb{D}$ be any sequence converging nontangentially to σ. By Proposition 2.2.7(iii), we can find $R > 0$, independent of k, such that for any $k \in \mathbb{N}$ there is $t_k \in (0,1)$ such that $\omega(w_k, t_k\sigma) \le R$. Clearly, $t_k \to 1^-$ as $k \to +\infty$; without loss of generality, we can assume $1 > t_k \ge 1/2$ for all $k \in \mathbb{N}$. For each k, choose $\nu_k \in \mathbb{N}$ such that

$$\frac{t_k - \frac{1}{2}}{1 - \frac{1}{2}t_k} \le |a_{\nu_k}| \iff \frac{t_k - |a_{\nu_k}|}{1 - |a_{\nu_k}| t_k} \le \frac{1}{2};$$

it follows that $|\gamma^{\nu_k}(t_k\sigma)| \le \frac{1}{2}$, i. e., $\omega(0, \gamma^{\nu_k}(t_k\sigma)) \le \frac{1}{2}\log 3$. Clearly, $\nu_k \to +\infty$ as $k \to +\infty$. Putting these estimates together, we have

$$\begin{aligned}\omega(0, \gamma^{\nu_k}(w_k)) &\le \omega(0, \gamma^{\nu_k}(t_k\sigma)) + \omega(\gamma^{\nu_k}(t_k\sigma), \gamma^{\nu_k}(w_k)) \\ &\le \omega(0, \gamma^{\nu_k}(t_k\sigma)) + \omega(t_k\sigma, w_k) \le \frac{1}{2}\log 3 + R;\end{aligned}$$

in particular, the whole sequence $\{\gamma^{\nu_k}(w_k)\}$ is contained in a fixed compact subset of $\mathbb{D}$. Then

$$\lim_{k\to+\infty} \arg \frac{\sigma - \psi(w_k)}{\sigma - w_k} = \lim_{k\to+\infty} \arg \frac{\sigma - \psi(\gamma^{-\nu_k}(\gamma^{\nu_k}(w_k)))}{\sigma - \gamma^{-\nu_k}(\gamma^{\nu_k}(w_k))} = 0,$$

by (4.63); by the arbitrariness of $\{w_k\}$, we have proved the first equality in (iii). The second equality follows from Corollary 2.4.10.

To prove (iv), choose a sequence $\{M_\nu\}$ increasing to $+\infty$. Since $\arg\psi'$ has vanishing nontangential limit at σ, for each ν we can find ε_ν such that $|\arg\psi'| < \pi/4$

in $K(\sigma, M_\nu) \cap D(\sigma, \varepsilon_\nu)$; clearly, we can assume that the sequence $\{\varepsilon_\nu\}$ is decreasing. Considering the convex hull of the points obtained as intersection of the boundary of $K(\sigma, M_\nu) \cap D(\sigma, \varepsilon_\nu)$ with the boundary of $K(\sigma, M_{\nu+1}) \cap D(\sigma, \varepsilon_{\nu+1})$ for $\nu \geq 0$, we get a convex domain $\Omega \subset \mathbb{D}$ with an inner tangent at σ (see Figure 4.3 and recall Definition 2.4.8) and such that $|\arg \psi'| < \pi/4$ on Ω. In particular, $\operatorname{Re} \psi' > 0$ in Ω; hence $\psi|_\Omega$ is injective by Proposition 4.5.1.

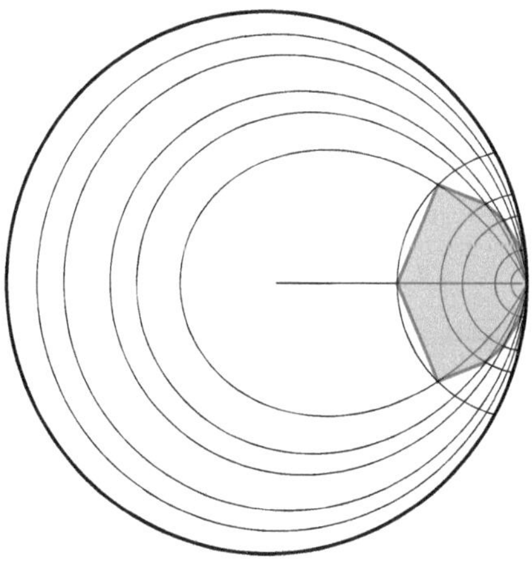

Figure 4.3: The domain $\Omega \subset \mathbb{D}$ from Theorem 4.9.12.

To complete the proof of (iv), it suffices to show that $\psi(\Omega)$ still has an inner tangent at σ. Fix $M > 1$ and choose $\beta > \arctan \sqrt{M^2 - 1}$. Since Ω has an inner tangent at σ, we can find $\delta > 0$ such that $U = A(\sigma, \beta) \cap D(\sigma, \delta) \subset \Omega$. Since $\psi|_\Omega$ is injective and (iii) holds, Lemma 2.2.15 implies that $\psi(U)$ is a Jordan domain whose boundary at σ consists in two curves forming an angle 2β symmetric with respect to the radius ending at σ. In particular, we can find $\arctan \sqrt{M^2 - 1} < \beta_1 < \beta$ and $\varepsilon > 0$ such that

$$K(\sigma, M) \cap D(\sigma, \varepsilon) \subset A(\sigma, \beta_1) \cap D(\sigma, \varepsilon) \subset \psi(\Omega),$$

where we have used (2.37), as needed.

We are left with the uniqueness. Assume that $\psi_1, \psi_2 \in \operatorname{Hol}(\mathbb{D}, \mathbb{D})$ satisfy (i)–(iii), and let Ω_1, Ω_2 be the corresponding domains as in (iv); notice that their existence follows from (iii) alone. Fix a compact subset $K \subset\subset \mathbb{D}$. Since Ω_1 has an inner tangent at σ, we can find $\nu_0 \in \mathbb{N}$ such that $\nu \geq \nu_0$ implies $\gamma^{-\nu}(K) \subset \Omega_1$. By (iii), also $\psi_1 \circ \gamma^{-\nu}$ converges to σ nontangentially uniformly on K; therefore, we can find $\nu_1 \geq \nu_0$ such that $\nu \geq \nu_1$ implies $\psi_1(\gamma^{-\nu}(K)) \subset \psi_2(\Omega_2)$. Put

$$F|_K = \gamma^\nu \circ \psi_2^{-1} \circ \psi_1 \circ \gamma^{-\nu}|_K.$$

Clearly, $F|_K$ is injective. Moreover, since (ii) yields

$$\gamma^\nu \circ \psi_2^{-1} \circ \psi_1 \circ \gamma^{-\nu} = \gamma^\nu \circ \psi_2^{-1} \circ f \circ \psi_1 \circ \gamma^{-\nu-1} = \gamma^{\nu+1} \circ \psi_2^{-1} \circ \psi_1 \circ \gamma^{-\nu-1}$$

it follows that F does not depend on ν. Thus, taking K larger and larger, we can define F on the whole of $\mathbb{D}$ and it is everywhere injective. Moreover, given $w_0 \in \mathbb{D}$ choose $\nu \in \mathbb{N}$ large enough so that $\psi_2 \circ \gamma^{-\nu}(w_0) \in \psi_1(\Omega_1)$; let $z_1 \in \Omega_1$ be such that $\psi_2 \circ \gamma^{-\nu}(w_0) = \psi_1(z_1)$ and put $z_0 = \gamma^\nu(z_1)$. Then $w_0 = F(z_0)$. So F is surjective, too; it follows that $F \in \mathrm{Aut}(\mathbb{D})$.

By construction, we have

$$F \circ \gamma^{-1} = \gamma^{-1} \circ \gamma^{\nu+1} \circ \psi_2^{-1} \circ \psi_1 \circ \gamma^{-\nu-1} = \gamma^{-1} \circ F;$$

therefore, F commutes with γ. Finally,

$$\psi_2 \circ F = \psi_2 \circ \gamma^\nu \circ \psi_2^{-1} \circ \psi_1 \circ \gamma^{-\nu} = f^\nu \circ \psi_2 \circ \psi_2^{-1} \circ \psi_1 \circ \gamma^{-\nu} = \psi_1 \circ \gamma^\nu \circ \gamma^{-\nu} = \psi_1,$$

and we are done. □

Having the model available, we can then show that backward orbits converge along a precise slope, and we can even find an explicit relation among the slope, the bounded step and the angular derivative.

Corollary 4.9.13. *Given $f \in \mathrm{Hol}(\mathbb{D}, \mathbb{D})$, let $\sigma \in \partial\mathbb{D}$ be a boundary repelling fixed point of f with $1 < f'(\sigma) < +\infty$. Let $\{z_\nu\}$ be a backward orbit with bounded step $1 < A < +\infty$ converging to σ. Then there is $\tau \in \partial\mathbb{D} \setminus \{\pm i\sigma\}$ such that*

$$\lim_{\nu\to+\infty} \frac{\sigma - z_\nu}{|\sigma - z_\nu|} = \tau. \tag{4.64}$$

Conversely, given $\tau = e^{i\theta}\sigma \in \partial\mathbb{D} \setminus \{\pm i\sigma\}$ there is a backward orbit $\{z_\nu\}$ with bounded step $1 < A < +\infty$ converging to σ such that (4.64) holds. Furthermore, A, $f'(\sigma)$ and θ are related by

$$A = \frac{|f'(\sigma) + e^{-2i\theta}| + f'(\sigma) - 1}{|f'(\sigma) + e^{-2i\theta}| - f'(\sigma) + 1}. \tag{4.65}$$

Proof. Let $\{z_\nu\}$ be a backward orbit with bounded step converging to σ. Let $\psi \in \mathrm{Hol}(\mathbb{D}, \mathbb{D})$, $\gamma \in \mathrm{Aut}(\mathbb{D})$ and $\Omega \subset \mathbb{D}$ be given by Theorem 4.9.12. Since, by Theorem 4.9.6, $z_\nu \to \sigma$ nontangentially, without loss of generality we can assume that $z_\nu \in \psi(\Omega)$ for all $\nu \in \mathbb{N}$. Put $w_\nu = \psi^{-1}(z_\nu)$. Then

$$\gamma(w_{\nu+1}) = \gamma \circ \psi^{-1}(z_{\nu+1}) = \psi^{-1} \circ f(z_{\nu+1}) = \psi^{-1}(z_\nu) = w_\nu;$$

therefore, $\{w_\nu\}$ is a backward orbit for γ, i. e., $w_\nu = \gamma^{-\nu}(w_0)$. By Lemma 4.9.10, it follows that there exists $\tau \in \partial\mathbb{D} \setminus \{\pm i\sigma\}$ such that

$$\lim_{\nu\to\infty} \frac{\sigma - w_\nu}{|\sigma - w_\nu|} = \tau, \tag{4.66}$$

and (4.64) follows from Theorem 4.9.12(iii) and Lemma 2.2.15.

Conversely, given $\tau \in \partial\mathbb{D} \setminus \{\pm i\sigma\}$ use Lemma 4.9.10 to choose a backward orbit $\{w_\nu\} \subset \Omega$ for γ converging to σ satisfying (4.66) and put $z_\nu = \psi(w_\nu)$. Then

$$f(z_{\nu+1}) = f \circ \psi(w_{\nu+1}) = \psi \circ \gamma(w_{\nu+1}) - \psi(w_\nu) - z_\nu;$$

thus $\{z_\nu\}$ is a backward orbit for f, converging to σ because $w_\nu \to \sigma$ nontangentially and σ is a boundary fixed point of ψ. Furthermore,

$$\omega(z_\nu, z_{\nu+1}) \leq \omega(w_\nu, w_{\nu+1}) = \omega(w_0, w_1) < +\infty,$$

and so $\{z_\nu\}$ has bounded step. Finally, (4.64) follows from (4.66), Theorem 4.9.12(iii) and Lemma 2.2.15.

Finally, (4.65) follows immediately from Corollary 2.3.10. □

Notes to Section 4.9

The material presented in this section has mainly be proven by Bracci [76] and Poggi-Corradini [325, 327] around 2003. More precisely, the results up to Theorem 4.9.6 come from [76, 327]; most of the rest in contained in [325], with the exception of Corollary 4.9.13, which is again in [327].

We have seen that, in the hyperbolic case, every backward orbit with bounded step converges to a boundary repelling fixed point along a precise direction, and conversely, that for any boundary repelling fixed point and any nontangential direction we can find a backward orbit converging to that point along that direction. If f is parabolic, this still holds at boundary repelling fixed points, but the situation is more complicated at the Wolff point. First of all, we already noticed (see Example 4.9.9) that there might not exist a backward orbit converging to the Wolff point. Assume anyway that $\{z_\nu\}$ is a backward orbit with bounded step converging to the Wolff point $\tau_f \in \partial\mathbb{D}$ of a parabolic map $f \in \mathrm{Hol}(\mathbb{D}, \mathbb{D})$; it turns out that we have two cases to consider. Using the notation and the terminology introduced in the notes to Section 4.6, the Julia lemma implies that the sequence $\{h_{\tau_f}(z_\nu)\}$ is not decreasing; we shall say that $\{z_\nu\}$ has *finite height* (resp., *infinite height*) if the limit of this sequence is finite (resp., infinite). If f has a positive hyperbolic step, then ([327, Theorem 1.21]) all backward orbits with bounded step converging to the Wolff point are of finite height. In [327], it is also possible to find examples of: parabolic maps with a positive hyperbolic step and finite height without backward orbits converging to the Wolff point; parabolic maps with a positive hyperbolic step and infinite height without backward orbits converging to the Wolff point; parabolic maps with zero hyperbolic step and infinite height without backward orbits converging to the Wolff point; parabolic maps with a positive hyperbolic step and finite height having a backward orbit with bounded step and finite height converging to the Wolff point; parabolic maps with zero hyperbolic step and infinite height having a backward orbit with bounded step and finite height converging to the Wolff point; parabolic maps with zero hyperbolic step and infinite height having a backward orbit with bounded step and infinite height converging to the Wolff point. Since parabolic maps with a zero hyperbolic step necessarily have infinite height (see [329]), the only example missing is that of a parabolic map with a positive hyperbolic step and finite height having a backward orbit with bounded step and infinite height converging to the Wolff point. It is also open the question of whether a given parabolic map with a zero hyperbolic step can have at the same time backward orbits with bounded step and finite height and backward orbits with bounded step and infinite height.

An example of backward orbit with unbounded step converging to a boundary fixed point with a (necessarily) infinite angular derivative can be obtained using the characterization of a superrepelling fixed point of first type for one-parameter semigroups (see Definition 5.2.1) given in [79]. An example is the following: take $D = \{w \in \mathbb{C} \mid \operatorname{Im} w > -1/|\operatorname{Re} w|\} \cup \{w \in \mathbb{C} \mid \operatorname{Re} w = 0\}$ with the map $F \in \operatorname{Hol}(D, D)$ given by $F(w) = w + i$. Since D is simply connected, we can find a biholomorphism $\psi\colon \mathbb{D} \to D$ allowing us to define $f = \psi^{-1} \circ F \circ \psi \in \operatorname{Hol}(\mathbb{D}, \mathbb{D})$. Then it is possible to prove (see [84]) that $\{z_\nu = \psi^{-1}(-i\nu)\}$ is a backward orbit converging to a boundary repelling fixed point $\sigma \in \partial\mathbb{D}$ with $f'(\sigma) = +\infty$; in particular, $\{z_\nu\}$ cannot be of bounded step.

As far as I know, it is an open problem whether there exists a backward orbit with unbounded step not converging to a boundary fixed point with infinite angular derivative; it cannot happen for one-parameter semigroups (see [80, Lemma 13.1.5]).

4.10 Commuting functions

In this section, we shall examine again the relationships between commuting functions and common fixed points using what we know now about the dynamics on $\mathbb{D}$.

In Theorem 3.1.20, we saw that a commuting family of holomorphic maps continuous up to the boundary necessarily have a common fixed point. However, the continuity at the boundary was used just to take care of functions without fixed points in the interior. In the unit disk $\mathbb{D}$ we have developed a lot of material about boundary behavior of functions without fixed points; using it, we can remove the hypothesis of boundary continuity with just a bit more effort obtaining the *Behan theorem.*

Theorem 4.10.1 (Behan, 1973). *Let $f, g \in \operatorname{Hol}(\mathbb{D}, \mathbb{D})$, with $g \not\equiv \mathrm{id}_{\mathbb{D}}$, be such that $f \circ g = g \circ f$. Assume that f has no fixed points. Let $\tau_f \in \partial\mathbb{D}$ be the Wolff point of f. Then g also has no fixed points and*

$$K\text{-}\lim_{z\to\tau_f} g(z) = \tau_f = K\text{-}\lim_{z\to\tau_f} f(z).$$

Proof. If g would have a fixed point $z_0 \in \mathbb{D}$, then by Lemma 3.1.18, z_0 would be a fixed point for f, impossible. By Corollary 2.5.5, we just have to show that g has non-tangential limit τ_f at τ_f. To do so, it will suffice, by the Lindelöf Theorem 2.4.2, to construct a continuous curve $\eta\colon [0,1) \to \mathbb{D}$ with $\eta(t) \to \tau_f$ as $t \to 1$ such that $g(\eta(t)) \to \tau_f$ as $t \to 1$.

For $0 \le t < 1$, let $\nu(t)$ be the greatest integer less than or equal to $-\log_2(1-t)$. Let $z_0 = f(0)$ and for $t \in [0,1)$ define

$$\eta(t) = f^{\nu(t)}(2[1 - 2^{\nu(t)}(1-t)]z_0). \tag{4.67}$$

Since $\eta([1-2^{-\nu}, 1-2^{-\nu-1}])$ is the image under f^ν of the segment S from 0 to $f(0)$, it is easily checked that η is continuous. Moreover, $\eta(t) \to \tau_f$ as $t \to 1$, because $f^\nu \to \tau_f$ uniformly on S.

Now, $f^\nu \to \tau_f$ also uniformly on $g(S)$; since $g(f^\nu(S)) = f^\nu(g(S))$, we have $g(\eta(t)) \to \tau_f$ as $t \to 1$ and we are done. □

Theorem 4.10.1 raises a natural question: if $f, g \in \mathrm{Hol}(\mathbb{D}, \mathbb{D})$ are without fixed points and commute, do they have the same Wolff point?

In general, the answer is negative: if γ is a hyperbolic automorphism of $\mathbb{D}$ then γ and γ^{-1} commute but $\tau_\gamma \neq \tau_{\gamma^{-1}}$. Our aim now is to show that this is the only exception.

We shall need an arithmetic lemma.

Lemma 4.10.2. (i) *Let $a, b, c, d \in (-\infty, +\infty]$ be such that*

$$a + b > 0, \quad c + d \geq 0, \quad a \leq 0, \quad b > 0, \quad c > 0 \quad \text{and} \quad d \leq 0.$$

Then there are positive integers $h, k \in \mathbb{N}^*$ such that $ha + kc > 0$ and $hb + kd > 0$.

(ii) *Assume that $0 < \theta \leq 1$ and $-1 \leq \phi \leq 1$. Then there exists $k \in \mathbb{Z}$ such that $-1/2 \leq k\theta + \phi \leq 1/2$.*

Proof. (i) If b (resp., c) is equal to $+\infty$ we take $h = 1$ and k large (resp., $k = 1$ and h large). If d (resp., a) is zero, we again take $h = 1$ and k large (resp., $k = 1$ and h large). So assume a, b, c and d finite and not zero. From $a + c|a/c| = 0$, we infer

$$0 < (a + b) + (c + d)|a/c| = b + d|a/c|.$$

Hence $|a/c| < |b/d|$. Then $h, k \in \mathbb{N}$ such that $|a/c| < k/h < |b/d|$ are exactly as we need.

(ii) If $-1/2 \leq \phi \leq 1/2$, then $k = 0$ works. Assume then $-1 \leq \phi < -1/2$; the case $1/2 < \phi \leq 1$ can be treated similarly just changing signs. We need to find $k \in \mathbb{Z}$ such that $1/2 \leq k\theta \leq 1$. Take $k = \lfloor \frac{1}{\theta} \rfloor$, the largest integer less than or equal to $1/\theta$, so that $\frac{1}{\theta} - 1 < k \leq \frac{1}{\theta}$; this immediately yields $k\theta \leq 1$. For the other inequality, if $1/2 < \theta \leq 1$ we get $k = 1$, and thus it works. If $0 < \theta \leq 1/2$, we have $k\theta > 1 - \theta \geq 1/2$, and thus k works in this case, too. □

With this result we can prove the theorem we announced above.

Theorem 4.10.3. *Let $f, g \in \mathrm{Hol}(\mathbb{D}, \mathbb{D}) \setminus \{\mathrm{id}_\mathbb{D}\}$ be such that $f \circ g = g \circ f$. Then:*

(i) *if f is not a hyperbolic automorphism of $\mathbb{D}$ then $\tau_f = \tau_g$;*

(ii) *otherwise, g also is a hyperbolic automorphism of $\mathbb{D}$, with the same fixed points as f, and either $\tau_f = \tau_g$ or $\tau_{f^{-1}} = \tau_g$.*

Proof. If f or g have a fixed point in $\mathbb{D}$, which is their Wolff point, the assertion follows from Lemma 3.1.18; so we can assume that f and g have no fixed points in $\mathbb{D}$. For the same reason, all composition of iterates $f^h \circ g^k$ have no fixed points in $\mathbb{D}$.

If f is a hyperbolic automorphism of $\mathbb{D}$ the assertion follows by Theorem 2.6.1; so suppose that f is not a hyperbolic automorphism. Let $\sigma = \tau_f$ and $\tau = \tau_g$; hence, by Theorem 4.10.1 $f(\sigma) = g(\sigma) = \sigma$ and $f(\tau) = g(\tau) = \tau$. In particular, by the Julia–Wolff–Carathéodory Theorem 2.3.2, $f'(\sigma)$, $f'(\tau)$, $g'(\sigma)$ and $g'(\tau)$ are real and positive (possibly equal to $+\infty$).

Assume, by contradiction, $\tau \neq \sigma$. Since f is not a hyperbolic automorphism of $\mathbb{D}$, neither g is (by Theorem 2.6.1), and thus Corollary 2.3.16 implies

$$f'(\tau)f'(\sigma) > 1 \quad \text{and} \quad g'(\tau)g'(\sigma) > 1.$$

Moreover, $f'(\sigma) \leq 1$, $f'(\tau) > 1$, $g'(\sigma) > 1$ and $g'(\tau) \leq 1$. If we apply Lemma 4.10.2(i) with $a = \log f'(\sigma)$, $b = \log f'(\tau)$, $c = \log g'(\sigma)$ and $d = \log g'(\tau)$, we obtain two positive integers h and k such that

$$(f^h \circ g^k)'(\sigma) > 1 \quad \text{and} \quad (f^h \circ g^k)'(\tau) > 1,$$

where we used Proposition 2.4.7. In particular, the Wolff point $\eta \in \partial\mathbb{D}$ of $f^h \circ g^k$ is neither σ nor τ. Since both f and g commute with $f^h \circ g^k$, Theorem 4.10.1 shows that $f(\eta) = g(\eta) = \eta$. Hence we should have $f'(\eta) > 1$ and $g'(\eta) > 1$; in particular, using again Proposition 2.4.7 we find

$$(f^h \circ g^k)'(\eta) = (f')^h(\eta) \cdot (g')^k(\eta) > 1,$$

whereas we should have $(f^h \circ g^k)'(\eta) \leq 1$ because η is the Wolff point of $f^h \circ g^k$, contradiction. □

So, commuting holomorphic maps have (with rare exceptions) the same Wolff point. What about repelling boundary fixed points? It turns out that they are not necessarily fixed, but at least they are permuted.

Theorem 4.10.4 (Bracci, 2003). *Let $f, g \in \mathrm{Hol}(\mathbb{D}, \mathbb{D})$ be different from the identity and not constant. Assume that $f \circ g = g \circ f$. Let $\sigma \in \partial\mathbb{D}$ be a boundary fixed point of f with $f'(\sigma) < +\infty$. Then g has nontangential limit $g(\sigma) \in \partial\mathbb{D}$ at σ and $g(\sigma)$ too is a boundary fixed point of f, with $f'(g(\sigma)) \leq f'(\sigma)$.*

Proof. If $\sigma = \tau_f$, then the assertion follows trivially from Theorem 4.10.1. So, we can assume that σ is not the Wolff point of f (and, for the same reason, of g); in particular, $f'(\sigma) > 1$. Notice that the existence of σ implies that f cannot be an elliptic automorphism.

By Theorem 4.9.8, we can find a backward f-orbit $\{z_\nu\}$ converging to σ and with bounded step $f'(\sigma)$. We claim that we can assume that $g(z_0) \neq \tau_f$. If $\tau_f \in \partial\mathbb{D}$, this is obvious. If instead $\tau_f \in \mathbb{D}$, let (ψ, γ, Ω) by given by Theorem 4.9.12. Since ψ and g are not constant, we can find $z_0 \in \psi((0, 1/2)\sigma)$ such that $g(z_0) \neq \tau_f$. Let $t_0 \in (0, 1/2)$ be such that $z_0 = \psi(t_0\sigma)$ and put $z_\nu = \psi(\gamma^{-\nu}(t_0\sigma))$. Since $\gamma^{-\nu}(t_0\sigma) \to \sigma$ nontangentially and σ is a boundary fixed point of ψ, it follows that $z_\nu \to \sigma$. Furthermore,

$$f(z_\nu) = f(\psi(\gamma^{-\nu}(t_0\sigma))) = \psi(\gamma^{-\nu+1}(t_0\sigma)) = z_{\nu-1},$$

and hence $\{z_\nu\}$ is a backward f-orbit. Finally, setting $\gamma_0(z) = (z - t_0\sigma)/(1 - t_0\overline{\sigma}z)$ we have $\gamma_0 \circ \gamma^{-1} = \gamma^{-1} \circ \gamma_0$ and

$$\begin{aligned}\omega(z_\nu, z_{\nu+1}) &= \omega(\psi(\gamma^{-\nu}(t_0\sigma)), \psi(\gamma^{-\nu-1}(t_0\sigma)))\\&\le \omega(\gamma^{-\nu}(t_0\sigma), \gamma^{-\nu-1}(t_0\sigma)) = \omega(t_0\sigma, \gamma^{-1}(t_0\sigma))\\&= \omega(\gamma_0(t_0\sigma), \gamma_0\circ\gamma^{-1}(t_0\sigma)) = \omega(\gamma_0(t_0\sigma), \gamma^{-1}\circ\gamma_0(t_0\sigma))\\&= \omega(0, \gamma^{-1}(0)) = \frac{1}{2}\log f'(\sigma),\end{aligned}$$

and thus $\{z_\nu\}$ is as required, by Theorem 4.9.4.

Now put $z_\nu^1 = g(z_\nu)$. We claim that $\{z_\nu^1\}$ is a backward f-orbit with bounded step less than or equal to $f'(\sigma)$. Indeed,

$$f(z_\nu^1) = f(g(z_\nu)) = g(f(z_\nu)) = g(z_{\nu-1}) = z_{\nu-1}^1$$

and

$$\omega(z_\nu^1, z_{\nu+1}^1) \le \omega(z_\nu, z_{\nu+1}) \le \frac{1}{2}\log f'(\sigma). \tag{4.68}$$

By Theorem 4.9.4, it follows that $\{z_\nu^1\}$ converges to a boundary fixed point σ^1 (either repelling or equal to the Wolff point) of f, because $z_0^1 \neq \tau_f$ by construction. Moreover, Proposition 2.1.21 and Theorem 2.3.2 yield

$$\begin{aligned}\frac{1}{2}\log f'(\sigma^1) &\le \liminf_{\nu\to+\infty}[\omega(0, z_\nu^1) - \omega(0, f(z_\nu^1))]\\&\le \lim_{\nu\to+\infty}\omega(z_\nu^1, z_{\nu-1}^1) \le \frac{1}{2}\log f'(\sigma),\end{aligned}$$

where the last inequality follows from (4.68).

So, to conclude the proof it suffices to show that g has nontangential limit σ^1 at σ. Let (ψ, γ, Ω) by given by Theorem 4.9.12. Since $z_\nu \to \sigma$ nontangentially by Theorem 4.9.6, without loss of generality we can assume that $z_\nu \in \psi(\Omega)$ for all $\nu \in \mathbb{N}$; in particular, $z_\nu = \psi(w_\nu)$ for a suitable $w_\nu \in \Omega$. Arguing as in the proof of Corollary 4.9.13, we see that $\{w_\nu\}$ is a backward orbit of γ converging to σ; in particular, $\{w_\nu\}$ is contained (see Lemma 4.9.10) in the image of a γ-invariant geodesic $\ell\colon [0,1) \to \mathbb{D}$ with $\ell(t) \to \sigma$ as $t \to 1^-$; let $a_0 \in [0,1)$ be such that $w_0 = \ell(a_0)$. We claim that $g(\psi(\ell(t))) \to \sigma^1$; since ψ has nontangential limit σ at σ, the Lindelöf Theorem 2.4.2 then implies that g has nontangential limit σ^1 at σ and we are done.

Assume, by contradiction, that this is not the case and let $\{t_\nu\} \subset [0,1)$ with $t_\nu \to 1^-$ be such that $g(\psi(\ell(t_\nu))) \to \sigma_0 \in \overline{\mathbb{D}}$ with $\sigma_0 \neq \sigma^1$. Since

$$\ell([0,1)) = \bigcup_{k\in\mathbb{N}} \gamma^{-k}(\ell([0,a_0])) + \infty,$$

we can find $s_\nu \in [0, a_0]$ and $k_\nu \in \mathbb{N}$ such that $\ell(t_\nu) = \gamma^{-k_\nu}(\ell(s_\nu))$. Therefore,

$$\omega(g(\psi(\ell(t_\nu))), g(z_{k_\nu})) \le \omega(\psi(\ell(t_\nu)), z_{k_\nu}) = \omega(\psi(\gamma^{-k_\nu}(\ell(s_\nu))), \psi(\gamma^{-k_\nu}(w_0)))$$
$$\le \omega(\ell(s_\nu), \ell(a_0)) \le \omega(0, a_0).$$

Since $g(z_\nu) \to \sigma^1$, Proposition 1.7.27 implies that $g(\psi(\ell(t_\nu))) \to \sigma^1$, contradiction, and we are done. □

In general, it is not true that commuting maps have the same boundary fixed points.

Example 4.10.5. Let $f \in \mathrm{Hol}(\mathbb{D}, \mathbb{D})$ be given by

$$f(z) = \left(\frac{z + 1/2}{1 + z/2}\right)^2$$

and $g = f \circ f$. Then f has three fixed points $\{1, \sigma, \overline{\sigma}\}$, where $\sigma = e^{2\pi i/3}$, with $f'(1) = 2/3$ and $f'(\sigma) = f'(\overline{\sigma}) = 2$. But g has two more fixed points $\sigma_\pm = \frac{1}{8}(-7 \pm i\sqrt{15})$, both with finite angular derivative ($g'(\sigma_\pm) = 16$), permuted by f because $f(\sigma_\pm) = \sigma_\mp$.

We end this section describing another application of dynamics to the construction of fixed points of particular families of holomorphic functions.

By Theorem 2.6.9 and Proposition 1.6.38, the automorphism group of a multiply connected domain of regular type is always compact. On the other hand, we know that $\mathrm{Aut}(\mathbb{D})$ is not compact, though the isotropy group of one point is. Actually, it turns out that a subgroup of $\mathrm{Aut}(\mathbb{D})$ is relatively compact if and only if it has a fixed point. For the sake of generality, we shall prove something more.

Theorem 4.10.6. *Let $G \subset \mathrm{Hol}(\mathbb{D}, \mathbb{D})$ be a group under composition. Then G has a fixed point in $\mathbb{D}$ if and only if G is relatively compact in $\mathrm{Hol}(\mathbb{D}, \mathbb{D})$. Moreover, in this case $\overline{G}$ is either: a single constant function; or, finite cyclic; or, the isotropy group of its fixed point. In all cases, G is Abelian.*

Proof. Let $e \in G$ be the identity of G. If e is constant, then for all $f \in G$ we have $f = e \circ f = e$, and thus $G = \{e\}$.

If e is not constant, then $e(\mathbb{D})$ is open in $\mathbb{D}$; from $e = e \circ e$ we infer that $e \equiv \mathrm{id}_{\mathbb{D}}$ on the open set $e(\mathbb{D})$, and hence everywhere. This implies that G is a subgroup of $\mathrm{Aut}(\mathbb{D})$.

If G has a fixed point, then it is relatively compact by Corollary 1.7.21. Conversely, assume that $\overline{G} \subset \mathrm{Hol}(\mathbb{D}, \mathbb{D})$ is compact. Take $f \in \overline{G}$. By Proposition 1.4.10, f must have a fixed point in $\mathbb{D}$. Up to conjugation, we can assume $f(0) = 0$ and thus $f(z) = e^{i\pi\theta}z$ for some $0 < \theta < 2$. Replacing f by f^{-1} if necessary, we can assume $0 < \theta \le 1$.

By compactness, there is $g_0 \in \overline{G}$ maximizing $|g(0)|$. Write

$$g_0(z) = e^{i\pi\phi}\frac{z + a}{1 + \overline{a}z},$$

with $|a| = |g_0(0)|$ and $-1 \le \phi < 1$. Lemma 4.10.2(ii) yields $k \in \mathbb{Z}$ such that $\mathrm{Re}(e^{i(k\theta+\phi)}) \ge 0$; then, letting $\tau = e^{i(k\theta+\phi)}$, we have

$$\left|1+\tau|a|^2\right|^2 < |1+\tau|^2.$$

Now, $(f^k \circ g_0)^2 \in \overline{G}$; since $f^k \circ g_0(0) = \tau a$, we obtain

$$(f^k \circ g_0)^2(0) = \tau a \frac{1+\tau}{1+\tau|a|^2}.$$

Hence we should have $a = 0$ since otherwise $|(f^k \circ g_0)^2| > |g_0(0)|$ would contradict the maximality of $|g_0(0)|$. It follows that $\overline{G}$ is contained in the isotropy group of 0 and the rest of the assertion follows from Proposition 1.4.13. □

Corollary 4.10.7. *Let $G \subset \mathrm{Hol}(\mathbb{D}, \mathbb{D})$ be a group under composition. If there exists a compact set $K \subset \mathbb{D}$ invariant under G, then G has a fixed point in $\mathbb{D}$.*

Proof. In fact, G is clearly relatively compact in $\mathrm{Hol}(\mathbb{D}, \mathbb{D})$. □

Notes to Section 4.10

Theorems 4.10.1 and 4.10.3 are due to Behan [53].

Each holomorphic self-map obviously commutes with its iterates; a natural question is whether there are other possibilities. Cowen in [129] has studied this problem by using the models we introduced in Section 4.5. Let $f \in \mathrm{Hol}(\mathbb{D}, \mathbb{D})$ be with Wolff point $\tau_f \in \overline{\mathbb{D}}$. Assume that $f'(\tau_f) \neq 0$ and let (X^o, ψ, Φ) be the model given by Theorem 4.6.8. We say that $g \in \mathrm{Hol}(\mathbb{D}, \mathbb{D})$ belongs to the *pseudo-iteration semigroup* of f if there is a linear fractional transformation Ξ commuting with Φ such that $\psi \circ g = \Xi \circ \psi$. If instead $f'(\tau_f) = 0$ (and necessarily $\tau_f \in \mathbb{D}$), let $m \geq 2$ be the multiplicity of $f - \tau_f$ at τ_f and φ the local biholomorphism given by Theorem 4.2.2. Then we shall say that $g \in \mathrm{Hol}(\mathbb{D}, \mathbb{D})$ belongs to the *pseudo-iteration semigroup* of f if there are $k \in \mathbb{N}^*$ and $\lambda \in \mathbb{S}^1$ such that $\lambda^{m-1} = 1$ and $\psi \circ g(z) = \lambda \psi(z)^k$ in a neighborhood of τ_f.

Using these definitions, Cowen [129] has shown that if $|f'(\tau_f)| < 1$ and g commutes with f then g is in the pseudo-iteration semigroup of f; if instead $f'(\tau_f) = 1$ and g commutes with f then both f and g are in the pseudo-iteration semigroup of $f \circ g$. Conversely, if g is in the pseudo-iteration semigroup of f then there exists $k \in \mathbb{N}$ such that $f^k \circ g$ and f commute; if $f'(\tau_f) = 0$, we can take $k = 1$. [129] also contains examples showing that if $f'(\tau_f) \neq 0$ then it is not necessarily true that maps in the pseudo-iteration semigroup of f commute with f. Even if it is not very explicit, this almost characterization suffices to show, for instance, that if f is elliptic (resp., superattracting, hyperbolic, parabolic) then any map commuting with f is elliptic (resp., superattracting, hyperbolic, parabolic). Vlacci [406] has shown that if f and g (not automorphisms) have the same Wolff point $\tau \in \partial\mathbb{D}$ and they both admit orbits converging to τ nontangentially then they commute if and only if g belongs to the pseudo-iteration semigroup of f.

When $f \in \mathrm{Hol}(\mathbb{D}, \mathbb{D})$ is such that $f(0) = 0$ and $0 < |f'(0)| < 1$, let $\mathcal{F} = \{g \in \mathrm{Hol}(\mathbb{D}, \mathbb{D}) \mid f \circ g = g \circ f\}$. If $g \in \mathcal{F}$, we know that $g(0) = 0$; put $\Gamma = \{g'(0) \mid g \in \mathcal{F}\} \subseteq \overline{\mathbb{D}}$. Then Pranger [343] has proved that Γ is closed as a subset of $\overline{\mathbb{D}}$ and under multiplication, $0, 1 \in \Gamma$ but $\Gamma \cap \mathbb{D} \neq \{0\}$ and $\overline{\mathbb{D}} \setminus \Gamma$ is connected. Conversely, any $\Gamma \subseteq \overline{\mathbb{D}}$ satisfying these properties can be realized in this way.

It is possible to completely characterize commuting finite Blaschke products. Assume that $f, g \in \mathrm{Hol}(\mathbb{D}, \mathbb{D})$ are commuting finite Blaschke products respectively of degree $m \geq n \geq 2$. If f has a fixed point $z_0 \in \mathbb{D}$, then Chalendar and Mortini [110] proved that two possibilities may arise: either there exist $p, q \in \mathbb{N}^*$ such that $f^p \equiv g^q$; or, there exist $\gamma \in \mathrm{Aut}(\mathbb{D})$ and $n_0 \in \mathbb{N}^*$ such that $\gamma^{-1} \circ f \circ \gamma(z) = z^m$ and $\gamma^{-1} \circ g^{n_0} \circ \gamma(z) = z^{nn_0}$. Moreover, if $f'(z_0) \neq 0$ and $f^p \equiv g^q$ then f and g commute. This latter assertion has been proved by Arteaga [23, Theorem B(b)]. Actually, in Arteaga's statement the condition $f'(z_0) \neq 0$ was missing, but it is used in the proof. Indeed, when $f'(z_0) = 0$ this result is false, as seen taking for instance $f(z) = e^{2\pi i/3}z^2$ and $g(z) = z^2$; then $f^2 \equiv g^2$ but f and g do not commute.

If instead the Blaschke products f and g have no fixed points in $\mathbb{D}$, then f and g commute if and only if there exist $p, q \in \mathbb{N}^*$ such that $f^p \equiv g^q$. This result has been stated in [23], but Arteaga's proof works only when at least one of the two functions is of positive hyperbolic step; the complete proof has been given by Basallote, Contreras, and Hernández-Mancera in [40].

The rational functions commuting with a given rational function have been completely characterized by Julia [215] and Ritt [353].

In [85] and [388], Bracci, Tauraso, and Vlacci gave conditions ensuring when two commuting holomorphic self-maps f and g of $\mathbb{D}$ actually coincide. They proved that:

(i) if there exist $z_0 \in \mathbb{D}$ and $k \in \mathbb{N}^*$ such that $f(z_0) = g(z_0) = z_0$, $f^{(m)}(z_0) = g^{(m)}(z_0) = 0$ for $1 \leq m < k$ and $f^{(k)}(z_0) = g^{(k)}(z_0) \neq 0$, then $f \equiv g$;

(ii) if g is a hyperbolic automorphism of $\mathbb{D}$ with fixed point $\tau \in \partial\mathbb{D}$ and we have $f'(\tau) = g'(\tau)$, then $f \equiv g$;

(iii) if g is a parabolic automorphism of $\mathbb{D}$ with fixed point $\tau \in \partial\mathbb{D}$ and we have $f'(\tau) = g'(\tau)$ and $f \in C^2(\tau)$ with $f''(\tau) = g''(\tau)$, then $f \equiv g$;

(iv) if $g \equiv \mathrm{id}_{\mathbb{D}}$ and there exists $\tau \in \partial\mathbb{D}$ with $f \in C^3(\tau)$ such that $f(\tau) = \tau$, $f'(\tau) = \tau$ and $f''(\tau) = f'''(\tau) = 0$, then $f \equiv \mathrm{id}_{\mathbb{D}}$;

(v) if $f, g \notin \mathrm{Aut}(\mathbb{D})$ have Wolff point $\tau \in \partial\mathbb{D}$ with $f'(\tau) = g'(\tau) < 1$, then $f \equiv g$;

(vi) if $f, g \notin \mathrm{Aut}(\mathbb{D})$ have Wolff point $\tau \in \partial\mathbb{D}$ with $f, g \in C^2(\tau)$, $f'(\tau) = g'(\tau) = 1$ and $f''(\tau) = g''(\tau) \neq 0$, then $f \equiv g$;

(vii) if $f, g \notin \mathrm{Aut}(\mathbb{D})$ have Wolff point $\tau \in \partial\mathbb{D}$ with $f, g \in C^3(\tau)$, $f'(\tau) = g'(\tau) = 1$, $f''(\tau) = g''(\tau) = 0$ and $f'''(\tau) = g'''(\tau)$, then $f \equiv g$.

Part (i) follows by using Theorem 4.1.2 when $k = 1$ and Theorem 4.2.2 when $k > 1$. Part (iii) follows from Proposition 2.6.11. Part (iv) is a slight generalization of Theorem 2.7.4. Parts (v), (vi), and (vii) are proved by using the models constructed in Section 4.5 and, for part (vi) and (vii), the techniques introduced in Section 4.7.

Theorem 4.10.4 comes from [76]. The final part of the proof showing that g has nontangential limit σ^1 at σ can be drastically simplified by using the discrete Lindelöf theorem mentioned at the end of the Notes to Section 2.4. Indeed, in the previous part of the proof we have constructed a sequence $\{z_\nu\}$ converging to σ with $\{\omega(z_\nu, z_{\nu+1})\}$ bounded and such that $g(z_\nu) \to \sigma^1$; the quoted discrete Lindelöf theorem then immediately gives that g has nontangential limit σ^1 at σ.

Example 4.10.5 comes from [110]. In [76], it is also proved the following more detailed statement: if $\mathrm{Fix}_A(f)$ is not empty and finite and g commutes with f, then one and only one of the following alternatives occur: either there exist $m \geq 1$, distinct $\sigma_0, \dots, \sigma_{m-1} \in \mathrm{Fix}_A(f)$ and $1 < A' \leq A$ such that $g(\sigma_j) = \sigma_{j+1}$, $f'(\sigma_j) = A'$ and $1 < |g'(p_j)| < +\infty$ for $j = 0, \dots, m-1$ (where $\sigma_m = \sigma_0$); or, there exists $m \geq 1$ such that $g^m(\sigma) = \tau_f$ for all $\sigma \in \mathrm{Fix}_A(f)$. In this case, $(g^m)'(\sigma) = +\infty$ for all $\sigma \in \mathrm{Fix}_A(f) \setminus \{\tau_f\}$.

Cowen in [129] stated three conjectures about fixed points of commuting maps. Conjecture 1: If f is hyperbolic, then every g that commutes with f has the same fixed point set as f in $\overline{\mathbb{D}}$. Conjecture 2: If f is not superattracting elliptic, g commutes with f and f and g have two fixed points in common then the fixed point sets of f and g are the same. Conjecture 3: If f is elliptic but not superattracting, then there is an integer n such that for every g that commutes with f the fixed point sets of f and g^n are the same. However, [110] contains examples showing that these conjectures, as stated, are false. Indeed, Example 4.10.5 shows that Conjectures 1 and 2 are false; a counterexample to Conjecture 3 is given by $f(z) = z(z + 1/2)(1 + z/2)^{-1}$ and $g = f \circ f$ (we have $\mathrm{Fix}(f) = \{0, 1\}$ while $\{0, 1, -1, -1/2\} \subseteq \mathrm{Fix}(g^n)$ for all $n \geq 1$). On a more positive tone, Bracci [76] has shown that Conjectures 1 and 3, and Conjecture 2 in the elliptic case, hold when f and g are injective with bounded derivative.

Theorem 4.10.6 was first proved by Mitchell [295] in 1979, with quite different arguments involving elementary facts of semigroup theory. Our proof is taken from [89].

An important consequence of Theorem 4.10.6 is that a compact group acting holomorphically on $\mathbb{D}$ has a fixed point. This should be compared with the fact that there are examples of compact Lie groups acting on topological spaces homeomorphic to $\mathbb{R}^n$ without fixed points (see, e. g., [309]).

5 Continuous dynamics on Riemann surfaces

In this chapter, we shall look at dynamics from another point of view. Let X be a Riemann surface; then $\mathrm{Hol}(X,X)$, endowed, as usual, with the compact-open topology, is a *topological semigroup with identity*, i. e., the operation given by the composition $(f,g) \mapsto f \circ g$ is continuous, associative, and has an identity. From this point of view, a semigroup homomorphism $\Phi\colon \mathbb{N} \to \mathrm{Hol}(X,X)$ is the same thing as the sequence of iterates of the single function $\Phi(1)$. In other words, in the previous chapters we have actually studied semigroup homomorphisms of $\mathbb{N}$ into $\mathrm{Hol}(X,X)$.

From this point of view, a natural generalization of the sequence of iterates is a *one-parameter semigroup*, i. e., a continuous semigroup homomorphism $\Phi\colon \mathbb{R}^+ \to \mathrm{Hol}(X,X)$. In this chapter, we shall thoroughly study these objects, aiming toward a complete classification. This will be possible because on Riemann surfaces with non-Abelian fundamental group every one-parameter semigroup Φ is trivial, i. e., $\Phi_t = \mathrm{id}_X$ for all $t \geq 0$. Furthermore, the one-parameter semigroups on other Riemann surfaces different from the disk can be classified (Section 5.3); so the main problem is the description of one-parameter semigroups on $\mathbb{D}$.

We shall actually provide several different descriptions of one-parameter semigroups on $\mathbb{D}$, useful in different contexts. We shall show how to relate one-parameter semigroups to Cauchy problems and ordinary differential equations, proving that a semigroup is completely determined by a holomorphic function $F\colon \mathbb{D} \to \mathbb{C}$, its *infinitesimal generator*. We shall give both a differential characterization and a completely explicit description of infinitesimal generators. Finally, we shall show how to replace $\mathbb{D}$ by another simply connected domain (in essentially a unique way) so to express a generic one-parameter semigroup in a particularly simple form; in a sense we shall transfer the analytic intricacies of one-parameter semigroups in a geometrically simple domain as $\mathbb{D}$ to the geometrical intricacies of a domain of definition for analytically very simple one-parameter semigroups, expressed in terms of affine maps.

More in detail, Section 5.1 summarizes a few well-known facts about algebraic semigroup homomorphisms that we shall need later. Section 5.2 introduces the notion of one-parameter semigroups of holomorphic maps establishing their basic properties. In Section 5.3, we classify all possible one-parameter semigroups on Riemann surfaces different from the unit disk. In Section 5.4, we introduce the notion of infinitesimal generator of a one-parameter semigroup, obtaining a strong link between our theory and the theory of ordinary differential equations. Section 5.5 contains a continuous version of the Wolff–Denjoy theorem. In Section 5.6, we shall give a precise formula for infinitesimal generators that shall be instrumental in proving, in Section 5.7, that one-parameter semigroups on $\mathbb{D}$ are conjugated to affine one-parameter semigroups defined on simply connected domains with special geometrical properties.

The material presented here is just an introduction to this beautiful theory. If we were to follow the pattern used in discussing discrete dynamics, we would need an-

https://doi.org/10.1515/9783110601978-005

other chapter entitled "Continuous dynamics on the unit disk." However, such a chapter already exists and it is actually a whole book [80]; the interested reader is strongly invited to read that book after finishing this chapter.

5.1 Algebraic semigroup homomorphisms

In this section, we collect some well-known facts about algebraic semigroups homomorphism of $\mathbb{R}^+$ into other groups or semigroups that we shall need later. In this section, as operation on $\mathbb{R}^+$ we shall always consider the sum, that makes $\mathbb{R}^+$ in a semigroup but of course not a group. Moreover, we shall put $\mathbb{R}^+_* = (0, +\infty)$, so that $(\mathbb{R}^+_*, \cdot)$ is a topological group.

Definition 5.1.1. Let G be a semigroup with identity element e. A function $\Phi\colon \mathbb{R}^+ \to G$ is a *semigroup homomorphism* if $\Phi(0) = e$ and $\Phi(t+s) = \Phi(t)\circ\Phi(s)$ for all $t, s \geq 0$, where $\circ$ denotes the operation in G. In the sequel, we shall often write Φ_t instead of $\Phi(t)$.

Lemma 5.1.2. *Let G be a group. Then:*
(i) *every semigroup homomorphism $\Phi\colon \mathbb{R}^+ \to G$ can be extended in a unique way to a group homomorphism $\tilde{\Phi}\colon \mathbb{R} \to G$; in particular, if G is a topological group and Φ is continuous, then $\tilde{\Phi}$ is continuous too;*
(ii) *if G is finite, then every semigroup homomorphism $\Phi\colon \mathbb{R}^+ \to G$ is trivial.*

Proof. (i) The (unique) extension is obviously given by

$$\tilde{\Phi}(t) = \begin{cases} \Phi(t) & \text{if } t \geq 0; \\ [\Phi(-t)]^{-1} & \text{if } t \leq 0, \end{cases}$$

where $[\cdot]^{-1}$ denotes the inverse operator in G. The continuity of $\tilde{\Phi}$ follows immediately from the continuity of the group operations.

(ii) Extend Φ to a group homomorphism $\tilde{\Phi}\colon \mathbb{R} \to G$; let K be the kernel of $\tilde{\Phi}$. If n is the order of G, we have $nt \in K$ for all $t \in \mathbb{R}$. Hence $\mathbb{R} = n\mathbb{R} \subseteq K \subseteq \mathbb{R}$, i. e., Φ is trivial. □

Lemma 5.1.3. *Let $\Phi\colon \mathbb{R}^+ \to (\mathbb{C}, +)$ be a continuous semigroup homomorphism of $\mathbb{R}^+$ in $(\mathbb{C}, +)$. Then $\Phi(t) = at$ for some $a \in \mathbb{C}$. Analogously, if $\Phi\colon \mathbb{R}^+ \to (\mathbb{R}, +)$ is a continuous homomorphism of $\mathbb{R}^+$ in $(\mathbb{R}, +)$ then $\Phi(t) = at$ for some $a \in \mathbb{R}$.*

Proof. Fix $t_0 > 0$. Then $\Phi(nt_0) = n\Phi(t_0)$ for all $n \in \mathbb{N}$. Hence for any $q \in \mathbb{N}^*$ we have

$$\Phi(1) = \Phi(q/q) = q\Phi(1/q),$$

that is $\Phi(1/q) = \Phi(1)/q$. Therefore, $\Phi(r) = r\Phi(1)$ for all $r \in \mathbb{Q}^+$ and, by continuity, for all $r \in \mathbb{R}^+$. The assertions follow setting $a = \Phi(1)$. □

Lemma 5.1.4. *Let $\Phi\colon \mathbb{R}^+ \to (\mathbb{R}^+_*, \cdot)$ be a continuous semigroup homomorphism. Then $\Phi(t) = e^{at}$ for some $a \in \mathbb{R}$.*

Proof. Since Φ takes values in $\mathbb{R}^+_*$, the logarithm $\log \Phi$ is a well-defined continuous semigroup homomorphism from $\mathbb{R}^+$ to $(\mathbb{R}, +)$. By Lemma 5.1.3, $\log \Phi(t) = at$ for some $a \in \mathbb{R}$ and the assertion follows. □

Lemma 5.1.5. *Let $\Phi\colon \mathbb{R}^+ \to (\mathbb{S}^1, \cdot)$ be a continuous semigroup homomorphism. Then $\Phi(t) = e^{ibt}$ for some $b \in \mathbb{R}$.*

Proof. Let $\pi\colon \mathbb{R} \to \mathbb{S}^1$ be the covering map $\pi(x) = e^{ix}$; notice that π is a group homomorphism from $(\mathbb{R}, +)$ to $(\mathbb{S}^1, \cdot)$. Then Φ lifts to a continuous semigroup homomorphism $\tilde{\Phi}\colon \mathbb{R}^+ \to (\mathbb{R}, +)$ such that $\Phi = \pi \circ \tilde{\Phi}$. Then the assertion follows from Lemma 5.1.3. □

Lemma 5.1.6. *Let $\Phi\colon \mathbb{R}^+ \to (\mathbb{C}^*, \cdot)$ be a continuous semigroup homomorphism. Then $\Phi(t) = e^{\lambda t}$ for some $\lambda \in \mathbb{C}$.*

Proof. Starting from Φ, we can construct two new homomorphisms: $|\Phi|\colon \mathbb{R}^+ \to (\mathbb{R}^+_*, \cdot)$ and $\Phi/|\Phi|\colon \mathbb{R}^+ \to (\mathbb{S}^1, \cdot)$. Then Lemmas 5.1.4 and 5.1.5 imply $\Phi(t) = e^{(a+ib)t}$ for suitable $a, b \in \mathbb{R}$. □

Later on we shall need a measurable version of Lemma 5.1.4. We start with a measure theory lemma.

Lemma 5.1.7. *Let $A \subseteq \mathbb{R}$ be a measurable subset of $\mathbb{R}$ with $\lambda(A) > 0$, where λ denotes the Lebesgue measure. Then there exists $\delta > 0$ such that*

$$(-\delta, \delta) \subset A - A = \{a_1 - a_2 \mid a_1, a_2 \in A\}.$$

Proof. By inner regularity, there exists a compact subset $K \subset A$ such that $\lambda(K) > 0$. On the other hand, by outer regularity applied to K, we can find an open neighborhood $U \supset K$ such that $\lambda(U) < 2\lambda(K)$. Since K is compact and U is open, we have that $\delta = d(K, \mathbb{R} \setminus U) > 0$, where d is the Euclidean distance. In particular, $(-\delta, \delta) + K \subset U$.

We now claim that $(t + K) \cap K \neq \emptyset$ for every $t \in (-\delta, \delta)$. Assume, by contradiction, that this is not the case; hence we can find $t_0 \in (-\delta, \delta)$ such that $(t_0 + K) \cap K = \emptyset$. Then since $t_0 + K \subset U$ by construction, we have

$$2\lambda(K) = \lambda(K) + \lambda(t_0 + K) = \lambda(K \cup (t_0 + K)) \le \lambda(U),$$

against the choice of U.

Therefore, $(t + K) \cap K \neq \emptyset$ for every $t \in (-\delta, \delta)$ and this implies that $(-\delta, \delta) \subseteq K - K \subset A - A$, as claimed. □

Proposition 5.1.8. (i) *Let $\Phi\colon \mathbb{R}^+ \to (\mathbb{R}, +)$ be a Lebesgue measurable semigroup homomorphism. Then there exists $a \in \mathbb{R}$ such that $\Phi(t) = at$ for all $t \in \mathbb{R}^+$.*

(ii) *Let* $\Psi\colon \mathbb{R}^+ \to (\mathbb{R}^+_*, \cdot)$ *be a Lebesgue measurable semigroup homomorphism. Then there exists* $a \in \mathbb{R}$ *such that* $\Psi(t) = e^{at}$ *for all* $t \in \mathbb{R}^+$.

Proof. (i) We start by proving that Φ is continuous at 0. Since $\Phi(0) = 0$, we have to show that for every $\varepsilon > 0$ there is $\delta > 0$ such that $|\Phi(t)| < \varepsilon$ if $t \in [0, \delta)$.

Fix $\varepsilon > 0$ and set $I = (-\varepsilon/2, \varepsilon/2)$. Let $\{q_n\}$ be an enumeration of the rational numbers. Since $\mathbb{R} = \bigcup_n (q_n + I)$, we have $\mathbb{R}^+ = \bigcup_n \Phi^{-1}(q_n + I)$. Since Φ is measurable, each $\Phi^{-1}(q_n + I)$ is measurable; therefore, there must exist $N \in \mathbb{N}$ such that $A = \Phi^{-1}(q_N + I)$ has positive Lebesgue measure. By Lemma 5.1.7, there is $\delta > 0$ such that $(-\delta, \delta) \subset A - A$; in particular, $[0, \delta) \subset (A - A) \cap \mathbb{R}^+$. Take $t \in [0, \delta)$, Then there must exist $a_1, a_2 \in A$ with $a_1 \geq a_2$ such that $t = a_1 - a_2$ and $\Phi(a_1), \Phi(a_2) \in q_N + I$. Since $\Phi(a_1 - a_2) + \Phi(a_2) = \Phi(a_1)$ because Φ is a semigroup homomorphism, it follows that

$$\Phi(t) = \Phi(a_1) - \Phi(a_2) \in I - I = (-\varepsilon, \varepsilon),$$

i. e., $|\Phi(t)| < \varepsilon$, as claimed.

Now we show that Φ is continuous everywhere. Fix $t_0 > 0$ and choose a sequence $\{t_\nu\}$ converging to t_0 from above; in particular $t_\nu - t_0 \geq 0$ always. Since $t_\nu - t_0 \to 0$ and we already shown that Φ is continuous at 0, we have $\Phi(t_\nu - t_0) \to 0$, and hence $\Phi(t_\nu) = \Phi(t_\nu - t_0) + \Phi(t_0) \to \Phi(t_0)$ as $\nu \to +\infty$. If instead $\{t_\nu\}$ converges to t_0 from below, we write $\Phi(t_\nu) = \Phi(t_0) - \Phi(t_0 - t_\nu)$ and arguing as above, we get again $\Phi(t_\nu) \to \Phi(t_0)$ as $\nu \to +\infty$.

Finally, since Φ is continuous we can apply Lemma 5.1.3 and we are done.

(ii) Put $\Phi(t) = \log \Psi(t)$. Then Φ is a Lebesgue measurable semigroup homomorphism from $\mathbb{R}^+$ to $(\mathbb{R}, +)$ and the assertion follows from (i). □

***Notes to Section* 5.1**

The content of this section is classical. Our proof of Proposition 5.1.8 is modeled on the proof of [80, Theorem 8.1.11]. It should be remarked that Proposition 5.1.8 fails without assuming some regularity condition; see, e. g., [80, Example 8.1.13].

5.2 One-parameter semigroups

We can now officially define the main object of study of this chapter, one-parameter semigroups on Riemann surfaces.

Definition 5.2.1. Let X be a Riemann surface. A *one-parameter semigroup of holomorphic maps* (briefly, a *one-parameter semigroup*) on X is a continuous semigroup homomorphism Φ from $\mathbb{R}^+$ to $\mathrm{Hol}(X, X)$ endowed with the composition. A *one-parameter group* of holomorphic maps on X is a continuous group homomorphism from $(\mathbb{R}, +)$ to $\mathrm{Hol}(X, X)$. When $t \in \mathbb{R}^+$ and $z \in X$, we shall often write $\Phi_t(z)$ or $\Phi(t, z)$ instead of $\Phi(t)(z)$. The *trivial one-parameter semigroup* is the trivial homomorphism $\Phi_t \equiv \mathrm{id}_X$

for all $t \in \mathbb{R}^+$. Finally, we shall say that a nontrivial one-parameter semigroup is *periodic* if there exists $t_0 > 0$ such that $\Phi_{t_0} \equiv \mathrm{id}_X$.

Remark 5.2.2. The definition of one-parameter semigroup as a continuous map $\Phi\colon \mathbb{R}^+ \to \mathrm{Hol}(X,X)$ has as an immediate consequence the fact that also the map, still denoted by Φ, from $\mathbb{R}^+ \times X$ to X sending (t,z) in $\Phi_t(z)$ is continuous.

Remark 5.2.3. If $\Phi_{t_0} \equiv \mathrm{id}_X$, then $\Phi_{kt_0} \equiv \mathrm{id}_X$ for all $k \in \mathbb{N}$. Furthermore, if $t > t_0$, writing $t = s + kt_0$ with $k = \lfloor t/t_0 \rfloor \in \mathbb{N}$ and $s \in [0, t_0)$ we see that $\Phi_t \equiv \Phi_s$, and hence Φ is completely determined by $\Phi_{[0,t_0]}$.

Our first result shows that not every function can be imbedded in a one-parameter semigroup

Proposition 5.2.4. *Let $\Phi\colon \mathbb{R}^+ \to \mathrm{Hol}(X,X)$ be a one-parameter semigroup on a Riemann surface X. Then Φ_t is injective for all $t \geq 0$.*

Proof. First of all notice that, since $\Phi_t' \to 1$ as $t \to 0$, for t small enough every Φ_t is locally injective.

Assume, by contradiction, that $\Phi_{t_0}(z_1) = \Phi_{t_0}(z_2)$ for some $t_0 > 0$ and $z_1, z_2 \in X$, with $z_1 \neq z_2$. Then if $t > t_0$, we have

$$\Phi_t(z_1) = \Phi_{t-t_0}(\Phi_{t_0}(z_1)) = \Phi_{t-t_0}(\Phi_{t_0}(z_2)) = \Phi_t(z_2).$$

In other words, the two curves $t \mapsto \Phi_t(z_1)$ and $t \mapsto \Phi_t(z_2)$ start at distinct points, meet at $t = t_0$ and coincide thereafter. Let $\tilde{t}_0 > 0$ be the least $t > 0$ such that $\Phi_t(z_1) = \Phi_t(z_2)$ and set $z_0 = \Phi_{\tilde{t}_0}(z_1) = \Phi_{\tilde{t}_0}(z_2)$. Then no Φ_t can be injective in a neighborhood of z_0 and this is a contradiction. □

In particular, it may happen that $\Phi_{t_0} \in \mathrm{Aut}(X)$ for some $t_0 \geq 0$. In this case, Φ automatically extends to a one-parameter group.

Proposition 5.2.5. *Let $\Phi\colon \mathbb{R}^+ \to \mathrm{Hol}(X,X)$ be a one-parameter semigroup on a hyperbolic Riemann surface X. Assume $\Phi_{t_0} \in \mathrm{Aut}(X)$ for some $t_0 > 0$; then Φ is the restriction to $\mathbb{R}^+$ of a one-parameter group. In particular, this happens for periodic one-parameter semigroups and in this case, if Φ is nontrivial, setting*

$$\ell = \inf\{t_0 \in \mathbb{R}^+ \mid \Phi_{t_0} \equiv \mathrm{id}_X\}, \tag{5.1}$$

we have $\ell > 0$.

Proof. Since $(\Phi_{t_0/n})^n = \Phi_{t_0} \in \mathrm{Aut}(X)$ for all $n \in \mathbb{N}^*$, we clearly have $\Phi_{rt_0} \in \mathrm{Aut}(X)$ for all $r \in \mathbb{Q}^+$. By continuity, $\Phi_{rt_0} \in \mathrm{Aut}(X)$ for all $r \in \mathbb{R}^+$ (because $\mathrm{Aut}(X)$ is closed in $\mathrm{Hol}(X,X)$, by Corollary 1.7.21) and the assertion follows from Lemma 5.1.2(i).

If Φ is periodic, we have $\Phi_{t_0} \equiv \mathrm{id}_X \in \mathrm{Aut}(X)$ for some $t_0 > 0$, and hence Φ extends to a one-parameter group. Assume now that $\ell = 0$; in particular we have a decreasing sequence $\{t_\nu\} \subset \mathbb{R}^+$ converging to 0 such that $\Phi_{t_\nu} \equiv \mathrm{id}_X$ for all $\nu \in \mathbb{N}$. Fix $t > 0$. Arguing as in Remark 5.2.3, we see that for every $\nu \in \mathbb{N}$ we can find $s_\nu \in [0, t_\nu]$ such that $\Phi_t \equiv \Phi_{s_\nu}$. But $t_\nu \to 0$ forces $s_\nu \to 0$; then continuity yields $\Phi_t = \lim_{\nu\to+\infty} \Phi_{s_\nu} = \mathrm{id}_X$, and thus Φ is trivial. □

Definition 5.2.6. Let Φ be a nontrivial periodic one-parameter group on a hyperbolic Riemann surface X. The number $\ell > 0$ defined by (5.1) is the *period* of Φ.

Remark 5.2.7. In the next section, we shall show that when X is a nonhyperbolic Riemann surface then all one-parameter semigroups on X extend to one-parameter groups and all nontrivial periodic one-parameter groups have a positive period (see Propositions 5.3.2–5.3.5).

As usual, a main role in our theory will be played by the fixed points.

Definition 5.2.8. A *fixed point* of a one-parameter semigroup Φ on a Riemann surface X is a point $z_0 \in X$ such that $\Phi_t(z_0) = z_0$ for all $t \geq 0$. If Φ has no fixed points, we say that it is *fixed point free*.

When we studied discrete dynamics, we saw that a fixed point of an iterate of a function f not necessarily is fixed by f itself. In the present context, the situation is much simpler.

Proposition 5.2.9. *Let $\Phi\colon \mathbb{R}^+ \to \mathrm{Hol}(X, X)$ be a one-parameter semigroup on a Riemann surface X. Assume that $\Phi_{t_0} \not\equiv \mathrm{id}_X$ has a fixed point $z_0 \in X$ for some $t_0 > 0$. Then z_0 is a fixed point of Φ.*

Proof. For any $t > 0$, we have

$$\Phi_{t_0}(\Phi_t(z_0)) = \Phi_{t+t_0}(z_0) = \Phi_t(\Phi_{t_0}(z_0)) = \Phi_t(z_0).$$

Therefore, $t \mapsto \Phi_t(z_0)$ is a curve issuing from z_0 contained in the fixed point set of Φ_{t_0}. Since $\Phi_{t_0} \not\equiv \mathrm{id}_X$, its fixed point set is discrete; therefore, $\Phi_t(z_0) = z_0$ for all $t \geq 0$, as claimed. □

Let $\Phi\colon \mathbb{R}^+ \to \mathrm{Hol}(X, X)$ be a one-parameter semigroup on a Riemann surface X with a fixed point $z_0 \in X$. Then we can define a continuous semigroup homomorphism $\mu\colon \mathbb{R}^+ \to \mathbb{C}^*$ setting $\mu(t) = \Phi_t'(z_0)$; the derivative is well-defined because z_0 is a fixed point of Φ_t for all $t \geq 0$ (see Definition 3.1.8). Notice that $\mu(t) \neq 0$ by Proposition 5.2.4; then, by Lemma 5.1.6, $\mu(t) = e^{-\lambda t}$ for some $\lambda \in \mathbb{C}$. Let us introduce a definition.

Definition 5.2.10. Let $\Phi\colon \mathbb{R}^+ \to \mathrm{Hol}(X, X)$ be a one-parameter semigroup on a Riemann surface X with a fixed point $z_0 \in X$. The *spectral value* of Φ at z_0 is the unique $\lambda \in \mathbb{C}$ such that $\Phi_t'(z_0) = e^{-\lambda t}$ for all $t \geq 0$. Notice that the spectral value is invariant under conjugation, because the derivative is.

Remark 5.2.11. Assume that X is a hyperbolic Riemann surface. By Theorem 3.1.10, the spectral value λ of a one-parameter semigroup Φ at a fixed point satisfies $\operatorname{Re}\lambda \geq 0$; moreover, $\operatorname{Re}\lambda = 0$ if and only if Φ is a one-parameter group and $\lambda = 0$ if and only if Φ is trivial.

Notes to Section 5.2

It is difficult to date precisely the birth of the concept of one-parameter semigroups of holomorphic functions. In 1917, Tricomi (see, for instance, [393]) dealt with problems somehow regarding the asymptotic behavior of one-parameter semigroups. Slightly later, Löwner [273] introduced the theory now known as *Löwner theory* to study extremal problems in complex analysis; as shown for instance by Pommerenke [336], Löwner theory is strictly related to holomorphic one-parameter semigroups with a fixed point. In 1943, Kufarev [253] found a way of extending Löwner theory, and the corresponding relationships with one-parameter semigroups, to the fixed-point-free case. A modern exposition of these results can be found in [78].

We assumed the continuity of one-parameter semigroups; as we shall see in Section 5.4 this will imply that one-parameter semigroups actually are real-analytic. Conditions ensuring the continuity of a one-parameter semigroup are discussed in [80, Section 8.1].

A related problem consists in embedding a given holomorphic self-map into a one-parameter semigroup, sometimes described as the problem of *fractional iteration*. Loosely stated, given $f \in \operatorname{Hol}(X, X)$, the problem consists in finding a reasonable way for defining the t-th iterate f^t of f for any $t > 0$, at least in some subset of X. Typically, it is also required to have $f^t \circ f^s = f^{t+s}$, and thus another way to state this problem is whether there exists a (local or global) one-parameter semigroup Φ such that $f = \Phi_1$.

In Proposition 5.2.4, we saw that the injectivity of f is a necessary condition for the existence of a global one-parameter semigroup Φ such that $f = \Phi_1$; the problem consists in finding necessary and sufficient conditions. When $X = \mathbb{D}$ (that, as we shall see in the next section, is the only interesting case), a solution expressed in terms of properties of the models we introduced in the previous chapter has been found; see, e. g., [385, 126, 230] and [140, pp. 96–99]. However, a solution expressed only in terms of analytic properties of f is not yet known. Other papers devoted to this topics are, for instance, [264, 179, 236, 205, 386, 33, 222, 123, 84].

5.3 One-parameter semigroups on Riemann surfaces

The aim of this section is to thoroughly investigate one-parameter semigroups on Riemann surfaces different from the unit disk, postponing the study of one-parameter semigroups on $\mathbb{D}$ to the remaining sections of this chapter.

Our task is made possible by the following.

Proposition 5.3.1. *Let $\Phi\colon \mathbb{R}^+ \to \operatorname{Hol}(X, X)$ be a one-parameter semigroup on a Riemann surface X with non-Abelian fundamental group. Then Φ is trivial.*

Proof. By Theorem 2.6.2, we should have $\Phi_t \equiv \mathrm{id}_X$ for small t, and hence for all t. □

So, we are left with just a few cases to investigate; let us start with the Riemann sphere.

Proposition 5.3.2. *Let $\Phi\colon \mathbb{R}^+ \to \mathrm{Hol}(\widehat{\mathbb{C}}, \widehat{\mathbb{C}})$ be a nontrivial one-parameter semigroup on the Riemann sphere $\widehat{\mathbb{C}}$. Then Φ extends to a one-parameter group, still denoted by Φ, and there is $\gamma \in \mathrm{Aut}(\widehat{\mathbb{C}})$ such that either:*
(i) $\gamma^{-1} \circ \Phi_t \circ \gamma(z) = z + at$ *for some* $a \in \mathbb{C}^*$, *or*
(ii) $\gamma^{-1} \circ \Phi_t \circ \gamma(z) = e^{-bt}z$ *for some* $b \in \mathbb{C}^*$.

In case (i), *Φ has a unique fixed point with spectral value* 0 *and it is never periodic. In case* (ii), *Φ has two distinct fixed points with spectral value respectively $\pm b$; moreover, Φ is periodic if and only if $b \in \mathbb{R}^* i$ and then it has period $2\pi/|b|$.*

Proof. By Propositions 5.2.4 and 5.2.5, Φ extends to a one-parameter group, because the compactness of $\widehat{\mathbb{C}}$ implies that any injective holomorphic self-map of $\widehat{\mathbb{C}}$ is also surjective, and hence an automorphism.

By Corollary 1.6.21, Φ_1 admits either two distinct fixed points or one (double) fixed point. In the latter case, up to conjugation by a suitable $\gamma \in \mathrm{Aut}(\widehat{\mathbb{C}})$ we can assume that the unique fixed point of Φ_1 is ∞, and so we have $\Phi_1(z) = z + a$ for a suitable $a \in \mathbb{C}$. Since Φ_t for $t \neq 1$ is an automorphism of $\widehat{\mathbb{C}}$ commuting with Φ_1, Corollary 1.6.21 shows that we should have $\Phi_t(z) = z + \alpha(t)$, where $\alpha\colon \mathbb{R}^+ \to (\mathbb{C}, +)$ is a continuous semigroup homomorphism with $\alpha(1) = a$. Then Lemma 5.1.3 yields $\alpha(t) = at$ and we are in case (i). In particular, an easy computation shows that the spectral value is 0 and that $\Phi_t \not\equiv \mathrm{id}_{\widehat{\mathbb{C}}}$ for all $t > 0$; in particular, Φ is not periodic.

If Φ_1 has two distinct fixed points, up to conjugation by a suitable $\gamma \in \mathrm{Aut}(\widehat{\mathbb{C}})$, we can assume that the fixed points are 0 and ∞, and thus we can write $\Phi_1(z) = e^{-b}z$, for a suitable $b \in \mathbb{C}$. Using again Corollary 1.6.21, we see that an automorphism ϕ of $\widehat{\mathbb{C}}$ commuting with Φ_1 should be of the form $\phi(z) = \lambda(\phi)z^{\varepsilon(\phi)}$, where $\lambda(\phi) \in \mathbb{C}^*$ and $\varepsilon(\phi) = \pm 1$.

The map $\varepsilon\colon \mathbb{R}^+ \to \mathbb{Z}_2$ given by $\varepsilon(t) = \varepsilon(\Phi_t)$ is a semigroup homomorphism; by Lemma 5.1.2, we must have $\varepsilon \equiv 1$. Hence $\Phi_t(z) = \lambda(t)z$ for all $t \geq 0$, where $\lambda\colon \mathbb{R}^+ \to (\mathbb{C}^*, \cdot)$ is a continuous semigroup homomorphism with $\lambda(1) = e^{-b}$. By Lemma 5.1.6, we have $\lambda(t) = e^{-bt}$ and we are in case (ii).

A standard computation yields the spectral values. Furthermore, we have $\Phi_{t_0} = \mathrm{id}_{\widehat{\mathbb{C}}}$ for some $t_0 > 0$ if and only if $bt_0 \in 2\pi\mathbb{Z}i$. In particular, if Φ is periodic then $b \in t_0^{-1}2\pi\mathbb{Z}i \subset \mathbb{R}^* i$. Conversely, if $b \in \mathbb{R}^* i$ then taking $t_0 = 2\pi/|b| > 0$ we have $\Phi_{t_0} \equiv \mathrm{id}_{\widehat{\mathbb{C}}}$, and thus Φ is periodic. Finally, if $b \in \mathbb{R}^* i$ then $\Phi_t \equiv \mathrm{id}_{\widehat{\mathbb{C}}}$ if and only if $t \in |b|^{-1}2\pi\mathbb{N}$, and thus the period is $\ell = 2\pi/|b|$. □

We consider now the complex plane.

Proposition 5.3.3. *Let $\Phi\colon \mathbb{R}^+ \to \mathrm{Hol}(\mathbb{C}, \mathbb{C})$ be a nontrivial one-parameter semigroup on $\mathbb{C}$. Then Φ extends to a one-parameter group, still denoted by Φ, and either:*
(i) $\Phi_t(z) = z + at$ *for some* $a \in \mathbb{C}^*$, *or*
(ii) $\Phi_t(z) = e^{-bt}z + a(1 - e^{-bt})$ *for some* $a \in \mathbb{C}$ *and* $b \in \mathbb{C}^*$.

In case (i), Φ *has no fixed points and it is never periodic. In case* (ii), Φ *has the unique fixed point a where it has spectral value b; moreover,* Φ *is periodic if and only if* $b \in \mathbb{R}^* i$ *and then* Φ *has period* $\ell = 2\pi/|b|$.

Proof. By Proposition 5.2.4, every Φ_t is an injective entire function, i. e., a linear polynomial. Write $\Phi_t(z) = \beta(t)z + \alpha(t)$, where $\beta\colon \mathbb{R}^+ \to \mathbb{C}^*$ and $\alpha\colon \mathbb{R}^+ \to \mathbb{C}$ are continuous and satisfy

$$\begin{aligned} \beta(s+t) &= \beta(s)\beta(t), \\ \alpha(s+t) &= \beta(s)\alpha(t) + \alpha(s). \end{aligned} \tag{5.2}$$

The first formula, by Lemma 5.1.6, implies $\beta(t) = e^{-bt}$ for some $b \in \mathbb{C}$. If $b = 0$, the second one implies $\alpha(t) = at$ for some $a \in \mathbb{C}^*$, by Lemma 5.1.3, and we are in case (i).

If $b \neq 0$, fix $t_0 > 0$ such that $e^{-bt_0} \neq 1$. Then, setting $s = t_0$ in the second formula of (5.2) and recalling that $\alpha(t_0 + t) = \alpha(t + t_0)$ we get

$$\alpha(t) = \frac{\alpha(t_0)}{1 - e^{-bt_0}}(1 - e^{-bt}),$$

and we are in case (ii). The last part of the assertion follows arguing as in the proof of the previous proposition. □

The next case is $\mathbb{C}^*$.

Proposition 5.3.4. *Let* $\Phi\colon \mathbb{R}^+ \to \mathrm{Hol}(\mathbb{C}^*, \mathbb{C}^*)$ *be a nontrivial one-parameter semigroup on* $\mathbb{C}^*$. *Then* Φ *extends to a one-parameter group, still denoted by* Φ, *and* $\Phi_t(z) = e^{-bt}z$ *for some* $b \in \mathbb{C}^*$; *in particular,* Φ *has no fixed points in* $\mathbb{C}^*$. *Moreover,* Φ *is periodic if and only if* $b \in \mathbb{R}^* i$ *and then* Φ *has period* $\ell = 2\pi/|b|$.

Proof. By Proposition 5.2.4, every Φ_t is injective; moreover, it is also surjective, because every holomorphic function from $\mathbb{C}^*$ to $\mathbb{C}$ missing two points is constant by Proposition 1.6.27. Hence we can apply Proposition 1.6.38(i) to write $\Phi_t(z) = \beta(t)z^{\varepsilon(t)}$ for suitable continuous maps $\beta\colon \mathbb{R}^+ \to \mathbb{C}^*$ and $\varepsilon\colon \mathbb{R}^+ \to \mathbb{Z}_2$. The assertions then follow arguing exactly as in the last part of the proof of Proposition 5.3.2. □

The next case is a torus.

Proposition 5.3.5. *Let* $\Phi\colon \mathbb{R}^+ \to \mathrm{Hol}(X, X)$ *be a nontrivial one-parameter semigroup on a torus* $X = \mathbb{C}/\Gamma_\tau$. *Then* Φ *extends to a one-parameter group. Moreover, there is a one-parameter group* $\tilde{\Phi}\colon \mathbb{R}^+ \to \mathrm{Hol}(\mathbb{C}, \mathbb{C})$ *of the form* $\tilde{\Phi}_t(z) = z + at$ *for a suitable* $a \in \mathbb{C}^*$ *such that* $\Phi_t \circ \pi = \pi \circ \tilde{\Phi}_t$ *for all* $t \in \mathbb{R}^+$, *where* $\pi\colon \mathbb{C} \to X$ *is the universal covering map. Furthermore,* Φ *has no fixed points, and it is periodic if and only if* $a \in \mathbb{R}^+\Gamma_\tau$. *In this case* Φ *has period* $\ell = |a|^{-1}|\gamma_0| > 0$, *where* $\gamma_0 \in \Gamma_\tau$ *is the nonzero element of* Γ_τ *proportional to a with least module.*

Proof. Since X is compact, every injective holomorphic self-map of X is also surjective. Propositions 5.2.4 and 5.2.5 then imply that Φ extends to a one-parameter group; in particular, the image of Φ is contained in the connected component at the identity of $\mathrm{Aut}(X)$. In Proposition 1.6.38(ii), we saw that the connected component at the identity of the automorphism group of X is isomorphic to X itself, endowed with the operation induced by the sum in $\mathbb{C}$ (see Remark 1.6.25). In particular, under this isomorphism, $\Phi\colon \mathbb{R}^+ \to X$ lifts to a continuous semigroup homomorphism $\widehat{\Phi}\colon \mathbb{R}^+ \to \mathbb{C}$ such that $\pi \circ \widehat{\Phi} = \Phi$. By Lemma 5.1.3, we have $\widehat{\Phi}(t) = at$, and thus $\tilde{\Phi}_t(z) = z + at$ is a lifting of Φ, as claimed.

Now, we have $\Phi_t(\pi(z)) = \pi(z)$ if and only if $\tilde{\Phi}_t(z) - z \in \Gamma_\tau$ if and only if $at \in \Gamma_\tau$. In particular, Φ_t has a fixed point if and only if $\Phi_t \equiv \mathrm{id}_X$; thus Φ has no fixed points. Furthermore, Φ is periodic if and only if $a \in \mathbb{R}^+\Gamma_\tau$. Writing $a = r\gamma_0$ with $r > 0$ and $\gamma_0 \in \Gamma_\tau$, we have $at \in \Gamma_\tau$ if and only if $t \in \mathbb{R}^+$ and $t = r^{-1}(\gamma/\gamma_0)$ for some $\gamma \in \Gamma_\tau$. Hence γ is proportional to γ_0 and the formula for the period follows. □

Finally, we are left with the doubly connected hyperbolic domains, i. e., $\mathbb{D}^*$ and the annuli $A(r,1) = \{z \in \mathbb{C} \mid r < |z| < 1\}$ for $0 < r < 1$. We begin with $A(r,1)$.

Proposition 5.3.6. *Let $\Phi\colon \mathbb{R}^+ \to \mathrm{Hol}(A(r,1), A(r,1))$ be a nontrivial one-parameter semigroup on an annulus $A(r,1)$ with $0 < r < 1$. Then $\Phi_t(z) = e^{ibt}z$ for some $b \in \mathbb{R}^*$ and Φ has no fixed points in $A(r,1)$. Furthermore, Φ is always periodic with period $\ell = 2\pi/|b|$.*

Proof. By Corollary 2.6.7 and Proposition 5.2.5, the image of Φ is contained in the automorphism group of $A(r,1)$. Hence by Proposition 1.6.38(iv), we can write

$$\Phi_t(z) = e^{i\theta(t)}\, r^{(1-\varepsilon(t))/2} z^{\varepsilon(t)}$$

for suitable continuous maps $\theta\colon \mathbb{R}^+ \to \mathbb{R}$ and $\varepsilon\colon \mathbb{R}^+ \to \mathbb{Z}_2$. Since $\mathbb{Z}_2$ is discrete and $\varepsilon(0) = 1$ it follows that $\varepsilon \equiv 1$; then θ is a continuous homomorphism and the assertion follows from Lemma 5.1.3. □

We end this section with $\mathbb{D}^*$.

Proposition 5.3.7. *Let $\Phi\colon \mathbb{R}^+ \to \mathrm{Hol}(\mathbb{D}^*, \mathbb{D}^*)$ be a nontrivial one-parameter semigroup on $\mathbb{D}^*$. Then there is a nontrivial one-parameter semigroup $\tilde{\Phi}$ on $\mathbb{D}$ with fixed point 0 such that $\Phi_t = \tilde{\Phi}_t|_{\mathbb{D}^*}$ for all $t \geq 0$. In particular, Φ has no fixed points in $\mathbb{D}^*$ and it is periodic if and only if $\tilde{\Phi}$ is, with the same period.*

Proof. Every Φ_t has a removable singularity in 0, and hence is the restriction of a function $\tilde{\Phi}_t \in \mathrm{Hol}(\mathbb{D}, \mathbb{D})$. Obviously, $\tilde{\Phi}\colon \mathbb{R}^+ \to \mathrm{Hol}(\mathbb{D}, \mathbb{D})$ is a one-parameter semigroup on $\mathbb{D}$; it remains to show that $\tilde{\Phi}_t(0) = 0$ for all $t > 0$, for then the last assertion follows from Corollary 1.1.14.

Assume, by contradiction, that $\tilde{\Phi}_{t_0}(0) \neq 0$ for some $t_0 > 0$. Since $\tilde{\Phi}_{t_0}$ sends $\mathbb{D}^*$ into itself, $\widetilde{\Phi}_{t_0}$ has no zeroes in $\mathbb{D}$. The same argument works for any $\tilde{\Phi}_{t_0/n}$ (where $n \geq 1$), because $\tilde{\Phi}_{t_0/n}(0) = 0$ implies $\tilde{\Phi}_{t_0}(0) = (\tilde{\Phi}_{t_0/n})^n(0) = 0$. But $\tilde{\Phi}_{t_0/n} \to \mathrm{id}_{\mathbb{D}}$ as $n \to +\infty$ and a Hurwitz theorem (Corollary A.1.4) provides a contradiction. □

So, the study of one-parameter semigroups on Riemann surfaces is reduced to the study of one-parameter semigroups on $\mathbb{D}$, a study that we shall begin in earnest in the next section.

Notes to Section 5.3

Most of this section comes from [190], where also the case of discontinuous one-parameter semigroups is considered. See also [209] for a different approach.

5.4 The infinitesimal generator

In the previous section, we saw that the theory of one-parameter semigroups on Riemann surfaces is interesting only on $\mathbb{D}$; we shall devote the rest of this chapter to describe the foundations of the theory in this case.

Our first main goal is to relate one-parameter semigroups and ordinary differential equations, thus making our work easier calling in an already well-established (and quite powerful) theory. We shall repeatedly use a basic existence theorem, the *Cauchy–Kovalevskaya theorem.*

Theorem 5.4.1. *Let Ω be an open subset of $\mathbb{R}^n$ and $F\colon \Omega \to \mathbb{R}^n$ a real analytic map. Then for any compact subset K of Ω there are $\delta > 0$, a neighborhood $U \subset \Omega$ of K and a real analytic map $u\colon (-\delta, \delta) \times U \to \Omega$ such that*

$$\begin{cases} \dfrac{\partial u}{\partial t}(t,x) = F(u(t,x)), \\ u(0,x) = x. \end{cases} \tag{5.3}$$

The solution of the Cauchy problem (5.3) *is unique in the sense that if there are $\delta' > 0$, another neighborhood $U' \subset \Omega$ of K and another map $u'\colon (-\delta', \delta') \times U' \to \Omega$ satisfying* (5.3), *then $u \equiv u'$ on $((-\delta,\delta)\times U)\cap((-\delta',\delta')\times U')$. Finally, if Ω is a domain in $\mathbb{C}^n$ and $F\colon \Omega \to \mathbb{C}^n$ is holomorphic, then for every $t \in (-\delta, \delta)$ the map $u(t,\cdot)\colon U \to \Omega$ is holomorphic.*

For a proof see, e. g., [300, Section 1.8].

Using the uniqueness statement of the latter theorem, we can now introduce the link between one-parameter semigroups and ordinary differential equations.

Corollary 5.4.2. *Let Ω be an open subset of $\mathbb{R}^n$, $F\colon \Omega \to \mathbb{R}^n$ a real analytic map, and K a compact subset of Ω. Choose $\delta > 0$ and a neighborhood $U \subset \Omega$ of K such that there is a real analytic solution $u\colon (-\delta, \delta) \times U \to \Omega$ of the Cauchy problem* (5.3). *Then for every $s, t \in (-\delta, \delta)$ and $x \in K$ such that $s + t \in (-\delta, \delta)$ and $u(t,x) \in U$ we have*

$$u(s, u(t,x)) = u(s+t, x). \tag{5.4}$$

Proof. Fix $t_0 \in (-\delta, \delta)$ and $x_0 \in K$ such that $u(t_0, x_0) \in U$ and take $0 < \delta' \leq \delta - |t_0|$. Now define $v_1, v_2 \colon (-\delta', \delta') \to \Omega$ setting $v_1(s) = u(s, u(t_0, x_0))$ and $v_2(s) = u(s + t_0, x_0)$. Then v_1 and v_2 are two real-analytic solutions of

$$\begin{cases} \dfrac{dv}{ds} = F \circ v, \\ v(0) = u(t_0, x_0). \end{cases}$$

By the uniqueness statement in Theorem 5.4.1, it follows that $v_1 \equiv v_2$ and (5.4) is proved. □

In other words, the solution of the Cauchy problem (5.3) is locally a one-parameter semigroup. In particular, if $F \in \mathrm{Hol}(\mathbb{D}, \mathbb{C})$ is such that the Cauchy problem

$$\begin{cases} \dfrac{\partial \Phi}{\partial t} = F \circ \Phi \\ \Phi(0, z) = z \end{cases} \tag{5.5}$$

has a *global* solution $\Phi \colon \mathbb{R}^+ \times \mathbb{D} \to \mathbb{D}$ then Φ is automatically a one-parameter semigroup, holomorphic in z and real analytic in t.

Remark 5.4.3. More precisely, the map $F \in \mathrm{Hol}(\mathbb{D}, \mathbb{C})$ in (5.5) should be thought of as a holomorphic vector field on $\mathbb{D}$. However, since we are interested only in one-parameter semigroups on $\mathbb{D}$ and not on general Riemann surfaces, we shall interpret F just as a standard holomorphic function, thanks to the canonical isomorphism $T\mathbb{D} \cong \mathbb{D} \times \mathbb{C}$.

The main result of this section is that every one-parameter semigroup on $\mathbb{D}$ is obtained in this way.

Theorem 5.4.4 (Berkson–Porta, 1978). *Let $\Phi \colon \mathbb{R}^+ \to \mathrm{Hol}(\mathbb{D}, \mathbb{D})$ be a one-parameter semigroup. Then there is a unique holomorphic function $F \colon \mathbb{D} \to \mathbb{C}$ such that*

$$\frac{\partial \Phi}{\partial t} = F \circ \Phi. \tag{5.6}$$

In particular, for every $z_0 \in \mathbb{D}$ the function $\Phi(\cdot, z_0)$ is real analytic.

Proof. Let K be a compact convex subset of $\mathbb{D}$. We can choose $\alpha \in (0, 1)$ such that the convex hull $\widehat{K}$ of $\Phi([0, \alpha] \times K)$ is still contained in $\mathbb{D}$. Take $\delta \in (0, \alpha]$ such that

$$\sup_{z \in \widehat{K}} |\Phi_t'(z) - 1| \leq 2 - 2^{2/3}$$

for all $t \leq \delta$. Hence for all $t \in [0, \delta]$ and $z \in K$ we have

$$|\Phi_{2t}(z) - 2\Phi_t(z) + z| = \left| \int_z^{\Phi_t(z)} \frac{d}{d\zeta} [\Phi_t(\zeta) - \zeta] \, d\zeta \right| \leq (2 - 2^{2/3}) |\Phi_t(z) - z|, \tag{5.7}$$

where the integration path is the segment from z to $\Phi_t(z)$, which is contained in $\widehat{K}$.

Since

$$2\left|\Phi_t(z) - z\right| - \left|\Phi_{2t}(z) - z\right| \le \left|\Phi_{2t}(z) - 2\Phi_t(z) + z\right|, \tag{5.8}$$

it follows that for all $t \in [0, \delta]$ and $z \in K$ we have

$$\left|\Phi_t(z) - z\right| \le 2^{-2/3}\left|\Phi_{2t}(z) - z\right|. \tag{5.9}$$

Let $k \in \mathbb{N}$ be such that $2^k\delta \ge 1$ and put

$$M = 2^{2k/3} \sup\{\left|\Phi_t(z) - z\right| \mid z \in K,\ t \in [2^{-k}, 1]\}.$$

Then (5.9) implies

$$\left|\Phi_t(z) - z\right| \le Mt^{2/3} \tag{5.10}$$

for all $t \in [0, 1]$ and all $z \in K$. Now we can repeat the same argument on a compact convex subset $K_1 \subset \mathbb{D}$ containing $\widehat{K}$ in its interior, obtaining a constant $M_1 > 0$ such that

$$\left|\Phi_t(z) - z\right| \le M_1 t^{2/3}$$

for all $t \in [0, 1]$ and all $z \in K_1$. Then the Cauchy inequalities produce a constant $\widetilde{M} > 0$ such that

$$\left|\Phi_t'(z) - 1\right| \le \widetilde{M}t^{2/3} \tag{5.11}$$

for all $t \in [0, 1]$ and all $z \in \widehat{K}$. If we plug (5.10) and (5.11) in (5.7), we find that for all $t \in [0, \alpha]$ and $z \in K$,

$$\left|\Phi_{2t}(z) - 2\Phi_t(z) + z\right| \le \widetilde{M}t^{2/3}\left|\Phi_t(z) - z\right| \le M\widetilde{M}t^{4/3}.$$

Thus (5.8) yields

$$\left|\frac{\Phi_{2t}(z) - z}{2t} - \frac{\Phi_t(z) - z}{t}\right| \le \frac{M\widetilde{M}}{2}t^{1/3}, \tag{5.12}$$

for $z \in K$ and $t \in (0, \alpha]$.

Let $j_0 \ge 1$ be such that $2^{-j_0+1} < \alpha$. Then for every $n \ge j_0$, we have

$$\frac{\Phi(2^{-n}, z) - z}{2^{-n}} = \frac{\Phi(2^{-j_0+1}, z) - z}{2^{-j_0+1}} + \sum_{j=j_0}^{n}\left[\frac{\Phi(2^{-j}, z) - z}{2^{-j}} - \frac{\Phi(2^{-j+1}, z) - z}{2^{-j+1}}\right].$$

Since the telescopic series on the right-hand side is convergent, by (5.12), it follows that

$$\lim_{n\to\infty} \frac{\Phi(2^{-n}, z) - z}{2^{-n}} = F(z) \tag{5.13}$$

exists uniformly on K. Since every compact of $\mathbb{D}$ can be covered by a finite number of compact convex subsets of $\mathbb{D}$, it follows that the limit (5.13) exists uniformly on compact subsets of $\mathbb{D}$, defining a holomorphic function $F\colon \mathbb{D} \to \mathbb{C}$.

For $z_0 \in \mathbb{D}$ and $t_0 > 0$, the set $\Phi([0, t_0] \times \{z_0\})$ is a compact subset of $\mathbb{D}$. Hence, as $n \to +\infty$, the function

$$\frac{\Phi(t + 2^{-n}, z_0) - \Phi(t, z_0)}{2^{-n}} = \frac{\Phi(2^{-n}, \Phi(t, z_0)) - \Phi(t, z_0)}{2^{-n}}$$

tends uniformly to $F(\Phi(t, z_0))$ for $t \in [0, t_0]$. This implies that

$$\int_0^s \frac{\Phi(t + 2^{-n}, z_0) - \Phi(t, z_0)}{2^{-n}}\, dt \longrightarrow \int_0^s F(\Phi_t(z_0))\, dt$$

as $n \to +\infty$. But on the other hand, a standard calculus argument yields

$$\int_0^s \frac{\Phi(t + 2^{-n}, z_0) - \Phi(t, z_0)}{2^{-n}}\, dt \longrightarrow \Phi_s(z_0) - z_0;$$

therefore, we get

$$\Phi_s(z) = z + \int_0^s F(\Phi_t(z))\, dt$$

for all $z \in \mathbb{D}$ and $s \in \mathbb{R}^+$, and (5.6) is proved.

The uniqueness of F is clear: if (5.6) holds, then $F(z) = \frac{\partial \Phi(\cdot, z)}{\partial t}\Big|_{t=0}$. Finally, the assertion about the regularity of $\Phi(\cdot, z_0)$ follows from Theorem 5.4.1. □

Definition 5.4.5. Let $\Phi\colon \mathbb{R}^+ \to \mathrm{Hol}(\mathbb{D}, \mathbb{D})$ be a one-parameter semigroup. The holomorphic function $F\colon \mathbb{D} \to \mathbb{C}$ satisfying (5.6) is the *infinitesimal generator* of Φ. Note that Φ is completely determined by its infinitesimal generator.

Remark 5.4.6. Recalling what we said in Remark 5.4.3, an infinitesimal generator $F \in \mathrm{Hol}(\mathbb{D}, \mathbb{C})$ of a one-parameter semigroup is sometimes called a *semicomplete vector field*; if Φ is actually defined on $\mathbb{R} \times \mathbb{D}$ (and hence it is a one-parameter group), then F is a *complete vector field.*

So, every one-parameter semigroup is the solution of a Cauchy problem and there is a one-to-one correspondence between infinitesimal generators and one-parameter semigroups. In particular, the classification of one-parameter semigroups is reduced to the classification of their infinitesimal generators. This can be summarized as follows.

Corollary 5.4.7. *A map $F \in \mathrm{Hol}(\mathbb{D}, \mathbb{C})$ is an infinitesimal generator of a one-parameter semigroup on $\mathbb{D}$ if and only if* (5.5) *has a global solution $\Phi: \mathbb{R}^+ \times \mathbb{D} \to \mathbb{D}$.*

Proof. This follows from Corollary 5.4.2 and Theorem 5.4.4, □

As we shall see, this approach will eventually provide us with a complete classification of one-parameter semigroups on $\mathbb{D}$. For the moment, we show how to recover fixed points and spectral values using infinitesimal generators.

Proposition 5.4.8. *Let $\Phi: \mathbb{R}^+ \to \mathrm{Hol}(\mathbb{D}, \mathbb{D})$ be a one-parameter semigroup and let $F \in \mathrm{Hol}(\mathbb{D}, \mathbb{C})$ be its infinitesimal generator. Then:*
(i) *$z_0 \in \mathbb{D}$ is a fixed point of Φ if and only if $F(z_0) = 0$;*
(ii) *if Φ has a fixed point $z_0 \in \mathbb{D}$, then its spectral value at z_0 is $-F'(z_0)$.*

Proof. If $z_0 \in \mathbb{D}$ is a fixed point of Φ, then (5.6) immediately yields $F(z_0) = 0$. Conversely, assume $F(z_0) = 0$, and set $\varphi(t) = \Phi(t, z_0)$. Then φ solves the Cauchy problem

$$\begin{cases} \dfrac{d\psi}{dt} = F \circ \psi, \\ \psi(0) = z_0. \end{cases}$$

Since $F(z_0) = 0$, the constant map $\psi \equiv z_0$ is also a solution, and thus, by Theorem 5.4.1, it is the only solution. Hence $\varphi \equiv z_0$ and z_0 is a fixed point of Φ.

Finally, assume $z_0 \in \mathbb{D}$ is a fixed point of Φ. If λ is the spectral value of Φ at z_0, we have

$$\begin{aligned} F'(z_0)e^{-\lambda t} &= F'(z_0)\Phi_t'(z_0) = \frac{\partial}{\partial z}(F \circ \Phi)(t, z_0) \\ &= \frac{\partial^2 \Phi}{\partial z \partial t}(t, z_0) = \frac{\partial}{\partial t}\Phi_t'(z_0) = -\lambda e^{-\lambda t} \end{aligned}$$

and so $F'(z_0) = -\lambda$, as claimed. □

A natural question now is whether it is possible to characterize the maps $F \in \mathrm{Hol}(\mathbb{D}, \mathbb{C})$ that are infinitesimal generators. This is possible, and there are many different characterizations. Some of them are collected in the following.

Theorem 5.4.9. *Let $F \in \mathrm{Hol}(\mathbb{D}, \mathbb{C})$. Then the following statements are equivalent:*
(i) *F is an infinitesimal generator;*
(ii) *$d\omega_{(z,w)}(F(z), F(w)) \le 0$ for all $z \ne w \in \mathbb{D}$;*

(iii) *for all* $z \neq w \in \mathbb{D}$, *we have*

$$\operatorname{Re}\left(\frac{F(z)-F(w)}{z-w}+\frac{\overline{F(z)}w+\bar{z}F(w)}{1-\bar{z}w}\right)\leq 0;$$

(iv) *for all* $z \in \mathbb{D}$, *we have*

$$\operatorname{Re}[2\bar{z}F(z)+(1-|z|^2)F'(z)]\leq 0; \tag{5.14}$$

(v) $d(\kappa_{\mathbb{D}}\circ F)_z(F(z))\leq 0$ *for all* $z\in\mathbb{D}$ *such that* $F(z)\neq 0$, *where the function* $\kappa_{\mathbb{D}}\circ F$ *is defined by* $(\kappa_{\mathbb{D}}\circ F)(z)=\kappa_{\mathbb{D}}(z;F(z))$.

Proof. Assume that F is an infinitesimal generator of a one-parameter semigroup $\Phi\colon\mathbb{R}^+\to\operatorname{Hol}(\mathbb{D},\mathbb{D})$. If $s\geq t$, we have $\Phi_s=\Phi_{s-t}\circ\Phi_t$; the Schwarz–Pick lemma (Theorem 1.2.3) then implies that the function $g(t)=\omega(\Phi_t(z),\Phi_t(w))$ is nonincreasing for all $z,w\in\mathbb{D}$. Therefore,

$$0\geq\frac{dg}{dt}(0)=d\omega_{(\Phi_t(z),\Phi_t(w))}\left(\frac{\partial\Phi}{\partial t}(t,z),\frac{\partial\Phi}{\partial t}(t,z)\right)\bigg|_{t=0}=d\omega_{(z,w)}(F(z),F(w)),$$

and (ii) is proved.

Assume now that (ii) holds. To prove (iii), consider the curves $\gamma_1(t)=z+tF(z)$ and $\gamma_2(t)=w+tF(w)$, issuing respectively from z and w and with support in $\mathbb{D}$ for t small enough, and set $g(t)=\omega(\gamma_1(t),\gamma_2(t))$. Since $\gamma_1'(0)=F(z)$ and $\gamma_2'(0)=F(w)$, we have

$$0\geq d\omega_{(z,w)}(F(z),F(w))=\frac{dg}{dt}\bigg|_{t=0}.$$

Putting $\tilde{g}(t)=\left|\frac{\gamma_1(t)-\gamma_2(t)}{1-\overline{\gamma_2(t)}\gamma_1(t)}\right|$, we have $g=\tanh^{-1}\circ\tilde{g}$, and hence $g'(0)=(\tanh^{-1})'(\tilde{g}(0))\tilde{g}'(0)$; since $\tanh^{-1}$ is strictly increasing we then get $\tilde{g}'(0)\leq 0$.

To compute $\tilde{g}'(0)$, first of all we remark that $\tilde{g}(t)^2=k(t)\overline{k(t)}$, where

$$k(t)=\frac{\gamma_1(t)-\gamma_2(t)}{1-\overline{\gamma_2(t)}\gamma_1(t)}.$$

Therefore, $\tilde{g}\tilde{g}'=\operatorname{Re}(k'\bar{k})$; since $\tilde{g}(t)\geq 0$ always it follows that $\tilde{g}'(0)\leq 0$ if and only if $\operatorname{Re}(k'(0)\overline{k(0)})\leq 0$. Now,

$$k'(0)=\frac{F(z)-F(w)}{1-\bar{w}z}+\frac{z-w}{(1-\bar{w}z)^2}[\overline{F(w)}z+\bar{w}F(z)];$$

hence

$$k'(0)\overline{k(0)}=\left[\frac{F(z)-F(w)}{1-\bar{w}z}+\frac{z-w}{(1-\bar{w}z)^2}[\overline{F(w)}z+\bar{w}F(z)]\right]\frac{\bar{z}-\bar{w}}{1-w\bar{z}}$$

$$= \left[\frac{F(z) - F(w)}{z - w} + \frac{\overline{F(w)}z + \overline{w}F(z)}{1 - \overline{w}z}\right]\left|\frac{z - w}{1 - \overline{w}z}\right|^2$$

and we have proved (iii).

(iv) follows from (iii) just letting w converge to z.

Assume now (iv) holds. Fix $z_0 \in \mathbb{D}$ such that $F(z_0) \neq 0$ and let $\gamma: [0, \varepsilon) \to \mathbb{D}$ be the solution of the Cauchy problem $\gamma'(t) = F(\gamma(t))$ with $\gamma(0) = z_0$. Then

$$\begin{aligned}\frac{d}{dt}\kappa_{\mathbb{D}}(\gamma(t); \gamma'(t))^2 &= \frac{d}{dt}\frac{|\gamma'(t)|^2}{(1 - |\gamma(t)|^2)^2} \\ &= \frac{2|F(\gamma(t))|^2}{(1 - |\gamma(t)|^2)^3}\left[2\overline{\gamma(t)}F(\gamma(t)) + (1 - |\gamma(t)|^2)F'(\gamma(t))\right] \leq 0.\end{aligned}$$

In particular, the function $t \mapsto \kappa_{\mathbb{D}}(\gamma(t); \gamma'(t))$ is nonincreasing; hence

$$0 \geq \frac{d}{dt}\kappa_{\mathbb{D}}(\gamma(t); \gamma'(t))\bigg|_{t=0} = d(\kappa_{\mathbb{D}} \circ F)_z(F(z))$$

and (v) is proved.

Finally, assume that (v) holds; by Corollary 5.4.7 we need to show that (5.5) admits a solution defined on $\mathbb{R}^+ \times \mathbb{D}$ with values in $\mathbb{D}$. Fix $z_0 \in \mathbb{D}$, and let $\phi: (-\delta', \delta) \to \mathbb{D}$ be the maximal solution of the Cauchy problem

$$\begin{cases} \dfrac{d\phi}{dt} = F \circ \phi, \\ \phi(0) = z_0, \end{cases} \tag{5.15}$$

where $\delta', \delta \in (0, +\infty]$; we have to prove that $\delta = +\infty$.

If $F(z_0) = 0$, then the maximal solution is the constant function $\phi \equiv z_0$ that, in particular, it is defined on $\mathbb{R}$; so if $F(z_0) = 0$ we are done.

Assume then $F(z_0) \neq 0$. Then $\phi'(t) = F(\phi(t))$ is always different from 0, by the uniqueness of the solution of a Cauchy problem. Therefore, the function $g(t) = \kappa_{\mathbb{D}}(\phi(t); \phi'(t))$ is differentiable in $t \in [0, \delta)$ and we have

$$\frac{d}{dt}\kappa_{\mathbb{D}}(\phi(t); \phi'(t)) = d(\kappa_{\mathbb{D}} \circ F)_{\phi(t)}(F(\phi(t))) \leq 0$$

by (v). In particular, g is nonincreasing. Take $s \in [0, \delta)$. By Proposition 1.9.14(i), we have

$$\begin{aligned}\omega(z_0, \phi(s)) &\leq \int_0^s \kappa_{\mathbb{D}}(\phi(t); \phi'(t))\,dt \\ &\leq \int_0^s \kappa_{\mathbb{D}}(\phi(0); \phi'(0))\,dt = s\,\kappa_{\mathbb{D}}(z_0; F(z_0)).\end{aligned}$$

Therefore, if $\delta < +\infty$ the support of ϕ is contained in a compact subset K of $\mathbb{D}$. Let $\delta_1 > 0$ and $u\colon (-\delta_1, \delta_1) \times K \to \mathbb{D}$ be given by Theorem 5.4.1 applied to K and choose $t_0 \in [0, \delta)$ such that $\delta - t_0 < \delta_1$. Then the uniqueness statement of Theorem 5.4.1 shows that the function $\psi\colon [0, \delta_1 + t_0) \to \mathbb{D}$ given by

$$\psi(t) = \begin{cases} \phi(t) & \text{if } t < \delta, \\ u(t - t_0, \phi(t_0)) & \text{if } t > t_0, \end{cases}$$

is still a solution of (5.15), against the maximality of δ. Therefore, $\delta = +\infty$ and F is an infinitesimal generator, as claimed. □

In particular, infinitesimal generators are characterized by an ordinary differential inequality. A first consequence is the following.

Corollary 5.4.10. *The set of all infinitesimal generators of one-parameter semigroups on* $\mathbb{D}$ *is a real convex cone in* $\mathrm{Hol}(\mathbb{D}, \mathbb{C})$ *with vertex at* 0.

Proof. Indeed, if $F, G \in \mathrm{Hol}(\mathbb{D}, \mathbb{C})$ satisfy (5.14) then also $F+G$ and tF with $t \ge 0$ satisfy the same differential inequality, and hence are infinitesimal generators. □

At this point, the knowledgeable reader may wonder whether the infinitesimal generators of one-parameter groups are maybe characterized by a differential equation; after all, $\mathrm{Aut}(\mathbb{D})$ is a Lie group and so we can apply the third Lie theorem, and... Indeed, the knowledgeable reader is right, and the differential equation is easily found.

Corollary 5.4.11. *A holomorphic function* $F\colon \mathbb{D} \to \mathbb{C}$ *is the infinitesimal generator of a one-parameter group on* $\mathbb{D}$ *if and only if for all* $z \in \mathbb{D}$ *we have*

$$\mathrm{Re}\big[2\bar{z}F(z) + (1 - |z|^2)F'(z)\big] = 0. \tag{5.16}$$

Proof. Assume first that F is the infinitesimal generator of a one-parameter group Φ on $\mathbb{D}$. For every $t \ge 0$, set $\Phi_t^+ = \Phi_t$ and $\Phi_t^- = \Phi_{-t}$. Then Φ^+ and Φ^- are two one-parameter semigroups, with infinitesimal generators F and $-F$, respectively; so Theorem 5.4.9(iv) yields (5.16).

Conversely, assume (5.16) holds; then, by Theorem 5.4.9(iv) and Corollary 5.4.7, (5.5) has a global solution Φ on $\mathbb{R} \times \mathbb{D}$. Hence we can quote Corollary 5.4.2 to infer that Φ is a one-parameter group with infinitesimal generator F. □

We end this section with two other formulas relating one-parameter semigroups and infinitesimal generators.

Proposition 5.4.12. *Let* $\Phi\colon \mathbb{R}^+ \to \mathrm{Hol}(\mathbb{D}, \mathbb{D})$ *be a one-parameter semigroup with infinitesimal generator* $F \in \mathrm{Hol}(\mathbb{D}, \mathbb{C})$. *Then for all* $z \in \mathbb{D}$ *and* $t \ge 0$, *we have*

$$\frac{\partial \Phi_t}{\partial t}(z) = F(z)\Phi_t'(z) \tag{5.17}$$

and

$$\Phi_t'(z) = \exp\left(\int_0^t F'(\Phi_s(z))\,ds\right). \tag{5.18}$$

Proof. Differentiating with respect to s the identity $\Phi_{s+t}(z) = \Phi_t(\Phi_s(z))$, we get

$$\frac{\partial \Phi_{s+t}(z)}{\partial s} = \Phi_t'(\Phi_s(z))\frac{\partial \Phi_s(z)}{\partial s} = \Phi_t'(\Phi_s(z))F(\Phi_s(z));$$

putting $s = 0$ we get (5.17).

Now differentiating (5.6) with respect to z, we obtain

$$\frac{\partial \Phi_t'(z)}{\partial t} = \frac{\partial}{\partial z}\frac{\partial \Phi_t(z)}{\partial t} = \frac{\partial}{\partial z}(F \circ \Phi_t)(z) = F'(\Phi_t(z))\Phi_t'(z).$$

Fix $z \in \mathbb{D}$. Since Φ_t is always injective, $\Phi_t'(z)$ never vanishes; therefore, we can define in a continuous way on $\mathbb{R}^+$ the argument $\arg(t)$ of $\Phi_t'(z)$ using as starting value $\arg(0) = 0$. Then if we put $f(t) = \log|\Phi_t'(z)| + i\arg(t)$, we have $f(0) = 0$ and $\Phi_t'(z) = e^{f(t)}$. In particular,

$$f'(t) = \frac{1}{\Phi_t'(z)}\frac{\partial \Phi_t'(z)}{\partial t} = F'(\Phi_t(z)).$$

Integrating and then taking the exponential we obtain (5.18), as claimed. □

Notes to Section 5.4

Possibly the first one to notice the existence of relationships between one-parameter semigroups and differential equations of the form (5.6) has been Wolff in 1938 [420], even though that paper is devoted to a slightly different topic and does not contain an explicit proof of Theorem 5.4.4.

Most of this section comes from the 1978 paper by Berkson and Porta [61], which has possibly been the first paper devoted to a systematic study of holomorphic one-parameter semigroups in $\mathbb{D}$. The only exceptions are Theorem 5.4.9 and Proposition 5.4.12. To be precise, in Theorem 5.4.9 the equivalence of (i), (ii), and (iii) comes from [346, 77], the equivalence of (i) and (iv) is in [61] and the equivalence between (i) and (v) is taken from [6]. On the other hand, Proposition 5.4.12 comes from [80, Proposition 10.1.8], where (5.17) is called *Kolmogorov backward equation*.

It should be remarked that, interpreting infinitesimal generators as holomorphic vector fields, a version of Theorem 5.4.4 can be proved for any Riemann surface, but the results described in the previous section make such a generalization useless.

5.5 The continuous Wolff–Denjoy theorem

In the next section, we shall be able to give more explicit characterizations of the infinitesimal generators of one-parameter semigroups on $\mathbb{D}$, but to do so we need a continuous version of the Wolff–Denjoy theorem.

Let $\Phi: \mathbb{R}^+ \to \mathrm{Hol}(\mathbb{D}, \mathbb{D})$ be a nontrivial one-parameter semigroup. If Φ has a fixed point $z_0 \in \mathbb{D}$, then (Remark 5.2.11) the spectral value λ of Φ at z_0 satisfies $\operatorname{Re}\lambda \geq 0$, and $\operatorname{Re}\lambda = 0$ if and only if Φ is a one-parameter group of elliptic automorphisms (see Theorem 3.1.10 and Proposition 5.2.5). Furthermore, in the latter case Φ_t has no limit in $\mathrm{Hol}(\mathbb{D}, \mathbb{C})$ as $t \to +\infty$ (by Lemma 5.1.5). Recalling the Wolff–Denjoy Theorem 3.2.1 it is now clear what is going on and we can get a *continuous Wolff–Denjoy theorem.*

Theorem 5.5.1. *Let $\Phi: \mathbb{R}^+ \to \mathrm{Hol}(\mathbb{D}, \mathbb{D})$ be a nontrivial one-parameter semigroup on $\mathbb{D}$. Assume Φ is not a one-parameter group of elliptic automorphisms. Then Φ_t converges as $t \to +\infty$ to a constant $\tau \in \overline{\mathbb{D}}$, the Wolff point of Φ_1.*

Proof. By assumption, Φ either is fixed point free or has a fixed point with spectral value having positive real part. By the Wolff–Denjoy Theorem 3.2.1, in both cases the sequence $\{\Phi_k = \Phi_1^k\}$ converges as $k \to +\infty$ to a point $\tau \in \overline{\mathbb{D}}$, the Wolff point of Φ_1.

Fix $z_0 \in \mathbb{D}$, and let $K = \Phi([0, 1] \times \{z_0\})$. The set K is a compact subset of $\mathbb{D}$; hence for every $\varepsilon > 0$ there is $k_0 \in \mathbb{N}$ such that for any $k \geq k_0$ and $z \in K$ we have $|\Phi_k(z) - \tau| < \varepsilon$. Therefore, for every $t \in [0, 1]$ and every $k \geq k_0$ we have

$$\left|\Phi_{k+t}(z_0) - \tau\right| < \varepsilon.$$

This means that $\lim_{t\to+\infty} \Phi_t(z_0) = \tau$. But z_0 was arbitrary; hence, by Theorem 1.7.19, we have proved that $\Phi_t \to \tau$ as $t \to +\infty$. □

Definition 5.5.2. Let $\Phi: \mathbb{R}^+ \to \mathrm{Hol}(\mathbb{D}, \mathbb{D})$ be a nontrivial one-parameter semigroup. The *Wolff point* $\tau_\Phi \in \overline{\mathbb{D}}$ of Φ is the Wolff point of Φ_1. In particular, $\Phi_t \to \tau_\Phi$ as $t \to +\infty$ unless Φ is a group of elliptic automorphisms.

As a consequence, the Wolff point of all functions in a one-parameter semigroup is always the same (cf. Theorems 3.1.20 and 4.10.1).

Corollary 5.5.3. *Let $\Phi: \mathbb{R}^+ \to \mathrm{Hol}(\mathbb{D}, \mathbb{D})$ be a nontrivial one-parameter semigroup on $\mathbb{D}$. Then the Wolff point of Φ_t is independent of t.*

Proof. If Φ has a fixed point $z_0 \in \mathbb{D}$, then z_0 is the Wolff point of all Φ_t. If Φ is fixed point free, for any $t_0 > 0$ we have

$$\tau_\Phi = \lim_{t\to+\infty} \Phi_t = \lim_{k\to+\infty} \Phi_{kt_0} = \tau_{\Phi_{t_0}}$$

and again $\tau_{\Phi_{t_0}}$ does not depend on t_0. □

It turns out that we can always define the spectral value at the Wolff point of a one-parameter semigroup.

Proposition 5.5.4. *Let $\Phi\colon \mathbb{R}^+ \to \mathrm{Hol}(\mathbb{D},\mathbb{D})$ be a nontrivial one-parameter semigroup on $\mathbb{D}$ without fixed points and let $\tau_\Phi \in \partial\mathbb{D}$ be its Wolff point. Then there is $\lambda \geq 0$ such that*

$$\beta_{\Phi_t}(\tau_\Phi) = e^{-\lambda t}$$

for all $t \in \mathbb{R}^+$, where $\beta_{\Phi_t}(\tau_\Phi)$ is the boundary dilation of Φ_t at τ_Φ.

Proof. First of all, since τ_Φ is the Wolff point of each Φ_t, we have $0 < \beta_{\Phi_t}(\tau_\Phi) \leq 1$ for all $t \in \mathbb{R}^+$ (Corollary 2.5.5). Furthermore, Proposition 2.4.7 implies that $\beta_{\Phi_{s+t}}(\tau_\Phi) = \beta_{\Phi_s}(\tau_\Phi)\beta_{\Phi_t}(\tau_\Phi)$ for all $s,t \in \mathbb{R}^+$. So, $\beta(t) = \beta_{\Phi_t}(\tau_\Phi)$ is a semigroup homomorphism from $\mathbb{R}^+$ to $(\mathbb{R}^+_*,\cdot)$ with image contained in $(0,1]$. Moreover, by the Weierstrass theorem, for each $n > 0$ the function $t \mapsto \Phi_t'((1-\frac{1}{n})\tau_\Phi)$ is continuous; since $\beta(t) = \lim_{n\to+\infty}\Phi_t'((1-\frac{1}{n})\tau_\phi)$ and the pointwise limit of continuous functions is Lebesgue measurable, it follows that β is Lebesgue measurable. Then we can apply Proposition 5.1.8 and we are done. □

The following definitions come now naturally.

Definition 5.5.5. Let $\Phi\colon \mathbb{R}^+ \to \mathrm{Hol}(\mathbb{D},\mathbb{D})$ be a one-parameter semigroup with Wolff point $\tau_\Phi \in \partial\mathbb{D}$. The number $\lambda \geq 0$ such that $\beta_{\Phi_t}(\tau_\Phi) = e^{-\lambda t}$ is the *spectral value* of Φ at τ_Φ.

Definition 5.5.6. Let $\Phi\colon \mathbb{R}^+ \to \mathrm{Hol}(\mathbb{D},\mathbb{D})$ be a nontrivial one-parameter semigroup with Wolff point $\tau_\Phi \in \overline{\mathbb{D}}$. We say that:
(i) Φ is *elliptic* if $\tau_\Phi \in \mathbb{D}$;
(ii) Φ is *hyperbolic* if $\tau_\Phi \in \partial\mathbb{D}$ and its spectral value is strictly positive;
(iii) Φ is *parabolic* if $\tau_\Phi \in \partial\mathbb{D}$ and its spectral value is zero.

We end this section with an easy observation concerning how hyperbolic one-parameter semigroups converge to their Wolff point.

Proposition 5.5.7. *Let $\Phi\colon \mathbb{R}^+ \to \mathrm{Hol}(\mathbb{D},\mathbb{D})$ be a hyperbolic one-parameter semigroup with Wolff point $\tau_\Phi \in \partial\mathbb{D}$. Then for every $z_0 \in \mathbb{D}$ the curve $t \mapsto \Phi_t(z_0)$ converges nontangentially to τ_Φ as $t \to +\infty$.*

Proof. Fix $z_0 \in \mathbb{D}$ and put $K = \Phi([0,1]\times\{z_0\})$. Since Φ is continuous, K is a compact subset of $\mathbb{D}$. Given $t > 0$, write $t = \lfloor t\rfloor + \{t\}$, with $\lfloor t\rfloor \in \mathbb{N}$ and $\{t\} \in [0,1]$. Then $\Phi_t(z_0) = \Phi_1^{\lfloor t\rfloor}(\Phi_{\{t\}}(z_0))$. The assertion then follows because $\Phi_{\{t\}}(z_0) \in K$ and $\Phi_1^n \to \tau_\Phi$ nontangentially uniformly on compact subsets (Proposition 4.3.2). □

***Notes to Section* 5.5**

Theorems 5.5.1 has been proved in 1978 by Berkson and Porta [61], using different arguments. Our proof is due to Edoardo Vesentini (personal communication).

5.6 The Berkson–Porta formula

In this section, we shall solve explicitely (5.14). To express our results, we need a notation.

Definition 5.6.1. Put

$$\mathcal{P} = \{p \in \mathrm{Hol}(\mathbb{D}, \mathbb{C}) \mid \operatorname{Re} p \geq 0\} \setminus \{0\}.$$

Notice that, by the minimum principle for harmonic functions, if $p \in \mathcal{P}$ then either $\operatorname{Re} p > 0$ on $\mathbb{D}$ or $p \equiv ib$ for some $b \in \mathbb{R}^*$.

Then we can prove the *Berkson–Porta formula.*

Theorem 5.6.2 (Berkson–Porta, 1978). *Let $\Phi\colon \mathbb{R}^+ \to \mathrm{Hol}(\mathbb{D}, \mathbb{D})$ be a nontrivial one-parameter semigroup on $\mathbb{D}$ with Wolff point $\tau \in \overline{\mathbb{D}}$. Then the infinitesimal generator of Φ is of the form*

$$F(z) = (\overline{\tau} z - 1)(z - \tau)p(z), \tag{5.19}$$

for a suitable $p \in \mathcal{P}$. In particular, Φ has a fixed point if and only if $\tau \in \mathbb{D}$, the fixed point is exactly τ, and the spectral value is $(1 - |\tau|^2)p(\tau)$. Conversely, given $p \in \mathcal{P}$ and $\tau \in \overline{\mathbb{D}}$, the function $F\colon \mathbb{D} \to \mathbb{C}$ given by (5.19) is the infinitesimal generator of a one-parameter semigroup on $\mathbb{D}$ with Wolff point τ.

Proof. Let $\Phi\colon \mathbb{R}^+ \to \mathrm{Hol}(\mathbb{D}, \mathbb{D})$ be a nontrivial one-parameter semigroup on $\mathbb{D}$, with Wolff point τ and infinitesimal generator F. Using Corollary 2.5.4, we see that the function

$$t \mapsto \frac{|1 - \overline{\tau}\Phi_t(z)|^2}{1 - |\Phi_t(z)|^2}$$

is nonincreasing for any $z \in \mathbb{D}$. Hence its derivative in 0 is nonpositive and we get

$$\operatorname{Re}\left[(1 - \tau\overline{z})(\overline{z} - \overline{\tau})F(z)\right] \leq 0. \tag{5.20}$$

Let $p(z) = (\overline{\tau} z - 1)^{-1}(z - \tau)^{-1}F(z)$; the quotient is well-defined because if $\tau \in \mathbb{D}$ then $F(\tau) = 0$ by Proposition 5.4.8. By (5.20), we have $p \in \mathcal{P}$ and F is of the form (5.19) as claimed. If $\tau \in \mathbb{D}$, that is if Φ has a fixed point, then by Proposition 5.4.8 the spectral value of Φ at τ is $-F'(\tau) = (1 - |\tau|^2)p(\tau)$.

Conversely, let $F \in \mathrm{Hol}(\mathbb{D}, \mathbb{C})$ be of the form (5.19); we should prove that F satisfies (5.14). If $p \equiv ib$ for some $b \in \mathbb{R}^*$, then (5.14) is easily verified. Assume instead that $\operatorname{Re} p(z) > 0$ for all $z \in \mathbb{D}$. The Schwarz–Pick lemma (Corollary 1.1.16) applied to $(p-1)/(p+1)$ yields

$$\frac{|p'(z)|}{2\operatorname{Re} p(z)} \leq \frac{1}{1-|z|^2} \tag{5.21}$$

for all $z \in \mathbb{D}$. Now

$$2\bar{z}F(z) + (1-|z|^2)F'(z) = (1-|z|^2)(\overline{\tau}z-1)(z-\tau)p'(z) - \big[|\overline{\tau}z-1|^2 + |z-\tau|^2\big]p(z);$$

hence (5.21) yields

$$\begin{aligned}
&\operatorname{Re}\big[2\bar{z}F(z) + (1-|z|^2)F'(z)\big]\\
&\quad = (1-|z|^2)\operatorname{Re}\big[(\overline{\tau}z-1)(z-\tau)p'(z)\big] - \big[|\overline{\tau}z-1|^2 + |z-\tau|^2\big]\operatorname{Re} p(z)\\
&\quad \leq (1-|z|^2)|\overline{\tau}z-1|\,|z-\tau|\,|p'(z)| - \big[|\overline{\tau}z-1|^2 + |z-\tau|^2\big]\operatorname{Re} p(z)\\
&\quad \leq -\big(|\overline{\tau}z-1| - |z-\tau|\big)^2 \operatorname{Re} p(z) \leq 0
\end{aligned} \tag{5.22}$$

and we are done. □

A consequence of this formula is that we can compute the spectral value using the infinitesimal generator also when the Wolff point is at the boundary.

Corollary 5.6.3. *Let $\Phi\colon \mathbb{R}^+ \to \operatorname{Hol}(\mathbb{D},\mathbb{D})$ be a one-parameter semigroup with Wolff point $\tau_\Phi \in \partial\mathbb{D}$. Let $\lambda \in \mathbb{R}^+$ be the spectral value of Φ at τ_Φ and let $F \in \operatorname{Hol}(\mathbb{D},\mathbb{C})$ be the infinitesimal generator of Φ. Then*

$$K\text{-}\lim_{z\to\tau_\Phi} F(z) = 0 \quad \textit{and} \quad K\text{-}\lim_{z\to\tau_\Phi} F'(z) = K\text{-}\lim_{z\to\tau_\Phi} \frac{F(z)}{z-\tau_\Phi} = -\lambda.$$

Proof. Theorem 5.6.2 implies that we can write

$$F(z) = -(1-\overline{\tau_\Phi}z)(z-\tau_\Phi)p(z),$$

for a suitable $p \in \mathcal{P}$. A classical result on holomorphic functions with nonnegative real part (see Proposition A.3.4) shows that there exists $\beta \geq 0$ such that

$$K\text{-}\lim_{z\to\tau_\Phi} \frac{F(z)}{z-\tau_\Phi} = -K\text{-}\lim_{z\to\tau_\Phi}(1-\overline{\tau_\Phi}z)p(z) = -\beta \leq 0.$$

From this, it immediately follows that the nontangential limit of F at τ_Φ is 0 and Corollary 2.4.11 shows that $-\beta$ is also the nontangential limit of F'. To conclude the proof, we have to show that $\beta = \lambda$.

Fix $t > 0$. Recalling (5.18) and its proof, we see that for each $r \in (0,1)$ we have

$$\log \Phi_t'(r\tau_\Phi) = \int_0^t F'(\Phi_s(r\tau_\Phi))\,ds.$$

Since τ_Φ is the common Wolff point of all Φ_s, we have $\beta_{\Phi_s}(\tau_\Phi) \le 1$ for all $s \ge 0$. Moreover, the set $\{\Phi_s(0)\}_{s\in[0,t]}$ is compact in $\mathbb{D}$; therefore, we can invoke Proposition 2.2.9 to infer that we can find $M > 1$ such that all the curves $r \mapsto \Phi_s(r\tau_\Phi)$ for $s \in [0,t]$ are contained in $K(\tau_\Phi, M)$. Then the Lebesgue dominated convergence theorem implies

$$-\lambda t = \lim_{r\to 1^-} \log \Phi_t'(r\tau_\Phi) = \int_0^t \lim_{r\to 1^-} F'(\Phi_s(r\tau_\Phi))\, ds = -\int_0^t \beta\, ds = -\beta t$$

and we are done. □

We end this section discussing one-parameter groups of automorphisms of $\mathbb{D}$. A one-parameter group $\Phi\colon \mathbb{R} \to \operatorname{Aut}(\mathbb{D})$ clearly acts on $\overline{\mathbb{D}}$; therefore, by Proposition 1.4.12, the fixed point set of Φ_t is independent of $t > 0$. Thus we have the following version of Definition 5.5.6.

Definition 5.6.4. Let $\Phi\colon \mathbb{R}^+ \to \operatorname{Aut}(\mathbb{D})$ be a nontrivial one-parameter group. We say that:

(i) Φ is *elliptic* if it has a fixed point in $\mathbb{D}$ (and then all Φ_t are elliptic automorphisms);
(ii) Φ is *parabolic* if it has a unique common fixed point in $\partial\mathbb{D}$ (and then all Φ_t are parabolic automorphisms);
(iii) Φ is *hyperbolic* if it has exactly two common fixed points in $\partial\mathbb{D}$ (and then all Φ_t are hyperbolic automorphisms).

Since, by Remark 1.4.5, the group of all elliptic (resp., parabolic, hyperbolic) automorphisms with given fixed point is isomorphic to $(\mathbb{S}^1,\cdot)$ (resp., to $(\mathbb{R},+)$, $(\mathbb{R}^+_*,\cdot)$), the one-parameter groups on $\mathbb{D}$ are determined by Lemmas 5.1.4, 5.1.3, and 5.1.6.

Corollary 5.6.5. *Let $\Phi\colon \mathbb{R} \to \operatorname{Aut}(\mathbb{D})$ be a nontrivial one-parameter group of automorphisms of $\mathbb{D}$. Then:*

(i) *if Φ is elliptic, then there exists $\gamma \in \operatorname{Aut}(\mathbb{D})$ and $b \in \mathbb{R}^*$ such that*

$$\gamma^{-1} \circ \Phi_t \circ \gamma(z) = e^{ibt} z;$$

(ii) *if Φ is parabolic, then there exists $\gamma \in \operatorname{Aut}(\mathbb{D})$ and $a \in \mathbb{R}$ such that*

$$\gamma^{-1} \circ \Phi_t \circ \gamma(z) = \frac{(2 + iat)z - iat}{2 - iat + iatz};$$

(iii) *if Φ is hyperbolic, then there exists $\gamma \in \operatorname{Aut}(\mathbb{D})$ and $a \in \mathbb{R}$ such that*

$$\gamma^{-1} \circ \Phi_t \circ \gamma(z) = \frac{(e^{at} + 1)z + e^{at} - 1}{(e^{at} - 1)z + e^{at} + 1}.$$

Proof. (i) Up to conjugating by an automorphism of $\mathbb{D}$, we can assume that the fixed point of Φ is the origin. This means that $\Phi_t(z) = \alpha(t)z$ for some function $\alpha\colon \mathbb{R}^+ \to \mathbb{S}^1$.

Since Φ is continuous, also α is continuous; since Φ is a one-parameter group, α is a semigroup homomorphism. The assertion then follows from Lemma 5.1.5.

(ii) Up to conjugating by an automorphism of $\mathbb{D}$, we can assume that the only fixed point of Φ is 1. Conjugating by the Cayley transform, we obtain a one-parameter group $\Psi: \mathbb{R} \to \mathrm{Hol}(\mathbb{H}^+, \mathbb{H}^+)$ with infinity as unique fixed point; therefore, $\Psi_t(w) = w + \beta(t)$ for some $\beta: \mathbb{R} \to \mathbb{R}$. Arguing as in (i), we see that β is a continuous group homomorphism and hence Lemma 5.1.3 implies that $\beta(t) = at$ for some $a \in \mathbb{R}^*$. Using the Cayley transform again to go back to $\mathbb{D}$, we get the assertion.

(iii) Up to conjugating by an automorphism of $\mathbb{D}$, we can assume that the two fixed points of Φ are ± 1. Conjugating by the Cayley transform we obtain a one-parameter group $\Psi: \mathbb{R} \to \mathrm{Hol}(\mathbb{H}^+, \mathbb{H}^+)$ with 0 and ∞ as fixed points; therefore, $\Psi_t(w) = \alpha(t)w$ for some $\alpha: \mathbb{R} \to \mathbb{R}^+$. Arguing as in (i), we see that α is a continuous group homomorphism and hence Lemma 5.1.4 implies that $\alpha(t) = e^{at}$ for some $a \in \mathbb{R}^*$. Using the Cayley transform again to go back to $\mathbb{D}$, we get the assertion. □

Finally, we explicitly describe the infinitesimal generators of one-parameter groups in $\mathbb{D}$.

Corollary 5.6.6. *Let $\Phi: \mathbb{R} \to \mathrm{Aut}(\mathbb{D})$ be a nontrivial one-parameter group on $\mathbb{D}$ with Wolff point $\tau_\Phi \in \overline{\mathbb{D}}$. Let $F \in \mathrm{Hol}(\mathbb{D}, \mathbb{C})$ be its infinitesimal generator.*

(i) *If Φ is elliptic, then*

$$F(z) = (\overline{\tau_\Phi} z - 1)(z - \tau_\Phi) ia, \tag{5.23}$$

for some $a \in \mathbb{R}^*$, and the spectral value of Φ at τ_Φ is $(1 - |\tau_\Phi|^2) ia$;

(ii) *if Φ is parabolic, then*

$$F(z) = (z - \tau_\Phi)^2 \overline{\tau_\Phi} ia, \tag{5.24}$$

for some $a \in \mathbb{R}^*$, and the spectral value of Φ at τ_Φ is 0;

(iii) *if Φ is hyperbolic, then*

$$F(z) = (z^2 - \tau_\Phi^2) \overline{\tau_\Phi} a + ib \overline{\tau_\Phi} (z - \tau_\Phi)^2, \tag{5.25}$$

for some $a > 0$ and $b \in \mathbb{R}$, the other fixed point is

$$\sigma = \tau_\Phi \frac{-a + ib}{a + ib} \in \partial \mathbb{D}$$

and the spectral value of Φ at τ_Φ is $-2a$.

Conversely, every function of the form (5.23) (resp., (5.24), (5.25)) is the infinitesimal generator of an elliptic (resp., parabolic, hyperbolic) one-parameter group with Wolff point τ_Φ.

Proof. The fact that Φ is an one-parameter group implies that both F and $-F$ must be infinitesimal generators. In cases (i) and (ii), the Wolff point associated to $\Phi|_{\mathbb{R}^+}$ and to $\Phi|_{\mathbb{R}^-}$ is the same; therefore, (5.6) implies that

$$F(z) = (\overline{\tau_\Phi} z - 1)(z - \tau_\Phi)ia$$

for some $a \in \mathbb{R}^*$. Recalling that in case (ii) we have $\tau_\Phi \in \partial\mathbb{D}$, we obtain (5.23) and (5.24).

Assume now that Φ is hyperbolic, with fixed points $\tau_\Phi, \sigma \in \partial\mathbb{D}$. By Corollary 5.4.11, all the inequalities in (5.22) must be equalities. The last one just says that $\tau_\Phi \in \partial\mathbb{D}$. The previous one implies that (5.21) is an equality, too; the first one implies that $p'(z) = 2a\tau_\Phi(z-\tau_\Phi)^{-2}$ for a suitable $a \geq 0$. It follows that $p(z) = 2a\tau_\Phi(\tau_\Phi - z)^{-1} + c$ for some $c \in \mathbb{C}$; moreover, the equality in (5.21) forces $\operatorname{Re} c = -a$ and so

$$p(z) = a\,\frac{\tau_\Phi + z}{\tau_\Phi - z} + ib,$$

for some $b \in \mathbb{R}$.

Recalling (5.19), we see that F is of the form (5.25), as desired. Notice that if a were 0 then F would have only one zero, against the assumption that Φ is hyperbolic; so $a > 0$ and the other fixed point of Φ is obtained by solving $F(\sigma) = 0$ with $\sigma \neq \tau_\Phi$. The spectral value is computed by using Corollary 5.6.3.

Conversely, it is easy to check that every F of the form (5.23), (5.24), or (5.25) satisfies (5.16); furthermore, since (5.23) has only one zero which is in $\mathbb{D}$, (5.24) has a unique zero which is in $\partial\mathbb{D}$ and (5.25) has two distinct zeroes on $\partial\mathbb{D}$, the assertion follows. □

Notes to Section 5.6

Theorem 5.6.2 has been proved by Berkson and Porta [61] in 1978.

There are many other characterizations of infinitesimal generators. For instance, it is possible to prove (see, e. g., [346, 14, 77]) that for a function $F \in \operatorname{Hol}(\mathbb{D}, \mathbb{C})$ the following statements are equivalent:

(i) F is an infinitesimal generator;

(ii) there exist $a \in \mathbb{C}$ and $p \in \mathcal{P}$ such that $F(z) = a - \overline{a}z^2 - zp(z)$;

(iii) for all $\sigma \in \partial\mathbb{D}$, we have

$$\limsup_{z\to\sigma} \operatorname{Re}\big[\overline{z}F(z)\big] \leq 0;$$

(iv) $\operatorname{Re}[\overline{z}F(z)] \leq (1-|z|^2)\operatorname{Re}[\overline{z}F(0)]$ for all $z \in \mathbb{D}$;

(v) for all $z, w \in \mathbb{D}$, we have

$$\operatorname{Re}\left[\frac{\overline{z}F(z)}{1-|z|^2} + \frac{\overline{w}F(w)}{1-|w|^2}\right] \leq \operatorname{Re}\left[\frac{\overline{z}F(w) + w\overline{F(z)}}{1-\overline{z}w}\right].$$

Characterizations of infinitesimal generators of one-parameter semigroups having more than one common fixed point in the boundary, or satisfying other conditions, can be found, e. g., in [170, 171, 28, 172, 377, 177].

5.7 One-parameter semigroups on the unit disk

In this final section, we shall give a geometric realization of one-parameter semigroups on $\mathbb{D}$, relying on two particular kind of domains.

Definition 5.7.1. Let $D \subset \mathbb{C}$ be a domain and $\lambda \in \mathbb{C}^*$ with $\operatorname{Re}\lambda \geq 0$. Then
(a) if $0 \in D$ and $e^{-\lambda t}z \in D$ for every $z \in D$ and $t \geq 0$ we say that D is *λ-spirallike*; in particular, if $\lambda \in \mathbb{R}^+$ we say that D is *starlike*;
(b) if $z + it \in D$ for every $z \in D$ and $t \geq 0$ we say that D is *starlike at infinity*.

The point is that λ-spirallike domains and domains starlike at infinity admit natural one-parameter semigroups, given respectively by $\Phi_t(z) = e^{-\lambda t}z$ and $\Phi_t(z) = z + it$. If moreover the domain is simply connected, we can then use a Riemann mapping to obtain a one-parameter semigroup on $\mathbb{D}$.

For instance, let $D \subset \mathbb{C}$ be a simply connected domain starlike at infinity and fix a biholomorphism $g\colon \mathbb{D} \to D$. Then we can define $\Phi\colon \mathbb{R}^+ \to \operatorname{Hol}(\mathbb{D}, \mathbb{D})$ by

$$\Phi_t(z) = g^{-1}(g(z) + it).$$

The map Φ clearly is a one-parameter semigroup without fixed points, with infinitesimal generator $F(z) = i/g'(z)$. The idea is that every fixed-point-free one-parameter semigroup on $\mathbb{D}$ is obtained in this way.

Theorem 5.7.2. *Every one-parameter semigroup $\Phi\colon \mathbb{R}^+ \to \operatorname{Hol}(\mathbb{D}, \mathbb{D})$ without fixed points is of the form*

$$\Phi_t(z) = g^{-1}(g(z) + it), \tag{5.26}$$

where g is a biholomorphism between $\mathbb{D}$ and a simply connected domain $D \subset \mathbb{C}$ starlike at infinity; g is uniquely determined up to an additive constant. Furthermore, Φ is a one-parameter group if and only if D is either a vertical strip or a vertical half-plane.

Proof. Let Φ be a fixed point free one-parameter semigroup with infinitesimal generator F. By Theorem 5.6.2, we know that $F(z) = \overline{\tau_\Phi}(z - \tau_\Phi)^2 p(z)$ for suitable $\tau_\Phi \in \partial\mathbb{D}$ and $p \in \mathcal{P}$. Since we are assuming Φ without fixed points, $p(z) \neq 0$ for all $z \in \mathbb{D}$.

Let $\Psi\colon \mathbb{D} \to \mathbb{H}^+$ be given by

$$\Psi(z) = \frac{i}{2}\frac{\tau_\Phi + z}{\tau_\Phi - z}.$$

Then Ψ is a biholomorphism between $\mathbb{D}$ and $\mathbb{H}^+$ such that $\Psi'(z) = i\tau_\Phi(z - \tau_\Phi)^{-2}$. Let $h \in \operatorname{Hol}(\mathbb{H}^+, \mathbb{C})$ be such that $h'(z) = 1/p(\Psi^{-1}(z))$; by Lemma 4.5.1, h is injective. Therefore, $g = h \circ \Psi\colon \mathbb{D} \to \mathbb{C}$ is injective, too, and furthermore,

$$g'(z) = \frac{i}{F(z)}.$$

Given $z \in \mathbb{D}$, we have

$$\frac{\partial}{\partial t}(g \circ \Phi_t)(z_0) = g'(\Phi_t(z_0)) \cdot F(\Phi_t(z_0)) \equiv i;$$

therefore,

$$g(\Phi_t(z)) = g(z) + it$$

for all $z \in \mathbb{D}$ and $t \geq 0$. This means that $D = g(\mathbb{D})$ is starlike at infinity and that Φ is given by (5.26).

For the uniqueness, let $g_1 \colon \mathbb{D} \to D_1$ be another biholomorphism between $\mathbb{D}$ and a domain $D_1 \subset \mathbb{C}$ starlike at infinity such that (5.26) holds. Differentiating

$$g^{-1}(g(z) + it) = g_1^{-1}(g_1(z) + it)$$

with respect to t at $t = 0$ we get $g'(z) = g_1'(z)$ for all $z \in \mathbb{D}$; in particular, $g - g_1$ is constant. Conversely, it is easy to check that $g + c$ satisfies (5.26) for every $c \in \mathbb{C}$.

Finally, Φ is a one-parameter group if and only if for every $t \in \mathbb{R}$ the function $z \mapsto z + it$ is an automorphism of D, that is if and only if D is either a vertical strip or a vertical half-plane. □

Remark 5.7.3. If g satisfies (5.26), we can use it to determine whether Φ is hyperbolic or parabolic. Indeed, if F is the infinitesimal generator of Φ we know that $F(z) = i/g'(z)$; recalling Corollary 5.6.3 we see that Φ is parabolic if and only if

$$K\text{-}\lim_{z\to\tau)\Phi} |(z - \tau_\Phi)g'(z)| = +\infty,$$

where $\tau_\Phi \in \partial\mathbb{D}$ is the Wolff point of Φ.

An analogous result holds for one-parameter groups with a fixed point. Given $\lambda \in \mathbb{C}^*$ with $\operatorname{Re}\lambda \geq 0$, let $D \subset \mathbb{C}$ be a simply connected λ-spirallike domain (different from $\mathbb{C}$). Let $g \colon \mathbb{D} \to D$ be a biholomorphism such that $g(0) = 0$. Then $\Phi \colon \mathbb{R}^+ \to \operatorname{Hol}(\mathbb{D}, \mathbb{D})$ given by

$$\Phi_t(z) = g^{-1}(e^{-\lambda t} g(z))$$

is a one-parameter semigroup on $\mathbb{D}$ with fixed point 0 and spectral value λ at 0. In this way, we have recovered all elliptic one-parameter semigroups of $\mathbb{D}$.

Theorem 5.7.4. *Every nontrivial one-parameter semigroup $\Phi \colon \mathbb{R}^+ \to \operatorname{Hol}(\mathbb{D}, \mathbb{D})$ with fixed point $\tau_\Phi \in \mathbb{D}$ is of the form*

$$\Phi_t(z) = g^{-1}(e^{-\lambda t} g(z)) \tag{5.27}$$

where $\lambda \in \mathbb{C}^*$ with $\operatorname{Re}\lambda \geq 0$ and g is a biholomorphism between $\mathbb{D}$ and a λ-spirallike domain $D \subset \mathbb{C}$ so that $g(\tau_\Phi) = 0$. The constant λ is uniquely determined and g is uniquely determined up to a multiplicative constant. Furthermore, Φ is a one-parameter group if and only if $\operatorname{Re}\lambda = 0$ and D is a disk.

Proof. By (5.19), the infinitesimal generator of Φ is of the form

$$F(z) = -(1 - \overline{\tau_\Phi} z)(z - \tau_\Phi)p(z)$$

for a suitable $p \in \mathcal{P}$; note that $0 \notin p(\mathbb{D})$ because Φ is nontrivial. Let $\gamma_0 \in \operatorname{Aut}(\mathbb{D})$ be given by $\gamma_0(z) = (z - \tau_\Phi)/(1 - \overline{\tau_\Phi} z)$; set $p_0 = (1 - |\tau_\Phi|^2)(p \circ \gamma_0^{-1})$ and $\tilde{p} = p_0 \circ \pi \colon \mathbb{H}^+ \to \mathbb{C}$, where $\pi \colon \mathbb{H}^+ \to \mathbb{D}^*$ is the universal covering map $\pi(w) = e^{2\pi i w}$. Choose a holomorphic function $\tilde{h} \colon \mathbb{H}^+ \to \mathbb{C}$ such that $\tilde{h}' = 1/\tilde{p}$. Now, $\tilde{p}(w+1) = \tilde{p}(w)$ for all $w \in \mathbb{H}^+$; hence there is $\mu \in \mathbb{C}$ such that $\tilde{h}(w+1) = \tilde{h}(w) + \mu$ for all $w \in \mathbb{H}^+$. Since $\tilde{h}$ is injective (Lemma 4.5.1), necessarily $\mu \neq 0$; set $h = \mu^{-1}\tilde{h}$. The map h is a biholomorphism between $\mathbb{H}^+$ and $\widetilde{D} = h(\mathbb{H}^+)$; moreover, $h' = \mu^{-1}/\tilde{p}$ and $h(w+1) = h(w) + 1$ for all $w \in \mathbb{H}^+$. This implies that h factorizes through π, defining a biholomorphism $\hat{g} \colon \mathbb{D}^* \to D^* = \pi(\widetilde{D}) \subset \mathbb{C}^*$ such that $\hat{g} \circ \pi = \pi \circ h$. In particular, $\hat{g}$ satisfies

$$z\hat{g}'(z) = \frac{\lambda \hat{g}(z)}{p_0(z)}$$

for all $z \in \mathbb{D}^*$, where $\lambda = \mu^{-1}$.

Now set $g_0 = \hat{g} \circ \gamma_0$. Then g_0 satisfies

$$g_0'(z) = -\frac{\lambda g_0(z)}{F(z)}$$

for all $z \in \mathbb{D} \setminus \{\tau_\Phi\}$. Therefore, for every $z_0 \in \mathbb{D} \setminus \{\tau_\Phi\}$ we have

$$\frac{\partial}{\partial t} g_0(\Phi_t(z_0)) = g_0'(\Phi_t(z_0)) \cdot F(\Phi_t(z_0)) = -\lambda g_0(\Phi_t(z_0)),$$

and so

$$g_0(\Phi_t(z_0)) = e^{-\lambda t} g_0(z_0). \tag{5.28}$$

Now, g_0 extends to a holomorphic map $g \colon \mathbb{D} \to D = D^* \cup \{w_0\} \subset \widehat{\mathbb{C}}$ such that $g(\tau_\Phi) = w_0$ (by the big Picard Theorem 1.7.25). By (5.28), we must have either $w_0 = 0$ or $w_0 = \infty$; in particular, g is everywhere injective (because $D^* \subset \mathbb{C}^*$), and hence a biholomorphism.

We claim that, up to possibly replace g by $1/g$ and λ by $-\lambda$, we can assume $w_0 = 0$ and $\operatorname{Re}\lambda \geq 0$. As soon as we prove this, it immediately follows that D is λ-spirallike and that g satisfies (5.27).

Assume first $\operatorname{Re}\lambda > 0$. In this case, (5.28) shows that $g_0(\Phi_t(z_0)) \to 0$ as $t \to +\infty$ for any $z_0 \in \mathbb{D} \setminus \{\tau_\Phi\}$; since $\Phi_t(z_0) \to \tau_\Phi$ as $t \to +\infty$ it follows that $w_0 = 0$, and in this case we are done.

If $\operatorname{Re}\lambda = 0$ and $w_0 = 0$, there is nothing to do. If $\operatorname{Re}\lambda = 0$ and $w_0 = \infty$, then it suffices to replace λ by $-\lambda$ and g by $1/g$ (notice that $0 \notin D^*$).

Finally, if $\operatorname{Re}\lambda < 0$, (5.28) shows that $g_0(\Phi_t(z_0)) \to \infty$ as $t \to +\infty$ for any $z_0 \in \mathbb{D} \setminus \{\tau_\Phi\}$; hence $w_0 = \infty$ and replacing λ by $-\lambda$ and g by $1/g$ we are again done.

It remains to prove the last two assertions. Assume that $\mu \in \mathbb{C}^*$ and $g_1 : \mathbb{D} \to D_1$ are such that $g_1(\tau_\Phi) = 0$ and

$$g^{-1}(e^{-\lambda t} g(z)) = g_1^{-1}(e^{-\mu t} g_1(z))$$

for all $z \in \mathbb{D}$ and for all $t \geq 0$. Differentiating with respect to z at $z = \tau_\Phi$, we immediately get $\mu = \lambda$. Differentiating with respect to t at $t = 0$, we get $g g_1' - g_1 g' = 0$, i. e., $g_1 = cg$ for some constant $c \in \mathbb{C}^*$. Conversely, it is easy to check that cg satisfies (5.27) for every $c \in \mathbb{C}^*$.

Finally, assume Φ is a one-parameter group. Then $\Phi_t(z_0)$ cannot converge as $t \to +\infty$ for any $z_0 \in \mathbb{D} \setminus \{\tau_\Phi\}$; this implies, by (5.27), $\operatorname{Re}\lambda = 0$. In particular, D is a simply connected domain invariant under rotations, that is a disk. □

Notes to Section 5.7

Theorems 5.7.2 and 5.7.4 have been proved by Heins [190] in 1981; see also [379]. They are the starting point of the so-called geometrical theory of one-parameter semigroups on $\mathbb{D}$, where analytic properties of the semigroup Φ are deduced from geometrical properties of the domain D. For instance, assume that Φ is fixed point free. Then it is possible to show (see, e. g., [64] and [80, Theorem 9.4.10]) that Φ is hyperbolic if and only if D is contained in a vertical strip of finite width; moreover, if $d > 0$ is the width of the smallest vertical strip containing D then the spectral value of Φ is π/d. For much more on this beautiful theory see, e. g., [80], and references therein.

A Appendix

In this Appendix, we collected a few classical results in (real and) complex analysis that we used in the book and that are not always covered by standard courses in complex analysis of one variable.

A.1 The Hurwitz theorems

In this section, we shall prove a list of results collectively known as *Hurwitz theorems*. We shall deduce them from a simplified version of the *Rouché theorem*.

Theorem A.1.1 (Rouché, 1862). *Let f and g be functions holomorphic in a neighborhood of a closed disk $\overline{D(z_0,r)} \subset \mathbb{C}$ and such that $|f-g| < |g|$ on $\partial D(z_0,r)$. Then f and g have the same number of zeroes in $D(z_0,r)$, counted with multiplicities.*

Proof. Set $f_\lambda = g + \lambda(f-g)$ for $\lambda \in [0,1]$. Then on $\partial D(z_0,r)$ we have

$$0 < |g(z)| - |f(z) - g(z)| \le |g(z)| - \lambda|f(z) - g(z)| \le |f_\lambda(z)|.$$

Let a_λ denote the number of zeroes of f_λ in $D(z_0,r)$, counted with multiplicities. The logarithmic indicator theorem yields

$$a_\lambda = \frac{1}{2\pi i} \int\limits_{\partial D(z_0,r)} \frac{f_\lambda'(\zeta)}{f_\lambda(\zeta)}\, d\zeta = \frac{1}{2\pi i} \int\limits_{\partial D(z_0,r)} \frac{g'(\zeta) + \lambda(f'(\zeta) - g'(\zeta))}{g(\zeta) + \lambda(f(\zeta) - g(\zeta))}\, d\zeta.$$

It follows that a_λ is an integer depending continuously on λ; thus it must be constant. In particular a_0, the number of zeroes of g, is equal to a_1, the number of zeroes of f, both counted with multiplicities. □

As a consequence we get the following *Hurwitz theorems*.

Corollary A.1.2 (Hurwitz, 1889). *Let $\{f_\nu\}$ be a sequence of functions holomorphic in a given neighborhood of a closed disk $\overline{D(z_0,r)} \subset \mathbb{C}$. Assume that $\{f_\nu\}$ converges uniformly on $\overline{D(z_0,r)}$ to a function f holomorphic in a neighborhood of $\overline{D(z_0,r)}$ and vanishing nowhere on $\partial D(z_0,r)$. Then for ν large enough f_ν has the same number of zeroes as f in $D(z_0,r)$, counted with multiplicities.*

Proof. Let $m = \inf\{|f(z)| \mid z \in \partial D(z_0,r)\} > 0$. Since $f_\nu \to f$ uniformly, for ν large enough we have $|f_\nu - f| < m \le |f|$ on $\partial D(z_0,r)$. Hence we can apply Theorem A.1.1 to f_ν and f. □

Corollary A.1.3. *Let $\{f_\nu\} \subset \mathrm{Hol}(X,Y)$ be a sequence of holomorphic functions between two Riemann surfaces X and Y. Assume that $\{f_\nu\}$ converges, uniformly on compact sets, to a nonconstant function $f \in \mathrm{Hol}(X,Y)$. Choose $w_0 \in f(X)$ and assume that $f^{-1}(w_0)$ contains at least $\ell \ge 1$ distinct points. Then for all ν large enough $f_\nu^{-1}(w_0)$ contains at least ℓ distinct points.*

https://doi.org/10.1515/9783110601978-006

Proof. Let $z_1, \dots, z_\ell \in f^{-1}(w_0)$ be distinct points. Choose a neighborhood $V \subset\subset Y$ of w_0 and $r_0 > 0$ such that there is a biholomorphism $\varphi: V \to D(0, r_0)$ with $\varphi(w_0) = 0$. Since $f_\nu \to f$ uniformly on compact sets, and f is not constant, for all $j = 1, \dots, \ell$ we can find a neighborhood $U_j \subset\subset X$ of z_j such that

(a) there are $r_j > 0$ and a biholomorphism $\psi_j: D(0, 2r_j) \to U_j$ with $\psi_j(0) = z_j$;
(b) $f_\nu(U_j) \subset V$ for all ν large enough;
(c) $U_j \cap f^{-1}(w_0) = \{z_j\}$ for all $j = 1, \dots, \ell$;
(d) $U_h \cap U_k = \emptyset$ if $h \neq k$.

Then Corollary A.1.2 applied to $\varphi \circ f_\nu \circ \psi_j: D(0, r_j) \to D(0, r_0)$ shows that $f_\nu^{-1}(w_0) \cap U_j$ contains at least one point for ν large enough (and all j) and we are done. □

Corollary A.1.4. *Let $\{f_\nu\} \subset \mathrm{Hol}(X, Y)$ be a sequence of holomorphic functions between two Riemann surfaces X and Y converging, uniformly on compact sets, to $f \in \mathrm{Hol}(X, Y)$. Then:*

(i) *if $w_0 \in Y$ is such that $w_0 \notin f_\nu(X)$ for all ν, then either $w_0 \notin f(X)$ or $f \equiv w_0$.*
(ii) *if all f_ν are injective, then f is either injective or constant.*

Proof. (i) If f is constant, the assertion is obvious. If f is not constant, by Corollary A.1.3 $w_0 \in f(X)$ would contradict the fact that $w_0 \notin f_\nu(X)$ for all ν.

(ii) Assume, by contradiction, that $f^{-1}(w_0)$ contains at least two points for some $w_0 \in Y$. Then, by Corollary A.1.3, either $f \equiv w_0$ or $f_\nu^{-1}(w_0)$ contains at least two points for every ν large enough, impossible. □

Corollary A.1.5. *Let X and Y be two Riemann surfaces, $D \subseteq Y$ a domain, and $\{f_\nu\} \subset \mathrm{Hol}(X, D)$ a sequence of holomorphic functions converging, uniformly on compact subsets, to a function $f \in \mathrm{Hol}(X, Y)$. Then either f is a constant belonging to ∂D or $f(X) \subseteq D$. In particular, $\overline{\mathrm{Hol}(X, D)} = \mathrm{Hol}(X, D) \cup \partial D$, where the closure is taken in $\mathrm{Hol}(X, Y)$.*

Proof. Clearly, $f(X) \subseteq \overline{D}$. If f is not constant, by Corollary A.1.3 every $w_0 \in f(X)$ must eventually belong to $f_\nu(X) \subseteq D$ and the assertion follows. □

Notes to Section A.1

Theorem A.1.1 is the easiest version of a general result proved by Rouché [357] in 1862, where the disk is replaced by a more general bounded domain of $\mathbb{C}$. Our proof follows [363] and it is inspired by [117].

The original versions of Corollaries A.1.2, A.1.3, and A.1.4 are in [203].

A.2 The Fatou uniqueness theorem

In this section, we shall prove a version of the Fatou uniqueness theorem that we shall obtain as a consequence of the following result, the so-called *two constants theorem.*

Theorem A.2.1 (Two constants theorem; Ostrowski, 1922). *Let $f: \mathbb{D} \to \mathbb{C}$ be holomorphic and such that $|f| \le M$ for a suitable $M > 0$. Assume that on a given arc $A \subset \partial\mathbb{D}$ of length α there exists $m \in (0, M]$ such that for all $\tau \in A$ we have*

$$\limsup_{z\to\tau} |f(z)| \le m.$$

For all $0 < r < 1$, *set*

$$\lambda(r) = \frac{\alpha}{2\pi}\,\frac{1-r}{1+r}.$$

Then for any $r \in (0,1)$ *and* $z \in \overline{D(0,r)}$, *we have*

$$|f(z)| \le m^{\lambda(r)} M^{1-\lambda(r)}. \tag{A.1}$$

Proof. Assume first that f extends continuously to $\overline{\mathbb{D}}$, and define $\varphi, h\colon \overline{\mathbb{D}} \to \mathbb{R}$ by

$$\varphi(z) = \log(\max\{|f(z)|, m\}),$$

and

$$h(z) = \frac{1}{2\pi i}\int_{\partial\mathbb{D}} P(z,\zeta)\varphi(\zeta)\,d\zeta,$$

where

$$P(z,\zeta) = \operatorname{Re}\left(\frac{\zeta+z}{\zeta-z}\right) = \frac{1-|z|^2}{|\zeta-z|^2}$$

is the Poisson kernel. Since φ is subharmonic on $\mathbb{D}$, h is harmonic on $\mathbb{D}$, and $\varphi = h$ on $\partial\mathbb{D}$, we have $\varphi \le h$ on $\overline{\mathbb{D}}$.

After a rotation, we can assume that $\overline{A} = \{e^{i\theta} \mid 0 \le \theta \le \alpha\}$. On A, we have $|f(z)| \le m$ and so $\varphi(z) = \log m$. Therefore,

$$\begin{aligned} h(z) &\le \frac{1}{2\pi}\int_0^{\alpha} P(z,e^{i\theta})\log m\,d\theta + \frac{1}{2\pi}\int_{\alpha}^{2\pi} P(z,e^{i\theta})\log M\,d\theta \\ &= \log M - \psi(z)\,[\log M - \log m], \end{aligned}$$

where

$$\psi(z) = \frac{1}{2\pi}\int_0^{\alpha} P(z,e^{i\theta})\,d\theta \ge \frac{1}{2\pi}\int_0^{\alpha} \frac{1-|z|}{1+|z|}\,d\theta = \lambda(|z|).$$

Hence

$$h(z) \le \log M - \lambda(|z|)[\log M - \log m] = \lambda(|z|)\log m + (1-\lambda(|z|))\log M.$$

Since $|f| \le \max\{|f|, m\} = e^{\varphi} \le e^{h}$, we obtain

$$|f(z)| \le m^{\lambda(|z|)} M^{1-\lambda(|z|)}, \tag{A.2}$$

that is (A.1) in this case.

If f is not necessarily continuous on $\overline{\mathbb{D}}$, for every $s \in (0,1)$ define $f_s\colon \overline{\mathbb{D}} \to \mathbb{C}$ by $f_s(z) = f(sz)$. Then (A.2) together with the maximum principle yields

$$|f(sz)| \le \Big(\sup_{\tau \in \overline{A}} |f_s(\tau)|\Big)^{\lambda(r)} M^{1-\lambda(r)}$$

for all $z \in \overline{D(0,r)}$ as soon as $r < s$. Letting $s \to 1$, we obtain (A.1), as claimed. □

Then our version of *Fatou uniqueness theorem* is the following.

Corollary A.2.2 (Fatou uniqueness theorem; Fatou, 1908). *Let $D \subset X$ be a domain in a Riemann surface X such that ∂D is a Jordan curve, Y another Riemann surface, and $f\colon D \to Y$ holomorphic. Assume there is a nonvoid open arc $A \subset \partial D$ and $y_0 \in Y$ such that*

$$\lim_{z \to \tau} f(z) = y_0$$

for all $\tau \in A$. Then $f \equiv y_0$.

Proof. Fix a neighborhood U of y_0 in Y such that there is a biholomorphism $\phi\colon U \to \mathbb{D}$ with $\phi(y_0) = 0$. By continuity, we can find a neighborhood V in X of a point $\tau_0 \in A$, a biholomorphism $\psi\colon V \to \mathbb{D}$ and a simply connected subdomain $D_1 \subset D$ bounded by a Jordan curve such that $D_1 \subset\subset V$, $\partial D_1 \cap \partial D \subset A$, and $f(D_1) \subset U$.

By Theorem 1.6.32, there is a biholomorphism $\eta\colon \psi(D_1) \to \mathbb{D}$, which extends to a homeomorphism of the closures. Then if we apply Theorem A.2.1 to $\phi \circ f \circ (\eta \circ \psi)^{-1}$, letting $m \to 0$ we find $f \equiv y_0$ on D_1, and thus everywhere. □

Notes to Section A.2

The two constants Theorem A.2.1 is due to Ostrowsky [314]; our proof is taken from [88, p. 152].

The original Fatou uniqueness theorem [145] states that if $f\colon \mathbb{D} \to \mathbb{C}$ is a bounded holomorphic function such that $\lim_{r\to 1} f(re^{i\theta}) = 0$ for all $e^{i\theta}$ belonging to a subset of $\partial\mathbb{D}$ of positive measure then $f \equiv 0$. A modern proof can be found, e. g., in Rudin [360, Theorem 15.19].

A.3 Holomorphic functions with nonnegative real part

In this section, we collect a few results about holomorphic functions with nonnegative real part, mostly obtained as consequences of the Poisson formula. The first one, the simple but useful *Zygmund inequality*, involves the modulus of the real part.

Proposition A.3.1 (Zygmund, 1945). *Let $f \in \mathrm{Hol}(\mathbb{D}, \mathbb{C}) \cap C^0(\overline{\mathbb{D}})$. Then*

$$(1 - |z|^2)|f'(z)| \le 2U(z)$$

for all $z \in \mathbb{D}$, where $U\colon \overline{\mathbb{D}} \to \mathbb{R}^+$ is the harmonic extension of $|\operatorname{Re} f|_{\partial\mathbb{D}}|$.

Proof. The Cauchy–Schwarz version of the Poisson formula (see, e. g., [360, Theorem 11.12]) says that

$$f(z) = i\,\mathrm{Im}\, f(0) + \frac{1}{2\pi}\int_{\partial\mathbb{D}} \frac{\zeta+z}{\zeta-z}\,\mathrm{Re}\, f(\zeta)\,d\zeta.$$

Therefore,

$$f'(z) = \frac{1}{2\pi}\int_{\partial\mathbb{D}} \frac{\partial}{\partial z}\left(\frac{\zeta+z}{\zeta-z}\right)\mathrm{Re}\, f(\zeta)\,d\zeta = \frac{1}{2\pi}\int_{\partial\mathbb{D}} \frac{2\zeta}{(\zeta-z)^2}\,\mathrm{Re}\, f(\zeta)\,d\zeta.$$

It follows that

$$|f'(z)| \le \frac{2}{1-|z|^2}\frac{1}{2\pi}\int_{\partial\mathbb{D}} \frac{1-|z|^2}{|\zeta-z|^2}|\mathrm{Re}\, f(\zeta)|\,d\zeta = \frac{2}{1-|z|^2}U(z)$$

and we are done. □

Now we specialize to the case of holomorphic functions with nonnegative real part. The first formula we prove is the *Harnack inequality*.

Theorem A.3.2 (Harnack, 1886). *Let $p\colon \mathbb{D} \to \mathbb{C}$ be holomorphic with nonnegative real part. Then*

$$\frac{1+|z|}{1-|z|}\,\mathrm{Re}\, p(0) \ge \mathrm{Re}\, p(z) \ge \frac{1-|z|}{1+|z|}\,\mathrm{Re}\, p(0)$$

for all $z \in \mathbb{D}$.

Proof. Fix $0 < r < 1$. Then the Poisson formula applied to $p_r(z) = p(rz)$ yields

$$\mathrm{Re}\, p_r(z) = \frac{1}{2\pi}\int_0^{2\pi} \frac{1-|z|^2}{|e^{i\theta}-z|^2}\,\mathrm{Re}\, p_r(e^{i\theta})\,d\theta.$$

Now, for all $z \in \mathbb{D}$ and $\theta \in \mathbb{R}$ we have

$$\frac{1+|z|}{1-|z|} \ge \frac{1-|z|^2}{|e^{i\theta}-z|^2} \ge \frac{1-|z|}{1+|z|};$$

therefore, recalling the mean value property of harmonic functions we get

$$\frac{1+|z|}{1-|z|}\,\mathrm{Re}\, p(0) \ge \mathrm{Re}\, p_r(z) \ge \frac{1-|z|}{1+|z|}\,\mathrm{Re}\, p(0).$$

The assertion follows letting $r \to 1^-$. □

The next result is the *Herglotz representation formula* for holomorphic functions with nonnegative real part. Our proof uses a bit of measure theory; the relevant facts can be found, e. g., in [360, Chapter 2].

Theorem A.3.3 (Herglotz, 1911). *Let $p\colon \mathbb{D} \to \mathbb{C}$ be holomorphic with nonnegative real part. Then there exists a positive Borel measure μ on $\partial\mathbb{D}$ such that for all $z \in \mathbb{D}$ we have*

$$p(z) = \int_{\partial\mathbb{D}} \frac{\zeta + z}{\zeta - z}\, d\mu(\zeta) + i \operatorname{Im} p(0). \tag{A.3}$$

Proof. Let $\varphi = \operatorname{Re} p$, and set $\varphi_r(z) = \varphi(rz)$ for $0 < r < 1$. Every φ_r is harmonic in $\mathbb{D}$, continuous in $\overline{\mathbb{D}}$ and nonnegative; thus

$$\operatorname{Re} p(0) = \varphi_r(0) = \frac{1}{2\pi i} \int_{\partial\mathbb{D}} \varphi_r(\zeta)\, d\zeta. \tag{A.4}$$

Let $\Lambda\colon C^0(\partial\mathbb{D}) \to \mathbb{R}$ be the $\mathbb{R}$-linear functional given by

$$\Lambda u = \limsup_{r\to 1} \frac{1}{2\pi i} \int_{\partial\mathbb{D}} u(\zeta)\varphi_r(\zeta)\, d\zeta.$$

The functional Λ is well-defined because $\{\varphi_r\}$ is uniformly bounded in $L^1(\partial\mathbb{D})$, thanks to (A.4). Since every φ_r is nonnegative, Λ gives rise to a positive measure μ on $\partial\mathbb{D}$ by

$$\Lambda u = \int_{\partial\mathbb{D}} u(\zeta)\, d\mu(\zeta);$$

in particular, $\mu(\partial\mathbb{D}) = \operatorname{Re} p(0)$.

Then the reproducing property of the Poisson kernel yields

$$\begin{aligned} \operatorname{Re} \int_{\partial\mathbb{D}} \frac{\zeta + z}{\zeta - z}\, d\mu(\zeta) &= \int_{\partial\mathbb{D}} P(z,\zeta)\, d\mu(\zeta) = \limsup_{r\to 1} \frac{1}{2\pi i} \int_{\partial\mathbb{D}} P(z,\zeta)\varphi_r(\zeta)\, d\zeta \\ &= \lim_{r\to 1} \varphi_r(z) = \varphi(z) = \operatorname{Re} p(z). \end{aligned} \tag{A.5}$$

Since two holomorphic functions with the same real part must differ by a purely imaginary additive constant, we are done. □

Notice that if p is continuous up to the boundary then μ is just $(\operatorname{Re} p)|_{\partial\mathbb{D}}\, d\zeta$, and thus it is absolutely continuous with respect to the standard Lebesgue measure on $\partial\mathbb{D}$.

We end this section using the Herglotz representation formula to prove the existence of a particular nontangential limit.

Proposition A.3.4. *Let $p\colon \mathbb{D} \to \mathbb{C}$ be a holomorphic function with nonnegative real part. Then for every $\sigma \in \partial\mathbb{D}$ the nontangential limit*

$$K\text{-}\lim_{z\to\sigma}(1-\overline{\sigma}z)p(z)$$

exists and belongs to $[0,+\infty)$.

Proof. If $\operatorname{Re} p(z_0)=0$ for some $z_0\in\mathbb{D}$, we have $p\equiv ia$ for some $a\in\mathbb{R}$ and the assertion is obvious. Assume then that $\operatorname{Re} p>0$ always and let μ be the positive Borel measure on $\partial\mathbb{D}$ associated to p by the Herglotz representation formula (Theorem A.3.3). Put $\lambda=\mu(\{\sigma\})\ge 0$ and $\mu_0=\mu-\lambda\delta_\sigma$, where δ_σ is the Dirac measure on $\partial\mathbb{D}$ concentrated at σ. The Herglotz formula (A.3) can then be written as

$$p(z)=\lambda\frac{\sigma+z}{\sigma-z}+\int_{\partial\mathbb{D}}\frac{\zeta+z}{\zeta-z}\,d\mu_0(\zeta)+i\operatorname{Im}p(0).$$

In particular, since

$$\lim_{z\to\sigma}(1-\overline{\sigma}z)\frac{\sigma+z}{\sigma-z}=2,$$

to get the assertion it suffices to prove that

$$K\text{-}\lim_{z\to\sigma}(1-\overline{\sigma}z)\int_{\partial\mathbb{D}}\frac{\zeta+z}{\zeta-z}d\mu_0(\zeta)=0. \tag{A.6}$$

Since $\mu_0(\{\sigma\})=0$, for every $\varepsilon>0$ we can find a $\delta>0$ such that if $C\subset\partial\mathbb{D}$ is an arc of length δ centered at σ then $\mu_0(C)<\varepsilon$. Notice that if $\zeta\in\partial\mathbb{D}\setminus C$ then $|\zeta-z|$ stays bounded away from 0 as $z\to\sigma$; this implies that

$$\lim_{z\to\sigma}(1-\overline{\sigma}z)\int_{\partial\mathbb{D}\setminus C}\frac{\zeta+z}{\zeta-z}\,d\mu_0(\zeta)=0.$$

On the other hand, fix $M>1$ and take $z\in K(\sigma,M)$. Then we have

$$\left|(1-\overline{\sigma}z)\int_C\frac{\zeta+z}{\zeta-z}\,d\mu_0(\zeta)\right|\le 2\frac{|\sigma-z|}{1-|z|}\left|\int_C d\mu_0(\zeta)\right|\le 2M\varepsilon.$$

Since this holds for every M and ε is arbitrary, (A.6) holds and we are done. □

***Notes to Section* A.3**

Proposition A.3.1 comes from [431]. Theorem A.3.2 has originally been proved in [180]. Theorem A.3.3 was first proved by Herglotz [193] for nonnegative harmonic functions.

A.4 Sequences

In this section, we collect a few more or (sometimes) less well-known facts about sequences and series of real numbers.

Lemma A.4.1. *Let $\{a_\nu\} \subset (0, +\infty)$ be a sequence of positive real numbers. Then*

$$\liminf_{\nu\to+\infty} \frac{a_{\nu+1}}{a_\nu} \le \liminf_{\nu\to+\infty} (a_\nu)^{1/\nu}$$

and

$$\limsup_{\nu\to+\infty} \frac{a_{\nu+1}}{a_\nu} \ge \limsup_{\nu\to+\infty} (a_\nu)^{1/\nu}.$$

Proof. If $\liminf_{\nu\to+\infty}(a_{\nu+1}/a_\nu) = 0$, the first assertion is obvious. Otherwise, choose $\varepsilon > 0$ such that $0 < \varepsilon < \liminf_{\nu\to+\infty}(a_{\nu+1}/a_\nu)$; in particular, we have $a_{\nu+1} \ge \varepsilon a_\nu$ for all $\nu \ge \nu_0$ sufficiently large. This implies $a_{\nu_0+\mu} \ge \varepsilon^\mu a_{\nu_0}$ for all $\mu \ge 0$, and hence

$$(a_{\nu_0+\mu})^{\frac{1}{\nu_0+\mu}} \ge \varepsilon \left(\frac{a_{\nu_0}}{\varepsilon^{\nu_0}}\right)^{\frac{1}{\nu_0+\mu}}.$$

Hence

$$\liminf_{\nu\to+\infty}(a_\nu)^{1/\nu} = \liminf_{\mu\to+\infty}(a_{\nu_0+\mu})^{\frac{1}{\nu_0+\mu}} \ge \varepsilon \liminf_{\mu\to+\infty}\left(\frac{a_{\nu_0}}{\varepsilon^{\nu_0}}\right)^{\frac{1}{\nu_0+\mu}} = \varepsilon$$

and the first assertion follows by the arbitrariness of ε.

If $\limsup_{\nu\to+\infty}(a_{\nu+1}/a_\nu) = +\infty$, the second assertion is obvious. Otherwise, choose $M > \limsup_{\nu\to+\infty}(a_{\nu+1}/a_\nu)$; in particular, we have $a_{\nu+1} \le M a_\nu$ for all $\nu \ge \nu_0$ sufficiently large. This implies $a_{\nu_0+\mu} \le M^\mu a_{\nu_0}$ for all $\mu \ge 0$, and hence

$$(a_{\nu_0+\mu})^{\frac{1}{\nu_0+\mu}} \le M \left(\frac{a_{\nu_0}}{M^{\nu_0}}\right)^{\frac{1}{\nu_0+\mu}}.$$

Hence

$$\limsup_{\nu\to+\infty}(a_\nu)^{1/\nu} = \limsup_{\mu\to+\infty}(a_{\nu_0+\mu})^{\frac{1}{\nu_0+\mu}} \le M \limsup_{\mu\to+\infty}\left(\frac{a_{\nu_0}}{M^{\nu_0}}\right)^{\frac{1}{\nu_0+\mu}} = M$$

and the second assertion follows by the arbitrariness of M. □

The next result is a version of the well-known *Fekete lemma.*

Definition A.4.2. A sequence $\{a_\nu\}_{\nu\in\mathbb{N}} \subset \mathbb{R}$ is *subadditive* if for all $\mu, \nu \in \mathbb{N}$ we have

$$a_{\mu+\nu} \le a_\mu + a_\nu.$$

Lemma A.4.3 (Fekete, 1923). *Let $\{a_\nu\}_{\nu\in\mathbb{N}} \subset \mathbb{R}$ be a subadditive sequence of real numbers. Then the sequence $\{\frac{1}{\nu}a_\nu\}$ converges and we have*

$$\lim_{\nu\to+\infty} \frac{a_\nu}{\nu} = \inf_{\nu\in\mathbb{N}} \frac{a_\nu}{\nu}.$$

Proof. Put $\alpha = \inf\{a_\nu/\nu\} \in [-\infty, +\infty)$ and choose $a > \alpha$. Take $\nu_0 \in \mathbb{N}$ so that $\frac{a_{\nu_0}}{\nu_0} < a$ and put $M = \max\{a_\nu \mid \nu \in [2\nu_0, 3\nu_0]\}$. If $\nu > 3\nu_0$, we can find $\mu \in \mathbb{N}$ such that $\nu - \mu\nu_0 \in [2\nu_0, 3\nu_0]$, and hence

$$a_\nu \le a_{\mu\nu_0} + a_{\nu-\mu\nu_0} \le \mu a_{\nu_0} + M.$$

Dividing by ν, we get

$$\alpha \le \frac{a_\nu}{\nu} \le \frac{\mu\nu_0}{\nu}\frac{a_{\nu_0}}{\nu_0} + \frac{M}{\nu};$$

notice that, by construction $\nu - 3\nu_0 \le \mu\nu_0 \le \nu - 2\nu_0$. Letting $\nu \to +\infty$, we find

$$\alpha \le \liminf_{\nu\to+\infty}\frac{a_\nu}{\nu} \le \limsup_{\nu\to+\infty}\frac{a_\nu}{\nu} \le \frac{a_{\nu_0}}{\nu_0} < a,$$

because $\frac{\mu\nu_0}{\nu} \to 1$, and the assertion follows from the arbitrariness of $a > \alpha$. □

The next result is a version of the *Stolz–Cesàro theorem.*

Proposition A.4.4. *Let $\{a_\nu\}, \{b_\nu\} \subset \mathbb{R}$ be two sequences. Assume that $\{b_\nu\}$ is strictly increasing diverging to $+\infty$ and that the following limit exists:*

$$\lim_{\nu\to+\infty}\frac{a_\nu - a_{\nu-1}}{b_\nu - b_{\nu-1}} = \ell \in [-\infty, +\infty].$$

Then the sequence $\{a_\nu/b_\nu\}$ admits a limit and we have

$$\lim_{\nu\to+\infty}\frac{a_\nu}{b_\nu} = \ell. \tag{A.7}$$

Proof. Assume first that $\ell \in \mathbb{R}$. By assumption, for all $\varepsilon > 0$ we can find $\nu_0 \in \mathbb{N}$ such that for $\nu > \nu_0$ we have

$$\left|\frac{a_\nu - a_{\nu-1}}{b_\nu - b_{\nu-1}} - \ell\right| < \varepsilon.$$

Since $\{b_\nu\}$ is strictly increasing, we get

$$(\ell - \varepsilon)(b_\nu - b_{\nu-1}) < a_\nu - a_{\nu-1} < (\ell + \varepsilon)(b_\nu - b_{\nu-1})$$

for all $\nu > \nu_0$. Now write $a_\nu = (a_\nu - a_{\nu-1}) + \cdots + (a_{\nu_0+1} - a_{\nu_0}) + a_{\nu_0}$; then repeatedly summing the previous inequalities, we obtain

$$(\ell - \varepsilon)(b_\nu - b_{\nu_0}) + a_{\nu_0} < a_\nu < (\ell + \varepsilon)(b_\nu - b_{\nu_0}) + a_{\nu_0}.$$

Dividing by b_ν and recalling that $b_\nu \to +\infty$, we get

$$\ell - \varepsilon \leq \liminf_{\nu \to +\infty} \frac{a_\nu}{b_\nu} \leq \limsup_{\nu \to +\infty} \frac{a_\nu}{b_\nu} \leq \ell + \varepsilon$$

and (A.7) follows by the arbitrariness of ε.

Assume now $\ell = +\infty$ and fix $M > 1$. Then we can find $\nu_0 \in \mathbb{N}$ such that for $\nu > \nu_0$ we have

$$\frac{a_\nu - a_{\nu-1}}{b_\nu - b_{\nu-1}} > M,$$

i. e.,

$$a_\nu - a_{\nu-1} > M(b_\nu - b_{\nu-1}).$$

Arguing as before, we obtain

$$a_\nu > M(b_\nu - b_{\nu_0}) + a_{\nu_0}.$$

Hence dividing by b_ν and letting $\nu \to +\infty$, we find

$$\liminf_{\nu \to +\infty} \frac{a_\nu}{b_\nu} \geq M$$

and the arbitrariness of M yields (A.7). The case $\ell = -\infty$ is completely analogous. □

As a corollary, we get the *Cesàro means theorem.*

Corollary A.4.5. *Let $\{c_\nu\} \subset \mathbb{R}$ be a sequence converging to $\ell \in [-\infty, +\infty]$. Then*

$$\lim_{\nu \to +\infty} \frac{1}{\nu} \sum_{j=0}^{\nu-1} c_j = \ell.$$

Proof. It follows from Proposition A.4.4 by taking $a_\nu = \sum_{j=0}^{\nu-1} c_j$ and $b_\nu = \nu$. □

***Notes to Section* A.4**

Lemma A.4.3 comes from [150]. Proposition A.4.4 has been first proved by Stolz [382, pp. 173–175] and Cesàro [109].

A.5 Topological groups

In this last section, we prove a basic fact about Hausdorff topological group not easily found in the literature.

Proposition A.5.1. *A discrete subgroup Γ of a Hausdorff topological group G is always closed.*

Proof. Assume, by contradiction, that there exists $g \in G \setminus \Gamma$ such that every neighborhood of g contains infinitely many elements of Γ. Since Γ is discrete, we can find a neighborhood $U \subset G$ of the identity element e such that $U \cap \Gamma = \{e\}$. Since G is a topological group, we can find another neighborhood $V \subset G$ of e such that $V^{-1}V \subseteq U$, where $V^{-1} = \{h^{-1} \mid h \in V\}$. Now gV is a neighborhood of g; therefore, we can find $h_1, h_2 \in gV \cap \Gamma$ with $h_1 \neq h_2$. But then $e \neq h_1^{-1}h_2 \in V^{-1}V \cap \Gamma \subseteq U \cap \Gamma = \{e\}$, contradiction. □

Notes to Section A.5

A good introduction to the theory of topological groups, including the background needed for the proof of Proposition A.5.1, is [196].

Bibliography

[1] J. Aaronson. Ergodic theory for inner functions of the upper half plane, *Ann. Inst. Henri Poincaré Sect. B* **14** (1978), 233–253.

[2] J. Aaronson. A remark on the exactness of inner functions, *J. Lond. Math. Soc.* **23** (1981), 469–474.

[3] M. Abate. *Iteration theory of holomorphic maps on taut manifolds*. Commenda di Rende: Mediterranean Press, 1989.

[4] M. Abate. The Lindelöf principle and the angular derivative in strongly convex domains, *J. Anal. Math.* **54** (1990), 189–228.

[5] M. Abate. Iteration theory, compactly divergent sequences and commuting holomorphic maps, *Ann. Sc. Norm. Super. Pisa, Cl. Sci.* **18** (1991), 167–191.

[6] M. Abate. The infinitesimal generators of semigroups of holomorphic maps, *Ann. Mat. Pura Appl.* **161** (1992), 167–180.

[7] M. Abate. Iteration of holomorphic families, *Rend. Ist. Mat. Univ. Trieste* **26** (1994), 141–150.

[8] M. Abate. Dynamics in several complex variables. In: *Metrical and dynamical aspects in complex analysis*, ed. by L. Blanc-Centi, vol. 2195, Lecture Notes in Mathematics. Berlin: Springer Verlag, 2017, 25–54.

[9] M. Abate. Multipoint Julia theorems, *Rend. Lincei Mat. Appl.* **32** (2021), 593–625.

[10] M. Abate and A. Christodoulou. Random iteration on hyperbolic Riemann surfaces, *Ann. Mat. Pura Appl.* **201** (2022), 2021–2035.

[11] M. Abate and F. Tovena. *Curves and surfaces*. Milan: Springer, 2012.

[12] M. Abate and J.-P. Vigué. Common fixed points in hyperbolic Riemann surfaces and convex domains, *Proc. Am. Math. Soc.* **112** (1991), 503–512.

[13] N. H. Abel. Untersuchungen über die Reihe $1+\frac{m}{1}x+\frac{m(m-1)}{1\cdot 2}+\cdots$, *J. Reine Angew. Math.* **1** (1826), 311–339.

[14] D. Aharonov, M. Elin, S. Reich and D. Shoikhet. Parametric representation of semi-complete vector fields on the unit balls in $\mathbb{C}^n$ and in Hilbert space, *Rend. Mat. Acc. Lincei* **10** (1999), 229–253.

[15] L. V. Ahlfors. Untersuchungen zur Theorie der konformen Abbildung und der ganzen Funktionen, *Acta Soc. Sci. Fenn.* **1** (1930), 1–40.

[16] L. V. Ahlfors. An extension of Schwarz's lemma, *Trans. Am. Math. Soc.* **43** (1938), 359–364.

[17] L. V. Ahlfors. *Complex analysis*. Third edition. New York: McGraw-Hill, 1973.

[18] L. V. Ahlfors. *Conformal invariants*. New York: McGraw-Hill, 1979.

[19] J. M. Anderson and J. Rovnyak. On generalized Schwarz-Pick estimates, *Mathematika* **53** (2006), 161–168.

[20] J. M. Anderson and A. Vasil'ev. Lower Schwarz-Pick estimates and angular derivatives, *Ann. Acad. Sci. Fenn., Math.* **33** (2008), 101–110.

[21] A. Andreotti and E. Vesentini. On deformations of discontinuous groups, *Acta Math.* **112** (1964), 249–298.

[22] L. Arosio and F. Bracci. Canonical models for holomorphic iteration, *Trans. Am. Math. Soc.* **368** (2016), 3305–3339.

[23] C. Arteaga. On a theorem of Ritt for commuting finite Blaschke products, *Complex Var. Theory Appl.* **48** (2003), 671–679.

[24] C. Arzelà. Sulle funzioni di linee, *Mem. Accad. Sci. Ist. Bologna Cl. Sci. Fis. Mat.* **5** (1895), 55–74.

[25] G. Ascoli. Le curve limite di una varietà data di curve, *Atti R. Accad. Lincei Mem. Cl. Sci. Fis. Mat. Nat.* **18** (1883–1884), 521–586.

[26] G. Aumann and C. Carathéodory. Ein Satz über konforme Abbildung mehrfach zusammenhängender ebener Gebiete, *Math. Ann.* **109** (1934), 756–763.

https://doi.org/10.1515/9783110601978-007

[27] F. G. Avkhadiev and K.-J. Wirths. *Schwarz-Pick type inequalities*. Basel: Birkhäuser, 2009.

[28] I. Ba and V. V. Goryaĭnov. Semigroup of conformal mappings of the upper half-plane into itself with hydrodynamic normalization at infinity, *Ukr. Math. J.* **44** (1992), 1209–1217.

[29] I. N. Baker. The iteration of entire transcendental functions and the solutions of the functional equation $f\{f(z)\} = F(z)$, *Math. Ann.* **129** (1955), 174–180.

[30] I. N. Baker. Zusammensetzungen ganzer Funktionen, *Math. Z.* **69** (1958), 121–163.

[31] I. N. Baker. Fixpoints and iterates of entire functions, *Math. Z.* **71** (1959), 146–153.

[32] I. N. Baker. The existence of fixpoints of entire functions, *Math. Z.* **73** (1960), 280–284.

[33] I. N. Baker. Fractional iteration near a fixpoint of multiplier 1, *J. Aust. Math. Soc.* **4** (1964), 143–148.

[34] I. N. Baker and Ch. Pommerenke. On the iteration of analytic functions in a half-plane, II, *J. Lond. Math. Soc.* **20** (1979), 255–258.

[35] I. N. Baker and P. J. Rippon. Towers of exponents and other composite maps, *Complex Var. Theory Appl.* **12** (1989), 181–200.

[36] I. N. Baker and P. J. Rippon. On compositions of analytic self-mappings of a convex domain, *Arch. Math. (Basel)* **55** (1990), 380–386.

[37] K. Barański, N. Fagella, X. Jarque and B. Karpińska. Absorbing sets and Baker domains for holomorphic maps, *J. Lond. Math. Soc.* **92** (2014), 144–162.

[38] K. Barański, N. Fagella, X. Jarque and B. Karpińska. On the connectivity of the Julia sets of meromorphic functions, *Invent. Math.* **198** (2014), 591–636.

[39] L. Baribeau, P. Rivard and E. Wegert. On hyperbolic divided differences and the Nevanlinna-Pick problem, *Comput. Methods Funct. Theory* **9** (2009), 391–405.

[40] M. Basallote, M. D. Contreras and C. Hernández-Mancera. Commuting finite Blaschke products with no fixed points in the unit disk, *J. Math. Anal. Appl.* **359** (2009), 547–555.

[41] A. F. Beardon. Iteration of contractions and analytic maps, *J. Lond. Math. Soc.* **41** (1990), 141–150.

[42] A. F. Beardon. Iteration of analytic Euclidean contractions. In: *Computational methods and function theory 1994 (Penang)*, vol. 5. Ser. Approx. Decompos. River Edge, NJ: World Sci. Publ., 1995, 57–74.

[43] A. F. Beardon. The Schwarz-Pick lemma for derivatives, *Proc. Am. Math. Soc.* **125** (1997), 3255–3256.

[44] A. F. Beardon. Continued fractions, discrete groups and complex dynamics, *Comput. Methods Funct. Theory* **1** (2001), 535–594.

[45] A. F. Beardon. Repeated compositions of analytic maps, *Comput. Methods Funct. Theory* **1** (2001), 235–248.

[46] A. F. Beardon. Semi-groups of analytic maps, *Comput. Methods Funct. Theory* **1** (2001), 249–258.

[47] A. F. Beardon and T. K. Carne. A strengthening of the Schwarz-Pick inequality, *Am. Math. Mon.* **99** (1992), 216–217.

[48] A. F. Beardon and D. Minda. A multi-point Schwarz-Pick lemma, *J. Anal. Math.* **92** (2004), 81–104.

[49] A. F. Beardon and D. Minda, The hyperbolic metric and geometric function theory. In: *Quasiconformal mappings and their applications*. New Delhi: Narosa, 2007, 9–56.

[50] A. F. Beardon and D. Minda, Conformal automorphisms of finitely connected regions. In: *Transcendental dynamics and complex analysis*. Cambridge: Cambridge Univ. Press, 2008, 37–73.

[51] A. F. Beardon and D. Minda. Dieudonné points of holomorphic self-maps of regions, *Comput. Methods Funct. Theory* **8** (2008), 409–432.

[52] A. F. Beardon, T. K. Carne, D. Minda and T. W. Ng. Random iteration of analytic maps, *Ergod. Theory Dyn. Syst.* **24** (2004), 659–675.

[53] D. F. Behan. Commuting analytic functions without fixed points, *Proc. Am. Math. Soc.* **37** (1973), 114–120.
[54] S. R. Bell and S. G. Krantz. Smoothness to the boundary of conformal maps. *Rocky Mount. J. Math.* **17** (1983), 23–40.
[55] E. Beltrami. Saggio di interpretazione della geometria non-euclidea, *G. Mat. Battaglini* **6** (1868), 248–312.
[56] E. Beltrami. Teoria fondamentale degli spazi di curvatura costante, *Ann. Mat. Pura Appl.* **2** (1868), 232–255.
[57] C. Bénéteau, A. Dahlner and D. Khavinson. Remarks on the Bohr phenomenon, *Comput. Methods Funct. Theory* **4** (2004), 1–19.
[58] A. M. Benini, V. Evdoridou, N. Fagella, P. J. Rippon and G. M. Stallard. Classifying simply connected wandering domains, *Math. Ann.* **383** (2022). 1127–1178.
[59] A. M. Benini, V. Evdoridou, N. Fagella, P. J. Rippon and G. M. Stallard. The Denjoy-Wolff set for holomorphic sequences, non-autonomous dynamical systems and wandering domains. Preprint, ArXiv:2203.06235 (2022).
[60] W. Bergweiler. Singularities in Baker domains, *Comput. Methods Funct. Theory* **1** (2001), 41–49.
[61] E. Berkson and H. Porta. Semigroups of analytic functions and composition operators, *Mich. Math. J.* **25** (1978), 101–115.
[62] D. Betsakos. Hyperbolic geometric versions of Schwarz's lemma, *Conform. Geom. Dyn.* **17** (2013), 119–132.
[63] D. Betsakos. On the images of horodisks under holomorphic self-maps of the unit disk, *Arch. Math.* **102** (2014), 91–99.
[64] D. Betsakos. Geometric description of the classification of holomorphic semigroups, *Proc. Am. Math. Soc.* **144** (2016), 1595–1604.
[65] L. Bianchi. *Lezioni di geometria differenziale*. Bologna: Zanichelli, 1927.
[66] L. Bieberbach. Über einen Satz des Herrn Carathéodory, *Nachr. Ges. Wiss. Gött.* **1913** (1913), 552–560.
[67] C. Bisi and G. Gentili. Boundary construction of petals at the Wolff point in the parabolic case, *J. Anal. Math.* **104** (2008), 1–11.
[68] H. P. Boas. Julius and Julia: mastering the art of the Schwarz lemma, *Am. Math. Mon.* **117** (2010), 770–785.
[69] V. Bolotnikov, M. Elin and D. Shoikhet. Inequalities for angular derivatives and boundary interpolation, *Anal. Math. Phys.* **3** (2013), 63–96.
[70] P. Bonfert. On iteration in planar domains, *Mich. Math. J.* **44** (1997), 47–68.
[71] L. E. Böttcher. The principal laws of convergence of iterates and their application to analysis, *Izv. Kazan. Fiz.-Mat. Obshch.* **14** (1904), 155–234.
[72] P. S. Bourdon, V. Matache and J. H. Shapiro. On convergence to the Denjoy-Wolff point, *Ill. J. Math.* **49** (2005), 405–430.
[73] P. S. Bourdon and J. H. Shapiro. *Cyclic phenomena for composition operators*. Vol. 125. Mem. Amer. Math. Soc. Providence, RI: American Mathematical Society, 1997.
[74] W. M. Boyce. Commuting functions with no common fixed points, *Trans. Am. Math. Soc.* **137** (1969), 77–92.
[75] F. Bracci. Holomorphic dependence of the Heins map, *Compl. Var.* **47** (2002), 1115–1119.
[76] F. Bracci. Fixed points of commuting holomorphic mappings other than the Wolff point, *Trans. Am. Math. Soc.* **355** (2003), 2569–2584.
[77] F. Bracci, M. D. Contreras and S. Díaz-Madrigal. Pluripotential theory, semigroups and boundary behavior of infinitesimal generators in strongly convex domains, *J. Eur. Math. Soc.* **12** (2010), 23–53.

[78] F. Bracci, M. D. Contreras and S. Díaz-Madrigal. Evolution families and the Loewner equation I: the unit disc, *J. Reine Angew. Math.* **672** (2012), 1–37.
[79] F. Bracci, M. D. Contreras and S. Díaz-Madrigal. Topological invariants for semigroups of holomorphic self-maps of the unit disk, *J. Math. Pures Appl.* **107** (2017), 78–99.
[80] F. Bracci, M. D. Contreras, and S. Díaz-Madrigal. *Continuous semigroups of holomorphic self-maps of the unit disc*. Berlin: Springer, 2020.
[81] F. Bracci, G. Gentili and P. Poggi-Corradini. Valiron's construction in higher dimension, *Rev. Mat. Iberoam.* **26** (2010), 57–76.
[82] F. Bracci, D. Kraus and O. Roth. A new Schwarz-Pick lemma at the boundary and rigidity of holomorphic maps, Preprint, arXiv:2003.02019v1 (2020).
[83] F. Bracci and P. Poggi-Corradini, *On Valiron's theorem*. In: *Future trends in geometric function theory*,vol. 92. Rep. Univ. Jyväskylä Dep. Math. Stat. Jyväskylä: Univ. Jyväskylä, 2003, 39–55.
[84] F. Bracci and O. Roth. Semigroup-fication of univalent self-maps of the unit disc. To appear in *Ann. Inst. Fourier (Grenoble)*. Preprint arXiv:2002.08252 (2022).
[85] F. Bracci, R. Tauraso and F. Vlacci. Identity principles for commuting holomorphic self-maps of the unit disc, *J. Math. Anal. Appl.* **270** (2002), 451–473.
[86] H. Brolin. Invariant sets under iteration of rational functions, *Ark. Mat.* **6** (1965), 103–144.
[87] E. Bujalance, J. J. Etayo and E. Martínez. Automorphism groups of hyperelliptic Riemann surfaces, *Kodai Math. J.* **10** (1987), 174–181.
[88] R. B. Burckel. *An introduction to classical complex analysis*. New York: Academic Press, 1979.
[89] R. B. Burckel. Iterating self-maps of the disk, *Am. Math. Mon.* **88** (1981), 396–407.
[90] R. B. Burckel, D. E. Marshall, D. Minda, P. Poggi-Corradini and T. J. Ransford. Area, capacity and diameter versions of Schwarz's lemma, *Conform. Geom. Dyn.* **12** (2008), 133–152.
[91] D. M. Burns jr. and S. G. Krantz. Rigidity of holomorphic mappings and a new Schwarz lemma at the boundary, *J. Am. Math. Soc.* **7** (1994), 661–676.
[92] C. Carathéodory. Sur quelques généralisation du théorème de M. Picard, *C. R. Acad. Sci. Paris* **141** (1905), 1213–1215.
[93] C. Carathéodory. Untersuchungen über die konformen Abbildungen von festen und veränderlichen Gebieten, *Math. Ann.* **72** (1912), 107–144.
[94] C. Carathéodory. Über die gegenseitige Beziehung der Ränder bei der konformen Abbildung des Inneren einer Jordanschen Jurve auf einen Kreis, *Math. Ann.* **73** (1913), 305–320.
[95] C. Carathéodory. Über die Begrenzung einfach zusammenhängender Gebiete, *Math. Ann.* **73** (1913), 323–370.
[96] C. Carathéodory. Zur Ränderzuordnung bei konformer Abbildung, *Nachr. Ges. Wiss. Gött.* **1913** (1913), 509–518.
[97] C. Carathéodory. Über die Winkelderivierten von beschränkten analytischen Funktionen, *Sitzungsber. Preuss. Akad. Wiss. Berlin* **1929** (1929), 39–54.
[98] C. Carathéodory. Über die Abbildungen, die durch Systeme von analytischen Funktionen von mehreren Veränderlichen erzeugt werden, *Math. Z.* **34** (1932), 758–792.
[99] C. Carathéodory. *Theory of functions of a complex variable, II*. New York: Chelsea, 1960.
[100] C. Carathéodory. *Conformal representation*. Second edition. Mineola, NY: Dover, 1998.
[101] C. Carathéodory and E. Landau. Beiträge zur Konvergenz von Funktionenfolgen, *Sitzungsber. Kön. Preuss. Akad. Wiss. Berlin* **1911** (1911), 587–613.
[102] L. Carleson. An interpolation problem for bounded analytic functions, *Am. J. Math.* **80** (1958), 921–930.
[103] H. Cartan. Les trasformation analytiques des domaines cerclés les uns dans les autres, *C. R. Acad. Sci. Paris* **190** (1930), 718–720.
[104] H. Cartan. Sur le fonctions de deux variables complexes. Les trasformations d'un domaine borné *D* en un domaine intérieur à *D*, *Bull. Soc. Math. Fr.* **58** (1930), 199–219.

[105] H. Cartan. Les fonctions de deux variables complexes. L'itération des transformations intérieures d'un domaine borné, *Math. Z.* **35** (1932), 760–773.

[106] H. Cartan. *Sur les groupes de transformations analytiques*. Paris: Hermann, 1935.

[107] A.-L. Cauchy. Mémoire sur les fonctions complémentaires, *C. R. Acad. Sci. Paris* **19** (1844), 1377–1384.

[108] A. Cayley. Sur quelques propriétés des déterminants gauches, *J. Reine Angew. Math.* **32** (1846), 119–123.

[109] E. Cesàro. Sur la convergence des séries, *Nouv. Ann. Math.* **7** (1888), 49–59.

[110] I. Chalendar and R. Mortini. When do finite Blaschke products commute?, *Bull. Aust. Math. Soc.* **64** (2001), 189–200.

[111] D. Chelst. A generalized Schwarz lemma at the boundary, *Proc. Am. Math. Soc.* **129** (2001), 3275–3278.

[112] C. H. Chen, S.-Y. Cheng and K. H. Look. On the Schwarz lemma for complete Kähler manifolds, *Acta Sci. Sin.* **22** (1979), 1238–1247.

[113] H. Chen, On the Bloch constant. In: *Approximation, complex analysis, and potential theory*, ed. by N. Arakelian et al. Amsterdam: Kluwer, 2001, 129–161.

[114] K. H. Cho, S.-A. Kim and T. Sugawa. On a multi-point Schwarz-Pick lemma, *Comput. Methods Funct. Theory* **12** (2012), 483–499.

[115] A. Christodoulou and I. Short. A hyperbolic-distance inequality for holomorphic maps, *Ann. Acad. Sci. Fenn., Math.* **44** (2019), 293–300.

[116] A. Christodoulou and I. Short. Stability of the Denjoy-Wolff theorem, *Ann. Acad. Sci. Fenn., Math.* **46** (2021), 421–431.

[117] A. Cohn. Über die Anzahl der Wurzeln einer algebraischen Gleichung in einem Kreise, *Math. Z.* **14** (1922), 110–148.

[118] M. D. Contreras and S. Díaz-Madrigal. Analytic flows on the unit disk: angular derivatives and boundary fixed points, *Pac. J. Math.* **222** (2005), 253–286.

[119] M. D. Contreras, S. Díaz-Madrigal and Ch. Pommerenke, Iteration in the unit disk: the parabolic zoo. In: *Complex and harmonic analysis*, ed. by Anthony Carbery et al. Lancaster, PA: DEStech Publications, Inc., 63–91.

[120] M. D. Contreras, S. Díaz-Madrigal and Ch. Pommerenke. Some remarks on the Abel equation in the unit disk, *J. Lond. Math. Soc.* **75** (2007), 623–634.

[121] M. D. Contreras, S. Díaz-Madrigal and Ch. Pommerenke. Second angular derivatives and parabolic iteration in the unit disk, *Trans. Am. Math. Soc.* **362** (2010), 357–388.

[122] M. D. Contreras, S. Díaz-Madrigal and Ch. Pommerenke. Boundary behavior of the iterates of a self-map of the unit disk, *J. Math. Anal. Appl.* **396** (2012), 93–97.

[123] M. D. Contreras, S. Díaz-Madrigal, M. J. Martín and D. Vukotić, Holomorphic self-maps of the disk intertwining two linear fractional maps. In: *Topics in complex analysis and operator theory*, Vol. 561. Providence, RI: Amer. Math. Soc., 2012, 199–227.

[124] J. B. Conway. *Functions of one complex variable*. Second edition. Berlin: Springer, 1995.

[125] J. B. Conway. *Functions of one complex variable II*. Berlin: Springer, 1995.

[126] C. C. Cowen. Iteration and the solution of functional equations for functions analytic in the unit disk, *Trans. Am. Math. Soc.* **265** (1981), 69–95.

[127] C. C. Cowen. Analytic solutions of Böttcher's functional equation in the unit disk, *Aequ. Math.* **24** (1982), 187–194.

[128] C. C. Cowen. Composition operators on H^2, *J. Oper. Theory* **9** (1983), 77–106.

[129] C. C. Cowen. Commuting analytic functions, *Trans. Am. Math. Soc.* **283** (1984), 685–695.

[130] C. C. Cowen and Ch. Pommerenke. Inequalities for the angular derivative of an analytic function in the unit disk, *J. Lond. Math. Soc.* **26** (1982), 271–289.

[131] C. Craig jr. and A. J. Macintyre. Inequalities for functions regular and bounded in a circle, *Pac. J. Math.* **20** (1967), 449–454.

[132] H. Cremer. Zum Zentrumproblem, *Math. Ann.* **98** (1927), 151–163.

[133] H. Cremer. Über die Haüfigkeit der Nichtzentren, *Math. Ann.* **115** (1938), 573–580.

[134] S. Dai and Y. Pan. Note on Schwarz-Pick estimates for bounded and positive real part analytic functions, *Proc. Am. Math. Soc.* **136** (2008), 635–640.

[135] A. Denjoy. Sur l'itération des fonctions analytiques, *C. R. Acad. Sci. Paris* **182** (1926), 255–257.

[136] J. Dieudonné. Recherches sur quelques problèmes relatifs aux polynômes et aux fonctions bornées d'une variable complexe, *Ann. Sci. Éc. Norm. Supér.* **48** (1931), 247–358.

[137] C. I. Doering and R. Mañé. *The dynamics of inner functions*. Vol. 3. Ensaios Matemáticos. Rio de Janeiro: Soc. Brasileira de Mat., 1991.

[138] V. N. Dubinin. The Schwarz inequality on the boundary for functions regular in the disk, *J. Math. Sci.* **122** (2004), 3623–3629.

[139] P. L. Duren. *Univalent functions*. Berlin: Springer, 1983.

[140] M. Elin and D. Shoikhet. *Linearization models for complex dynamical systems. Topics in univalent functions, functional equations and semigroup theory*. Basel: Birkhäuser, 2010.

[141] M. Elin, F. Jacobzon, M. Levenshtein and D. Shoikhet, The Schwarz lemma: rigidity and dynamics. In: *Harmonic and complex analysis and its applications*, ed. by A. Vasil'ev. Berlin: Springer, 2014, 135–230.

[142] D. B. A. Epstein. Curves on 2-manifolds and isotopies, *Acta Math.* **115** (1966), 83–107.

[143] H. M. Farkas and I. Kra. *Riemann Surfaces*. Berlin: Springer, 1980.

[144] J. Farkas. Sur les fonctions itératives, *J. Math. Pures Appl.* **10** (1884), 101–108.

[145] P. Fatou. Séries trigonométriques et séries de Taylor, *Acta Math.* **30** (1906), 335–400.

[146] P. Fatou. Sur les équations fonctionnelles. Premier mémoire, *Bull. Soc. Math. Fr.* **47** (1919), 161–271.

[147] P. Fatou. Sur les équations fonctionnelles. Deuxième mémoire, *Bull. Soc. Math. Fr.* **48** (1920), 33–94.

[148] P. Fatou. Sur les équations fonctionnelles. Troisième mémoire, *Bull. Soc. Math. Fr.* **48** (1920), 208–314.

[149] P. Fatou. Sur l'itération des fonctions trascendantes entières, *Acta Math.* **47** (1926), 337–370.

[150] M. Fekete. Über die Verteilung der Wurzeln bei gewissen algebraischen Gleichungen mit ganzzahligen Koeffizienten, *Math. Z.* **17** (1923), 228–249.

[151] J. L. Fernández, M. V. Melián and D. Pestana. Quantitative mixing results and inner functions, *Math. Ann.* **337** (2007), 233–251.

[152] J. Ferrand. Sur l'itération des fonctions analytiques, *C. R. Acad. Sci. Paris* **212** (1941), 1068–1071.

[153] G. R. Ferreira. A note on forward iteration of inner functions, Preprint, ArXiv:2206.00374v1 (2022).

[154] O. Forster. *Lectures on Riemann surfaces*. New York: Springer, 1981.

[155] R. Fournier and O. Roth. Jack and Julia, *Contemp. Math.* **743** (2020), 213–216.

[156] T. Franzoni and E. Vesentini. *Holomorphic maps and invariant distances*. Vol. 69. Notas de Matematica. Amsterdam: North-Holland, 1980.

[157] M. Fréchet. Sur quelques points du calcul fonctionnel, *Rend. Circ. Mat. Palermo* **22** (1906), 1–74.

[158] A. Frovlova, M. Levenshtein, D. Shoikhet and A. Vasil'ev. Boundary distorsion estimates for holomorphic maps, *Complex Anal. Oper. Theory* **8** (2014), 1129–1149.

[159] D. Gans. *An introduction to non-euclidean geometry*. New York: Academic Press, 1973.

[160] J. Gerretsen and G. Sansone. *Lectures on the theory of functions of a complex variable, II*. Groningen: Wolters-Noordhoff, 1969.

[161] J. Gill. Compositions of analytic functions of the form $F_n(z) = F_{n-1}(f_n(z)), f_n(z) \to f(z)$, *J. Comput. Appl. Math.* **23** (1988), 179–184.

[162] J. Gill. Limit periodic iteration, *Appl. Numer. Math.* **4** (1988), 297–308.

[163] J. Gill. Inner composition of analytic mappings on the unit disk, *Int. J. Math. Math. Sci.* **14** (1991), 221–226.
[164] J. Gill. The use of the sequence $F_n(z) = f_n \circ \cdots \circ f_1(z)$ in computing fixed points of continued fractions, products, and series, *Appl. Numer. Math.* **8** (1991), 469–476.
[165] J. Gill. Outer compositions of hyperbolic/loxodromic linear fractional transformations, *Int. J. Math. Math. Sci.* **15** (1992), 819–822.
[166] J. Gill. Convergence of infinite compositions of complex functions, *Commun. Anal. Theory Contin. Fract.* **19** (2012), 1–27.
[167] J. Gill. A primer on the elementary theory of infinite compositions of complex functions, Preprint, https://www.researchgate.net/profile/John_Gill6/publications (2017).
[168] G. M. Goluzin. Some estimations of derivatives of bounded functions, *Mat. Sb.* **58** (1945), 295–306.
[169] G. M. Goluzin. *Geometric theory of functions of a complex variable*. Providence: American Mathematical Society, 1969.
[170] V. V. Goryaĭnov. Semigroups of conformal mappings, *Math. USSR Sb.* **57** (1987), 463–483.
[171] V. V. Goryaĭnov, One-parameter semigroups of analytic functions. In: *Geometric function theory and applications of complex analysis in mechanics: studies in complex analysis and its applications to partial differential equations 2*, ed. by R. Kühnau and W. Tutschke. Harlow: Longman Sci. Tech., 1991, 160–164.
[172] V. V. Goryaĭnov. Fractional iterates of functions that are analytic in the unit disk with given fixed points, *Math. USSR Sb.* **74** (1993), 29–46.
[173] M. Goto. On an arcwise connected subgroup of a Lie group, *Proc. Am. Math. Soc.* **20** (1969), 157–162.
[174] H. Grauert and H. Rieckziegel. Hermitesche Metriken und normale Familien holomorphen Abbildungen, *Math. Z.* **89** (1965), 108–125.
[175] L. Greenberg. Conformal transformations of Riemann surfaces, *Am. J. Math.* **82** (1960), 749–760.
[176] H. Grunsky. Zur konformen Abbildung von Gebieten, die in einer Richtung konvex sind, *J. Math. Anal. Appl.* **34** (1971), 685–701.
[177] P. Gumenyuk. Parametric representation of univalent functions with boundary regular fixed points, *Constr. Approx.* **46** (2017), 435–458.
[178] P. Gumenyuk and D. Prokhorov. Value regions of univalent self-maps with two boundary fixed points, *Ann. Acad. Sci. Fenn., Math.* **43** (2018), 451–462.
[179] J. Hadamard. Two works on iteration and related questions, *Bull. Am. Math. Soc.* **50** (1944), 67–75.
[180] A. Harnack. Existenzbeweise zur Theorie des Potentiales in der Ebene und im Raume, *Ber. Verh. K. Sächs. Ges. Wiss. Leipzig* **1886** (1886), 144–169.
[181] W. K. Hayman. Some remarks on Schottky's theorem, *Proc. Camb. Philos. Soc.* **43** (1947), 442–454.
[182] M. H. Heins. Extremal problems for functions analytic and single-valued in a doubly connected region, *Am. J. Math.* **62** (1940), 91–106.
[183] M. H. Heins. A generalization of the Aumann-Carathéodory Starrheitssatz, *Duke Math. J.* **8** (1941), 312–316.
[184] M. H. Heins. On the iteration of functions which are analytic and single-valued in a given multiply-connected region, *Am. J. Math.* **63** (1941), 461–480.
[185] M. H. Heins. On the number of 1-1 directly conformal maps, which a multiply plane connected region of finite connectivity $p > 2$ admits onto itself, *Bull. Am. Math. Soc.* **52** (1946), 454–457.
[186] M. H. Heins. On a problem of Heinz Hopf, *J. Math. Pures Appl.* **37** (1958), 153–160.
[187] M. H. Heins. On a class of conformal metrics, *Nagoya Math. J.* **21** (1962), 1–60.

[188] M. H. Heins. On the pseudoperiods of the Weierstrass zeta function, *SIAM J. Numer. Anal.* **3** (1966), 266–268.

[189] M. H. Heins. On the pseudoperiods of the Weierstrass zeta function, II, *Nagoya Math. J.* **30** (1967), 113–119.

[190] M. H. Heins, Semigroups of holomorphic maps of a Riemann surface into itself which are homomorphs of the set of positive reals considered additively. In: *E. B. Christoffel — The influence of his work on mathematics and the physical sciences.* Basel: Birkhäuser, 1981, 314–331.

[191] M. H. Heins, A theorem of Wolff-Denjoy type. In: *Complex analysis.* Basel: Birkhäuser, 1988, 81–86.

[192] D. A. Hejhal. Universal covering maps for variable regions, *Math. Z.* **137** (1974), 7–20.

[193] G. Herglotz. Über Potenzreihen mit positivem reellen, Teil im Einheitskreis, *Ber. Verh. K. Sächs. Ges. Wiss. Leipzig* **63** (1911), 501–511.

[194] M. Hervé. Quelques propriétés des transformations intérieurs d'un domaine borné, *Ann. Sci. Éc. Norm. Supér.* **68** (1951), 125–168.

[195] A. Herzig. Die Winkelderivierte und das Poisson-Stieltjes Integral, *Math. Z.* **46** (1940), 129–156.

[196] P. J. Higgins. *Introduction to topological groups.* Vol. 15. London Mathematical Society Lecture Notes Series. London: Cambridge University Press, 1974.

[197] D. Hilbert. Über das Dirichletsche Prinzip, *Math. Ann.* **59** (1904), 161–186.

[198] K. Hoffman. *Banach spaces of analytic functions.* Second edition. Englewood Cliffs, N.J: Prentice-Hall, 1990.

[199] W. E. Hornor. Invariant measures for univalent self-maps of the unit disc, *Complex Anal. Oper. Theory* **1** (2007), 549–569.

[200] P. Hriljac. Analytic dynamics in some pseudoconvex domains, *Nonlinear Anal.* **19** (1992), 717–730.

[201] H. Huber. Über analytische Abbildungen Riemannscher Flächen in sich, *Comment. Math. Helv.* **27** (1953), 1–72.

[202] J. P. Huneke. On common fixed points of commuting continuous functions on an interval, *Trans. Am. Math. Soc.* **139** (1969), 371–381.

[203] A. Hurwitz. Über die Nullstellen der Bessel'schen Funktionen, *Math. Ann.* **33** (1889), 246–266.

[204] A. Hurwitz. Über algebraische Gebilde mit eindeutigen Transformationen in sich, *Math. Ann.* **41** (1893), 403–442.

[205] E. Jabotinsky. Analytic iteration, *Trans. Am. Math. Soc.* **108** (1963), 457–477.

[206] I. S. Jack. Functions starlike and convex of order α, *J. Lond. Math. Soc.* **3** (1971), 469–474.

[207] M. Jacques and I. Short. Repeated compositions of Möbius transformations, *Ergod. Theory Dyn. Syst.* **40** (2020), 437–452.

[208] M. Jacques and I. Short. Semigroups of isometries of the hyperbolic plane, *Int. Math. Res. Not.* **2022** (2022), 6403–6463.

[209] F. Jafari, Z. Słodkowski and T. Tonev. A new proof for the description of holomorphic flows on multiply connected domains, *Int. J. Math. Math. Sci.* **2014** (2014). Art. ID 160579, 1–5.

[210] M. Jarnicki and P. Pflug. *Invariant distances and metrics in complex analysis.* Second edition. Berlin: de Gruyter, 2013.

[211] J. A. Jenkins and N. Suita. On analytic self-mappings of Riemann surfaces, *Math. Ann.* **202** (1973), 37–56.

[212] J. A. Jenkins and N. Suita. On analytic self-mappings of Riemann surfaces II, *Math. Ann.* **209** (1974), 109–115.

[213] F. B. Jones. Connected and disconnected plane sets and the functional equation $f(x) + f(y) = f(x + y)$, *Bull. Am. Math. Soc.* **48** (1942), 115–120.

[214] V. Jørgensen. On an inequality for the hyperbolic measure and its applications in the theory of functions, *Math. Scand.* **4** (1956), 113–124.

[215] G. Julia. Mémoire sur l'itération des fonctions rationnelles, *J. Math. Pures Appl.* **1** (1918), 47–245.

[216] G. Julia. Extension nouvelle d'un lemme de Schwarz, *Acta Math.* **42** (1920), 349–355.

[217] G. Julia. *Leçons sur les fonctions uniformes à point singulier essentiel isolé*. Paris: Gauthier-Villars, 1924.

[218] G. Julia. *Principes géométriques d'analyse, I*. Paris: Gounod, 1930.

[219] G. Julia. Sur la représentation conforme des aires multiplement connexes, *Ann. Sc. Norm. Super. Pisa* **1** (1932), 113–138.

[220] G. Julia. *Leçons sur la réprésentations conforme des aires multiplement connexes*. Paris: Gauthier-Villars, 1934.

[221] H. T. Kaptanoğlu. Some refined Schwarz-Pick lemmas, *Mich. Math. J.* **50** (2002), 649–664.

[222] S. Karlin and J. McGregor. Embedding iterates of analytic functions with two fixed points into continuous groups, *Trans. Am. Math. Soc.* **132** (1968), 137–145.

[223] L. Keen and N. Lakic, Accumulation points of iterated function systems. In: *Complex dynamics*, vol. 396. Providence, RI: Amer. Math. Soc., 2006, 101–113.

[224] L. Keen and N. Lakic. Random holomorphic iterations and degenerate subdomains of the unit disk, *Proc. Am. Math. Soc.* **134** (2006), 371–378.

[225] L. Keen and N. Lakic. Accumulation constants of iterated function systems with Bloch target domains, *Ann. Acad. Sci. Fenn., Math.* **32** (2007), 73–82.

[226] L. Keen and N. Lakic. *Hyperbolic geometry from a local viewpoint*. Cambridge: Cambridge University Press, 2007.

[227] J. L. Kelley. *General topology*. Princeton, NJ: Van Nostrand, 1955.

[228] D. Kellogg. Harmonic functions and Green's integral, *Trans. Am. Math. Soc.* **13** (1912), 109–132.

[229] P. Kelly and G. Matthews. *The non-euclidean, hyperbolic plane*. Berlin: Springer, 1981.

[230] V. Khatskevich, S. Reich and D. Shoikhet. Abel-Schröder equations for linear fractional mappings and the Koenigs embedding problem, *Acta Sci. Math. (Szeged)* **69** (2003), 67–98.

[231] P. J. Kiernan. Hyperbolically imbedded spaces and the big Picard theorem, *Math. Ann.* **204** (1973), 203–209.

[232] H. O. Kim and H. S. Kim. Iterations, Denjoy-Wolff points and ergodic properties of point-mass singular inner functions, *J. Math. Anal. Appl.* **175** (1993), 105–117.

[233] K.-T. Kim and S. G. Krantz. The automorphism groups of domains, *Am. Math. Mon.* **12** (2005), 585–601.

[234] T. Kimura. On the iteration of analytic functions, *Funkc. Ekvacioj* **14** (1971), 197–238.

[235] W. A. Kirk and M. A. Khamsi. *An introduction to metric spaces and fixed point theory*. New York: Wiley, 2001.

[236] H. Kneser. Reelle analytische Lösungen der Gleichung $\phi(\phi(x)) = e^x$ und verwandter Funktional-gleichungen, *J. Reine Angew. Math.* **187** (1949), 56–67.

[237] S. Kobayashi. *Hyperbolic manifolds and holomorphic mappings*. New York: Dekker, 1970.

[238] S. Kobayashi. Intrinsic distances, measures and geometric function theory, *Bull. Am. Math. Soc.* **82** (1976), 357–416.

[239] S. Kobayashi. *Hyperbolic complex spaces*. Berlin: Springer, 1998.

[240] P. Kœbe. Über die Uniformisierung beliebiger analytischen Kurven, I, *Nachr. Ges. Wiss. Gött.* **1907** (1907), 191–210.

[241] P. Kœbe. Über die Uniformisierung beliebiger analytischen Kurven, III, *Nachr. Ges. Wiss. Gött.* **1908** (1908), 337–358.

[242] P. Kœbe. Abhandlungen zur Theorie der konformen Abbildung, II: Die Fundamentalabbildung beliebiger mehrfach zusammenhängender schlichter Bereiche nebst einer Anwendung auf die Bestimmung algebraischer Funktionen zu gegebener Riemannscher Fläche, *Acta Math.* **40** (1916), 251–290.
[243] P. Kœbe. Abhandlungen zur Theorie der konformen Abbildung, IV: Abbildung mehrfach zusammenhängender schlichter Bereiche auf Schlitzbereiche, *Acta Math.* **41** (1918), 305–344.
[244] G. Kœnigs. Recherches sur les substitutions uniformes, *Bull. Sci. Math.* **7** (1883), 340–357.
[245] Y. Komatu. On angular derivative, *Kodai Math. Semin. Rep.* **13** (1961), 167–179.
[246] Y. Komatu and A. Mori. Conformal rigidity of Riemann surfaces, *J. Math. Soc. Jpn.* **4** (1952), 302–309.
[247] H. König. Conformal conjugacies in Baker domains, *J. Lond. Math. Soc.* **59** (1999), 153–170.
[248] A. Korányi. Harmonic functions on hermitian hyperbolic spaces, *Trans. Am. Math. Soc.* **135** (1969), 507–516.
[249] I. Kra. *Automorphic forms and Kleinian groups*. Reading, Mass.: Benjamin, 1972.
[250] S. G. Krantz. The Schwarz lemma at the boundary, *Complex Var. Elliptic Equ.* **56** (2011), 455–468.
[251] D. Kraus. Critical sets of bounded analytic functions, zero sets of Bergman spaces and nonpositive curvature, *Proc. Lond. Math. Soc.* **106** (2013), 931–956.
[252] D. Kraus and O. Roth. Maximal Blaschke products, *Adv. Math.* **241** (2013), 58–78.
[253] P. P. Kufarev. On one-parameter families of analytic functions, *Mat. Sb.* **13** (1943), 87–118.
[254] A. Kuznetsov. Semi-groups of analytic functions that contain the identity map, *Comput. Methods Funct. Theory* **7** (2007), 239–247.
[255] E. Landau and G. Valiron. A deduction from Schwarz's lemma, *J. Lond. Math. Soc.* **4** (1929), 162–163.
[256] H. J. Landau and R. Osserman. On analytic mappings of Riemann surfaces, *J. Anal. Math.* **7** (1959/60), 249–279.
[257] S. Lang. *An introduction to complex hyperbolic spaces*. New York: Springer, 1987.
[258] F. Lárusson. A Wolff-Denjoy theorem for infinitely connected Riemann surfaces, *Proc. Am. Math. Soc.* **124** (1996), 2745–2750.
[259] A. Lecko and B. Uzar. A note on Julia-Carathéodory theorem for functions with fixed initial coefficients, *Proc. Jpn. Acad., Ser. A, Math. Sci.* **89** (2013), 133–137.
[260] J. M. Lee. *Introducton to topological manifolds*. Berlin: Springer-Verlag, 2000.
[261] O. Lehto and K. I. Virtanen. Boundary behaviour and normal meromorphic functions, *Acta Math.* **97** (1957), 47–65.
[262] L. Lempert, Intrinsic distances and holomorphic retracts. In: *Complex analysis and applications '81, Varna*. Sofia: Bulgarian Academy of Sciences, 1984, 341–364.
[263] K. Leschinger. Über Fixpunkte holomorpher Automorphismen, *Manuscr. Math.* **25** (1978), 391–396.
[264] P. Lévy. Fonctions à croissance régulière et itération d'ordre fractionnaire, *Ann. Mat. Pura Appl.* **5** (1928), 269–298.
[265] J. Lewittes. An extension of hyperbolic geometry and Julia's theorem, *Duke Math. J.* **35** (1968), 777–782.
[266] K. Y. Li. Inequalities for fixed points of holomorphic functions, *Bull. Lond. Math. Soc.* **22** (1990), 446–452.
[267] E. Lindelöf. Mémoire sur certaines inégalités dans la théorie des fonctions monogènes et sur quelques propriétés nouvelles de ces fonctions dans le voisinage d'un point singulier essentiel, *Acta Soc. Sci. Fenn., Math.* **35** (1909), 3–35.
[268] E. Lindelöf. Sur un principe générale de l'analyse et ses applications à la theorie de la représentation conforme, *Acta Soc. Sci. Fenn., Math.* **46** (1915), 1–35.

[269] J. E. Littlewood. Mathematical notes (4): On a theorem of Fatou, *J. Lond. Math. Soc.* **2** (1927), 172–176.
[270] A. J. Lohwater and G. Piranian. The boundary behavior of functions analytic in the disk, *Ann. Acad. Sci. Fenn., Math.* **239** (1957), 1–17.
[271] L. Lorentzen. Compositions of contractions, *J. Comput. Appl. Math.* **32** (1990), 169–178.
[272] L. Lorentzen. A convergence question inspired by Stieltjes and by value sets in continued fraction theory, *J. Comput. Appl. Math.* **65** (1995), 233–251.
[273] K. Löwner. Untersuchungen über schlichte konforme Abbildungen des Einheitskreises. I, *Math. Ann.* **89** (1923), 103–121.
[274] B. D. MacCluer, K. Stroethoff and R. Zhao. Generalized Schwarz-Pick estimates, *Proc. Am. Math. Soc.* **131** (2003), 593–599.
[275] B. D. MacCluer, K. Stroethoff and R. Zhao. Schwarz-Pick type estimates, *Complex Var. Theory Appl.* **48** (2003), 711–730.
[276] R. Maehara. On a connected dense proper subgroup of $\mathbb{R}^2$ whose complement is dense, *Proc. Am. Math. Soc.* **97** (1986), 556–558.
[277] A. Marden and Ch. Pommerenke. Analytic self-mappings of infinite order of Riemann surfaces, *J. Anal. Math.* **37** (1980), 186–207.
[278] A. Marden, I. Richards and B. Rodin. Analytic self-mappings of Riemann surfaces, *J. Anal. Math.* **18** (1967), 197–225.
[279] D. E. Marshall. An elementary proof of the Pick-Nevanlinna interpolation theorem, *Mich. Math. J.* **21** (1974), 219–223.
[280] B. Maskit. The conformal group of a plane domain, *Am. J. Math.* **90** (1968), 718–722.
[281] R. A. McCoy and I. Ntantu. *Topological properties of spaces of continuous functions*. Vol. 1315. Lect. Notes Math. Berlin: Springer, 1988.
[282] P. R. Mercer. Sharpened versions of the Schwarz lemma, *J. Math. Anal. Appl.* **205** (1997), 508–511.
[283] P. R. Mercer. On a strengthened Schwarz-Pick inequality, *J. Math. Anal. Appl.* **234** (1999), 735–739.
[284] P. R. Mercer. Schwarz-Pick-type estimates for the hyperbolic derivative, *J. Inequal. Appl.* **2006** (2006), 1–6.
[285] P. R. Mercer. An improved Schwarz lemma at the boundary, *Open Math.* **16** (2018), 1140–1144.
[286] P. R. Mercer. Boundary Schwarz inequalities arising from Rogosinski's lemma, *J. Class. Anal.* **12** (2018), 93–97.
[287] J. Milnor. *Dynamics in one complex variable*. Third edition. Princeton, NJ: Princeton University Press, 2006.
[288] C. D. Minda. Fixed points of analytic self-mappings of Riemann surfaces, *Manuscr. Math.* **27** (1979), 391–399.
[289] C. D. Minda. The Aumann-Carathéodory rigidity constant for doubly connected regions, *Kodai Math. J.* **2** (1979), 420–426.
[290] C. D. Minda. The hyperbolic metric and coverings of Riemann surfaces, *Pac. J. Math.* **84** (1979), 171–182.
[291] C. D. Minda. A reflection principle for the hyperbolic metric and applications to geometric function theory, *Complex Var.* **8** (1987), 129–144.
[292] C. D. Minda. The strong form of Ahlfors' lemma, *Rocky Mt. J. Math.* **17** (1987), 457–461.
[293] C. Miranda. Sur un nouveau critère de normalité pour les families de fonctions holomorphes, *Bull. Soc. Math. Fr.* **63** (1935), 185–196.
[294] R. Miranda. *Algebraic curves and Riemann surfaces*. Providence, RI, USA: American Mathematical Society, 1994.
[295] T. Mitchell. Invariants means and analytic actions, *Pac. J. Math.* **85** (1979), 145–153.
[296] P. Montel. Sur les suites infinies de fonctions, *Ann. Sci. Éc. Norm. Supér.* **24** (1907), 233–334.

[297] P. Montel. Sur le familles de fonctions analytiques, qui admettent des valeurs exceptionnelles dans un domaine, *Ann. Sci. Éc. Norm. Supér.* **29** (1912), 487–535.

[298] P. Montel. Sur les fonctions méromorphes limites de fractions rationnelles à termes entrelacés, *Ann. Sci. Éc. Norm. Supér.* **50** (1933), 171–196.

[299] M. Nakai. The Martin and Royden compactifications of Tôki covering surfaces, *RIMS Kôkyûroku* **1783** (2012), 1–28.

[300] R. Narasimhan. *Analysis on real and complex manifolds*. Third printing. Amsterdam: North-Holland, 1985.

[301] R. Narasimhan and Y. Nievergelt. *Complex analysis in one variable*. Second edition. Berlin: Springer, 2001.

[302] Z. Nehari. A generalization of Schwarz's lemma, *Duke Math. J.* **14** (1947), 1035–1049.

[303] R. Nevanlinna. Über beschränkte Funktionen, die in gegebenen Punkten vorgeschriebene Werte annehmen, *Ann. Acad. Sci. Fenn., Math.* **13** (1919), 1–71.

[304] R. Nevanlinna. Kriterien für die Randwerte beschränkter Funktionen, *Math. Z.* **13** (1922), 1–9.

[305] R. Nevanlinna. Remarques sur le lemme de Schwarz, *C. R. Acad. Sci. Paris* **188** (1929), 1027–1029.

[306] R. Nevanlinna. Über beschränkte analytische Funktionen, *Ann. Acad. Sci. Fenn., Math.* **32** (1929), 1–75.

[307] L. Nirenberg, S. M. Webster and P. Yang. Local boundary regularity of holomorphic mappings, *Comm. Pure Appl. Math.* **33** (1980), 305–338.

[308] K. Noshiro. On the theory of schlicht functions, *J. Fac. Sci. Hokkaido Imp. Univ.* **2** (1934), 129–155.

[309] R. Oliver. Weight systems for **SO**(3)-actions, *Ann. Math.* **110** (1979), 227–241.

[310] W. F. Osgood. On the existence of the Green's function for the most general simply connected plane region, *Trans. Am. Math. Soc.* **1** (1900), 310–314.

[311] W. F. Osgood and E. H. Taylor. Conformal trasformation on the boundaries of their regions of definition, *Trans. Am. Math. Soc.* **14** (1913), 277–298.

[312] R. Osserman. From Schwarz to Pick to Ahlfors and beyond, *Not. Am. Math. Soc.* **46** (1999), 868–873.

[313] R. Osserman. A sharp Schwarz inequality on the boundary, *Proc. Am. Math. Soc.* **128** (2000), 3513–3517.

[314] A. Ostrowsky. Über die Bedeutung der Jensenchen Formel für einige Fragen der komplexen Funktionentheorie, *Acta Litt. Sci. Univ. Hung.* **1** (1922), 80–87.

[315] P. Painlevé. Sur les lignes singulières des functions analytiques. Thèse, Gauthier-Villars, Paris, 1887.

[316] P. Painlevé. Sur la théorie de la représentation conforme. *C.R. Acad. Sci. Paris* **112** (1891), 653–657.

[317] J. E. Pascoe, M. Sargent and R. Tully-Doyle. A controlled tangential Julia-Carathéodory theory via averaged Julia quotients, *Anal. PDE* **14** (2021), 1773–1795.

[318] E. Peschl and M. Lehtinen. A conformal self-map which fixes three points is the identity, *Ann. Acad. Sci. Fenn., Math.* **4** (1978/1979), 85–86.

[319] E. Picard. Sur une propriété des fonctions entières, *C. R. Acad. Sci. Paris* **88** (1879), 1024–1027.

[320] E. Picard. Sur les fonctions analytiques uniformes dans le voisinage d'un point singulier essentiel, *C. R. Acad. Sci. Paris* **89** (1879), 745–747.

[321] E. Picard. Sur la représentation conforme des aires multiplement connexes, *Ann. Sci. Éc. Norm. Supér.* **30** (1913), 483–487.

[322] G. Pick. Über eine Eigenschaft der konformen Abbildung kreisförmiger Bereiche, *Math. Ann.* **77** (1915), 1–6.

[323] G. Pick. Über die Beschränkungen analytischer Funktionen, welche durch vorgegebene Funktionswerte bewirkt werden, *Math. Ann.* **77** (1915), 7–23.

[324] G. Pick. Über beschränkte Funktionen mit vorgeschriebenen Wertzuordnungen, *Ann. Acad. Sci. Fenn.* **15** (1920), 1–17.

[325] P. Poggi-Corradini. Canonical conjugations at fixed points other than the Denjoy-Wolff point, *Ann. Acad. Sci. Fenn., Math.* **25** (2000), 487–499.

[326] P. Poggi-Corradini. On the failure of a generalized Denjoy-Wolff theorem, *Conform. Geom. Dyn.* **6** (2002), 13–32.

[327] P. Poggi-Corradini. Backward-iteration sequences with bounded hyperbolic steps for analytic self-maps of the disk, *Rev. Mat. Iberoam.* **19** (2003), 843–970.

[328] P. Poggi-Corradini. On the uniqueness of classical semiconjugations for self-maps of the disk, *Comput. Methods Funct. Theory* **6** (2006), 403–421.

[329] P. Poggi-Corradini. Pointwise convergence on the boundary in the Denjoy-Wolff theorem, *Rocky Mt. J. Math.* **40** (2010), 1275–1288.

[330] P. Poggi-Corradini. Iteration in the disk and the ball: a survey of the role of hyperbolic geometry, *Anal. Math. Phys.* **1** (2011), 289–310.

[331] H. Poincaré. Theorie des groupes fuchsiens, *Acta Math.* **1** (1882), 1–62.

[332] H. Poincaré. Mémoire sur les groupes kleinéens, *Acta Math.* **3** (1883), 49–92.

[333] H. Poincaré. Sur les groupes des équations linéaires, *Acta Math.* **4** (1884), 201–312.

[334] H. Poincaré. Sur un théorème de M. Fuchs, *Acta Math.* **7** (1885), 1–32.

[335] H. Poincaré. Sur l'uniformisation des fonctions analytiques, *Acta Math.* **31** (1907), 1–64.

[336] Ch. Pommerenke. Über die subordination analytischer funktionen, *J. Reine Angew. Math.* **218** (1965), 159–173.

[337] Ch. Pommerenke. *Univalent functions.* Göttingen: Vandenhoeck & Ruprecht, 1975.

[338] Ch. Pommerenke. On the iteration of analytic functions in a half-plane, I, *J. Lond. Math. Soc.* **19** (1979), 439–447.

[339] Ch. Pommerenke. On asymptotic iteration of analytic functions in the disk, *Analysis* **1** (1981), 45–61.

[340] Ch. Pommerenke and A. Vasil'ev. On bounded univalent functions and the angular derivative, *Ann. Univ. Mariae Curie-Skłodowska* **54** (2000), 79–106.

[341] Ch. Pommerenke and A. Vasil'ev. Angular derivatives of bounded univalent functions and extremal partitions of the unit disk, *Pac. J. Math.* **206** (2002), 425–450.

[342] M. B. Porter. Concerning series of analytic functions, *Ann. Math.* **6** (1904), 190–192.

[343] W. A. Pranger jr.. Iteration of functions analytic on a disk, *Aequ. Math.* **4** (1970), 201–204.

[344] T. Radó. Über die konformen Abbildungen schlichter Gebiete, *Acta Litt. Sci. Univ. Hung.* **2** (1924), 47–60.

[345] R. Ratto, M. Rigoli and L. Véron. Conformal immersions of complete Riemannian manifolds and extensions of the Schwarz lemma, *Duke Math. J.* **74** (1994), 223–236.

[346] S. Reich and D. Shoikhet. Semigroups and generators on convex domains with hyperbolic metric, *Rend. Mat. Acc. Lincei* **9** (1997), 231–250.

[347] S. Reich and D. Shoikhet. *Nonlinear semigroups, fixed points, and geometry of domains in Banach spaces.* London: Imperial College Press, 2005.

[348] G. Ren and X. Wang. Extremal functions of boundary Schwarz lemma, Preprint, arXiv:1502.02369 (2015).

[349] B. Riemann, Grundlagen für eine allgemeine Theorie der Functionen einer veränderlichen complexen Grösse. (Inauguraldissertation, Göttingen 1851). In: *B. Riemann's Gesammelte Mathematische Werke und Wissenschaftlicher Nachlass.* Leipzig: Teubner, 1892, 3–43.

[350] B. Riemann, Über die Hypothesen, welche der Geometrie zu Grunde liegen. (Habilitationsschrift, Göttingen 1854). In: *B. Riemann's Gesammelte Mathematische Werke und Wissenschaftlicher Nachlass.* Leipzig: Teubner, 1892, 272–287.

[351] P. J. Rippon. A generalization of a theorem of Beardon on analytic contraction mappings, *J. Math. Anal. Appl.* **199** (1996), 157–161.
[352] J. F. Ritt. On the conformal mapping of a region into a part of itself, *Ann. Math.* **22** (1920–21), 157–160.
[353] J. F. Ritt. Permutable rational functions, *Trans. Am. Math. Soc.* **25** (1923), 399–448.
[354] A. M. Rockett and P. Szüsz. On tangential regions for power series, *Arch. Math.* **60** (1993), 446–450.
[355] W. Rogosinski. Zum Schwarzschen Lemma, *Jahresber. Dtsch. Math.-Ver.* **44** (1934), 258–261.
[356] O. Roth. The Nehari-Schwarz lemma and infinitesimal boundary rigidity of bounded holomorphic functions, *Stud. Univ. Babeş-Bolyai Math.* **67** (2022), 285–294.
[357] E. Rouché. Mémoire sur la série de Lagrange, *J. Éc. Polytech.* **22** (1862), 149–193.
[358] H. L. Royden. The Ahlfors-Schwarz lemma: the case of equality, *J. Anal. Math.* **46** (1986), 261–270.
[359] H. L. Royden. Hyperbolicity in complex analysis, *Ann. Acad. Sci. Fenn., Math.* **13** (1988), 387–400.
[360] W. Rudin. *Real and complex analysis*. New York: McGraw Hill, 1966.
[361] W. Rudin. *Function theory in the unit ball of* $\mathbb{C}^n$. Berlin: Springer, 1980.
[362] St. Ruscheweyh. Two remarks on bounded analytic functions, *Serdica* **11** (1985), 200–202.
[363] S. Saks and A. Zygmund. *Analytic functions*. New York: Elsevier, 1971.
[364] F. Schottky. Über die konforme Abbildung mehrfach zusammenhängender ebener Flächen, *J. Reine Angew. Math.* **83** (1877), 300–351.
[365] F. Schottky. Über die Wertschwankungen der harmonischen Funktionen zweier reellen Veränderlichen und der Funktionen eines complexen Arguments, *J. Reine Angew. Math.* **117** (1897), 225–253.
[366] F. Schottky. Über den Picard'schen Satz und die Borel'schen Ungleichungen, *Sitzungsber. Kön. Preuss. Akad. Wiss. Berlin* **1904** (1904), 1244–1262.
[367] F. Schottky. Über den Picard'schen Satz und die Borel'schen Ungleichungen, *Sitzungsber. Kön. Preuss. Akad. Wiss. Berlin* **1906** (1906), 32–36.
[368] E. Schröder. Über unendlich viele Algorithmen zur Auflösung der Gleichungen, *Math. Ann.* **2** (1870), 317–363.
[369] E. Schröder. Über iterierte Funktionen, *Math. Ann.* **3** (1871), 296–322.
[370] H. A. Schwarz. Über diejenigen algebraischen Gleichungen zwischen zwei veränderlichen Grössen, welche eine Schaar rationaler eindeutig umkehrbarer Transformationen in sich selbst zulassen, *J. Reine Angew. Math.* **87** (1879), 139–145.
[371] H. A. Schwarz, Zur Theorie der Abbildung. In: *Gesammelte Mathematische Abhandlungen, II*. Berlin: Springer, 1890, 108–132.
[372] W. Seidel. Über die Ränderzuordnung bei konformen Abbildungen. *Math. Ann.* **104** (1931), 182–243.
[373] J.-P. Serre. *A course in arithmetic*. Berlin: Springer Verlag, 1973.
[374] J. H. Shapiro. *Composition operators and classical function theory*. New York: Springer Verlag, 1993.
[375] A. L. Shields. On fixed points of commuting analytic functions, *Proc. Am. Math. Soc.* **15** (1964), 703–706.
[376] D. Shoikhet. Another look at the Burns-Krantz theorem, *J. Anal. Math.* **105** (2008), 19–42.
[377] D. Shoikhet. Rigidity and parametric embedding of semi-complete vector fields on the unit disk, *Milan J. Math.* **84** (2016), 159–201.
[378] C. L. Siegel. Iteration of analytic functions, *Ann. Math.* **43** (1942), 607–612.
[379] A. G. Siskasis. Semigroups of composition operators and the Cesaro operator on $H^p(\mathbb{D})$, Thesis (Ph.D.) University of Illinois at Urbana-Champaign, 1985.

[380] T. J. Stieltjes Jr. Recherches sur les fractions continues, *Ann. Fac. Sci. Toulouse* **8** (1894), 1–122.

[381] O. Stolz. Beweis einiger Sätze über Potenzreihen, *Z. Math. Phys.* **20** (1875), 369–376.

[382] O. Stolz. *Vorlesungen über allgemeine Arithmetik: nach den Neueren Ansichten*. Leipzig: Teubners, 1885.

[383] E. L. Stout. Bounded holomorphic functions on finite Riemann surfaces, *Trans. Am. Math. Soc.* **120** (1965), 255–285.

[384] N. Suita. Analytic mapping and harmonic length, *Kōdai Math. Sem. Rep.* **23** (1971), 351–356.

[385] G. Szekeres. Regular iteration of real and complex functions, *Acta Math.* **100** (1958), 203–258.

[386] G. Szekeres. Fractional iteration of entire and rational functions, *J. Aust. Math. Soc.* **4** (1964), 129–142.

[387] R. Tauraso. On fixed points of holomorphic maps of simply connected proper domains in $\mathbb{C}$, *Atti Accad. Naz. Lincei, Cl. Sci. Fis. Mat. Nat., Rend. Lincei* **5** (1994), 197–202.

[388] R. Tauraso. Commuting holomorphic maps of the unit disc, *Ergod. Theory Dyn. Syst.* **24** (2004), 945–953.

[389] R. Tauraso and F. Vlacci. Rigidity at the boundary for holomorphic self-maps of the unit disk, *Complex Var. Theory Appl.* **45** (2001), 151–165.

[390] K. Tavakoli. Boundary points as limit functions of iterated holomorphic function systems, *Proc. Am. Math. Soc.* **137** (2009), 1971–1976.

[391] H. Töpfer. Komplexe Iterationsindizes ganzer und rationaler Funktionen, *Math. Ann.* **121** (1949), 191–222.

[392] F. Tricomi. Sull'iterazione delle funzioni di una variabile complessa, *Atti Accad. Naz. Lincei, Rend. Cl. Sci. Fis. Mat. Natur.* **25** (1916), 156–162.

[393] F. Tricomi. Sull'iterazione delle funzioni di linee, *G. Mat. Battaglini* **55** (1917), 1–8.

[394] M. Troyanov. The Schwarz Lemma for non positively curved Riemannian surfaces, *Manuscr. Math.* **72** (1991), 251–256.

[395] M. Tsuji. *Potential theory in modern function theory*. Tokio: Maruzen, 1959.

[396] H. Unkelbach. Über die Randverzerrung bei konformer Abbildung, *Math. Z.* **43** (1938), 739–742.

[397] H. Unkelbach. Über die Randverzerrung bei schlichter konformer Abbildung, *Math. Z.* **46** (1940), 329–336.

[398] G. Valiron. Sur l'itération des fonctions holomorphes dans un demi-plan, *Bull. Sci. Math.* **55** (1931), 105–128.

[399] G. Valiron. *Fonctions analytiques*. Paris: Presses Universitaires de France, 1954.

[400] C. de la Vallée-Poussin. La représentation des aires multiplement connexes, *Ann. Sci. Éc. Norm. Supér.* **47** (1930), 267–309.

[401] C. de la Vallée-Poussin. Sur la représentation conforme des aires planes multiplement connexes, *Bull. Cl. Sci., Acad. R. Belg.* **17** (1931), 10–27.

[402] E. Vesentini. Su un teorema di Wolff e Denjoy, *Rend. Semin. Mat. Fis. Milano* **53** (1983), 17–25.

[403] J.-P. Vigué, Itérées et points fixes d'applications holomorphes. In: *Complex analysis and geometry (Trento, 1993)*, vol. 173. Lecture Notes in Pure and Appl. Math. New York: Dekker, 1996, 509–518.

[404] G. Vitali. Sopra le serie di funzioni analitiche, *Rend. R. Ist. Lomb. Sci. Lett.* **36** (1903), 772–774.

[405] G. Vitali. Sopra le serie di funzioni analitiche, *Ann. Mat. Pura Appl.* **10** (1904), 65–82.

[406] F. Vlacci. On commuting holomorphic maps in the unit disc of $\mathbb{C}$, *Complex Var. Theory Appl.* **30** (1996), 301–313.

[407] F. Vlacci. Iteration theory in hyperbolic domains, *J. Anal. Math.* **74** (1998), 51–66.

[408] S. E. Warschawski. Über das Randverhalten der Ableitung der Abbildungsfunktion bei konformer Abbildung. *Math. Z.* **35** (1932), 321–456.

[409] S. E. Warschawski. Über einen Satz von O. D. Kellogg. *Nachr. Ges. Wiss. Göttingen, Math.-Phys. Kl.* **1932** (1932), 73–86.
[410] S. E. Warschawski. On the higher derivatives at the boundary in conformal mappings. *Trans. Am. Math. Soc.* **38** (1935), 310–340.
[411] S. E. Warschawski. On differentiability at the boundary in conformal mapping. *Proc. Amer. Math. Soc.* **12** (1961), 614–620.
[412] S. M. Webster. Biholomorphic mappings and the Bergman kernel off the diagonal, *Invent. Math.* **51** (1979), 155–169.
[413] H. Weyl. *Die Idee der riemannschen Fläche*. Leipzig: Teubner, 1913.
[414] J. Wolff. Sur l'itération des fonctions holomorphes dans une région, et dont les valeurs appartiennent à cette région, *C. R. Acad. Sci. Paris* **182** (1926), 42–43.
[415] J. Wolff. Sur l'itération des fonctions bornées, *C. R. Acad. Sci. Paris* **182** (1926), 200–201.
[416] J. Wolff. Sur une généralisation d'un théorème de Schwarz, *C. R. Acad. Sci. Paris* **182** (1926), 918–920.
[417] J. Wolff. Sur une généralisation d'un théorème de Schwarz, *C. R. Acad. Sci. Paris* **183** (1926), 500–502.
[418] J. Wolff. Sur l'itération des fonctions holomorphes dans un demi-plan, *Bull. Soc. Math. Fr.* **57** (1929), 195–203.
[419] J. Wolff. L'intégrale d'une fonction holomorphe et à partie réelle positive dans un demi plan est univalente, *C. R. Acad. Sci. Paris* **198** (1934), 1209–1210.
[420] J. Wolff. L'équation différentielle $dz/dt = w(z)$ = fonction holomorphe à partie réelle positive dans un demi-plan, *Compos. Math.* **6** (1938), 296–304.
[421] J. Wolff. Sur les fonctions holomorphes dont l'esemble des valeurs est soumis à certaines restrictions, *Proc. Ned. Akad. Wet.* **43** (1940), 1018–1022.
[422] H. Wu. Normal families of holomorphic mappings, *Acta Math.* **119** (1967), 193–233.
[423] H. Yamabe. On an arcwise connected subgroup of a Lie group, *Osaka Math. J.* **2** (1950), 13–14.
[424] S. Yamashita. Conformality and semiconformality of a function holomorphic in the disk, *Trans. Am. Math. Soc.* **245** (1978), 119–138.
[425] S. Yamashita. The Pick version of the Schwarz lemma and comparison of the Poincaré densities, *Ann. Acad. Sci. Fenn., Math.* **19** (1994), 291–322.
[426] S. Yamashita. Goluzin's extension of the Schwarz-Pick inequality, *J. Inequal. Appl.* **1** (1997), 345–356.
[427] P. Yang. Holomorphic curves and boundary regularity of biholomorphic maps. Preprint (1978).
[428] S.-T. Yau. Remarks on conformal transformations, *J. Differ. Geom.* **8** (1973), 369–381.
[429] S.-T. Yau. A general Schwarz lemma for Kähler manifolds, *Am. J. Math.* **100** (1978), 197–203.
[430] M. Z. Zhang. Generalized Schwarz-Pick lemma, *Acta Math. Sinica (Chin. Ser.)* **49** (2006), 613–616.
[431] A. Zygmund. Smooth functions, *Duke Math. J.* **12** (1945), 47–76.

Index

https://doi.org/10.1515/9783110601978-008

De Gruyter Studies in Mathematics

Volume 88
Joel L. Schiff
Topics in Complex Analysis
ISBN 978-3-11-075769-9, e-ISBN 978-3-11-075782-8,
e-ISBN (ePUB) 978-3-11-075787-3

Volume 66
Igor V. Nikolaev
Noncommutative Geometry. A Functorial Approach, 2nd Edition, 2022
ISBN 978-3-11-078860-0, e-ISBN 978-3-11-078870-9,
e-ISBN (ePUB) 978 3 11 078881-5

Volume 87
René Erlin Castillo
Fourier Meets Hilbert and Riesz. An Introduction to the Corresponding Transforms, 2022
ISBN 978-3-11-078405-3, e-ISBN 978-3-11-078409-1,
e-ISBN (ePUB) 978-3-11-078412-1

Volume 86
Flavia Smarrazzo, Alberto Tesei
Measure Theory and Nonlinear Evolution Equations, 2022
ISBN 978-3-11-055600-1, e-ISBN 978-3-11-055690-2,
e-ISBN (ePUB) 978-3-11-055604-9

Volume 85
Shai Dekel
Pointwise Variable Anisotropic Function Spaces on $\mathbb{R}^n$, 2022
ISBN 978-3-11-076176-4, e-ISBN 978-3-11-076179-5,
e-ISBN (ePUB) 978-3-11-076187-0

Volume 84
Marko Kostić
Selected Topics in Almost Periodicity, 2021
ISBN 978-3-11-076322-5, e-ISBN 978-3-11-076352-2,
e-ISBN (ePUB) 978-3-11-076360-7

www.degruyter.com

Bei Fragen zur Produktsicherheit wenden Sie sich bitte an:
If you have any questions regarding product safety,
please contact:

Walter de Gruyter GmbH
Genthiner Straße 13
10785 Berlin
productsafety@degruyterbrill.com